FUNDAMENTALS OF COMPUTER ENGINEERING:
Logic Design and Microprocessors

FUNDAMENTALS OF COMPUTER ENGINEERING:
Logic Design and Microprocessors

Herman Lam/John O'Malley

JOHN WILEY & SONS
NEW YORK • CHICHESTER • BRISBANE • TORONTO • SINGAPORE

Library of Congress Cataloging in Publication Data:

Lam, Herman.
Fundamentals of computer engineering: logic design and
microprocessors/Herman Lam, John O'Malley.
p. cm.
Bibliography: p.
ISBN 0-471-60501-8
1. Logic design. 2. Microprocessors. I. O'Malley, John.
II. Title.
TK7868.L6L35 1988 87-31724
621.395—dc19 CIP

10 9 8 7 6 5 4 3 2 1

To my family
H.L.

To my granddaughters, Jenni, Alicia, Tess, and Ashley
J.O.

PREFACE

The top-down structured approach of digital design is the principal subject of this book. In this approach the preliminary design of a digital system is partitioned into more detailed designs of less complex subsystems, which in turn are partitioned until eventually the detailed designs of the resulting digital circuits are in forms suitable for ready physical implementations. At some point of the design process, a decision must be made on the implementation methods to be used for the various digital subsystems. The types of implementation methods can be classified as follows:

1. Combinational/sequential circuit design using catalogued SSI/MSI/LSI circuit elements.
2. Microprocessor-based design.
3. Design using bit-slice circuit elements.
4. Custom/semicustom LSI/VLSI circuit design.

Each implementation method requires a different set of structured design techniques and design principles. The choice of implementation method, of course, depends on the application and performance constraints of the digital subsystem being designed, along with factors such as cost, physical size limitations, and quantities to be produced.

In this book we are primarily concerned with the structured design techniques required for implementation methods 1 and 2. For this reason, this book is divided into two main parts. The first part of the book (Chapters 1–7) provides the foundation for structured design techniques using catalogued SSI, MSI, and LSI circuit elements. The remainder of the book (Chapters 8–13) provides the fundamentals of microprocessors (μP) and μP-based design.

More specifically, Chapter 1 presents number system basics, along with details of binary, octal, and hexadecimal number systems. Chapter 2 introduces some digital design fundamentals, including the elements of Boolean algebra and Karnaugh map minimization. Emphasis is on currently relevant concepts, with some de-emphasis of less relevant but traditional material. Chapter 3 contains the fundamentals of the design of digital circuits with SSI (small-scale integrated) circuit elements. The implementations are based on the three logic conventions: positive logic, negative logic, and mixed logic, with emphasis on mixed logic. Chapters 4 and 5 present the commonly used MSI (medium-scale integrated) circuit elements as digital building blocks, with combinational circuit elements in Chapter 4 and sequential ones in Chapter 5. The presentation is generally top-down, with the functional description of each circuit element being described first, then the internal realization of it, followed by applications of the circuit element. Chapter 6 presents the commonly used LSI (large-scale integrated) circuit elements, including the arithmetic logic unit (ALU), carry look-ahead adder, programmable logic array (PLA), and programmable array logic (PAL). Also included in Chapter 6 are the various types of random access memories, including the read-write memories (static and dynamic RAM) and read-only memory (ROM, PROM, and EPROM). Chapter 7 integrates the material of the preceding chapters by presenting the general principles for structured digital design, using catalogued SSI/MSI/LSI circuit elements. Key concepts emphasized in this chapter include top-down and modular design, controller and controlled circuit elements, and the algorithmic state machine (ASM). Design examples are also presented to illustrate the principles of structured design.

Chapter 8 is a transitional chapter. It presents the life cycle of a digital system and also provides an introduction to microprocessor-based design.

The remainder of the book (Chapters 9–13) concentrates on μP-based design fundamentals. Throughout these chapters we use a specific microprocessor, the Intel 8085, and its supporting chips, to illustrate the various μP concepts. We believe that the use of a commercially available microprocessor as a unifying thread throughout the discussion is a more effective approach than the use of a nonexistent ''generic'' microprocessor. The 8085 is not, of course, representative of the state of the art in μP technology. But, this is a problem with any microprocessor. Even the most current of the 16-bit or 32-bit microprocessors will be out of date in a relatively short time. Our intention in this book is not to present the specific details of a current microprocessor, but rather to use a microprocessor that is excellent for illustrating those concepts and fundamentals that will remain current. For this purpose, an 8-bit microprocessor such as the 8085 is ideal for an introductory text. The 8085 is powerful enough to illustrate the important μP concepts, but simple enough to avoid obscuring these concepts with complex component details. Finally, it is our experience that once a specific microprocessor and its applications are mastered, one can easily become proficient, with some training, in the use of other microprocessors and their applications.

Chapters 9 and 10 explore the software concepts that are essential to fully exploiting the power of a μP system. Topics include assembly language programming, addressing modes, program assembly, stacks, and subroutine linkage. Emphasis is placed on producing not simply a working program, but a ''good'' one, using structured programming techniques. Chapters 11 and 12 present μP hardware and interfacing concepts. Topics include a model of a μP system, μP bus structure, μP machine cycles/timing, and techniques for interfacing the components (memory and I/O) of a μP system. Emphasis is placed on the intimate hardware/software interaction of a μP system. Chapter 13 completes our introduction to the microprocessor and μP-based design. Topics include the history and state of the art of the microprocessor, selection criteria for a microprocessor, hardware/software trade-offs, a μP system development procedure, and a survey of the μP system development tools (both hardware and software) that are available.

There is a trend nationwide in electrical engineering (EE) and computer science (CS) departments to incorporate more and more computer engineering courses into their curricula. No longer is it sufficient to have a single logic design course and a single microprocessor course. Consequently, a first computer course must provide a firm foundation for subsequent computer courses.

At the University of Florida we begin the electrical engineering and the computer science curricula with a required course on the fundamentals of computer engineering. In such an introductory course it is important to have a good overview of computer engineering for general EE or CS students, while also providing a solid foundation for computer engineering students. For these reasons, we believe that it is essential in such a course to include an introduction to the μP along with the traditional logic design material. This book is written with the same belief.

The first half of this introductory course is devoted to logic design using SSI/MSI circuit elements, with some exposure to state machines. The second half is devoted to an introduction to microprocessor and μP-based design fundamentals. While providing an overview of computer engineering, this introductory course also serves in our undergraduate curriculum as a foundation for two subsequent sequences in the area of computer engineering. One of these sequences is in digital design; the other is in microprocessors.

In our digital design sequence, the first course is a projects-oriented course that allows the students to learn the finer points of state-machine design. Also emphasized are computer-aided design (CAD) and CAD tools for digital design. The second course in this sequence is on computer design and architecture.

The first course in the microprocessor sequence is also a projects-oriented course and is structured on a Hewlett-Packard 64000 Universal Development System. Because of the introductory microprocessor material in the prerequisite course, applications of various μP families can be presented more rapidly. The second course in this sequence deals with advanced microprocessor concepts and applications.

Finally, a pervasive theme that will become apparent in the reading of this book is the importance of structured design, both in hardware and software. The design methods, techniques, and notation presented throughout this book are consistent with this theme. Note particularly the use of mixed logic, the algorithmic state machine (ASM), and the structured programming techniques with well-documented programs.

CONTENTS

Chapter 1

Number Systems and Representations

1.1 INTRODUCTION

The major topics of this chapter are the number systems and the number representations that are useful in the study of computer engineering. First introduced is the binary number system, which is the number system used in digital computers and other digital applications. After this introduction to the binary number system we will consider the foundation for a positional number system, such as the binary number system and the more familiar decimal number system. Then we will relate the binary and decimal number systems. Next we will briefly consider the octal and hexadecimal number systems since they are used as shorthand systems for the binary number system. After that we will consider three different ways of representing signed binary numbers. Finally, we will consider the binary-coded decimal (BCD) representation.

We are well familiar with the decimal number system. Probably its popularity results from humans having ten fingers. If God had endowed us with, say, eight fingers instead, then perhaps the octal number system would have been the one we studied in grade school. In other words, the extensive use of the decimal number system is mostly a matter of chance and not because it is best for calculations.

Only one number system—the binary number system—is well suited, at present, for direct use in digital applications, including digital computers. In an electronic digital computer application, each different symbol of a number system is represented by a different physical quantity such as a different voltage level. Consequently, the direct use of, say, the decimal number system requires components that utilize ten different physical quantities. Other considerations require these components to be small, light, inexpensive, and also very fast in operation. Components satisfying these latter requirements seldom possess the necessary ten distinct physical quantities. In fact, many of these components have just two distinct physical quantities and therefore are binary components.

To better appreciate the need for the binary number system for digital applications, consider some of the important components used in electronic digital systems. The flip-flop, a storage circuit, is popular because its output voltages can be changed very rapidly. Inherently, a flip-flop produces just two voltage levels and thus is a binary component. Another even more basic component, the transistor, is usually operated in digital applications in just two conditions—saturation and cutoff—and hence it is a binary device. In a magnetic storage element, typically there are just two physical conditions—the two possible directions of residual magnetic flux.

Although the binary number system is best for digital systems, it has the disadvantage of requiring more digits than the decimal number system—generally, more than three times as many. Fortunately, though, having more binary digits is not a serious problem, and can be avoided for hand calculations by using an octal or hexadecimal shorthand. Using octal reduces the number of digits by a factor of approximately three, and using hexadecimal reduces it by a factor of approximately four. As will be seen, using either of these number systems for a shorthand is convenient because it is very easy to convert a binary number into either its octal or hexadecimal equivalent.

1.2 BINARY NUMBER SYSTEM

The binary number system is simpler, in most respects, than the decimal number system because it has only two distinct symbols, 0 and 1, which is much fewer than the ten distinct symbols: 0, 1, 2, 3, 4, 5, 6, 7, 8, and 9 of the decimal number system. Table 1.1 has the first 18 nonnegative binary integers and the corresponding decimal integers.

TABLE 1.1 DECIMAL-BINARY EQUIVALENCE

Decimal	Binary	Decimal	Binary
0	0	9	1001
1	1	10	1010
2	10	11	1011
3	11	12	1100
4	100	13	1101
5	101	14	1110
6	110	15	1111
7	111	16	10000
8	1000	17	10001

We can derive the binary contents of this table by starting with 0, and continually adding 1, using the rules for binary addition given in Table 1.2. Starting with 0, we add 1 to get the next binary entry of Table 1.1:

$$\begin{array}{r} 0 \\ +1 \\ \hline 1 \end{array}$$

Then, add 1 to get the next number.

$$\begin{array}{r} 1 \\ +1 \\ \hline 2 \end{array} \quad \text{(wrong!)}$$

Wrong! In the binary system, the only valid digits are 0 and 1. From the rules for binary addition in Table 1.2, we see that the next number should be

$$\begin{array}{r} 1 \\ +1 \\ \hline 10 \end{array} \quad \text{(sum of 0 and a carry of 1)}$$

Continuing counting in this manner, we obtain more table entries.

$$\begin{array}{r} 10 \\ +\ 1 \\ \hline 11 \\ +\ 1 \\ \hline 100 \\ +\ 1 \\ \hline 101 \end{array}$$

And so forth.

TABLE 1.2 BINARY ADDITION

a	*b*	Sum	Carry
0	0	0	0
0	1	1	0
1	0	1	0
1	1	0	1

$$\begin{array}{r} a \\ +b \\ \hline \end{array}$$

In order to avoid any ambiguity, we will often use a subscript to specify the number system of a particular number. As an illustration,

$$13_{10} = 1101_2$$

The subscript 2 is used for the binary system, 10 for the decimal system, 8 for the octal system, and 16 for the hexadecimal system. Alternatively, a letter B, D, O, or H can be used, respectively, instead of a subscript.

1.2.1 Addition and Subtraction

Addition in binary is quite simple, as is evident from Table 1.2. Some examples follow.

$$\begin{array}{r} 1010\ (10_{10}) \\ +\ 100\ (4_{10}) \\ \hline 1110\ (14_{10}) \end{array} \qquad \begin{array}{r} ^{1} \\ 101\ (5_{10}) \\ +101\ (5_{10}) \\ \hline 1010\ (10_{10}) \end{array} \qquad \begin{array}{r} ^{1\,1} \\ 111\ (7_{10}) \\ +\ 11\ (3_{10}) \\ \hline 1010\ (10_{10}) \end{array} \qquad \begin{array}{r} ^{1\,1} \\ ^{1}110\ (6_{10}) \\ 11\ (3_{10}) \\ +\ 111\ (7_{10}) \\ \hline 10000\ (16_{10}) \end{array} \qquad \begin{array}{r} ^{1\ \ 111\ 11} \\ 11011.101\ (27.625_{10}) \\ +\ 1010.111\ (10.875_{10}) \\ \hline 100110.100\ (38.5_{10}) \end{array}$$

TABLE 1.3 BINARY SUBTRACTION

a	*b*	Difference	Borrow
0	0	0	0
0	1	1	1
1	0	1	0
1	1	0	0

$$\begin{array}{rl} a & \text{(minuend)} \\ \underline{-b} & \text{(subtrahend)} \end{array}$$

Notice in the third example that the second column has the addition of two 1s plus a 1 carry from the first column of addition. The sum of $11_2 = 3_{10}$ is shown as a 1 in the sum plus a 1 carry in the third column from the right. The fourth example shows that one difficulty in adding binary numbers is the large number of carries that often occur. In this example, the carry from the second column of addition ($1 + 1 + 1 + 1 = 100$) extends over two places. Such an extension is not unusual in the addition of more than two binary numbers, but is rare in the addition of decimal numbers. The last example shows that the addition rules are the same for numbers that are not integers. In other words, the presence of the binary point (not decimal point) has no effect on the addition rules.

Binary subtraction is more difficult than addition, but no more so than decimal subtraction is more difficult than decimal addition. Table 1.3 specifies the rules for binary subtraction. A simple example of binary subtraction is the following:

$$\begin{array}{r} 1110\ (14_{10}) \\ -\ \ \underline{100}\ (4_{10}) \\ 1010\ (10_{10}) \end{array}$$

Just as in decimal subtraction, the principal difficulty in binary subtraction occurs when a borrow is required. This happens in the subtraction of a 1 from a 0. Then, it is necessary to borrow a 1 from the "next" digit. If this digit is a 1, then it is changed to a 0, and the borrow taken, as in the following example:

$$\begin{array}{rl} 1101 & (13_{10}) \\ \underline{-\ \ 10} & (2_{10}) \end{array} \quad \rightarrow \quad \begin{array}{rl} 10^{1}01 & \text{borrow} \\ \underline{-\ \ \ 10} & \\ 1011 & (11_{10}) \end{array}$$

If the "next" digit in the minuend is a 0, then the borrow must extend over several digits, just as in decimal subtraction. As an illustration,

$$\begin{array}{r} 1100001 \\ \underline{-\ \ 100011} \end{array} \quad \rightarrow \quad \begin{array}{rl} 10111^{1}01 & \text{borrow} \\ \underline{-\ \ 1000\ 11} & \\ 1111\ 10 & \end{array}$$

Note that the second, third, fourth, and fifth digits of the minuend change in the generation of the borrow for the sixth digit.

1.2.2 Multiplication and Division

Table 1.4 specifies the rules for binary multiplication. As is evident, 0 times a digit is 0, and 1 times a digit is that digit. Not shown by the table is the fact that the rule for

TABLE 1.4 BINARY MULTIPLICATION

a	*b*	Product
0	0	0
0	1	0
1	0	0
1	1	1

$$\begin{array}{r} a \\ \times\, b \\ \hline \end{array}$$

positioning the binary point in a product is the same as that for the decimal point in the decimal number system.

EXAMPLE 1.1

Multiply 1101.1 by 1010.1.

Solution.

```
     1101.1   multiplicand
   × 1010.1   multiplier
   --------
      11011
     00000
    11011
   00000
  11011
-----------
10001101.11   product
```

■■

From Example 1.1, we see that, in general, for each 1 in the multiplier we write the multiplicand and then shift left one position before writing again. For each 0 in the multiplier, we just shift left.

Division in binary is much simpler than in decimal. The process of binary division is best illustrated by an example. We will check the result of the last example by dividing the multiplier into the product to see whether the quotient is the original multiplicand.

EXAMPLE 1.2

Divide 10001101.11 by 1010.1.

Solution.

```
                         1101.1   quotient
divisor   1010.1)10001101.11   dividend
                 10101
                 -----
                  11100
                  10101
                  -----
                   11111
                   10101
                   -----
                    10101
                    10101
                    -----
                    00000
```

■■

In general, the first step in binary division is to mark off, starting from the left in the dividend, a number of digits equal to the number of digits in the divisor. If the number marked off is larger, then the divisor divides into this number exactly once. If the number marked off is smaller, we include another digit of the dividend. Then, the divisor always divides into this number exactly once. We subtract, and then continue the process just as in decimal division. We position the binary point of the quotient in the same way as in decimal division.

1.3 POSITIONAL NUMBER SYSTEMS

Now that we have studied the binary number system, we will provide a foundation for what we have done by briefly considering *positional number systems,* starting with the familiar decimal number system.

As we know, the decimal number system has ten different symbols: 0, 1, 2, 3, 4, 5, 6, 7, 8, and 9. The number of symbols of a number system is the *base* or *radix*. So, the base of the decimal system is 10. Note that the symbol for the base is a combination of the first two symbols; there is no single symbol for the 10 base of the decimal system.

In a *positional* number system (and all the number systems we will study are positional number systems), the value of a number depends not only on the symbols used but also on the *positions* of the symbols. For example, 3674_{10} and 3746_{10} contain the same symbols but have entirely different values. The significance of the positions is that they correspond to different powers of the base, even though we do not show these powers because we write these numbers in contracted form. As an illustration, 3674_{10} is a contraction of $3 \times 10^3 + 6 \times 10^2 + 7 \times 10^1 + 4 \times 10^0$. Graphically, it can be shown as follows:

```
3674
\\\└4 × 10^0
 \\└7 × 10^1
  \└6 × 10^2
   └3 × 10^3
```

Although we do not show the radix point (e.g., decimal point) for integer numbers, we assume it to be just to the right of the rightmost digit. Then, the first position to the *left* of the radix point corresponds to the base raised to the power of 0, the second position corresponds to the base raised to the power of 1, the third position corresponds to the base raised to the power of 2, and so on.

This concept of powers also applies to digits positioned to the *right* of a radix point, except the powers are *negative* powers of the base. The first position to the right corresponds to the base raised to the power of -1, the second position to the base raised to the power of -2, and so on. So, a radix point in a number designates the separation of the negative powers of the base from the nonnegative powers. As an illustration, consider the number 642.391_{10}, which is a contraction of $6 \times 10^2 + 4 \times 10^1 + 2 \times 10^0 + 3 \times 10^{-1} + 9 \times 10^{-2} + 1 \times 10^{-3}$. Graphically, it can be shown as follows:

642.391

1×10^{-3}
9×10^{-2}
3×10^{-1}
2×10^{0}
4×10^{1}
6×10^{2}

The decimal point in the contracted number is between digits 2 and 3—the multipliers of 10^0 and 10^{-1}, respectively. The digit just to the left of the decimal point is the multiplier of the zero power of the base, and the digit just to the right is the multiplier of the negative one power of the base.

This number representation applies not only to the decimal number system but also to all number systems we will consider. In general, a number $a_n a_{n-1} \cdots a_1 a_0 . a_{-1} a_{-2} \cdots$, in which the a_i's are digits of a positional number system with radix r, is representable as

$$a_n r^n + a_{n-1} r^{n-1} + \cdots + a_2 r^2 + a_1 r^1 + a_0 r^0 + a_{-1} r^{-1} + a_{-2} r^{-2} + \cdots$$

in which $0 \le a_i \le r - 1$. In using this representation in our following considerations of the binary, octal, and hexadecimal number systems, we will, for convenience, always express the r's and their powers in decimal.

Let us now see how this representation applies to the binary number system. The base or radix, r, is 2. Thus, this system has just two distinct digits 0 and 1 since

$$0 \le a_i \le r - 1$$
$$0 \le a_i \le 2 - 1$$
$$0 \le a_i \le 1$$

An example of a binary number is 1101.011, which is a contraction of

$$1 \times 2^3 + 1 \times 2^2 + 0 \times 2^1 + 1 \times 2^0 + 0 \times 2^{-1} + 1 \times 2^{-2} + 1 \times 2^{-3}$$

Graphically, it can be shown as

1101.011

1×2^{-3}
1×2^{-2}
0×2^{-1}
1×2^{0}
0×2^{1}
1×2^{2}
1×2^{3}

1.4 BINARY-TO-DECIMAL CONVERSION

In our study we will often want to convert a binary number into its decimal equivalent. One easy way of doing this is to express the binary number in expanded form as powers of the base 2 expressed in decimal, and then add in decimal. For example, 11101.011_2

is in decimal: $1 \times 2^4 + 1 \times 2^3 + 1 \times 2^2 + 0 \times 2^1 + 1 \times 2^0 + 0 \times 2^{-1} + 1 \times 2^{-2} + 1 \times 2^{-3} = 29.375$. Graphically,

$$
\begin{array}{l}
11101.011 \\
\quad 1 \times 2^{-3} = 1 \times \frac{1}{8} = 0.125 \\
\quad 1 \times 2^{-2} = 1 \times \frac{1}{4} = 0.25 \\
\quad 0 \times 2^{-1} = 0 \times \frac{1}{2} = 0 \\
\quad 1 \times 2^{0} = 1 \times 1 = 1 \\
\quad 0 \times 2^{1} = 0 \times 2 = 0 \\
\quad 1 \times 2^{2} = 1 \times 4 = 4 \\
\quad 1 \times 2^{3} = 1 \times 8 = 8 \\
\quad 1 \times 2^{4} = 1 \times 16 = \underline{16} \\
\qquad\qquad\qquad\qquad\quad 29.375
\end{array}
$$

Although some convenient algorithms exist for rapid conversion from binary to decimal, they are not worth the effort to learn unless we do a considerable amount of binary-to-decimal conversion.

1.5 DECIMAL-TO-BINARY CONVERSION

Frequently, we will need to convert a decimal integer into an equivalent binary integer. To do this, we repeatedly divide in decimal by the base 2, saving the remainders which will form the desired binary number. Specifically, using decimal division, we divide the decimal integer by 2. Next, we place the remainder of 0 or 1 to one side, and then divide the integer part of the quotient by 2. Again, we place the remainder from this second division to one side, and divide 2 into the integer part of the second quotient, and so on. We repeat this process until we obtain a zero quotient. Then, by arranging the remainders in reverse order, we obtain the desired binary equivalent.

EXAMPLE 1.3

Convert the decimal number 117 into its binary equivalent.

Solution.

	remainder
2 |117	
2 |58	1
2 |29	0
2 |14	1
2 |7	0
2 |3	1
2 |1	1
0	1

By arranging the remainders in the reverse order in which we obtained them, we have the result that 1110101 is the binary equivalent of decimal 117. ■■

The justification for this rule is that in converting a decimal integer into its binary equivalent, we are finding the a_i's of

$$\cdots a_3 \times 2^3 + a_2 \times 2^2 + a_1 \times 2^1 + a_0 \times 2^0$$

in which each a_i is either 0 or 1. The first division by 2 results in

$$\cdots a_3 \times 2^2 + a_2 \times 2^1 + a_1 + a_0/2 \quad \text{remainder}$$

With a little thought, we see that the part of the quotient including a_1 and to the left is an integer, and a_0 is the remainder of the division. Similarly, when we divide this integer part by 2, the remainder from this second division is a_1, and so on. So, the remainders from the repeated divisions are the a_i's with an order beginning with the least significant, which means that we must reverse this order to get the equivalent binary number.

We must use a different rule for converting the *fractional* part of a decimal number into an equivalent binary fraction. For it, we multiply by 2, instead of divide, and we save the integer parts of the products. Specifically, using decimal multiplication, we multiply the decimal fraction by 2. Next, we place the integer part of the product, which is 0 or 1, to one side, and then multiply the fractional part of the product by 2. Again, we place the resulting integer part to one side, and multiply the fractional part by 2, and so on. We repeat this procedure until the fractional part of the product is zero or until the number of binary digits is that desired. The integer parts of the products form the corresponding binary fraction, with the integers arranged in the order in which we obtained them.

EXAMPLE 1.4

Convert decimal 0.8125 to binary.

Solution.

	integer
$2 \times 0.8125 = 1.625$	1
$2 \times 0.625 = 1.25$	1
$2 \times 0.25 = 0.5$	0
$2 \times 0.5 = 1$	1

So, 0.8125 in decimal is 0.1101 in binary. ■■

This procedure is easy to justify. We convert a decimal fraction into binary by finding the a_i's of

$$a_{-1} \times 2^{-1} + a_{-2} \times 2^{-2} + a_{-3} \times 2^{-3} + \cdots$$

in which each a_i is either 0 or 1. Multiplying this expression by 2 results in

$$a_{-1} + \underbrace{a_{-2} \times 2^{-1} + a_{-3} \times 2^{-2} + \cdots}_{\text{fraction}}$$

Note that the portion of the product to the right of a_{-1} is less than 1 and so is a fraction. Also, a_{-1} is the integer part of the product. If the product has an integer part of 1, then a_{-1} is 1. If the integer part is 0, then a_{-1} is 0.

To find a_{-2}, we use only the fractional part of the first product:

$$a_{-2} \times 2^{-1} + a_{-3} \times 2^{-2} + \cdots$$

Multiplying this by 2 gives

$$a_{-2} + \underbrace{a_{-3} \times 2^{-1} + \cdots}_{\text{fraction}}$$

The portion of the product to the right of a_{-2} is the fraction part of the second product, and a_{-2} is the integer part. Repetition of this process gives the remainder of the binary digits. Incidentally, a terminating fraction in decimal may not be terminating in binary (for example, $0.6_{10} = 0.10011001100110011\cdots_2$), but a terminating fraction in binary is always terminating in decimal.

For a decimal number with both integer and fraction parts, we use the integer rule on the integer part and the fraction rule on the fraction part. Then, we combine parts into one number. For example, for the binary equivalent of 27.875_{10}, we apply the integer rule to obtain $11011_2 = 27_{10}$ and the fraction rule to obtain $0.111_2 = 0.875_{10}$, and then combine the binary parts to obtain $27.875_{10} = 11011.111_2$.

1.6 OCTAL NUMBER SYSTEM

Since the base of the octal number system is 8, this system has eight symbols, which are 0, 1, 2, 3, 4, 5, 6, and 7. Table 1.5 has the correspondences among the decimal, binary, and octal number systems for the first 20 nonnegative integers.

As to be expected, to count in the octal number system, we simply add 1 to the current number to obtain the next number:

$$\ldots, 5 + 1 = 6, 6 + 1 = 7, 7 + 1 = 10, 10 + 1 = 11, \ldots, 16 + 1 = 17, 17 + 1 = 20, 20 + 1 = 21, \ldots$$

TABLE 1.5 DECIMAL-BINARY-OCTAL EQUIVALENCE

Decimal	Binary	Octal	Decimal	Binary	Octal
0	0	0	10	1010	12
1	1	1	11	1011	13
2	10	2	12	1100	14
3	11	3	13	1101	15
4	100	4	14	1110	16
5	101	5	15	1111	17
6	110	6	16	10000	20
7	111	7	17	10001	21
8	1000	10	18	10010	22
9	1001	11	19	10011	23

Note that $7 + 1 = 10$. Since this is a base 8 system, adding 1 to 7 generates a carry to the next digit in an addition of two octal numbers. With this in mind, we can readily add and subtract in octal. In our use of octal as a binary shorthand, we will never have to multiply or divide in octal.

In octal-to-binary conversion, we replace each octal digit with its three-digit binary equivalent. In the resulting binary number, we can, of course, ignore any leading zeros in the integer part or trailing zeros in the fraction part.

EXAMPLE 1.5

Convert 345.5602_8 into its binary equivalent.

Solution.

$$\begin{array}{cccccccc} 3 & 4 & 5 & \cdot & 5 & 6 & 0 & 2 \\ \updownarrow & \updownarrow & \updownarrow & \updownarrow & \updownarrow & \updownarrow & \updownarrow & \updownarrow \\ 011 & 100 & 101 & \cdot & 101 & 110 & 000 & 010 \end{array}$$

So, $345.5602_8 = 11100101.10111000001_2$. ■■

The conversion from binary to octal is just the reverse of the above. Specifically, we group the binary digits (called *bits* for short) by threes from the right and from the left of the binary point. Then, we replace each group by its octal equivalent. In the binary fraction part, we add zeros, if needed, to complete the rightmost group.

EXAMPLE 1.6

Determine the octal equivalent of 11001110.0101101_2.

Solution. We group the bits by threes as follows:

$$011 \quad 001 \quad 110 \;.\; 010 \quad 110 \quad 100$$

We do not really need to add the leading zero in the integer part, but we do need to add the two trailing zeros to the fraction part to complete the last group. After making this grouping, we replace each group with its octal equivalent.

$$\begin{array}{ccccccc} 011 & 001 & 110 & . & 010 & 110 & 100 \\ \updownarrow & \updownarrow & \updownarrow & \updownarrow & \updownarrow & \updownarrow & \updownarrow \\ 3 & 1 & 6 & . & 2 & 6 & 4 \end{array}$$

So, the equivalent octal number is 316.264. If we had not added the two trailing zeros, we would have the erroneous 316.261. ■■

The justification for the conversion rules is that the grouping of the binary digits into groups of three forms, in effect, numbers times powers of 8. This is perhaps best understood from a specific example. Again, consider 011 001 110.010 110 100, which is a contraction of

$$011 \times 2^6 + 001 \times 2^3 + 110 \times 2^0 + 010 \times 2^{-3} + 110 \times 2^{-6} + 100 \times 2^{-9}$$

Converting this to octal, term by term, results in

$$3 \times 8^2 + 1 \times 8^1 + 6 \times 8^0 + 2 \times 8^{-1} + 6 \times 8^{-2} + 4 \times 8^{-3}$$

or 316.264 in contracted form. So, the grouping of bits by threes allows the powers of the binary base to be directly converted into powers of the octal base.

As is evident, octal numbers are a convenient shorthand for representing binary numbers because the equivalent octal numbers have only approximately one-third as many digits, and the conversion between binary and octal is easy and fast. There is another justification for this shorthand. In some computers, the basic binary numbers operated on have parts that are integer multiples of 3 bits. In other words, each part is either 3 bits, 6 bits, 9 bits, or some other integer multiple of 3 bits. So, conversion of these binary numbers to octal provides an exact conversion for each part, which is a convenience.

1.7 HEXADECIMAL NUMBER SYSTEM

In some computers the parts of binary numbers considered as units are multiples of 4 bits instead of 3 bits, making the octal shorthand unsuitable. But, the hexadecimal number system is useful because in converting from binary to hexadecimal, we group the bits by fours.

Since the base of the hexadecimal number system is 16, this system has 16 different symbols. These symbols are the ten decimal digits plus the first six letters of the alphabet: 0, 1, 2, 3, 4, 5, 6, 7, 8, 9, A, B, C, D, E, and F. Table 1.6 has the correspondences for the first 20 nonnegative integers of the decimal, binary, and hexadecimal number systems. By starting with 0 and adding 1 consecutively, we can generate the hexadecimal entries for this table. In doing this, note that since the hexadecimal number system is a base 16 system, a carry is not generated for the next digit until the digit sum exceeds F (15_{10}).

TABLE 1.6 DECIMAL-BINARY-HEXADECIMAL EQUIVALENCE

Decimal	Binary	Hexadecimal	Decimal	Binary	Hexadecimal
0	0	0	10	1010	A
1	1	1	11	1011	B
2	10	2	12	1100	C
3	11	3	13	1101	D
4	100	4	14	1110	E
5	101	5	15	1111	F
6	110	6	16	10000	10
7	111	7	17	10001	11
8	1000	8	18	10010	12
9	1001	9	19	10011	13

The conversion between binary and hexadecimal is similar to that for octal except that the grouping of bits is by fours instead of by threes. Justification for this conversion follows from that for the octal-binary conversion.

EXAMPLE 1.7

Convert 11100101101.1111010111_2 to hexadecimal.

Solution. We group the bits by fours to the right and to the left of the binary point, and then substitute the hexadecimal equivalent for each group.

0111	0010	1101	.	1111	0101	1100
↕	↕	↕	↕	↕	↕	↕
7	2	D	.	F	5	C

So, the equivalent hexadecimal number is 72D.F5C. ■■

EXAMPLE 1.8

Convert $B9A4.E6C_{16}$ to binary.

Solution. In the conversion from hexadecimal to binary, we convert each hexadecimal digit into its 4-bit equivalent.

B	9	A	4	.	E	6	C
↕	↕	↕	↕	↕	↕	↕	↕
1011	1001	1010	0100	.	1110	0110	1100

So, $B9A4.E6C_{16} = 1011100110100100.1110011011_2$. ■■

The easiest way to convert a hexadecimal number to a decimal number is to express the hexadecimal number in expanded form as powers of the base 16 and then add in decimal.

EXAMPLE 1.9

Convert $B63.4C_{16}$ to decimal.

Solution.

$$\begin{aligned} B63.4C_{16} &= 11 \times 16^2 + 6 \times 16^1 + 3 \times 16^0 + 4 \times 16^{-1} + 12 \times 16^{-2} \\ &= 2816 + 96 + 3 + 0.25 + 0.046875 \\ &= 2915.296875_{10} \end{aligned}$$

■■

The decimal-to-hexadecimal conversion parallels that of Examples 1.3 and 1.4 for decimal-to-binary conversion. But, of course, the division and multiplication are by 16 instead of 2, and each decimal intermediate result must be converted to a hexadecimal integer.

EXAMPLE 1.10

Convert $43{,}976.3046875_{10}$ to hexadecimal.

Solution.

	remainder
16 | 43,976	
16 | 2748	8
16 | 171	12→C
16 | 10	11→B
0	10→A

So, $43{,}976_{10} = \text{ABC8}_{16}$. Continuing,

	integer
$16 \times 0.3046875 = 4.875$	4
$16 \times 0.875 = 14.0$	14→E

So,

$$43{,}976.3046875_{10} = \text{ABC8.4E}_{16}$$

■■

1.8 REPRESENTING NEGATIVE BINARY NUMBERS

In a digital computer, or any other digital system, there are no positive (+) or negative (−) signs for representing positive and negative binary numbers. Instead, the signs of binary numbers must somehow be represented by 1s and 0s within the forms of the binary numbers. Three forms of signed binary numbers are popular: *signed magnitude, 1s complement,* and *2s complement*. We will consider these three forms as related to integer-type or fixed-point numbers in which the first bit in a number is the sign bit and each binary point is assumed to be immediately to the right of the least-significant digit. But the concepts apply as well to numbers that are not integers.

In the signed-magnitude form, a positive or negative binary number is represented by a sign bit followed by the magnitude in binary. For example, for an 8-bit representation, the decimal number 13 is represented by the binary number 00001101 and −5 by 10000101, with, as is conventional, a 0 bit representing a positive sign and a 1 bit a negative sign. Incidentally, there are two representations for the number zero, a positive zero and a negative zero, which for 8 bits are 00000000 and 10000000, respectively.

In the 1s-complement representation, positive numbers are the same as in the signed-magnitude representation, but negative numbers differ for they are represented in 1s-complement form. To find the 1s-complement representation of a negative number, all we have to do is to consider the number to be positive, and then change all 0s to 1s, including the sign bit, and all 1s to 0s. For example, to find the 8-bit representation of decimal number −13, we first find the 8-bit binary equivalent of +13, which is 00001101. Then we change the 0s to 1s and the 1s to 0s, and obtain 11110010, which is the desired complement representation.

In the 2s-complement representation, positive numbers are the same as in the signed-magnitude representation, but negative numbers differ; they are represented in 2s-complement form. The 2s complement of a negative number is simply the 1s complement plus 1. As an illustration, the 8-bit 1s complement of -13 is 11110010, as has been shown. Therefore, the 2s-complement representation is $11110010 + 1 = 11110011$. Applications of the 2s-complement representation, along with 1s-complement and signed-magnitude representations, will be demonstrated throughout this text.

1.9 BINARY-CODED DECIMAL (BCD) REPRESENTATION

The *binary-coded decimal (BCD)* representation is a code for decimal numbers. It is an alternative to converting a decimal number to its binary equivalent. Many applications require the inputting and/or displaying of decimal digits. In these applications, it is often convenient to store the data in the BCD representation.

In the BCD representation, each decimal digit of a decimal number is coded into binary. Since there are ten decimal digits, a binary representation of these digits requires 4 bits. Only 10 of the 16 combinations of the 4 bits are required for a BCD code. Of the many possible such codes, the most popular is the 8421 code illustrated in Table 1.7. In this code, each decimal digit is converted into its 4-bit binary equivalent, and then the groups of four bits are concatenated.

EXAMPLE 1.11

What is the representation for the decimal number 7963 in the 8421 code?

Solution. In the coding of 7963, we replace each digit by the corresponding 4 bits from Table 1.7. The result is

7	9	6	3	decimal number
↓	↓	↓	↓	
0111	1001	0110	0011	8421 code

Finally, we concatenate the four groups of four bits:

0111100101100011 ■■

The 8421 code is so popular as a BCD code that when a reference is made to the BCD code, we should assume that it is the 8421 code unless another BCD code is specified.

TABLE 1.7. 8421 BCD CODE

zero	0000	five	0101
one	0001	six	0110
two	0010	seven	0111
three	0011	eight	1000
four	0100	nine	1001

SUPPLEMENTARY READING (see Bibliography)

[Bartee 85], [Hill 81], [Mano 84], [McCluskey 75], [Roth 85]

PROBLEMS

1.1. Perform the indicated operations on the following binary numbers:

(a) 1101 + 110

(b)
```
   111.011
  1111.11
+ 111001.1
```

(c) 10011 − 110

(d) 1001.1101 − 11100.001

(e)
```
   111.01
  1111.1
  1011.11
+ 1101.111
```

(f) 11001.01 − 111.1

1.2. Repeat Problem 1.1 for

(a) 1101.1 + 111.101

(b)
```
    111.01
   1011.11
+ 11110.01
```

(c) 11011 − 101

(d) 1001.01 − 10110.101

(e)
```
     111.101
    1110.11
   11111.01
+ 101110.011
```

(f) 1100.01 − 111.1

1.3. Perform the indicated binary multiplications and divisions.

(a) 101 × 1101 (b) 11011 × 1100.1

(c) 111011.01 × 1111.001 (d) 10100 ÷ 100

(e) 100011.11 ÷ 101.1 (f) 101010001.1 ÷ 11011

1.4. Repeat Problem 1.3 for

(a) 1101 × 110011 (b) 11010.1 × 1101.01

(c) 1011101.011 × 10110.101 (d) 1001000 ÷ 110

(e) 10000000.01 ÷ 1101.1 (f) 110101001.11101 ÷ 110011.101

1.5. Convert the following binary numbers to decimal:

(a) 100111 (b) 10101.101 (c) 10001001.001

1.6. Repeat Problem 1.5 for

(a) 1100111 (b) 1110.111 (c) 1100110101.0111

1.7. Convert the following decimal numbers to binary:

(a) 146 (b) 1928.875

(c) 14.6 (d) $19\frac{2}{9}$

1.8. Repeat Problem 1.7 for

(a) 1689 (b) 1430.625 (c) 178.3

1.9. Perform the indicated operations on the following octal numbers:

(a) 324 + 564 (b) 710.145 + 217.633

(c) 352 − 163
(d) 1236.47 − 765.2377
(e) 40002.3 − 675.77
(f)
```
     46.23
    327.12
 + 4652.327
```
(g) 473.63 − 754.321

1.10. Repeat Problem 1.9 for
(a) 472 + 326
(b) 410.23 + 367.55
(c) 456 − 324
(d) 4723.636 − 724.647
(e)
```
     47.653
    327.734
  + 467.772
```
(f) 5346.72 − 600321.1

1.11. Convert the following octal numbers to decimal:
(a) 110011 **(b)** 1234.54 **(c)** 7006.302

1.12. Repeat Problem 1.11 for
(a) 30076 **(b)** 6403.2 **(c)** 400602.244

1.13. Convert the following decimal numbers to octal:
(a) 3094
(b) 2906.5625
(c) 250.3
(d) $37\frac{1}{9}$

1.14. Repeat Problem 1.13 for
(a) 1000
(b) 100
(c) 2345.46875
(d) $46\frac{3}{13}$

1.15. Convert the following binary numbers to octal:
(a) 101101
(b) 110010111011.01011011
(c) 1000111101.01

1.16. Repeat Problem 1.15 for
(a) 1101011
(b) 10111111011.0110111
(c) 110111.0111

1.17. Convert the following octal numbers to binary:
(a) 76002
(b) 773406.245

1.18. Repeat Problem 1.17 for
(a) 23077
(b) 432.7066

1.19. Convert the following hexadecimal numbers to decimal:
(a) 4A6
(b) ABD.C8

1.20. Repeat Problem 1.19 for
(a) CAB
(b) FE6.8C4

1.21. Convert the following binary numbers to hexadecimal:
(a) 110111001
(b) 110001101.1001

1.22. Repeat Problem 1.21 for
(a) 1001110101
(b) 1100111001.1000101

1.23. Convert the following hexadecimal numbers to binary:
(a) CD3F
(b) 8F2A.5EF

1.24. Repeat Problem 1.23 for
(a) ACD8
(b) 3FD2.ACD

1.25. Convert the following decimal numbers to hexadecimal:
(a) 329.5 **(b)** 7495.34 **(c)** $17\frac{1}{7}$

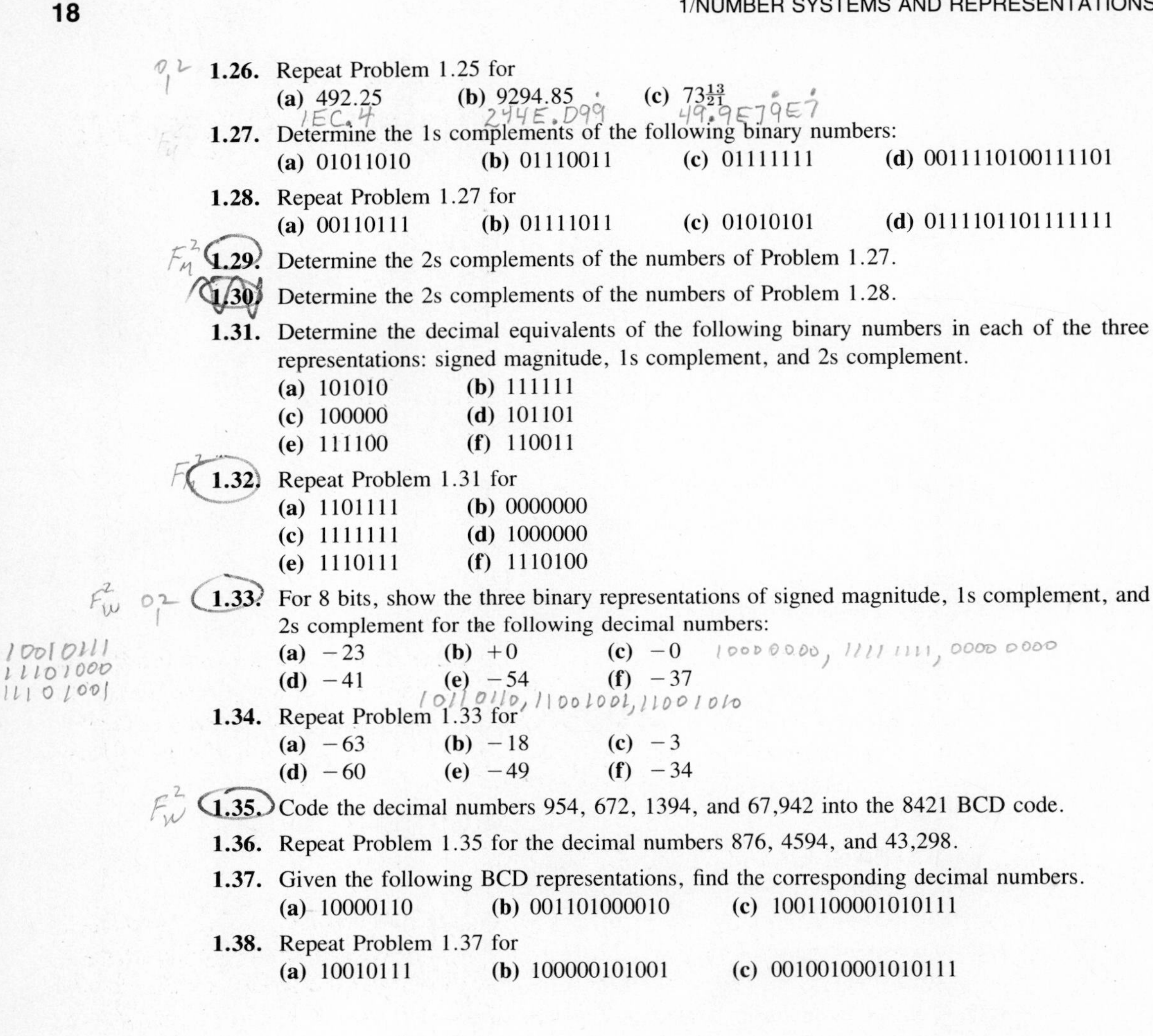

1.26. Repeat Problem 1.25 for
(a) 492.25 (b) 9294.85 (c) $73\frac{13}{21}$

1.27. Determine the 1s complements of the following binary numbers:
(a) 01011010 (b) 01110011 (c) 01111111 (d) 0011110100111101

1.28. Repeat Problem 1.27 for
(a) 00110111 (b) 01111011 (c) 01010101 (d) 0111101101111111

1.29. Determine the 2s complements of the numbers of Problem 1.27.

1.30. Determine the 2s complements of the numbers of Problem 1.28.

1.31. Determine the decimal equivalents of the following binary numbers in each of the three representations: signed magnitude, 1s complement, and 2s complement.
(a) 101010 (b) 111111
(c) 100000 (d) 101101
(e) 111100 (f) 110011

1.32. Repeat Problem 1.31 for
(a) 1101111 (b) 0000000
(c) 1111111 (d) 1000000
(e) 1110111 (f) 1110100

1.33. For 8 bits, show the three binary representations of signed magnitude, 1s complement, and 2s complement for the following decimal numbers:
(a) -23 (b) $+0$ (c) -0
(d) -41 (e) -54 (f) -37

1.34. Repeat Problem 1.33 for
(a) -63 (b) -18 (c) -3
(d) -60 (e) -49 (f) -34

1.35. Code the decimal numbers 954, 672, 1394, and 67,942 into the 8421 BCD code.

1.36. Repeat Problem 1.35 for the decimal numbers 876, 4594, and 43,298.

1.37. Given the following BCD representations, find the corresponding decimal numbers.
(a) 10000110 (b) 001101000010 (c) 1001100001010111

1.38. Repeat Problem 1.37 for
(a) 10010111 (b) 100000101001 (c) 0010010001010111

Chapter 2

Boolean Algebra

2.1 INTRODUCTION

This chapter presents *Boolean algebra,* which is the basic mathematics required for the study of the design of digital circuits. Since these circuits are also called switching circuits, the Boolean algebra used in their design is sometimes called *switching algebra*. Boolean algebra provides a systematic method for describing and simplifying the logic processes at the basic component level of digital design.

2.2 BOOLEAN VARIABLES AND LOGIC VALUES

A *Boolean variable,* unlike an ordinary algebraic variable, has only one of two values: true or false, called *logic values*. A Boolean variable can be viewed as representing a statement that can be only true or false. For example, the Boolean variable A may represent the statement ''Frank has red hair.'' Obviously, variable A can have only the logic value of either true or false. In other words, the statement ''Frank has red hair'' is either true (Frank does have red hair) or false (Frank does not have red hair). If the statement is false, Frank has another color hair or is bald.

A Boolean variable may be a function of other Boolean variables. Then, as will be seen in the next chapter, we will find it convenient to call the function variable an *output variable* and the other variables *input variables*. The value of the output variable depends not only on the values of the input variables, but also on the operations of the function.

2.3 FUNDAMENTAL OPERATIONS

Boolean algebra has three fundamental operations: AND, OR, and NOT. We will now consider each of them.

2.3.1 AND Operation

The AND operation is represented by the symbol "·" as in

$$Z = A \cdot B$$

which in words is "Z is equal to A AND B." This equation specifies that the Boolean variable Z is true if both A is true *and* B is true. Otherwise, Z is false.

The AND operation can be precisely defined in a *truth table* (also called a *logic table*), as follows:

A	B	$Z = A \cdot B$
F	F	F
F	T	F
T	F	F
T	T	T

A truth table is simply a listing of all possible values of the input variables (here, A and B) with the corresponding output variable values resulting from the operation (or operations). It is apparent from this table that Z is true (T) if and only if both A is true (T) *and* B is true (T). Otherwise, Z is false (F).

For a physical aid to the understanding of the AND operation, consider the circuit of Fig. 2.1, containing a voltage source, two switches A and B, and a light bulb Z. For this circuit, a variable A might represent the statement that "switch A is closed," a variable B the statement that "switch B is closed," and a variable Z the statement that "the light bulb Z is lit." Clearly, if either statement A or B is false, then statement Z is false, since both switches must be closed for the voltage source to energize the light bulb.

The AND operation of Boolean algebra is not related to the multiplication operation of ordinary arithmetic, although the symbols are the same. Generally, Boolean algebra has many of the same elements as ordinary algebra, and has many of the same symbols and terminology. But there are some important differences. Therefore, in working with Boolean algebra, we should not rely on our past study of ordinary algebra but rather rely solely on the definitions given here.

The AND operation applies to any number of input variables. For three input variables the AND truth table is

A	B	C	$Z = A \cdot B \cdot C$
F	F	F	F
F	F	T	F
F	T	F	F
F	T	T	F
T	F	F	F
T	F	T	F
T	T	F	F
T	T	T	T

Note that Z is false unless all input variables are true, which is a general property of the AND operation, regardless of the number of input variables.

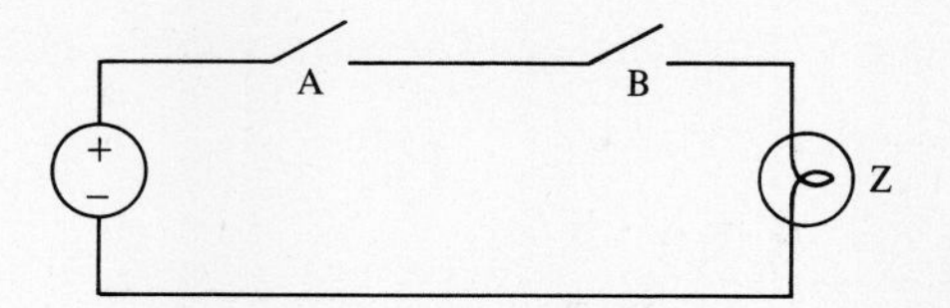

Figure 2.1 Illustrating AND.

For this truth table, as for all truth tables, there is a standard way of assigning the T's and F's in the input variable columns. All columns start with F. Then, in the rightmost input variable column, the F's and T's alternate. In the next column they alternate by twos, and in the first column they alternate by fours. If there were a fourth column, they would alternate by eights, and so on.

To have all possible combinations of input variable values, a truth table must have a number of rows equal to 2^n, in which n is the number of input variables. So, a two-variable truth table has 4 rows, a three-variable one has 8 rows, a four-variable one has 16 rows, and so on.

2.3.2 OR Operation

The OR operation is represented by the symbol "+" as in

$$Z = A + B$$

which in words is "Z is equal to A OR B." This equation specifies that Z is true if *either* A *or* B is true. Otherwise, Z is false. In a truth table, this specification is

A	B	Z = A + B
F	F	F
F	T	T
T	F	T
T	T	T

Although the symbols are the same, the OR operation of Boolean algebra is not related to the addition operation of ordinary arithmetic.

For an illustration of the OR operation, consider the circuit of Fig. 2.2. As with Fig. 2.1, let variable A correspond to the statement that "switch A is closed," variable B to the statement that "switch B is closed," and variable Z to the statement that "light bulb Z is lit." Obviously, if either statement A or B is true, then statement Z is true, since the closing of either switch completes a circuit from the voltage source to the light bulb.

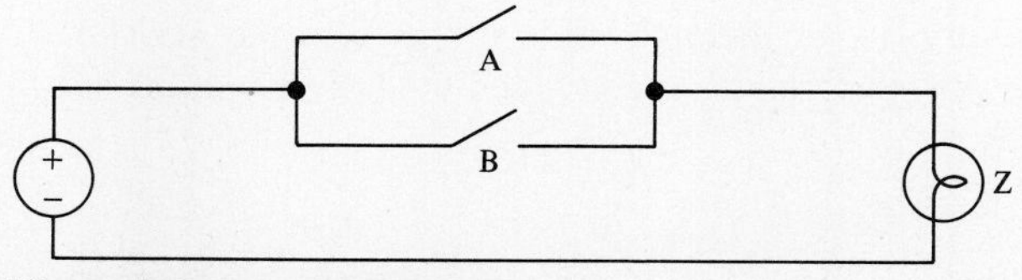

Figure 2.2 Illustrating OR.

The OR operation can extend to any number of input variables. For three input variables the OR truth table is

A	B	C	Z = A + B + C
F	F	F	F
F	F	T	T
F	T	F	T
F	T	T	T
T	F	F	T
T	F	T	T
T	T	F	T
T	T	T	T

Note that Z is true unless all input variables are false, which is a general property of the OR operation, regardless of the number of input variables.

2.3.3 NOT Operation

The NOT operation is designated by an overline, as in

$$Z = \overline{A}$$

which in words is that "Z is equal to A NOT." The NOT operation is defined as follows:

A	$Z = \overline{A}$
F	T
T	F

The NOT operation is a *complement* operation. Since a Boolean variable can have only one of two values, the complement of one logic value is the other value: $\overline{F} = T$ and $\overline{T} = F$.

The NOT operation can apply to more than one variable, and in fact to an entire expression. Then, the overline extends over more than one variable. As an illustration, the complement of $A \cdot \overline{B} + C$ has the designation $\overline{A \cdot \overline{B} + C}$.

2.3.4 Operation Hierarchy

To evaluate an expression, we have to know the priority of the operations. Suppose, for example, we want the value of the expression $\overline{A} + B \cdot C$ for A = F, B = T, and C = F. Since all three operations are present, we need to know the order in which to perform them. The priority or hierarchy rule is: perform the *individual* variable NOT operations first, the AND operations second, and the OR operations last. Therefore, this expression evaluates to

$$\overline{A} + B \cdot C = \overline{F} + T \cdot F = T + T \cdot F = T + F = T$$

We can override the priority rule by using parentheses, as in

$$(\overline{A} + B) \cdot C = (\overline{F} + T) \cdot F = (T + T) \cdot F = T \cdot F = F$$

The parentheses cause the OR operation to be performed before the AND operation.

If an expression is complemented, we must evaluate the expression before complementing; that is, we do not complement first.

The hierarchy of operations and the evaluation of expressions can best be understood from some examples.

EXAMPLE 2.1

Evaluate the following expressions for A = T, B = F, C = T, and D = T: (a) $A \cdot \overline{B} + C$, (b) $(\overline{A} + B) \cdot (C + \overline{B} \cdot D)$, and (c) $A \cdot \overline{B} \cdot (C + D \cdot \overline{E})$.

Solution.

(a) $A \cdot \overline{B} + C = T \cdot \overline{F} + T = T \cdot T + T = T + T = T$

We could have obtained the T result immediately after inserting the values, because T (from the C) OR anything is equal to T.

(b)
$$\begin{aligned}(\overline{A} + B) \cdot (C + \overline{B} \cdot D) &= (\overline{T} + F) \cdot (T + \overline{F} \cdot T)\\ &= (F + F) \cdot (T + T \cdot T)\\ &= (F + F) \cdot (T + T)\\ &= (F) \cdot (T) = F\end{aligned}$$

We could have stopped after finding that one factor is false, since F AND anything is F.

(c)
$$\begin{aligned}A \cdot \overline{B} \cdot (C + D \cdot \overline{E}) &= T \cdot \overline{F} \cdot (T + T \cdot \overline{E})\\ &= T \cdot T \cdot (T + T \cdot \overline{E}) = T \cdot T \cdot T = T\end{aligned}$$

Note that the quantity in parentheses is true because T + anything = T. ■■

2.3.5 Summary

A summary of the three basic Boolean operations is given in Table 2.1. Remember that a Boolean variable can have only one of two logic values: true (T) or false (F). For convenience, though, we will often represent the logic value true (T) by the symbol 1, and the logic value false (F) by the symbol 0, as in Table 2.1. *Important:* Although the symbol 1 *looks* like the numeric 1 and the symbol 0 *looks* like the numeric 0, they are *logic* values representing true and false, respectively. So we cannot add or subtract or perform any other ordinary arithmetic operations on the logic values 0 and 1.

TABLE 2.1

A	B	$A \cdot B$	$A + B$	$\overline{A}$	$\overline{B}$
0	0	0	0	1	1
0	1	0	1	1	0
1	0	0	1	0	1
1	1	1	1	0	0

2.4 LOGIC EXPRESSIONS FROM TRUTH TABLES

We have used truth tables in defining the AND, OR, and NOT operations. Truth tables are also useful in designing digital circuits. Specifically, in a design we form a truth table from the design specifications and then determine a logic expression from the truth table.

Consider the design of a simple digital circuit for controlling the operation of a light bulb Z with two switches A and B. Suppose the light bulb is to be on when both switches are on (closed), and also when both switches are off (open). Otherwise (one switch off and the other on), the light bulb is to be off.

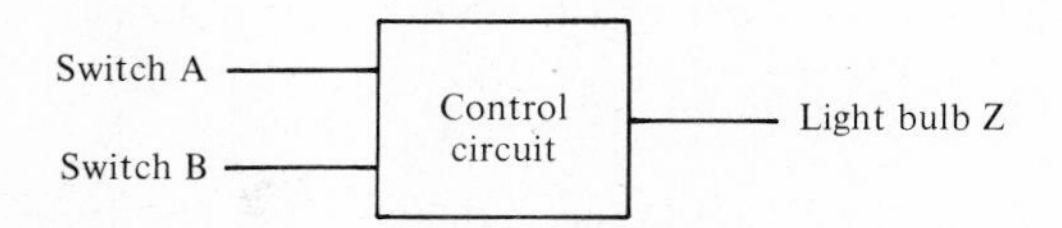

We can easily represent this description of the circuit operation with a truth table in which a 0 implies that a switch or the light bulb is off, and a 1 implies that a switch or the light bulb is on.

A	B	Z
0	0	1
0	1	0
1	0	0
1	1	1

The next step is to obtain a logic expression from this truth table. There are two methods we can use, and they give different types of logic expressions. One type is a sum-of-products (SOP) expression, and the other is a product-of-sums (POS) expression. An SOP expression comprises several "product" (AND) terms "summed" (ORed) together. For example, the following expression is in SOP form:

$$A \cdot \overline{B} + \overline{A} \cdot B + A \cdot B$$

A POS expression comprises several "sum" (OR) factors "multiplied" (ANDed) together, as in

$$(A + \overline{B}) \cdot (C + D) \cdot (\overline{A} + C)$$

2.4.1 SOP Expression from a Truth Table

To obtain an SOP expression for the truth table of the light control circuit, we need to derive a logic expression for Z in the form of

$$Z = f(A, B)$$

that will express the condition for which Z is true (the light bulb is on). In other words, what are the combinations of the inputs A and B for which Z is true (1)? From the truth table, we see that Z is true when A is false *and* B is false, *or* when A is true *and* B is true. Written in a formalized manner, this is

$$Z = \overline{A} \cdot \overline{B} + A \cdot B$$

We can generalize from this example and derive the following rules for obtaining an SOP expression from a truth table:

1. Identify each row in which the output function is *true* (1).
2. For each of these rows, write an AND term of all the input variables, applying complements such that the term evaluates to true for the variable values of that row. Incidentally, each of these terms is commonly called a *minterm*. (In general, a minterm is a term that contains all the input variables.)
3. OR together all the AND terms (minterms) found in step 2.

The resulting SOP expression is called a *canonical sum-of-products expression* (CSOP) or, more simply, a *minterm expansion*. Note that in a CSOP each minterm contributes one and only one true (1) value. Examples will help us better understand these rules.

EXAMPLE 2.2

Given the following truth table, find a CSOP expression for Z.

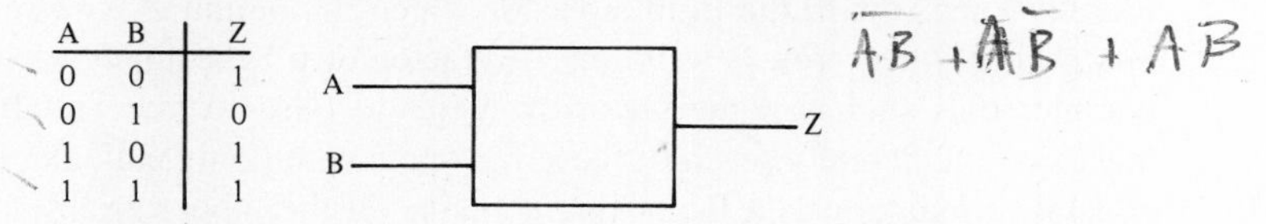

A	B	Z
0	0	1
0	1	0
1	0	1
1	1	1

Solution.

1. Rows 1, 3, and 4 have true (1) outputs.
2. The minterms corresponding to these rows are $\overline{A} \cdot \overline{B}$, $A \cdot \overline{B}$, and $A \cdot B$, respectively.
3. ORing these minterms results in $Z = \overline{A}\,\overline{B} + A\overline{B} + AB$.

Note that the AND operator symbol "·" is omitted in the final result. The AND operator is implied by the adjacent placement of the variables. We will often omit this symbol when its omission does not cause any confusion. ■■

EXAMPLE 2.3

Find a CSOP expression for the Z specified in the following truth table.

A	B	C	Z
0	0	0	1
0	0	1	0
0	1	0	1
0	1	1	0
1	0	0	0
1	0	1	0
1	1	0	1
1	1	1	1

A
B
C
Z

Solution.

1. Rows 1, 3, 7, and 8 have true (1) outputs.
2. The corresponding minterms are, respectively, $\overline{A}\,\overline{B}\,\overline{C}$, $\overline{A}B\overline{C}$, $AB\overline{C}$, and ABC.
3. ORing the minterms results in $Z = \overline{A}\,\overline{B}\,\overline{C} + \overline{A}B\overline{C} + AB\overline{C} + ABC$. ■■

2.4.2 POS Expression from a Truth Table

From the same truth table for the described light control circuit, we can also obtain a POS expression for Z. Recall that this truth table is

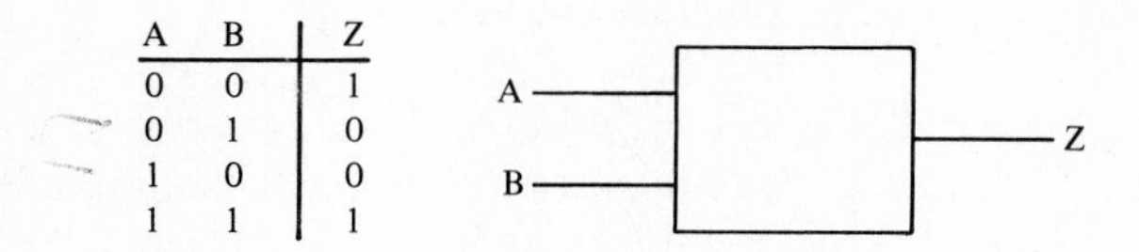

A	B	Z
0	0	1
0	1	0
1	0	0
1	1	1

The rules for obtaining a POS expression from a truth table are as follows:

1. Identify each row in the truth table for which the output Z is *false* (0).
2. For each of these rows, write an OR factor of all the input variables, using complements such that the factor evaluates to false for the variable values of that row. Incidentally, each of these factors is commonly called a *maxterm*. (In general, a maxterm is a factor that contains all the input variables.)
3. AND together all the OR factors (maxterms) found in step 2.

The resulting POS expression is called a *canonical product-of-sums expression* (CPOS) or, more simply, a *maxterm expansion*. Note that in a CPOS, each maxterm contributes one and only one false (0) value.

We will now apply these rules to the light control circuit example.

1. Z is false (0) for rows 2 and 3.
2. The corresponding maxterms are $A + \overline{B}$ and $\overline{A} + B$.
3. ANDing the maxterms, we obtain $Z = (A + \overline{B})(\overline{A} + B)$.

EXAMPLE 2.4

Find a CPOS expression for Z.

A	B	Z
0	0	1
0	1	0
1	0	1
1	1	1

Solution.

1. Row 2 is the only row with a false (0) output.
2. The maxterm corresponding to this row is $A + \overline{B}$.
3. Since there is only one maxterm, Z is equal to it: $Z = A + \overline{B}$. ■■

EXAMPLE 2.5

Find a CPOS expression for Z.

A	B	C	Z
0	0	0	1
0	0	1	0
0	1	0	1
0	1	1	0
1	0	0	0
1	0	1	0
1	1	0	1
1	1	1	1

Solution.

1. Rows 2, 4, 5, and 6 have false (0) outputs.
2. The corresponding maxterms are, respectively, $A + B + \overline{C}$, $A + \overline{B} + \overline{C}$, $\overline{A} + B + C$, and $\overline{A} + B + \overline{C}$.
3. By ANDing the maxterms, we obtain

$$Z = (A + B + \overline{C})(A + \overline{B} + \overline{C})(\overline{A} + B + C)(\overline{A} + B + \overline{C})$$ ■■

2.5 EQUIVALENT EXPRESSIONS

Two expressions are *equivalent* if for every set of values of the input variables, the two expressions evaluate to the *same* value. As an illustration, for the light control circuit we obtained a CSOP expression $\overline{A}\overline{B} + AB$ *and* an equivalent CPOS expression $(A + \overline{B})(\overline{A} + B)$. We can verify that these two expressions are equivalent by using a truth table.

A	B	$\overline{A} \cdot \overline{B} + A \cdot B$	$(A + \overline{B})(\overline{A} + B)$
0	0	$\overline{0} \cdot \overline{0} + 0 \cdot 0 = 1$	$(0 + \overline{0})(\overline{0} + 0) = 1$
0	1	$\overline{0} \cdot \overline{1} + 0 \cdot 1 = 0$	$(0 + \overline{1})(\overline{0} + 1) = 0$
1	0	$\overline{1} \cdot \overline{0} + 1 \cdot 0 = 0$	$(1 + \overline{0})(\overline{1} + 0) = 0$
1	1	$\overline{1} \cdot \overline{1} + 1 \cdot 1 = 1$	$(1 + \overline{1})(\overline{1} + 1) = 1$

Since the two expressions evaluate to the same values for each combination of input variable values, they are equivalent, or

$$Z = \overline{A}\overline{B} + AB = (A + \overline{B})(\overline{A} + B)$$

For another illustration, recall the truth table for the OR operation:

A	B	Z
0	0	0
0	1	1
1	0	1
1	1	1

The CSOP expression for Z is $AB + \overline{A}B + A\overline{B}$, and the CPOS expression is $A + B$. Obviously, the two expressions are equivalent. In other words,

$$Z = \overline{A}B + A\overline{B} + AB = A + B$$

Again, we can show this equivalence with a truth table. But there is another method of proof. We can *reduce* the expression $\overline{A}B + A\overline{B} + AB$ to the expression $A + B$ by using *Boolean identities:*

$$\begin{aligned} \overline{A}B + A\overline{B} + AB &= \overline{A}B + (A\overline{B} + AB) = \overline{A}B + A(\overline{B} + B) \\ &= \overline{A}B + A(1) = \overline{A}B + A = B + A = A + B \end{aligned}$$

Obviously, some of us may not fully comprehend this manipulation at this point since Boolean identities are discussed in the next section.

2.6 BOOLEAN IDENTITIES

Generally, we want to reduce any logic expression to its simplest form since a simpler expression usually requires fewer hardware components for its implementation. Boolean identities are often useful in a reduction.

Table 2.2 contains the Boolean identities most important to digital design. Identities 1–5 are the fundamental relations of Boolean algebra and provide the foundation for

TABLE 2.2 BOOLEAN IDENTITIES

	(a)	(b)
1.	$\overline{\overline{A}} = A$	$\overline{\overline{A}} = A$
2.	$A + \text{false} = A \quad (A + 0 = A)$	$A \cdot \text{true} = A \quad (A \cdot 1 = A)$
3.	$A + \text{true} = \text{true} \quad (A + 1 = 1)$	$A \cdot \text{false} = \text{false} \quad (A \cdot 0 = 0)$
4.	$A + A = A$	$A \cdot A = A$
5.	$A + \overline{A} = \text{true} \quad (A + \overline{A} = 1)$	$A \cdot \overline{A} = \text{false} \quad (A \cdot \overline{A} = 0)$
6.	$A + B = B + A$	$A \cdot B = B \cdot A$
7.	$A + B + C = (A + B) + C = A + (B + C)$	$A \cdot B \cdot C = (A \cdot B) \cdot C = A \cdot (B \cdot C)$
8.	$A \cdot (B + C) = A \cdot B + A \cdot C$	$A + B \cdot C = (A + B)(A + C)$
9.	$\overline{A + B} = \overline{A} \cdot \overline{B}$	$\overline{A \cdot B} = \overline{A} + \overline{B}$
10.	$A \cdot B + A \cdot \overline{B} = A$	$(A + B)(A + \overline{B}) = A$
11.	$A + A \cdot B = A$	$A(A + B) = A$
12.	$A(\overline{A} + B) = A \cdot B$	$A + \overline{A} \cdot B = A + B$
13.	$A \cdot B + \overline{A} \cdot C + B \cdot C = A \cdot B + \overline{A} \cdot C$	$(A + B)(\overline{A} + C)(B + C) = (A + B)(\overline{A} + C)$

Boolean manipulation. Identities 6, 7, and 8 are, respectively, the commutative, associative, and distributive laws of Boolean algebra. Except for 8(b), they are very similar to the corresponding laws of ordinary algebra. Identity 8(b) is unfamiliar only if we think in terms of "+" as addition and "·" as multiplication operators. Remember that "+" in Boolean algebra is the OR operator and so has properties that are not the same as the addition operator of ordinary algebra. One of these properties enables us to "multiply through" or "factor out," as in

$$A + BCD = (A + B)(A + C)(A + D)$$

Identities 9(a) and (b) are the well-known DeMorgan's laws. Note that both forms of DeMorgan's laws have the complement of an entire expression, and the effect of this complementing is to change each "+" to a "·" and each "·" to a "+" and to complement each *literal*. (A literal is a generic term for a variable or its complement.) Identities 10 through 13 are other identities frequently used in digital design. Identities 10(a) and (b) are the bases for several systematic Boolean simplification methods, one of which we will study in a following section. In each of these identities a variable may be replaced by an expression, and the identity will still be valid.

Using truth tables, we can prove all these Boolean identities, exhaustively. As an illustration, we will use this method to prove DeMorgan's law of 9(a) for three variables. The result is

A	B	C	$\overline{A + B + C}$	$\overline{A} \cdot \overline{B} \cdot \overline{C}$
0	0	0	1	1
0	0	1	0	0
0	1	0	0	0
0	1	1	0	0
1	0	0	0	0
1	0	1	0	0
1	1	0	0	0
1	1	1	0	0

Since the two output colums are identical, then $\overline{A + B + C} = \overline{A} \cdot \overline{B} \cdot \overline{C}$.

Another method of proving these identities is with Boolean manipulation, using other proven identities. As an illustration, we will prove identity 12(a) in this manner, with the assumption that the other identities we use have been proved.

$$A(\overline{A} + B) = A\overline{A} + AB \quad \text{by the distributive law 8(a)}$$
$$A\overline{A} + AB = 0 + AB \quad \text{by 5(b)}$$
$$0 + AB = AB \quad \text{by 2(a)}$$

So, we have proved that $A(\overline{A} + B) = AB$ by using Boolean manipulation to obtain the second expression from the first.

EXAMPLE 2.6

Using Boolean manipulation, prove identity 11(a), assuming that the preceding identities are valid.

Solution.

$$
\begin{aligned}
A + AB &= (A \cdot 1) + AB && 2(b)\\
(A \cdot 1) + AB &= A \cdot (1 + B) && 8(a)\\
A \cdot (1 + B) &= A \cdot (1) && 3(a)\\
A \cdot (1) &= A && 2(b)
\end{aligned}
$$

■ ■

To gain facility with the use of these identities, you should use Boolean manipulation to work through the proofs of some of the other identities of Table 2.2.

Note in Table 2.2 that the Boolean identities are divided into two columns. The identities in column (a) and the corresponding identities in column (b) are *duals* of each other. A dual of an identity is formed by replacing each AND operator with an OR operator, each OR operator with an AND operator, each true (1) with false (0), and each false with true. This procedure for finding a dual of an expression is identical to the application of DeMorgan's laws, except that there is no complementing of literals. In general, in Boolean algebra a dual of a theorem is another valid theorem.

2.7 SIMPLIFICATION USING BOOLEAN IDENTITIES

The primary use of Boolean manipulation is for simplifying Boolean expressions to obtain simpler expressions for digital design. A simpler expression usually results in a simpler implementation with digital devices.

In using Boolean manipulation we are usually trying to obtain either a *minimum sum of products* (MSOP) or a *minimum product of sums* (MPOS) that is equivalent to the original expression. An MSOP is an SOP that is equivalent to the original expression, and has no more terms or literals than any other equivalent SOP. In other words, in an MSOP, the number of terms and the number of literals are the minimum possible for any equivalent SOP. There may be several equivalent MSOP expressions. Once in a while, though, the requirements for a minimum number of terms and a minimum number of literals are mutually exclusive—they conflict. Then, there is no clear meaning of what an MSOP is.

The MSOP definition applies to an MPOS with the substitution of "factors" for "terms." An MPOS and an equivalent MSOP may have a different number of literals, as we will see for the expression of the next example. Also, the number of factors of the MPOS may be different than the number of terms of the MSOP.

EXAMPLE 2.7

Find an MSOP for $F = \overline{X}W + Y + \overline{Z}(Y + \overline{X}W)$.

Solution.

$$
\begin{aligned}
F &= \overline{X}W + Y + \overline{Z}(Y + \overline{X}W)\\
&= \overline{X}W + Y + \overline{Z}Y + \overline{Z}\overline{X}W && 8(a)\\
&= (\overline{X}W + \overline{Z}\overline{X}W) + (Y + \overline{Z}Y) && 6(a),\ 7(a)\\
&= \overline{X}W(1 + \overline{Z}) + Y(1 + \overline{Z}) && 8(a)
\end{aligned}
$$

$$= \overline{X}W(1) + Y(1) \qquad 3(a)$$
$$= \overline{X}W + Y \qquad 2(b)$$

So, $\overline{X}W + Y$ is an MSOP expression for F.

Sometimes we want an MPOS even though obtaining an MSOP is easier. We can often obtain an MPOS from an MSOP by taking the dual of the MSOP, multiplying out, and then taking the dual again, making obvious simplifications in multiplying out. Taking the dual twice gives us an equivalent expression. Doing this for the $\overline{X}W + Y$ of this example, we obtain $(\overline{X} + W)Y$ from taking the dual. Then, we multiply, getting $\overline{X}Y + WY$. Finally, we take the dual again, obtaining $(\overline{X} + Y)(W + Y)$, which is an MPOS expression for F. Note that this MPOS has one more literal than the MSOP. ■■

EXAMPLE 2.8

Find an MSOP for $F = \overline{W}XY + \overline{W}XZ + \overline{(Y + Z)}$.

Solution.

$$F = \overline{W}XY + \overline{W}XZ + \overline{(Y + Z)}$$
$$= \overline{W}X(Y + Z) + \overline{(Y + Z)} \qquad 8(a)$$
$$= \overline{W}X + \overline{(Y + Z)} \qquad 12(b)$$
$$= \overline{W}X + \overline{Y}\,\overline{Z} \qquad 9(a)$$

■■

EXAMPLE 2.9

Find an MSOP for $F = (\overline{X} + WY + \overline{Z})(\overline{X} + WY + Z)$.

Solution.

$$F = (\overline{X} + WY + \overline{Z})(\overline{X} + WY + Z)$$
$$= (\overline{X} + WY) + (\overline{Z} \cdot Z) \qquad 8(b)$$
$$= (\overline{X} + WY) + 0 \qquad 5(b)$$
$$= \overline{X} + WY \qquad 2(a)$$

Or, *directly* use 10(b) and obtain $F = \overline{X} + WY$ in one step. ■■

EXAMPLE 2.10

Find an MSOP for $F = V\overline{W}XY + VWYZ + V\overline{X}YZ$.

Solution.

$$F = V\overline{W}XY + VWYZ + V\overline{X}YZ$$
$$= VY(\overline{W}X + WZ + \overline{X}Z) \qquad 8(a)$$
$$= VY(\overline{W}X + Z(W + \overline{X})) \qquad 8(a)$$

Since $$(W + \overline{X}) = \overline{\overline{(W + \overline{X})}} = \overline{\overline{W}X} \qquad 1(a), 9(a)$$

$$F = VY(\overline{W}X + Z(\overline{\overline{W}X}))$$
$$= VY(\overline{W}X + Z) \qquad 12(b)$$
$$= VY\overline{W}X + VYZ \qquad 8(a)$$

■■

EXAMPLE 2.11

Find an MSOP for $F = W\overline{X}\,\overline{Y} + \overline{W}XY + W\overline{X}Z + WYZ + XYZ$.

Solution. Where do do we begin? By using identity 13(a), we can obtain

$$W\overline{X}Z + XYZ + WYZ = W\overline{X}Z + XYZ \quad (1)$$
$$W\overline{X}\,\overline{Y} + WYZ + W\overline{X}Z = W\overline{X}\,\overline{Y} + WYZ \quad (2)$$
$$WYZ + \overline{W}XY + XYZ = WYZ + \overline{W}XY \quad (3)$$

We can use equation (1) to eliminate the WYZ term. But with it eliminated, we cannot use equations (2) and (3) since this term is needed in these equations. Therefore, we should not use equation (1) and eliminate WYZ. Instead, we will use equations (2) and (3). Note that to do this, we use the same term (WYZ) twice, but that is acceptable because, from identity 4(a), $WYZ = WYZ + WYZ$. By using equations (2) and (3) we eliminate the $W\overline{X}Z$ and XYZ terms, obtaining $F = W\overline{X}\,\overline{Y} + \overline{W}XY + WYZ$. ■■

The difficulties with simplification we had in this last example should make us think that there must be a better way to simplify an expression than by using Boolean manipulation. With it, we have difficulty determining where to begin, how to proceed, and when to know that we are finished and have an MSOP or an MPOS. There are simply no definitive rules for Boolean manipulation. Consequently, we will use Boolean manipulation only for reducing simple Boolean expressions, or Boolean expressions that we cannot conveniently simplify with other methods. For simplifying expressions of up to six variables, we will usually prefer to use the graphical method of the next section.

2.8 KARNAUGH MAPS

A Karnaugh map (K-map) is a convenient graphical method for obtaining an MSOP or MPOS of a Boolean expression of three, four, or five variables—and possibly, though not conveniently, six or seven variables. All the information of a truth table of a function is contained in a K-map. In other words, there is a one-to-one correspondence between a truth table and its K-map.

A K-map contains squares, one for each row of a corresponding truth table. So, a two-variable K-map has 4 squares, a three-variable K-map has 8 squares, a four-variable K-map has 16 squares, and so on. The values of the input variables are arranged in a certain order along two edges of a K-map, and from these values we can determine the corresponding truth table row for each square. The function values are inserted into the squares.

Figure 2.3 illustrates truth tables and corresponding K-maps for some two-variable, three-variable, and four-variable functions. Note, for example, in Fig. 2.3(a), that the top left square corresponds to the first truth table row since both the A and B input values are 0 for this square. And, from the 1 inside the square, we see that $Z = 1$ for $A = 0$ and $B = 0$. Similarly, the bottom left square corresponds to the second row of the truth

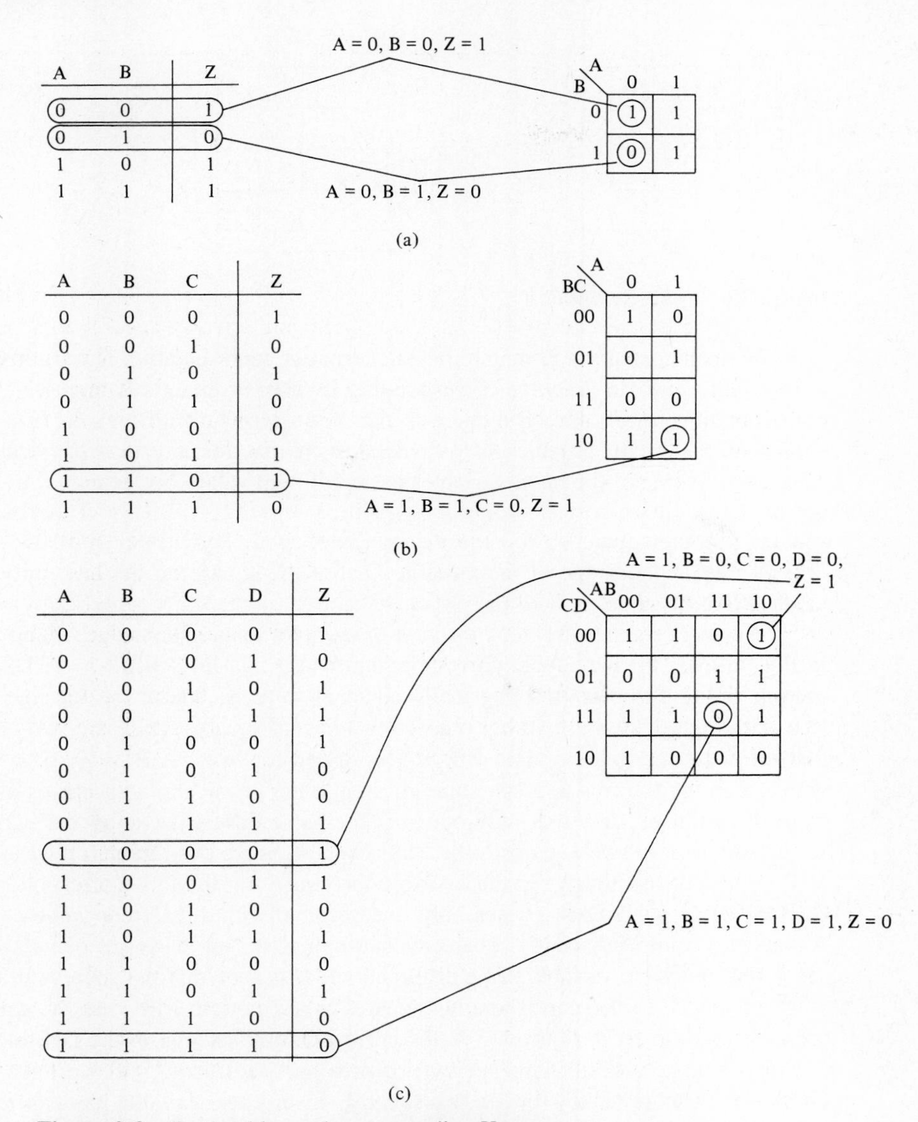

Figure 2.3 Truth tables and corresponding K-maps.

table, where A = 0, B = 1, Z = 0, and so on. In the three-variable K-map of Fig. 2.3(b), the bottom right square, for example, corresponds to the seventh row of the truth table, where A = 1, B = 1, C = 0, and Z = 1. In the four-variable K-map of Fig. 2.3(c), the top right corner square corresponds to the ninth row of the corresponding truth table, where A = 1, B = 0, C = 0, D = 0, and Z = 1. Also, the square in the third row and column of the K-map corresponds to the last row of the truth table, where A = 1, B = 1, C = 1, D = 1, and Z = 0, and so on. You should verify the other entries for each K-map.

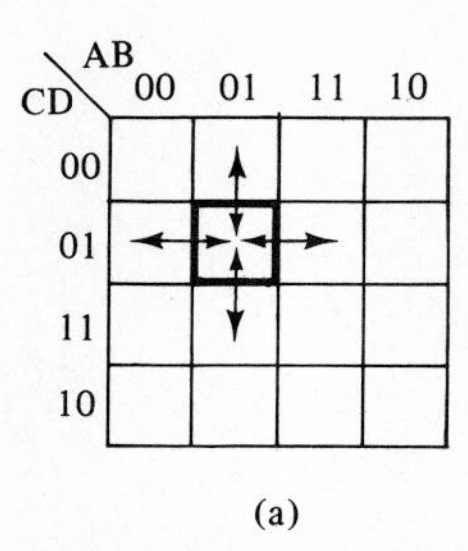

(a)

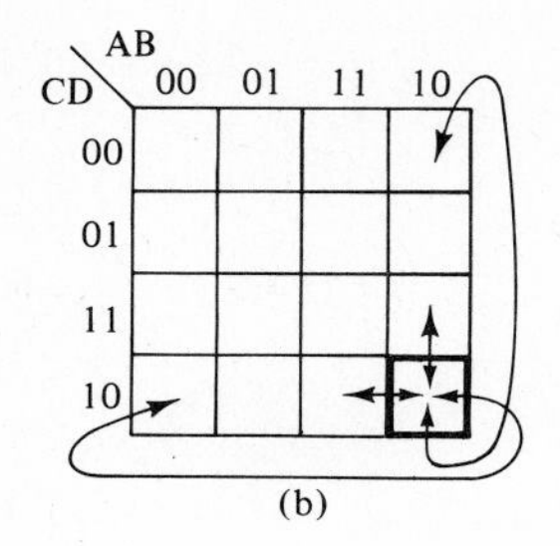

(b)

Figure 2.4 Adjacent squares.

We can consider a K-map to be an alternative representation of a truth table. Given a truth table, we can draw the corresponding K-map, or given a K-map we can form the corresponding truth table. But there is one very important difference: In a K-map the values of the input variables are arranged such that for any two physically adjacent squares, only *one* of the input variables has a different value. For example, in Fig. 2.4(a), consider the square corresponding to A = 0, B = 1, C = 0, D = 1, and any adjacent square. For the square on top the values are A = 0, B = 1, C = 0, D = 0, which differs only in the value of the variable D. For the square on the left, only variable B has a different value, and so on.

Adjacent squares are more difficult to see for a square at an edge. As an illustration, in Fig. 2.4(b), for the square corresponding to A = 1, B = 0, C = 1, D = 0, at the bottom right, there are two physically adjacent squares, one at the top and one on the left. Although there are no other physically adjacent squares, there are two other squares that differ in only one variable value. The square for A = 1, B = 0, C = 0, D = 0, which is in the top row and last column, though not physically adjacent, is also adjacent from the point of view that only one variable has a different value—the variable C. So we consider these two squares to be ''adjacent'' squares even though they are not physically adjacent. Similarly, in the last row the square on the left (corresponding to A = 0, B = 0, C = 1, D = 0) has only one variable with a different value—the variable A—and so is an ''adjacent'' square. Generalizing, for the purposes of adjacency in the sense that only one variable has a different value, we see that the squares in the top row are ''adjacent'' to the corresponding squares in the bottom row, and the squares in the leftmost column are ''adjacent'' to the corresponding squares in the rightmost column. If ever we have a doubt whether two squares are ''adjacent,'' all we have to do is to check the variable values for the two squares. If only one variable has a different value, then the two squares are ''adjacent.''

In Fig. 2.4, note the numbering of the variable values. If we interpret these values as being binary numbers, then the first column is numbered zero, the second column one, but the third column is numbered three, and the fourth column two. So, from a numeric sense, the numberings of the third and fourth columns are interchanged. The same is true for the third and fourth rows. It is this interchanging of numbering that gives the adjacencies needed to simplify Boolean functions entered on K-maps.

2.8.1 Obtaining an MSOP from a K-Map

As mentioned, a K-map is a specific graphical representation of a truth table. Consequently, we can obtain a canonical SOP expression (CSOP) directly from a K-map, just

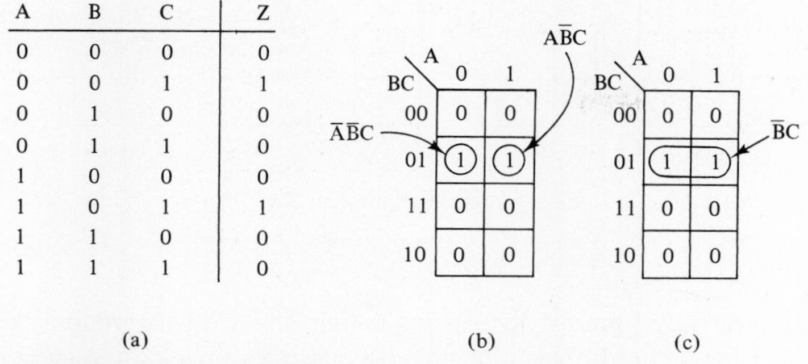

A	B	C	Z
0	0	0	0
0	0	1	1
0	1	0	0
0	1	1	0
1	0	0	0
1	0	1	1
1	1	0	0
1	1	1	0

Figure 2.5 Using a K-map.

as we can from a truth table. We will do this for the K-map of Fig. 2.5(b), which corresponds to the truth table of Fig. 2.5(a). Recall from Sec. 2.4.1, that to obtain a CSOP from a truth table, we identify each row for which the function (Z here) is 1. With this K-map, then, we must identify the squares in which 1s are entered for Z, and, for convenience, circle the 1s as shown in Fig. 2.5(b). Next, we obtain the minterm corresponding to each of these squares. Here, these minterms are $\overline{A}\,\overline{B}C$ and $A\overline{B}C$. Finally, we OR the minterms to obtain the CSOP for Z. For the function Z of Fig. 2.5, the result is $Z = \overline{A}\,\overline{B}C + A\overline{B}C$.

We can algebraically simplify this result by using Boolean identity 10(a) of Table 2.2:

$$Z = \overline{A}\,\overline{B}C + A\overline{B}C = \overline{B}C$$

Better yet, we can simplify *graphically* using the K-map, as illustrated in Fig. 2.5(c). For this, we just circle the two adjacent 1-squares and read the simplified expression directly from the K-map by keeping the literals that do not change and dropping the variable that changes in value—in this case, variable A. This graphical simplification is possible because of the K-map feature that for any two adjacent squares, only *one* input variable has a different value. So, this K-map simplification is a graphical application of Boolean identity 10(a): $AB + A\overline{B} = A$.

In a similar fashion, we can group four adjacent 1-squares, arranged in the form of a rectangle, to eliminate two variables, as illustrated in Fig. 2.6(a). For this group of four squares notice that two variables (A and B)—and only two—have different values. These are the variables that drop out. So, we can read the simplified expression for the

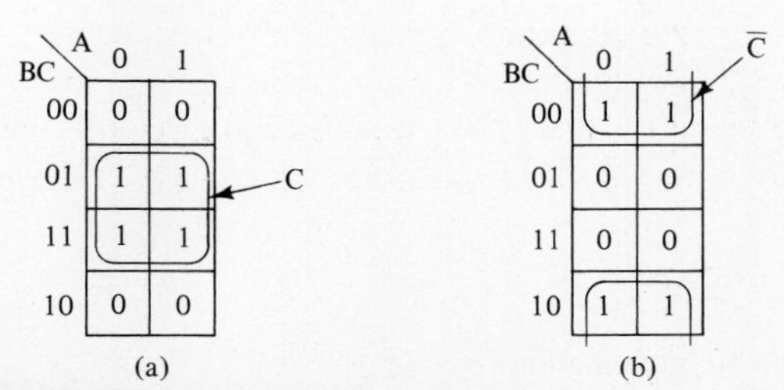

Figure 2.6 Grouping four 1-squares.

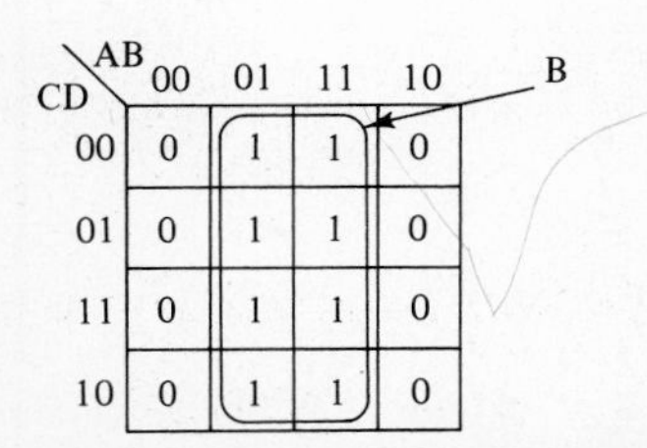

Figure 2.7 Grouping eight 1-squares.

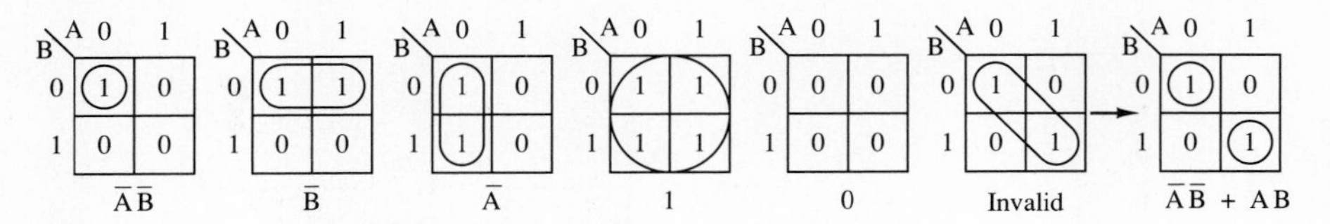

Figure 2.8 Two-variable K-map grouping examples.

circled group directly from the K-map and obtain Z = C. This grouping of four squares corresponds to applying Boolean identity 10(a) twice:

$$Z = (\overline{A}\overline{B}C + A\overline{B}C) + (\overline{A}BC + ABC) = \overline{B}C + BC = C$$

For this variation of variable values that allows dropping of two or more variables, the circled squares must be in the shape of a rectangle; however, the rectangle can extend from the top of the K-map to the bottom, as in Fig. 2.6(b). Also, the number of squares grouped must be a power of two: 2^n. Then, n variables drop out.

Figure 2.7 shows a rectangular grouping of eight adjacent squares. We can read the simplified expression directly from the K-map as Z = B. The justification for this simplification is that this graphical application corresponds to applying Boolean identity 10(a) three times, as you can prove to yourself.

The terms we have been obtaining from K-maps are called *prime implicants*. For every set of variable values that makes a prime implicant 1, the corresponding function is also 1. This is rather obvious, because to obtain a prime implicant, we circle only 1s of a function. Another feature of a prime implicant is that no literal can be deleted from it and yet have it remain a valid term of the function. Thus there is a sense of minimalness about the number of literals in a prime implicant.

Figure 2.8 shows some common groupings for two-variable K-maps. However, for the simplification of expressions of just two variables we will usually find it easier to use the identities of Table 2.2. Note the invalid grouping of the next-to-last K-map. We cannot validly circle the two 1s because both variables have different values for the two 1-squares. For a valid grouping of two 1s only one variable can and must have different values. Graphically, we should know that the grouping is invalid from the fact that the sides of the squares are not physically adjacent. Since we cannot group the two 1s, we cannot simplify the CSOP.

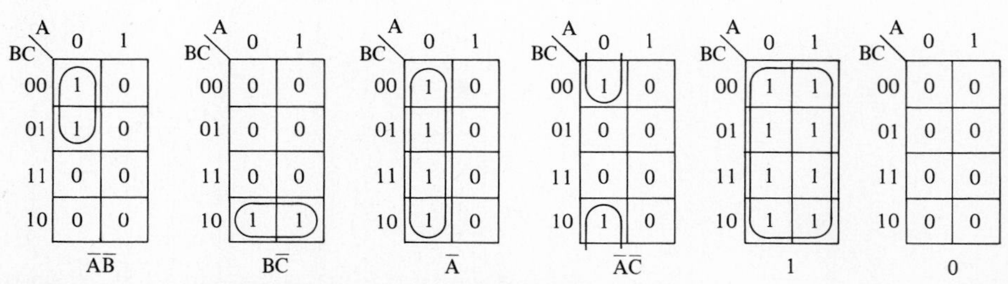

Figure 2.9 Three-variable K-map grouping examples.

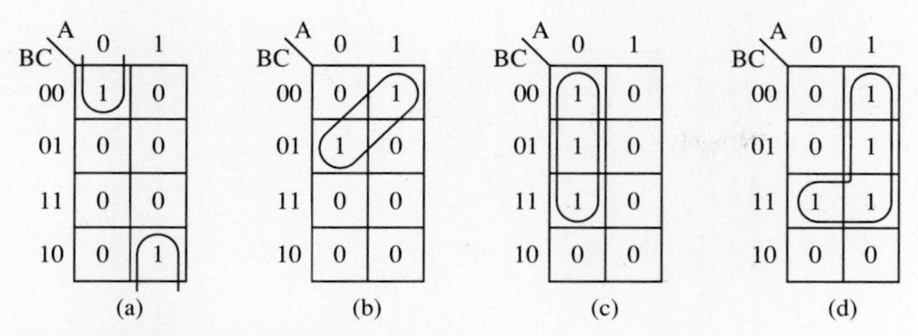

Figure 2.10 Invalid groupings.

Figure 2.9 shows some common groupings for three-variable K-maps, whereas Fig. 2.10 shows some invalid groupings for three-variable K-maps. In Fig. 2.10(a) the grouping is invalid because two variables (A and B), and not just one, have different values for the two 1-squares. We do not, however, have to check the variable values to know this. It is obvious from the lack of physical adjacency, even with the K-map rolled up to make the top edge and bottom edge join. Also, the grouping does not form a rectangle. For similar reasons, the grouping of Fig. 2.10(b) is invalid. The grouping of Fig. 2.10(c) is invalid because the number of grouped 1-squares (three) is not a power of two. The grouping of Fig. 2.10(d) is invalid, graphically speaking, because the grouped 1-squares do not form a rectangle. From an analytical point of view, the grouping is invalid because all three variables have different values for the grouped 1-squares, while for a valid grouping of four 1-squares, two and only two variables must have different values.

Figure 2.11 shows some common groupings for four-variable K-maps.

So far we have considered only a single grouping on each K-map. Almost always, though, more than one grouping is required to completely specify the function represented by the K-map, as shown by the following example.

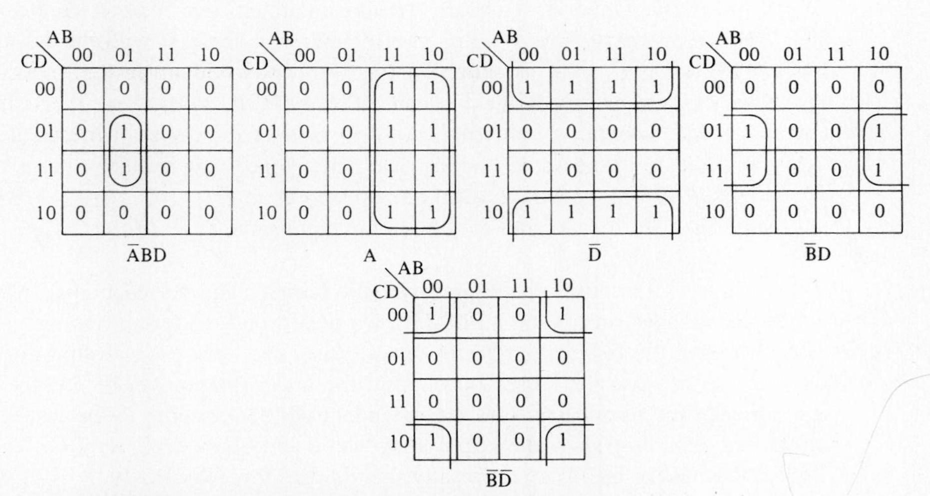

Figure 2.11 Four-variable K-map grouping examples.

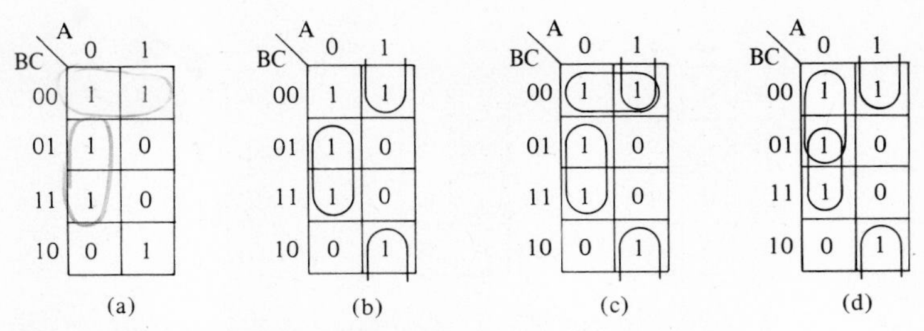

Figure 2.12 K-map illustration for Example 2.12.

EXAMPLE 2.12

Use a K-map to find an MSOP expression for Z.

A	B	C	Z
0	0	0	1
0	0	1	1
0	1	0	0
0	1	1	1
1	0	0	1
1	0	1	0
1	1	0	1
1	1	1	0

Solution. Our first step is to enter the function on a three-variable K-map, as shown in Fig. 2.12(a).

The next step is to find the 1-squares that we can circle only once. There are two of them: A = 1, B = 1, C = 0 and A = 0, B = 1, C = 1. We form groupings with them, as shown in Fig. 2.12(b). The corresponding prime implicants ($A\overline{C}$ and $\overline{A}C$) are called *essential prime implicants* since they must be included in an MSOP. Only one 1-square remains uncircled, that of A = 0, B = 0, C = 0. For it, we have two grouping choices. We can group it with the 1-square of A = 1, B = 0, C = 0, as shown in Fig. 2.12(c), to obtain prime implicant $\overline{B}\,\overline{C}$. Or, we can group it with the 1-square of A = 0, B = 0, C = 1, as shown in Fig. 2.12(d), to obtain prime implicant $\overline{A}\,\overline{B}$. Since the number of literals in these prime implicants is the same, we can use either one, which means there is no unique MSOP. From the Fig. 2.12(c) K-map we obtain $Z = \overline{A}C + A\overline{C} + \overline{B}\,\overline{C}$, and from the Fig. 2.12(d) K-map we obtain the equivalent $Z = \overline{A}C + A\overline{C} + \overline{A}\,\overline{B}$. ■■

Figure 2.13 shows some three-variable K-maps and corresponding MSOP expressions. Remember, in finding an MSOP, we want to use as few groupings as possible in order to have the fewest number of terms. Also, we want each grouping to be as large as possible because the larger a grouping, the fewer the number of literals in the corresponding prime implicant. Note the redundant dotted grouping in the last K-map. If we used this grouping also, the expression would be $A\overline{B} + \overline{A}C + \overline{B}C$. This is not an MSOP because it has more terms and literals than the MSOP $A\overline{B} + \overline{A}C$, which covers all the 1s.

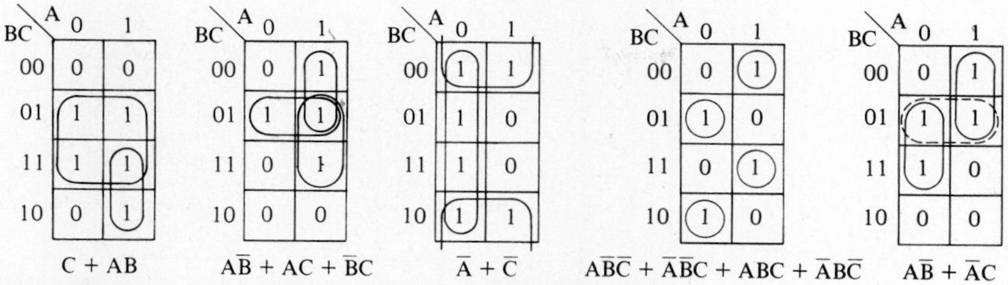

Figure 2.13 Three-variable K-map MSOP examples.

Obtaining an MSOP for a four-variable function is only slightly more difficult than for a three-variable function. Additionally, we must remember that 1-squares in the leftmost column are "adjacent" 1-squares in the corresponding rows in the rightmost column. Also, all four corner squares are "adjacent."

EXAMPLE 2.13

Use a K-map to find an MSOP expression for Z.

A	B	C	D	Z
0	0	0	0	1
0	0	0	1	1
0	0	1	0	1
0	0	1	1	0
0	1	0	0	0
0	1	0	1	1
0	1	1	0	0
0	1	1	1	1
1	0	0	0	0
1	0	0	1	0
1	0	1	0	1
1	0	1	1	0
1	1	0	0	0
1	1	0	1	0
1	1	1	0	0
1	1	1	1	0

Solution. Our first step is to enter the function on a four-variable K-map, as shown in Fig. 2.14(a). The next step is to find the essential prime implicant group-

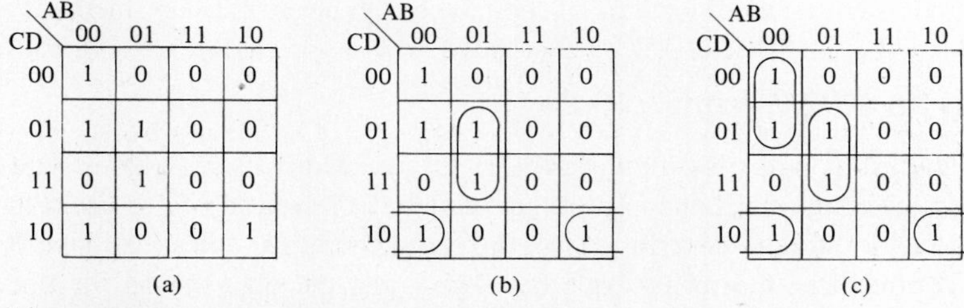

Figure 2.14 K-map illustration for Example 2.13.

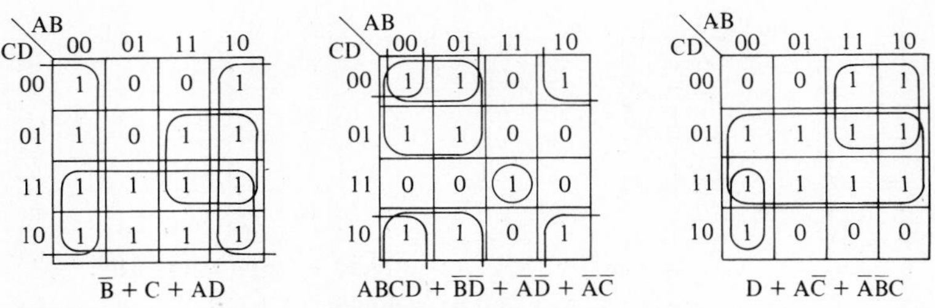

Figure 2.15 Four-variable K-map MSOP examples.

ings. There are just two of them since the only 1-squares we cannot circle more than once are those for A = 1, B = 0, C = 1, D = 0 and A = 0, B = 1, C = 1, D = 1. Figure 2.14(b) shows the essential prime implicant groupings. Two 1-squares remain to be circled. Since they are adjacent, we should obviously group them and obtain the complete grouping of Fig. 2.14(c). From it we obtain $Z = \overline{B}C\overline{D} + \overline{A}BD + \overline{A}\,\overline{B}\,\overline{C}$, which is a unique MSOP even though one of the terms is not an essential prime implicant. ■ ■

Figure 2.15 shows some four-variable K-maps and the corresponding MSOP expressions.

In summary, finding an MSOP from a K-map is an art. Although there are no definitive rules to follow to guarantee obtaining an MSOP, the following guidelines are helpful.

1. Circle every 1 at least once.
2. Circle a 1 more than once if this circling helps in making larger groupings, but do not circle any more times than necessary to circle all the 1s.
3. Make the groups as large as possible.
4. Use no more groups than necessary.
5. Start the circling with 1s that can be circled only once. In general, start with the 1s that are most difficult to group.

We could extend our study of K-maps to five-variable and six-variable K-maps. The principles are the same, but the adjacencies become progressively more difficult to visualize as the number of variables increases. Besides, there are computer programs for finding MSOPs for functions of many more variables than we can use K-maps for. These computer programs are not based on graphical techniques such as K-maps, but on tabular methods such as modifications of the tabular Quine-McCluskey method.

2.8.2 Obtaining an MPOS from a K-Map

We can use K-maps for finding MPOSs just as readily as for finding MSOPs. The rules for grouping are the same except that we circle 0s instead of 1s. From these groups we form factors instead of terms. Also, we complement variables that have 1 values and do not complement those that have 0 values—just the opposite as for finding literals of MSOPs.

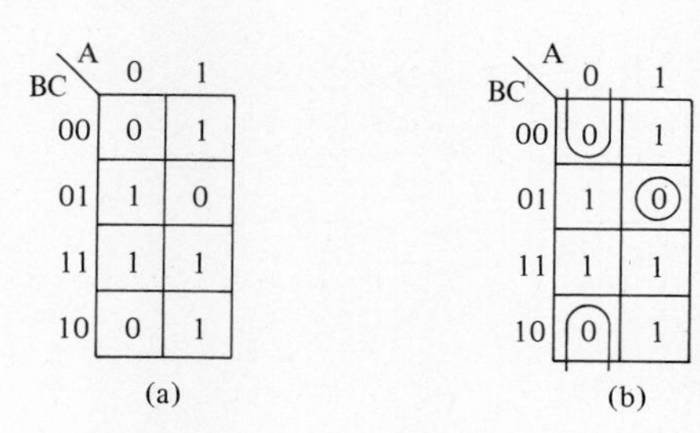

Figure 2.16 K-map illustration for Example 2.14.

EXAMPLE 2.14

Use a K-map to find an MPOS expression for Z.

A	B	C	Z
0	0	0	0
0	0	1	1
0	1	0	0
0	1	1	1
1	0	0	1
1	0	1	0
1	1	0	1
1	1	1	1

Solution. As usual, our first step is to enter the function on a K-map, as shown in Fig. 2.16(a). Then, we group the 0s, as shown in Fig. 2.16(b). Next, we find factors corresponding to the groups, remembering to complement if a variable has a 1 value. These factors are $(A + C)$ and $(\overline{A} + B + \overline{C})$. Finally, we AND these factors. The result is $Z = (A + C)(\overline{A} + B + \overline{C})$. ■■

Figure 2.17 shows some four-variable K-maps and corresponding MPOS expressions obtained from the shown groupings of 0s. Since K-maps can be used to find MSOPs or MPOSs, there arises the question of which type of minimum expression to solve for in a digital design. Actually, we may want to solve for both to determine which, possibly, is simpler. As an illustration, if for the function entered on the K-map of Fig. 2.16(a) we had solved for an MSOP, we would have obtained either $Z = \overline{A}C + A\overline{C} + BC$ or

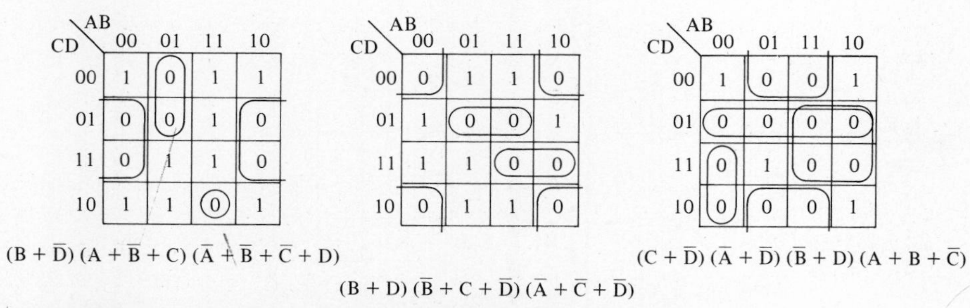

$(B + \overline{D})(A + \overline{B} + C)(\overline{A} + \overline{B} + \overline{C} + D)$

$(B + D)(\overline{B} + C + \overline{D})(\overline{A} + \overline{C} + \overline{D})$

$(C + \overline{D})(\overline{A} + \overline{D})(\overline{B} + D)(A + B + \overline{C})$

Figure 2.17 Four-variable K-map MPOS illustrations.

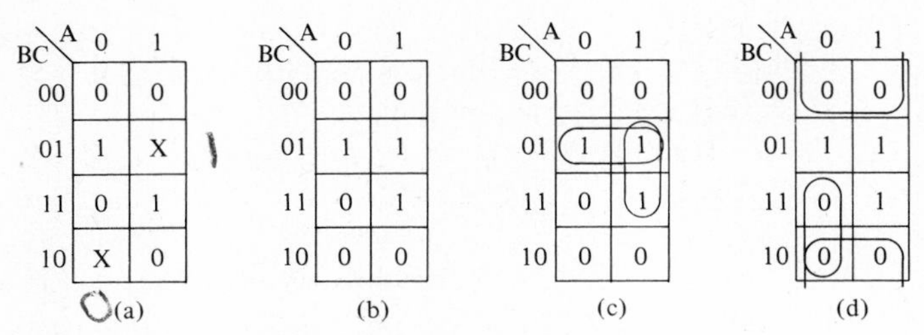

Figure 2.18 Three-variable don't-care illustrations.

$Z = \overline{A}C + A\overline{C} + AB$. Both MSOPs have three terms and six literals, while the MPOS $Z = (A + C)(\overline{A} + B + \overline{C})$ we found has just two factors and five literals, and so is simpler. Often, though, the MSOP and MPOS expressions are equally minimum. Another consideration is the type of hardware that will be used to implement the function. As shown in the next chapter, we want to use the minimum expression that more closely corresponds to the available hardware.

2.9 DON'T-CARE OUTPUTS

Often, in designing a digital system, a designer will know that certain components of the system will never have all possible combinations of inputs. A component with, say, inputs of A, B, and C, may never have inputs of $A = 1$, $B = 1$, $C = 0$ and $A = 1$, $B = 0$, $C = 0$, for example. And yet there are rows in the truth table and squares in the K-map for these input values. What then does the designer insert for the output for each of these inputs? The answer is a "don't care," identified by the symbol X. Clearly, for those inputs it does not matter what the outputs are, at least with regard to system operation. But it does make a difference with regard to minimization.

Since for don't-care outputs we are free to select either 0s or 1s, we select the values that are best for minimizing the output expression. In this, the selection for one don't care does not restrict us from selecting a different value for another don't care. In other words, the don't-care assignments do not have to be the same. By using K-maps we can readily determine which values for the don't cares result in minimum expressions.

Consider the K-map of Fig. 2.18(a), which has two don't cares. For an MSOP, obviously we want to replace the don't care of $A = 1$, $B = 0$, $C = 1$ with a 1 and the

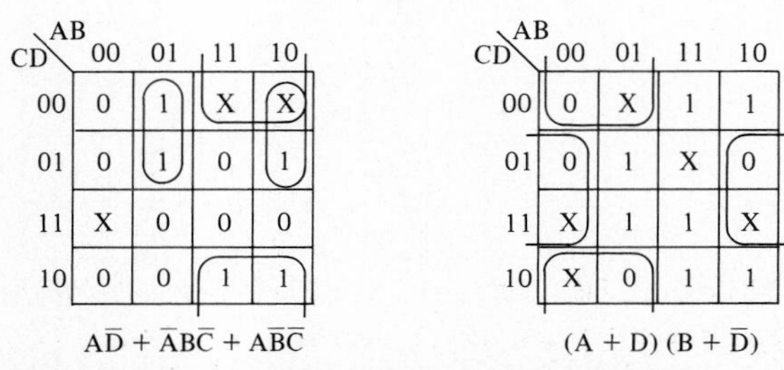

Figure 2.19 Four-variable don't-care illustrations.

don't care of A = 0, B = 1, C = 0 with a 0, as shown in Fig. 2.18(b). Then, with the grouping shown in Fig. 2.18(c), we obtain the MSOP expression $AC + \overline{B}C$ for Z. This don't care selection is also clearly best for an MPOS, which from Fig. 2.18(d) we see is $Z = C(A + \overline{B})$. But, the best don't-care selection for an MSOP is not always best for an MPOS. Figure 2.19 gives some additional don't-care illustrations.

2.10 MINIMIZATION SUMMARY

As has been shown, the K-map is a very convenient tool for simplifying logic functions of up to four variables—and, as mentioned, the K-map method can be extended to the simplifying of logic functions of five or more variables. However, the K-map method becomes increasingly more difficult to use with an increase in the number of variables, and eventually becomes unmanageable. Other systematic Boolean simplification methods have been developed to avoid this problem. These methods were important when hardware devices were the dominant cost of digital systems. In recent years, however, advances in microelectronic technology have been such that hardware costs are far from being the dominant costs. Consequently, our time will be better devoted to the study of concepts that are relevant to the enhancement of the overall digital system at the design level rather than in the study of sophisticated methods for eliminating a few logic gates.

2.11 OTHER COMMON LOGIC OPERATIONS

To complete our study of Boolean algebra, we will consider some other logic operations, that though not as fundamental as AND, OR, and NOT, are still important.

For two variables A and B, we can systematically generate 16 possible logic functions, as illustrated in Table 2.3. The three fundamental operations AND, OR, and NOT are represented by F_1, F_7, and F_{10} (and F_{12}), respectively. All other logic functions are combinations of these three fundamental operations. We will consider the most common of these.

2.11.1 NAND Operation

The NAND operation, represented by F_{14} in Table 2.3, is defined again as follows:

A	B	F_{14}
0	0	1
0	1	1
1	0	1
1	1	0

Note from the first three rows of the truth table that the Boolean variable F_{14} is true if *not* both A *and* B are true. From the last row of the truth table we get $F_{14} = \overline{A} + \overline{B}$, which by DeMorgan's laws also has the form $F_{14} = \overline{A \cdot B}$. From this second form we

TABLE 2.3

A	B	F_0	F_1	F_2	F_3	F_4	F_5	F_6	F_7	F_8	F_9	F_{10}	F_{11}	F_{12}	F_{13}	F_{14}	F_{15}
0	0	0	0	0	0	0	0	0	0	1	1	1	1	1	1	1	1
0	1	0	0	0	0	1	1	1	1	0	0	0	0	1	1	1	1
1	0	0	0	1	1	0	0	1	1	0	0	1	1	0	0	1	1
1	1	0	1	0	1	0	1	0	1	0	1	0	1	0	1	0	1
		↑	↑		↑		↑	↑	↑	↑	↑	↑		↑		↑	↑
		0	$A \cdot B$		A		B	$A \oplus B$	$A + B$	$\overline{A + B}$	$A \odot B$	$\overline{B}$		$\overline{A}$		$\overline{A \cdot B}$	1

see that NAND is a combination of NOT and AND—hence the name NAND. More specifically, the NAND operation is the complement of the AND operation; it is the AND operation followed by the NOT operation.

The NAND operation applies to any number of variables. For the three variables A, B, and C, the NAND truth table is

A	B	C	$\overline{A \cdot B \cdot C}$
0	0	0	1
0	0	1	1
0	1	0	1
0	1	1	1
1	0	0	1
1	0	1	1
1	1	0	1
1	1	1	0

Again, the only 0 output is for the last row, which is always the case, regardless of the number of input variables. In other words, the output variable is 1 unless all input variables are 1, in which case the output variable is 0.

2.11.2 NOR Operation

The NOR operation, represented by F_8 in Table 2.3, is defined again as follows:

A	B	F_8
0	0	1
0	1	0
1	0	0
1	1	0

Note from the first row of the truth table that the Boolean variable F_8 is true only if *neither* A *nor* B is true. Also, from this row we get that $F_8 = \overline{A} \cdot \overline{B}$, which by DeMorgan's laws also has the form $F_8 = \overline{A + B}$. From this second form we see that NOR is a combination of NOT and OR—hence the name NOR. More specifically, the NOR operation is the complement of the OR operation; it is the OR operation followed by the NOT operation.

The NOR operation applies to any number of variables. For the three variables A, B, and C, the NOR truth table is

A	B	C	$\overline{A + B + C}$
0	0	0	1
0	0	1	0
0	1	0	0
0	1	1	0
1	0	0	0
1	0	1	0
1	1	0	0
1	1	1	0

Again, the only 1 output is for the first row, which is always the case, regardless of the number of input variables. In other words, the output variable is 0 unless all input variables are 0, in which case the output variable is 1.

Although the NAND and NOR operations are by far the most popular of the nonfundamental Boolean operations, there are other important ones we should consider.

2.11.3 Exclusive OR Operation

The Exclusive OR (XOR) operation, represented by F_6 in Table 2.3, is commonly designated by the symbol $\oplus$ as in $F_6 = A \oplus B$. The XOR operation is defined by the following table:

A	B	F_6
0	0	0
0	1	1
1	0	1
1	1	0

Note that the Boolean variable F_6 is true if A is true or B is true, but not both. We can also view F_6 as being true if and only if $A \neq B$. Consequently this operation is useful in digital design for comparison purposes. Note also that this truth table differs from the two-variable OR truth table only in the last row, for both inputs of 1. From the second and third rows of the truth table we find that $F_6 = \overline{A}B + A\overline{B}$.

The XOR operation applies to any number of variables. For the three variables A, B, and C, the XOR truth table is

A	B	C	$A \oplus B \oplus C$
0	0	0	0
0	0	1	1
0	1	0	1
0	1	1	0
1	0	0	1
1	0	1	0
1	1	0	0
1	1	1	1

Note that for an odd number of 1 inputs, the output is 1. But for an even number of 1 inputs, the output is 0. This is generally true. Because of this, the XOR pattern on a K-map is a checkerboard, and so no terms combine.

2.11.4 Equivalence Operation

The equivalence operation, also called the coincidence operation, is represented by F_9 in Table 2.3. This operation is commonly designated by the symbol $\odot$, as in $F_9 = A \odot B$, and worded "F_9 is equal to A equivalence B." The equivalence operation is defined by the following table:

A	B	F_9
0	0	1
0	1	0
1	0	0
1	1	1

From the first and last rows, we see that the Boolean variable F_9 is true if and only if A = B. So $F_9 = \overline{A}\,\overline{B} + AB$. Also, from a comparison of this truth table and that for XOR, we see that *for two variables,* the equivalence and XOR operations are complements of one another, which means that the equivalence operation, like the XOR operation, is useful in digital design for comparison purposes.

SUPPLEMENTARY READING (see Bibliography)

[Bartee 85], [Boole 54], [Blakeslee 79], [Hill 81], [Karnaugh 53], [Mano 84], [McCluskey 75], [Peatman 80], [Roth 85], [Shannon 38]

PROBLEMS

2.1. Evaluate the following expressions for A = F, B = T, C = F, and D = T:

(a) $A\overline{B} + \overline{C}(\overline{A} + \overline{D})$
(b) $(A + B\overline{C})(A\overline{D} + \overline{A}D)$
(c) $\overline{AD}(A + \overline{B\overline{C}} + B\overline{D})$
(d) $\overline{A}B(A\overline{C} + \overline{B}D + \overline{\overline{A}B\overline{C}D})$
(e) $\overline{\overline{A\overline{B} + C} + C\overline{D}}$
(f) $(A + \overline{B\overline{C} + BC\overline{D}} + \overline{\overline{B} + C})(\overline{A}CD + B\overline{D})$

2.2. Evaluate the following expressions for A = F, B = F, C = T, and D = T:

(a) $\overline{A}B + \overline{B}CD(AB + \overline{C}D + \overline{A}D)$
(b) $\overline{A}BCD + A\overline{C}D + \overline{B}C(A + C\overline{D})$
(c) $(A + B + C)(\overline{A} + C + \overline{D})(A + B + C\overline{D})$
(d) $(\overline{A + B})(\overline{A + \overline{C}D + CD})$
(e) $(\overline{A\overline{B}C + \overline{C}D})(\overline{\overline{AC + \overline{D}} + BC\overline{D}})$
(f) $\overline{(\overline{\overline{A} + B\overline{C}})CD + B\overline{C}D}$

2.3. Repeat Problem 2.1 for A = 1, B = 1, C = 0, and D = 1.

2.4. Repeat Problem 2.2 for A = 1, B = 0, C = 1, and D = 0.

2.5. Find the CSOP expressions for the Z functions defined in the following truth tables:

ABC	Z_1		ABC	Z_2		ABC	Z_3		ABC	Z_4
000	0		000	1		000	1		000	0
001	1		001	1		001	0		001	0
010	1		010	1		010	1		010	1
011	0		011	0		011	1		011	1
100	1		100	0		100	0		100	1
101	1		101	1		101	0		101	0
110	0		110	1		110	1		110	1
111	1		111	0		111	1		111	0
(a)			(b)			(c)			(d)	

2.6. Find the CPOS expressions for the Z functions of Problem 2.5.

2.7. A function Z of four variables A, B, C, and D is 1 if and only if two of the four variables are 1. Express the function Z as a CSOP and also as a CPOS.

2.8. A function Z of four variables A, B, C, and D is 1 if and only if an odd number of the four variables is 1. Express the function Z as a CSOP and also as a CPOS.

2.9. A function Z of five variables A, B, C, D, and E is 1 if and only if an even number of the five variables is 1. Express the function Z as a CSOP and also as a CPOS.

2.10. Use truth tables to determine whether the following pairs of expressions are equivalent:

(a) $\bar{A}\bar{B} + A\bar{B}\bar{C} + AB + \bar{A}BC$ and $\bar{A}\bar{B}\bar{C} + \bar{A}C + A\bar{C}$

(b) $\bar{A}\bar{C} + BC + A\bar{B}\bar{C}$ and $\bar{B}\bar{C} + ABC + \bar{A}B$

(c) $AB + AC$ and $(A + \bar{C})(A + C)(B + C)$

(d) $(A + \bar{B} + C)(B + \bar{C})(\bar{A} + \bar{B})$ and $\bar{A}BC + \bar{B}\bar{C}$

(e) $(\bar{A} + C)(A + \bar{B} + \bar{C})(\bar{A} + \bar{B})$ and $(\bar{A} + \bar{B} + C)(\bar{B} + \bar{C})$

2.11. Repeat Problem 2.10 for the following.

(a) $\bar{B}\bar{C} + BCD + A\bar{B}C$ and $A\bar{B} + \bar{A}\bar{B}\bar{C} + \bar{A}BCD + ACD$

(b) $A\bar{D} + \bar{A}CD + C\bar{D}$ and $\bar{A}C + AC\bar{D} + AB\bar{C} + A\bar{B}\bar{C}\bar{D}$

(c) $BD + \bar{B}\bar{D}$ and $(A + B + \bar{D})(\bar{B} + D)(\bar{A} + B + \bar{D})$

(d) $\bar{B}D(\bar{A} + \bar{C})$ and $(\bar{B} + C)(B + D)(\bar{B} + \bar{C})$

(e) $B(\bar{A} + B + C + D)(\bar{B} + C)$ and $C(A + B)(\bar{A} + B + \bar{C})$

2.12. Repeat Problem 2.10 for the following:

(a) $\bar{A}\bar{C}\bar{D} + \bar{A}\bar{D}E + ABD + A\bar{B}\bar{C}D\bar{E} + ADE$ and $\bar{A}C\bar{D}E + \bar{A}\bar{C}\bar{D}\bar{E} + AD + \bar{A}\bar{C}\bar{D}E$

(b) $(A + E)(D + \bar{E})(A + \bar{D})(C + D)(B + C)$ and $A(C + BD)(D + C\bar{E})$

2.13. Use truth tables to determine whether the following pairs of expressions are complements of each other:

(a) $\bar{A}C + ABC + \bar{A}\bar{B}\bar{C}$ and $A\bar{B} + A\bar{C} + \bar{A}B\bar{C}$

(b) $\bar{C} + A\bar{B}C$ and $\bar{A}C + AB$

(c) $AB + \bar{A}BC + A\bar{B}\bar{C}$ and $\bar{A}\bar{C} + \bar{B}C + AB\bar{C}$

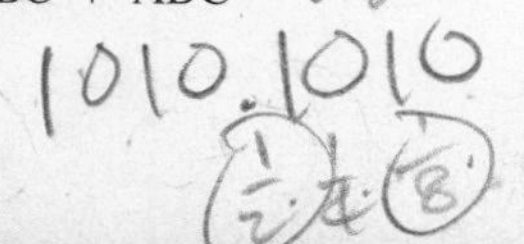

(d) $(A + B)(\overline{A} + C)$ and $\overline{A}\,\overline{B} + \overline{B}\,\overline{C}$
(e) $(\overline{A} + \overline{B})(A + \overline{C})$ and $(A + \overline{B} + C)(A + B + C)(\overline{A} + B)$
(f) $(\overline{A} + B)(\overline{A} + \overline{B} + \overline{C})(A + B)$ and $(A + \overline{B})(\overline{A} + \overline{B} + C)$

2.14. Repeat Problem 2.13 for
(a) $\overline{A}\,\overline{B} + CD + \overline{B}C$ and $B\overline{D} + B\overline{C} + A\overline{B}\,\overline{C}$
(b) $\overline{A}\,\overline{B}C + A\overline{D}$ and $\overline{A}B + AD + \overline{A}\,\overline{C}$
(c) $\overline{A}\,\overline{C} + AC$ and $(\overline{A} + \overline{C})(A + B + C + D)(A + \overline{B} + C)$
(d) $(B + \overline{C})(A + D)(\overline{A} + B)$ and $(\overline{A} + \overline{B})(\overline{B} + \overline{D})$
(e) $(\overline{A} + \overline{C})(\overline{A} + B)$ and $(A + D)(\overline{B} + C)(A + B)$

2.15. Using Boolean identities, simplify the following expressions:
(a) $A\overline{B}C + (\overline{A} + B)C\overline{D}$
(b) $A\overline{C}(B + D) + \overline{B}\,\overline{D} + C$
(c) $(A\overline{B} + C + D)(\overline{A} + B + C + D + E)$
(d) $(A\overline{B} + C\overline{D} + E + F)(\overline{A} + B + E + F)(\overline{C} + D + E + F)$

2.16. Repeat Problem 2.15 for
(a) $A + \overline{C}D + CDE + \overline{A}B$
(b) $AC\overline{D} + A\overline{B}\,\overline{C} + A\overline{B}D$
(c) $(A + C + D)(A + C + \overline{D} + E)$
(d) $(A + B + \overline{C})(A + \overline{C} + D)(\overline{B} + \overline{D})C$

2.17. Repeat Problem 2.15 for
(a) $ABC + A\overline{B}C + \overline{(\overline{\overline{A}} + \overline{B} + \overline{C})(\overline{A} + B + \overline{\overline{C}})}$
(b) $(A + B + C)(A + \overline{BC})(A + \overline{B} + C)(A + B + \overline{C})$
(c) $ABCD + AB\overline{C}D + \overline{A}BCD + \overline{A}B\overline{C}D$
(d) $(A + \overline{B})(\overline{B} + C)(\overline{B} + \overline{C})$

2.18. Find complements of the following expressions by using DeMorgan's laws and then simplify until DeMorgan's laws cannot be applied further.
(a) $A\overline{B} + C(\overline{D} + E)$
(b) $A\overline{B} + \overline{CD}(E + \overline{F}G)$
(c) $A\overline{B}(C + \overline{D}E) + \overline{A}\,\overline{B}C + \overline{B}(C + D)$

2.19. Repeat Problem 2.18 for
(a) $(A + \overline{C})(D + \overline{E}F) + A\overline{B}$
(b) $\overline{\overline{\overline{AB}(C + D\overline{E})}} + A\overline{B}(C + D)$
(c) $A\overline{C}D(E + \overline{F}G) + HI\overline{J}(K + \overline{\overline{L\overline{M}}})$

2.20. Using Boolean identities, convert each of the following expressions to a CSOP, and also to a CPOS:
(a) $\overline{A}\,\overline{B} + \overline{B}C$
(b) $A + \overline{B}C$
(c) $(A + B)(\overline{A} + \overline{B})$
(d) $(A + B)(\overline{B} + \overline{C})$

2.21. Repeat Problem 2.20 for
(a) $A\overline{B} + C$
(b) $A\overline{B} + \overline{C}D$
(c) $(A + \overline{B}C)(A + \overline{B}D)$
(d) $(A + \overline{B})(C + D)$

2.22. Find the K-maps for each function Z_1, Z_2, Z_3, and Z_4 defined as follows:

ABC	Z_1	Z_2
000	0	1
001	1	1
010	1	0
011	0	0
100	0	1
101	1	1
110	1	0
111	1	0

ABCD	Z_3	Z_4
0000	0	1
0001	1	0
0010	0	1
0011	1	1
0100	1	0
0101	0	0
0110	0	1
0111	1	0
1000	1	1
1001	0	1
1010	0	0
1011	0	1
1100	1	1
1101	1	0
1110	1	0
1111	0	1

2.23. Find the truth tables corresponding to the functions defined by the K-maps of Fig. 2.20.

(a) Z_1

BC \ A	0	1
00	1	1
01	1	0
11	1	1
10	0	0

(b) Z_2

BC \ A	0	1
00	0	1
01	0	1
11	1	0
10	1	1

(c) Z_3

CD \ AB	00	01	11	10
00	1	1	0	0
01	0	1	1	1
11	0	1	1	0
10	0	1	1	0

(d) Z_4

CD \ AB	00	01	11	10
00	1	0	0	1
01	1	0	0	1
11	1	1	1	1
10	0	1	0	0

Figure 2.20 K-maps for Problem 2.23.

2.24. Use K-maps to find an MSOP and an MPOS for each of the functions defined by the truth tables of Problem 2.22.

2.25. Find an MSOP and an MPOS for each of the functions defined by the K-maps of Fig. 2.20.

2.26. Obtain an MSOP and an MPOS for each of the functions defined by the K-maps of Fig. 2.21.

(a) Z_1

BC \ A	0	1
00	1	0
01	1	0
11	1	1
10	0	1

(b) Z_2

BC \ A	0	1
00	0	0
01	0	1
11	1	1
10	0	0

(c) Z_3

BC \ A	0	1
00	0	1
01	1	1
11	0	1
10	1	0

(d) Z_4

BC \ A	0	1
00	0	X
01	1	1
11	X	0
10	0	1

(e) Z_5

BC \ A	0	1
00	0	X
01	1	0
11	1	1
10	0	1

Figure 2.21 K-maps for Problem 2.26.

2.27. Obtain an MSOP and an MPOS for each of the functions defined by the K-maps of Fig. 2.22.

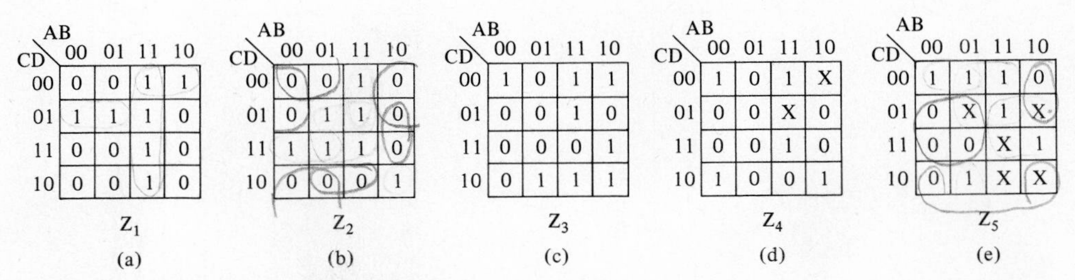

Figure 2.22 K-maps for Problem 2.27.

2.28. Use K-maps to obtain an MSOP and an MPOS for each of the following functions:

(a) $Z_1 = \overline{A}\overline{B}\overline{C} + ABC + \overline{A}\overline{B}C + A\overline{B}C$

(b) $Z_2 = \overline{A}\overline{B}\overline{C} + ABC + \overline{A}BC + A\overline{B}\overline{C}$

(c) $Z_3 = \overline{B}C + A\overline{B}\overline{C} + ABC$

(d) $Z_4 = (A + B + C)(\overline{A} + \overline{B} + \overline{C})(\overline{A} + \overline{B} + C)$

(e) $Z_5 = (\overline{A} + B + C)(A + B + \overline{C})(A + \overline{B} + C)$

(f) $Z_6 = (A + B + C)(\overline{A} + \overline{C})(A + \overline{B} + C)$

2.29. Repeat Problem 2.28 for

(a) $Z_1 = \overline{A}\overline{B}C + ABC + A\overline{B}\overline{C} + \overline{A}BC$ with a don't care for ABC = 101

(b) $Z_2 = A\overline{B}\overline{C} + ABC + \overline{A}B\overline{C} + A\overline{B}C$ with don't cares for ABC = 001 and 011

(c) $Z_3 = A\overline{B}\overline{C} + \overline{A}\overline{B}C + ABC$ with don't cares for ABC = 101 and 010

(d) $Z_4 = (A + B + C)(\overline{A} + \overline{B} + \overline{C})$ with don't cares for ABC = 100, 011, and 110

(e) $Z_5 = (\overline{A} + B + \overline{C})(A + \overline{B} + C)$ with don't cares for ABC = 001, 111, and 110

(f) $Z_6 = (\overline{A} + \overline{B})(B + \overline{C})(A + B + C)$ with a don't care for ABC = 011

2.30. Repeat Problem 2.28 for

(a) $Z_1 = \overline{A}\overline{B}\overline{C}\overline{D} + \overline{A}B\overline{C}D + A\overline{B}\overline{C}\overline{D} + \overline{A}BCD + AB\overline{C}D$

(b) $Z_2 = \overline{A}\overline{B}\overline{C}\overline{D} + \overline{A}\overline{B}CD + AB\overline{C}\overline{D} + \overline{A}\overline{B}C\overline{D} + ABC\overline{D} + A\overline{B}C\overline{D}$

(c) $Z_3 = \overline{A}B\overline{C} + \overline{A}\overline{B}CD + BCD + ABC\overline{D} + A\overline{C}D$

(d) $Z_4 = (A + B + \overline{C} + D)(A + \overline{B} + C + D)(\overline{A} + B + \overline{C} + \overline{D})(A + \overline{B} + \overline{C} + D)$
$(\overline{A} + \overline{B} + C + D)$

(e) $Z_5 = (\overline{A} + B + \overline{C} + D)(A + B)(C + D)(\overline{B} + C + \overline{D})(\overline{A} + B + C + \overline{D})$

(f) $Z_6 = (A + B + \overline{D})(\overline{B} + C + D)(\overline{A} + B + C + \overline{D})(\overline{A} + \overline{B} + \overline{C} + D)$
$(\overline{A} + B + \overline{C} + \overline{D})$

2.31. Repeat Problem 2.28 for

(a) $Z_1 = \overline{A}\overline{B}\overline{C}\overline{D} + \overline{A}BCD + AB\overline{C}D + \overline{A}\overline{B}\overline{C}D + AB\overline{C}\overline{D}$ with a don't care for ABCD = 0101

(b) $Z_2 = A\overline{B}CD + A\overline{B}\overline{C}\overline{D} + A\overline{B}C\overline{D} + \overline{A}\overline{B}C\overline{D} + \overline{A}BCD + \overline{A}\overline{B}CD$ with don't cares for ABCD = 0001 and 1111

(c) $Z_3 = AB\overline{C} + \overline{A}\overline{B}\overline{C}D + A\overline{B}C\overline{D} + \overline{A}\overline{B}CD$ with don't cares for ABCD = 0010 and 0101

03467 15 21 23 26 29
2, 8, 12, 14, 17, 24, 31
M 8, 9, A, B, 6, 7, 4

(d) $Z_4 = (\overline{A} + B)(A + \overline{B} + \overline{C})(A + \overline{B} + C + D)$ with a don't care for ABCD = 0101

(e) $Z_5 = (C + D)(A + \overline{B})(\overline{A} + \overline{C} + D)$ with don't cares for ABCD = 1001, 1101, and 1111

(f) $Z_6 = (C + D)(\overline{B} + D)(C + \overline{D})$ with don't cares for ABCD = 0111 and 1111

2.32. For two variables it has been stated that the equivalence and XOR operations are complements of one another. What about three variables? Compare the truth table of $A \odot B \odot C$ with that for $A \oplus B \oplus C$.

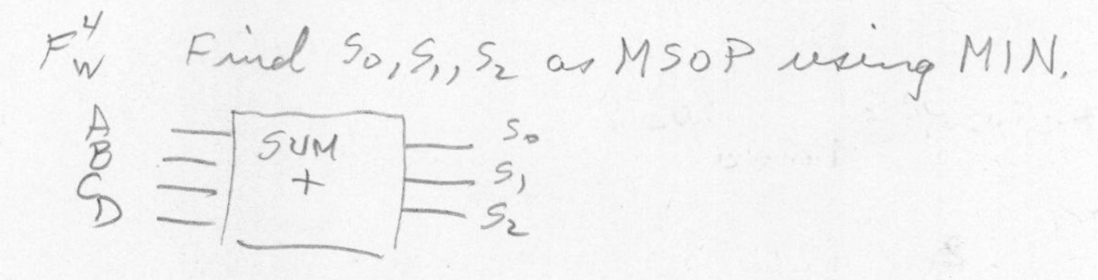

Chapter 3

Digital Design with Small-Scale Integrated Circuit Elements

3.1 INTRODUCTION

In Chapter 2 we considered the concept of Boolean variables (variables that can have only the logic values of true and false) and the three fundamental operations of Boolean algebra: AND, OR, and NOT. We proceeded to derive complex expressions based on these operations and discussed various methods of manipulating these expressions. With this background in mind, in this chapter we will study the design of digital systems using physical devices that actually perform Boolean operations, and with extraordinary speed. We will find that we can physically implement any imaginable digital system that is consistent with Boolean logic. Furthermore, the implementations will perform the basic Boolean operations in a nanosecond time frame. It is this fact that elevated Boolean algebra from an interesting concept derived by George Boole in the nineteenth century into the world of wizardry of modern-day digital computers.

Specifically, in this chapter we will study the design of digital systems with physical components called *small-scale integrated circuits* (SSI circuits). The term "small-scale" refers to the relatively small number of electronic gates in a single *integrated circuit*, often referred to as a *chip*. (A *gate* is an individual logic implementer.) Generally, integrated circuits with 12 or fewer equivalent logic gates per semiconductor chip are considered to be SSI circuits. Other types of integrated circuits include *medium-scale integrated circuits* (MSI, 13 to 99 logic gates), *large-scale integrated circuits* (LSI, 100 to 1000 logic gates), and *very large-scale integrated circuits* (VLSI, more than 1000 logic gates).

3.2 TRANSISTOR-TRANSISTOR LOGIC

For all our SSI implementations, we will use the SSI circuit family that has been most popular for years: the transistor-transistor logic (TTL) family. Currently, there are many series within the TTL family, including standard TTL, high-speed TTL (H-TTL), low-power TTL (L-TTL), Schottky TTL (S-TTL), low-power Schottky TTL (LS-TTL), advanced S-TTL (AS-TTL), and advanced LS-TTL (ALS-TTL).

Each TTL chip has a standard 74XY identifier label. The 74 specifies a commercial grade TTL product. The X identifies the series within the TTL family, and the Y, which is two or more digits, identifies the particular member of the series. For example, the 74ALS04 chip is a commercial grade TTL product in the advanced LS (ALS) series with part number 04. Under these labels, integrated-circuit manufacturers publish chip descriptions in their TTL data books. Included in each description is the functional operation of the chip components. We will use this information (in the form of voltage tables) in our design of digital systems.

The corresponding circuit elements of each of the TTL series are functionally identical, in regard to logic operations. For example, a 7400, a 74LS00, and a 74ALS00 are all two-input NAND gates. The differences among them are physical characteristics such as speed of operation, power dissipation, and so forth, all of which are unrelated to logic operations. Since in this chapter only the logic operations of the chips are of concern, we will adopt an apostrophe notation, such as in 74'00, rather than specify an actual series within the TTL family. A more detailed discussion of the TTL family is given in Sec. 4.8.

3.3 LOGIC CONVENTIONS

This section contains a unifying view of the various logic conventions that apply to digital circuits in general. In the discussion we will view a digital circuit as a network of digital *elements* energized by interconnecting digital *signals*. Each digital signal corresponds to a logic variable, and the voltage level of the signal represents the logic value of the variable. Also, each digital element, in the form of an integrated circuit, performs some logic function on input signals and produces corresponding output signals. We will begin our consideration of logic conventions with some definitions, terminology, and notational standards that will be used in this text.

For a digital signal, the logic values of true (T) and false (F) are represented by one of two voltage levels: high (H) or low (L). Therefore, there are just two possible assignments:

	Assignment	Terminology
(a)	$\left\{\begin{matrix} T \leftrightarrow H \\ F \leftrightarrow L \end{matrix}\right\}$	Active-high
(b)	$\left\{\begin{matrix} T \leftrightarrow L \\ F \leftrightarrow H \end{matrix}\right\}$	Active-low

For each signal of a digital circuit, we need to assign a voltage representation of either active-high or active-low. In the *positive-logic convention* of logic/voltage assignment, we assign all signals in the digital circuit to be active-high. Conversely, in the *negative-logic convention,* we assign all signals to be active-low. Finally, in the *mixed-logic convention,* we individually assign the voltage representation of each signal, which means that there can be a mixture of active-high and active-low signals in a single digital circuit. But if in the mixed-logic convention we happen to assign all signals to be active-high, then it is equivalent to the positive-logic convention. In other words, the positive-logic convention is a subset of the mixed-logic convention—as is the negative-logic convention.

To understand the concepts of active-high and active-low more fully, consider Fig. 3.1, which shows a fictitious device 74'XX with three input terminals (A, B, and C) and two output terminals (Y and Z). The small circles at terminals A and C indicate that they are active-low input terminals. In other words, the signals applied at terminals A and C will be interpreted by the 74'XX device as being active-low signals. The absence of a circle at terminal B indicates that it is an active-high input terminal and so any signal applied there will be interpreted by the device as being an active-high signal. Note the effects of such interpretations as illustrated by the "input voltage values" columns and the "input logic values" columns of the table shown in Fig. 3.1(b).

The device performs the logic function f on the signals applied at input terminals A, B, and C, and produces an output signal at output terminal Y. The circle at terminal Y designates that it is an active-low output terminal. Consequently, the signal that is produced there is active-low, as illustrated by the "output logic values" columns and the "output voltage values" columns of the table shown in Fig. 3.1(b). Similarly, the device also performs the logic function g and produces an active-high signal at the active-high output terminal Z (absence of a circle).

Generally, in order to make use of the function that is defined for a particular device, we must apply an active-low signal to an active-low input terminal, and an active-

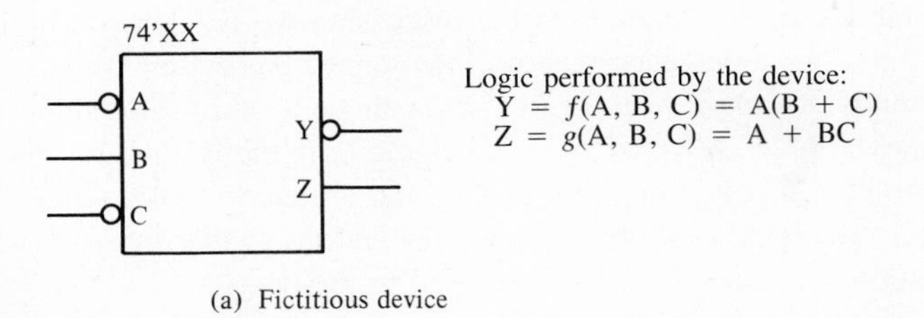

(a) Fictitious device

Input voltage values A	B	C	Input logic values A	B	C	Output logic values Y	Z	Output voltage values Y	Z
L	L	L	T	F	T	T	T	L	H
L	L	H	T	F	F	F	T	H	H
L	H	L	T	T	T	T	T	L	H
L	H	H	T	T	F	T	T	L	H
H	L	L	F	F	T	F	F	H	L
H	L	H	F	F	F	F	F	H	L
H	H	L	F	T	T	F	T	H	H
H	H	H	F	T	F	F	F	H	L

(b) Voltage and logic table

Figure 3.1 Illustration of active-high and active-low concepts.

high signal to an active-high input terminal. Also, we must accept the output signals as being either active-high or active-low, as defined for the output terminals. These facts are illustrated by the following example.

EXAMPLE 3.1

For the circuit shown in Fig. 3.2(a), find the logic expressions for S4 and S5 as functions of S1, S2, and S3.

Solution. Note from Fig. 3.2(a) that each signal, besides having a name label (e.g., S1), also has a voltage representation label of either .H for active-high or .L for active-low. We will consistently use this notation for the labeling of signals.

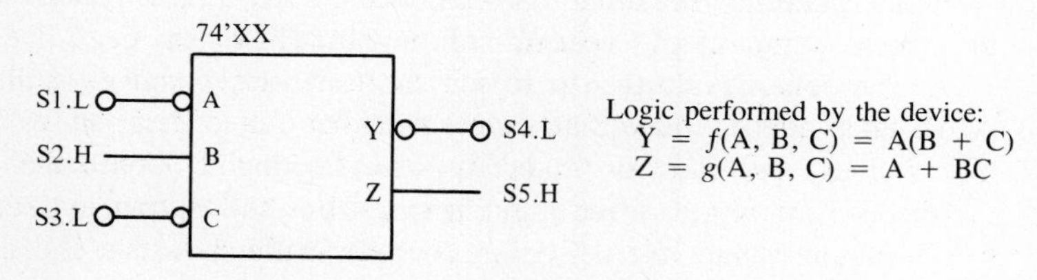

(a) Fictitious device

Input logic values			Input voltage values S1 S2 S3			Input logic values			Output logic values		Output voltage values Y Z		Output logic values	
S1	S2	S3	(A	B	C)	A	B	C	Y	Z	(S4	S5)	S4	S5
F	F	F	H	L	H	F	F	F	F	F	H	L	F	F
F	F	T	H	L	L	F	F	T	F	F	H	L	F	F
F	T	F	H	H	H	F	T	F	F	F	H	L	F	F
F	T	T	H	H	L	F	T	T	F	T	H	H	F	T
T	F	F	L	L	H	T	F	F	F	T	H	H	F	T
T	F	T	L	L	L	T	F	T	T	T	L	H	T	T
T	T	F	L	H	H	T	T	F	T	T	L	H	T	T
T	T	T	L	H	L	T	T	T	T	T	L	H	T	T

(b) Voltage and logic table

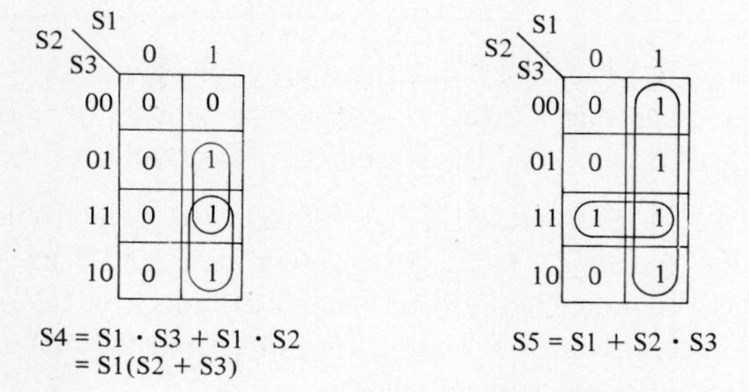

(c) Results

Figure 3.2 Illustration for Example 3.1.

Thus a signal is incompletely specified unless it is labeled with a name *and* a voltage representation. Also, to emphasize graphically that S1 is an active-low signal, a circle is associated with it (as with S3 and S4).

For this example, the input signals S1 and S3 and the output signal S4 are all active-low, whereas the input signal S2 and the output signal S5 are both active-high. Consequently, the logic/voltage assignments of these signals match those of the terminals of the device.

The desired logic expressions can be obtained in a systematic manner by using the table shown in Fig. 3.2(b). The "input logic values" columns show all the possible logic values for input signals S1, S2, and S3. And the "input voltage values" columns contain the corresponding voltage values for S1, S2, and S3 that are actually applied at the input terminals A, B, and C. These voltage values are interpreted by the input terminals of the device, resulting in the logic values shown in the "input logic values" columns for A, B, and C.

The device then performs the functions f and g on the input logic values, and produces the output logic values for Y and Z, as shown in the "output logic values" columns for Y and Z. Next, the device outputs the actual voltage values corresponding to these logic values, as is shown in the "output voltage values" columns. Note for active-low output terminal Y that the device outputs a low voltage value for a true logic value and a high voltage value for a false logic value. Finally, the voltage values generated by the device are interpreted by the output signals S4 and S5. The result is shown in the "output logic values" columns for S4 and S5.

With this table completed, it is a simple matter to derive the logic expressions for S4 and S5 by using the techniques presented in Chapter 2. The results are shown in Fig. 3.2(c). Note that they agree with the device logic expressions of Fig. 3.2(a), with S1, S2, and S3 substituted for A, B, and C, respectively. It is much easier, of course, to make these substitutions than to use the voltage and logic table approach.

Generalizing, we conclude from this example that to make use of the function that is defined for a particular device, we must make certain that the logic/voltage assignments of the input and output signals match those of the input and output terminals of the device. ■ ■

EXAMPLE 3.2

For the circuit shown in Fig. 3.3(a), find the logic expressions for S4 and S5 as functions of S1, S2, and S3. Note that the logic/voltage assignments of input signals S2 and S3 do not match those of input terminals B and C, respectively, of the device.

Solution. As in Example 3.1, the solution for this problem can be obtained systematically by using a table, as is shown in Fig. 3.3(b). Note specifically the difference between the "input logic values" columns for S2 and S3 and the "input logic values" columns for B and C, resulting from the difference in the interpretation of the same input voltage values.

With this table completed, it is again a simple matter to derive the logic expressions for S4 and S5 by using the techniques presented in Chapter 2. The

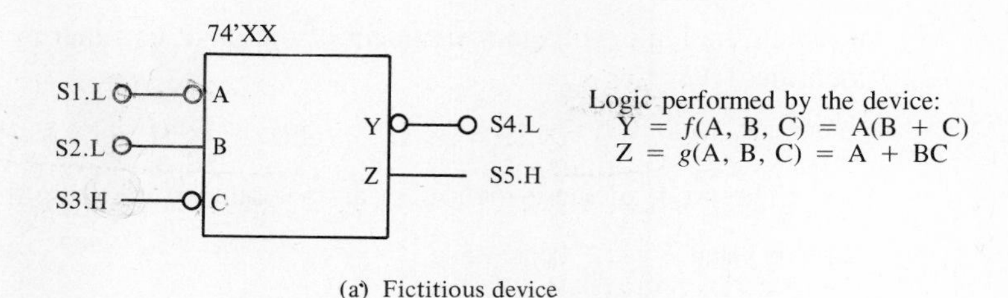

(a) Fictitious device

Input logic values			Input voltage values			Input logic values			Output logic values		Output voltage values		Output logic values	
S1	S2	S3	S1 (A	S2 B	S3 C)	A	B	C	Y	Z	Y (S4	Z S5)	S4	S5
F	F	F	H	H	L	F	T	T	F	T	H	H	F	T
F	F	T	H	H	H	F	T	F	F	F	H	L	F	F
F	T	F	H	L	L	F	F	T	F	F	H	L	F	F
F	T	T	H	L	H	F	F	F	F	F	H	L	F	F
T	F	F	L	H	L	T	T	T	T	T	L	H	T	T
T	F	T	L	H	H	T	T	F	T	T	L	H	T	T
T	T	F	L	L	L	T	F	T	T	T	L	H	T	T
T	T	T	L	L	H	T	F	F	F	T	H	H	F	T

(b) Voltage and logic table

$$S4 = S1 \cdot \overline{S2} + S1 \cdot \overline{S3} = S1(\overline{S2} + \overline{S3}) \qquad S5 = S1 + \overline{S2} \cdot \overline{S3}$$

(c) Results

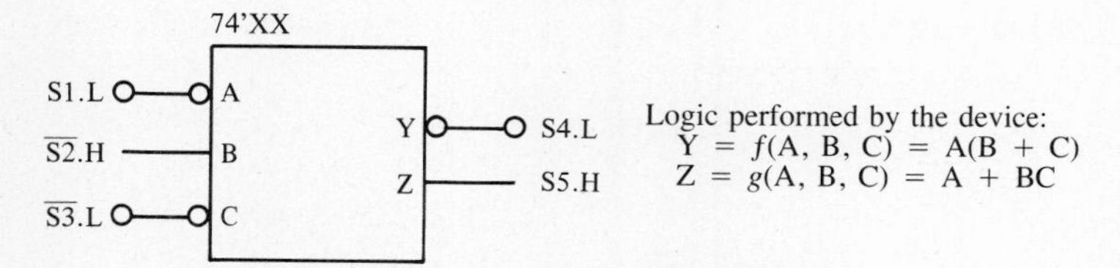

(d) Circuit diagram with transformed input signals

Figure 3.3 Illustration for Example 3.2.

result is shown in Fig. 3.3(c). From comparing these results with the logic expressions for Y and Z in Fig. 3.3(a), observe that in effect we have transformed the input signal S2.L to $\overline{S2}$.H so that it matches the logic/voltage assignment of input terminal B. Similarly, S3.H has been transformed to $\overline{S3}$.L to match the logic/voltage assignment of input terminal C. These transformations are shown in Fig. 3.3(d).

The conclusion that we can draw from this example is that when there is a mismatch of the logic/voltage assignment between an input signal and an input terminal, we need to transform the logic/voltage assignment of the signal to make it match the logic/voltage assignment of the input terminal by using the following identities.

$$\text{X.H} = \overline{\text{X}}\text{.L} \qquad \text{and} \qquad \overline{\text{X}}\text{.H} = \text{X.L}$$

The proof of these identities can be obtained by using the following table.

Voltage value X	Logic value X.H	X.L
L	F	T
H	T	F

Expressed in words, what this table shows is that for a low voltage (L), X.H "interprets" it as being false (F) and X.L "interprets" it as being true (T). Conversely, for a high voltage, X.H "interprets" it as being true and X.L "interprets" it as being false. Then from columns 2 and 3 of the table, we can determine that X.H = $\overline{\text{X}}$.L and $\overline{\text{X}}$.H = X.L. ■■

EXAMPLE 3.3

For the circuit shown in Fig. 3.4(a), determine the logic expressions for S4 and S5 as functions of S1, S2, and S3. Note that the logic/voltage assignments of the output signals S4 and S5 do not match those of the output terminals Y and Z, respectively, of the device.

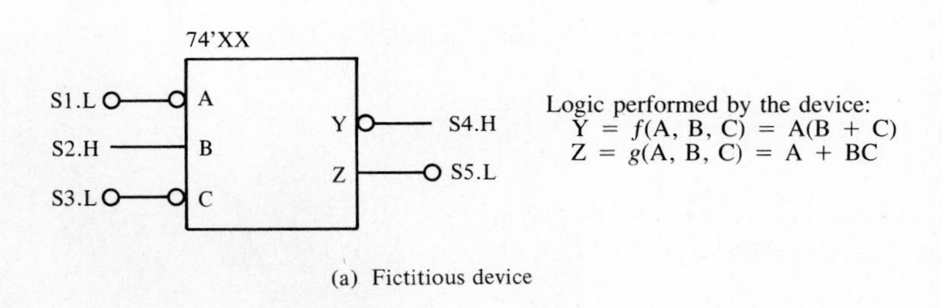

(a) Fictitious device

$\overline{\text{S4}} = \text{S1(S2 + S3)}$ $\qquad$ $\overline{\text{S5}} = \text{S1} + \text{S2} \cdot \text{S3}$

$\text{S4} = \overline{\text{S1(S2 + S3)}}$ $\qquad$ $\text{S5} = \overline{\text{S1} + \text{S2} \cdot \text{S3}}$

(b) Results

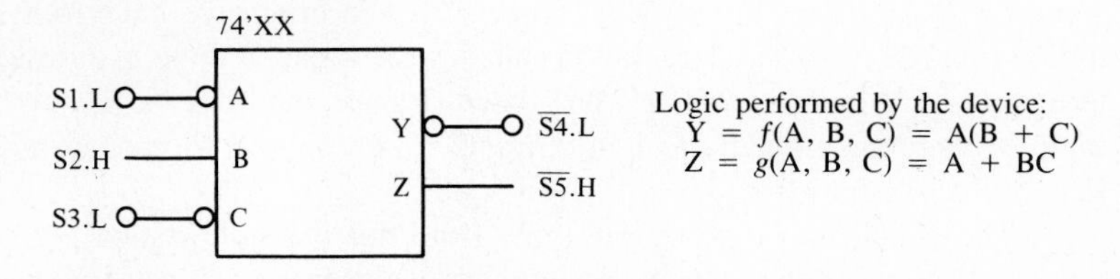

(c) Circuit diagram with transformed input signals

Figure 3.4 Illustration for Example 3.3.

Solution. As in Examples 3.1 and 3.2, the solution for this example [shown in Fig. 3.4(b)] can be systematically obtained by using a table like the one shown in Fig. 3.3(b). The solution, however, is left to the reader. (See Problem 3.3.) Observe that we can get the same result by transforming the output signal S4.H to $\overline{\text{S4}}$.L and S5.L to $\overline{\text{S5}}$.H to obtain matches with the logic/voltage assignments of output terminals Y and Z, respectively, as is shown in Fig. 3.4(c).

The conclusion that we can draw from this example is that when there is a mismatch of the logic/voltage assignment between an output signal and an output terminal, we need to transform the logic/voltage assignment of the signal to match the logic/voltage assignment of the output terminal by again using the following identities: X.H = $\overline{\text{X}}$.L and $\overline{\text{X}}$.H = X.L. ■■

We will now use the concepts of this section in a consideration of some specific SSI devices.

3.3.1 74'00

The 74'00 has two input terminals, which we will call terminals A and B, and one output terminal, which we will call terminal Z. From a TTL data book, we can determine the following voltage table for the 74'00.

A	B	Z
L	L	H
L	H	H
H	L	H
H	H	L

This table specifies how the 74'00 actually works, physically speaking. But what it does, logically speaking, depends on our logic/voltage assignment. Consider the assignment of

A—active-high B—active-high Z—active-low

For this assignment, the corresponding graphical representation is that of Fig. 3.5(a), and the 74'00 voltage table translates into the logic table of Fig. 3.5(b). From this table we see that Z = A · B, and so the 74'00 is an AND gate for this specific logic/voltage assignment.

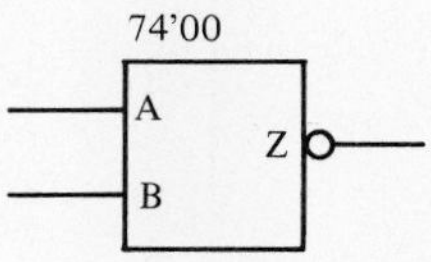

(a)

A	B	Z
F	F	F
F	T	F
T	F	F
T	T	T

(b)

Figure 3.5 74'00 AND gate.

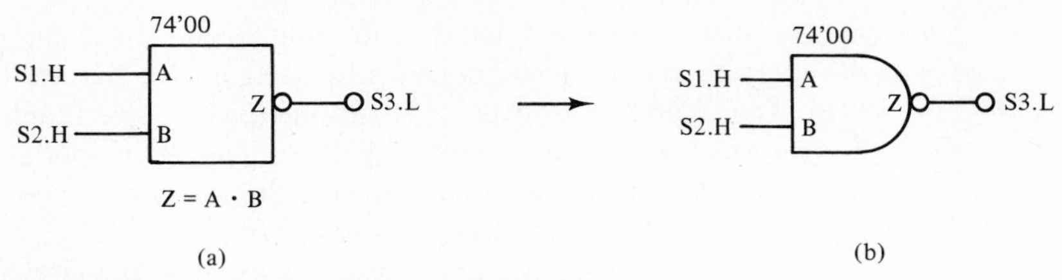

Figure 3.6 Use of the AND graphic symbol.

The commonly accepted graphic symbol for an AND gate is

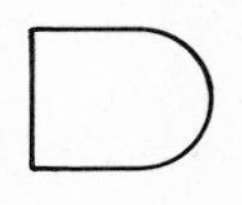

As illustrated in Fig. 3.6, we will use this symbol to replace the rectangular box for the 74'00. Then there is no need to specify the logic operation.

To recapitulate, from Fig. 3.6(b) we see that the 74'00 has two input terminals (A and B) and one output terminal (Z). Having no circles, both input terminals expect active-high signals. And, because of the circle at terminal Z, the output signal is active-low. As indicated by the shape of the symbol, the logic function performed on the signals applied at terminals A and B is the AND operation $Z = A \cdot B$, provided that the signals applied are active-high and provided that we interpret the output signal to be active-low. In this particular case, $S3 = S1 \cdot S2$, and the signal S3 is active-low.

Now, consider the logic/voltage assignment of

A—active-low B—active-low Z—active-high

For this assignment, the corresponding graphical representation is that of Fig. 3.7(a), and the 74'00 voltage table translates into the logic table of Fig. 3.7(b). Rearranging this logic table, we obtain the standard logic table of Fig. 3.7(c). From it we see that $Z = A + B$, and so the 74'00 is an OR gate for this specific logic/voltage assignment.

The commonly accepted graphic symbol for an OR gate is

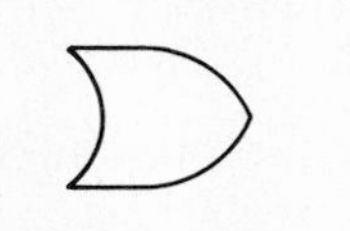

As illustrated in Fig. 3.8, we will use this symbol to replace the rectangular box for the 74'00. Then we do not have to specify the logic operation.

74'00
A
B
Z

(a)

A	B	Z
T	T	T
T	F	T
F	T	T
F	F	F

(b)

Rearrange →

A	B	Z
F	F	F
F	T	T
T	F	T
T	T	T

(c)

Figure 3.7 74'00 OR gate.

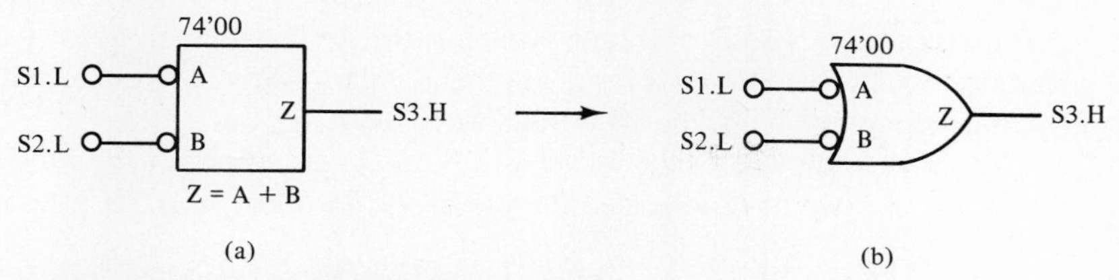

Figure 3.8 Use of the OR graphic symbol.

In Fig. 3.8(b) the two circles at both input terminals indicate that these terminals expect active-low input signals. And, because there is no circle at terminal Z, the output signal is active-high. As indicated by the shape of the symbol, the logic function performed on the signals applied at terminals A and B is the OR operation $Z = A + B$, provided that the applied signals are active-low and provided that we interpret the output signal to be active-high. In this particular case, $S3 = S1 + S2$, and the output signal S3 is active-high.

We have considered two of the eight possible logic/voltage assignments for the three terminals of a 74'00. All the possible assignments are

	Signals			
Assignments	A	B	Z	
(1)	.L	.L	.L	
(2)	.L	.L	.H	← OR
(3)	.L	.H	.L	
(4)	.L	.H	.H	
(5)	.H	.L	.L	
(6)	.H	.L	.H	
(7)	.H	.H	.L	← AND
(8)	.H	.H	.H	

We have shown that the 74'00 implements an OR gate for assignment (2), and an AND gate for assignment (7). The remaining six assignments do not implement any of the three fundamental Boolean operations (AND, OR, NOT), but rather a combination of them.

Of the remaining six assignments, the most popular one for the 74'00 is assignment (8), in which all signals are active-high. For this assignment, the corresponding graphical representation is that of Fig. 3.9(a), and the 74'00 voltage table translates into the logic table of Fig. 3.9(b). From this table we see that for an all active-high assignment, the

74'00
A
Z
B
$Z = \overline{A \cdot B}$

(a)

A	B	Z
F	F	T
F	T	T
T	F	T
T	T	F

(b)

Figure 3.9 74'00 NAND gate.

74'00 performs the NAND operation, which is the AND operation followed by the NOT operation. So, $Z = \overline{A \cdot B}$. As a result, in the popular positive-logic convention in which all signals are active high, the 74'00 is always a NAND gate.

To indicate the use of the 74'00 as a NAND gate, consider the following illustration. In Fig. 3.6 we have indicated the use of the 74'00 as an AND gate as follows:

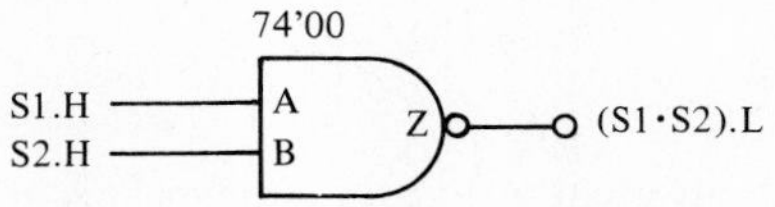

74'00
S1.H — A
S2.H — B
Z — (S1·S2).L

From the identity $X.L = \overline{X}.H$, this is equivalent to

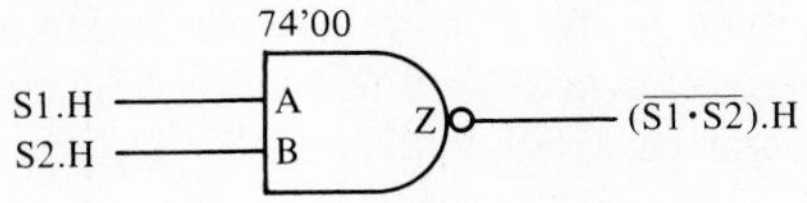

74'00
S1.H — A
S2.H — B
Z — $\overline{(S1 \cdot S2)}$.H

Therefore a mismatch of the voltage representation between the output terminal and the output signal results in a logic inversion of the function performed by the 74'00, causing it to go from an AND operation to a NAND operation.

We should remember that logic values (T, F) and the voltage representations of these logic values are two separate concepts. Also, although the voltage table of a device describes its physical behavior, the logic operation that the device performs in a particular digital circuit depends on our assignment of voltage representations to the logic values.

3.3.2 74'02

In a similar fashion we can consider the 74'02, which also has three terminals. But, we will omit some of the obvious details. The voltage table for the 74'02 is

A	B	Z
L	L	H
L	H	L
H	L	L
H	H	L

All possible logic/voltage assignments are, of course,

	Signals		
Assignments	A	B	Z
(1)	.L	.L	.L
(2)	.L	.L	.H
(3)	.L	.H	.L
(4)	.L	.H	.H
(5)	.H	.L	.L
(6)	.H	.L	.H
(7)	.H	.H	.L
(8)	.H	.H	.H

For assignment (7) of active-high input terminals and an active-low output terminal, we obtain the following logic table. From it we can see that for this assignment, the 74'02 is an OR gate.

A	B	Z
F	F	F
F	T	T
T	F	T
T	T	T

$Z = A + B$

74'02

S1.H — A; S2.H — B; Z —o (S1 + S2).L = S3.L

For assignment (2) of active-low input terminals and an active-high output terminal, we obtain the following logic table. From it we see that for this assignment, the 74'02 is an AND gate.

A	B	Z
T	T	T
T	F	F
F	T	F
F	F	F

rearrange →

A	B	Z
F	F	F
F	T	F
T	F	F
T	T	T

$Z = A \cdot B$

74'02

S1.L o— A; S2.L o— B; Z — $(S1 \cdot S2).H = S3.H$

For assignment (8), that of all active-high terminals, we obtain the following block diagram and NOR logic table.

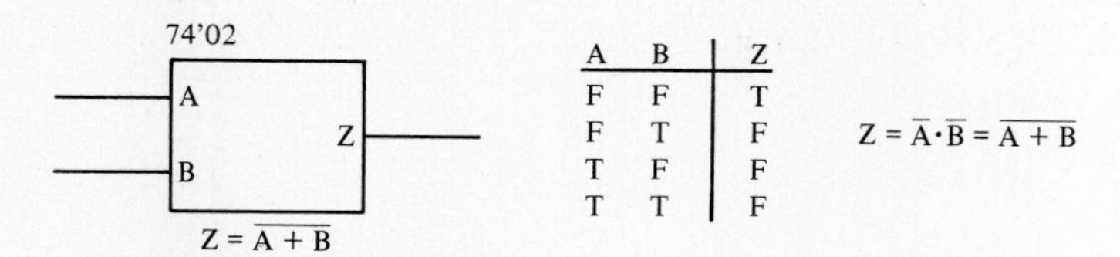

A	B	Z
F	F	T
F	T	F
T	F	F
T	T	F

$Z = \overline{A} \cdot \overline{B} = \overline{A + B}$

So, with the positive-logic convention in which all signals are active-high, the 74'02 is always a NOR gate.

The 74'02 performing as a NOR gate results from a mismatch of the voltage representation between the output terminal and the output signal, causing the operation to change from OR to NOR.

74'02

S1.H, S2.H — o $(\overline{S1 + S2}).H = S3.H$

3.3.3 74'04

We will now consider the 74'04, which has only one input terminal and one output terminal, which we will label A and Z, respectively. The 74'04 voltage table is

A	Z
L	H
H	L

Since the 74'04 has only two terminals, there are just four possible logic/voltage assignments:

	Signals	
Assignments	A	Z
(1)	.L	.L
(2)	.L	.H
(3)	.H	.L
(4)	.H	.H

For assignment (2) the graphical representation and logic table are as follows:

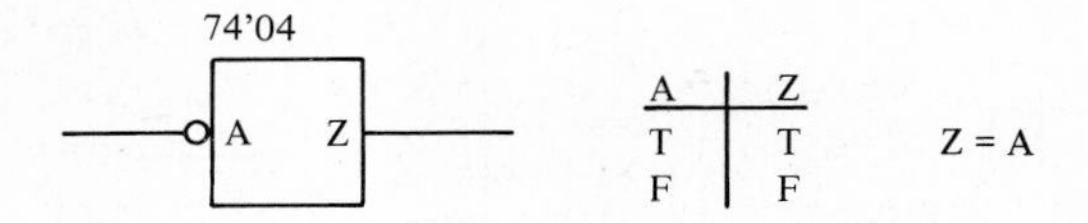

With this assignment, the 74'04 implements an IDENTITY logic operation, which means it does not do anything *logically*. But, of course, it does change the voltage representation from active-low to active-high, and so is useful in digital circuits that have mismatched voltage representations.

The commonly accepted symbol for an IDENTITY gate, also known as a *buffer,* is

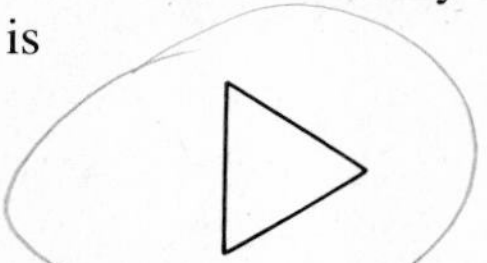

We will use this symbol to replace the rectangular box for the 74'04, as illustrated in Fig. 3.10.

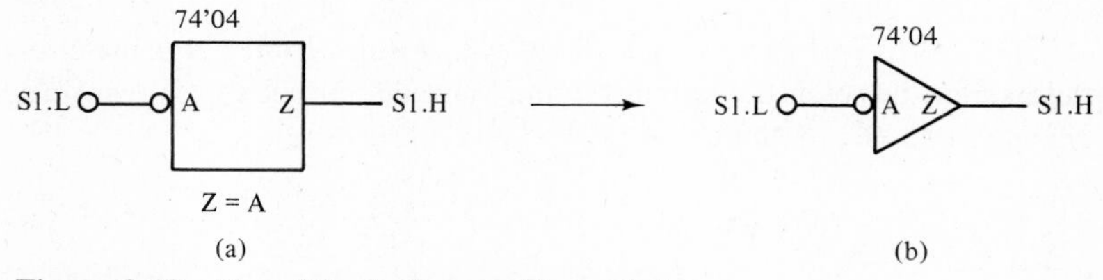

Figure 3.10 Use of the buffer graphic symbol.

In Fig. 3.10(b) the shape of the symbol tells us that the 74'04 performs the IDENTITY logic operation on A, making Z = A. Also, with a circle at the input but none at the output, the gate functions as a *voltage* inverter, transforming S1.L to S1.H for this assignment (2).

For assignment (3) the graphical representation and logic table are

74'04

A Z

A	Z
F	F
T	T

Z = A

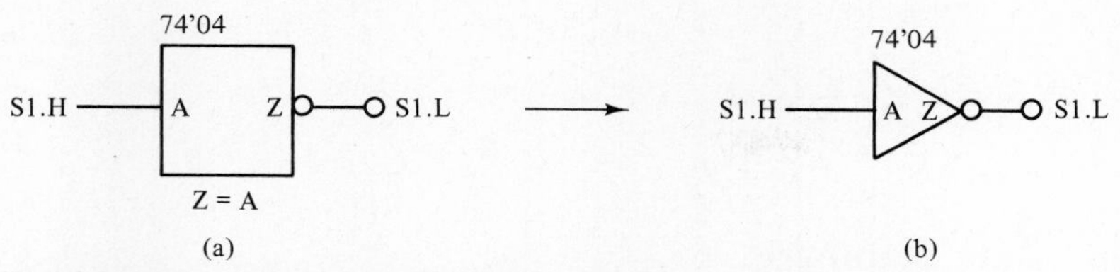

Figure 3.11 Active-high input voltage inverter.

Again, the 74'04 implements an IDENTITY operation. In this case, though, it changes the voltage representation from active-high to active-low. Also, as shown in Fig. 3.11, we can again use the buffer symbol instead of the rectangular box.

In Fig. 3.11(b) the shape of the symbol tells us that for assignment (3) the 74'04 performs the IDENTITY operation on A, making Z = A. Also, with no circle as the input but with one at the output, the gate functions as a *voltage* inverter, transforming S1.H to S1.L.

For assignment (4) of an active-high input and also output, the block diagram and the logic table are

74'04

A	Z
F	T
T	F

$Z = \overline{A}$

Therefore for the assignment of both signals active-high, the 74'04 functions as a *logic* inverter, but not a *voltage* inverter. In the positive-logic convention in which all signals are active-high, the 74'04 is always a logic inverter—a NOT gate.

In this case the logic inversion action of the 74'04 results from a mismatch of the voltage representation between the output terminal and the output signal, as follows:

74'04

S1.H → A Z → $\overline{\text{S1}}$.H

Finally, for assignment (1) of active-low input and also output, the block diagram and the logic table are

74'04

A	Z
T	F
F	T

$Z = \overline{A}$

from which we see that the 74'04 performs the logic NOT operation. So, with both signals active-low, the 74'04 again functions as a logic inverter, but not a voltage inverter. With the negative-logic convention, in which all signals are active-low, the 74'04 is again always a logic inverter. This logic inversion action of the 74'04 also results from a mismatch of the voltage representation between the output signal and output terminal, as follows:

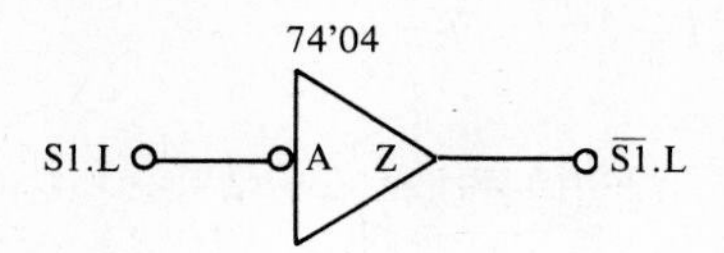

3.3.4 SSI Basic Gate Summary

Figure 3.12 summarizes the most important interpretations of some of the popular SSI devices described in TTL data books.

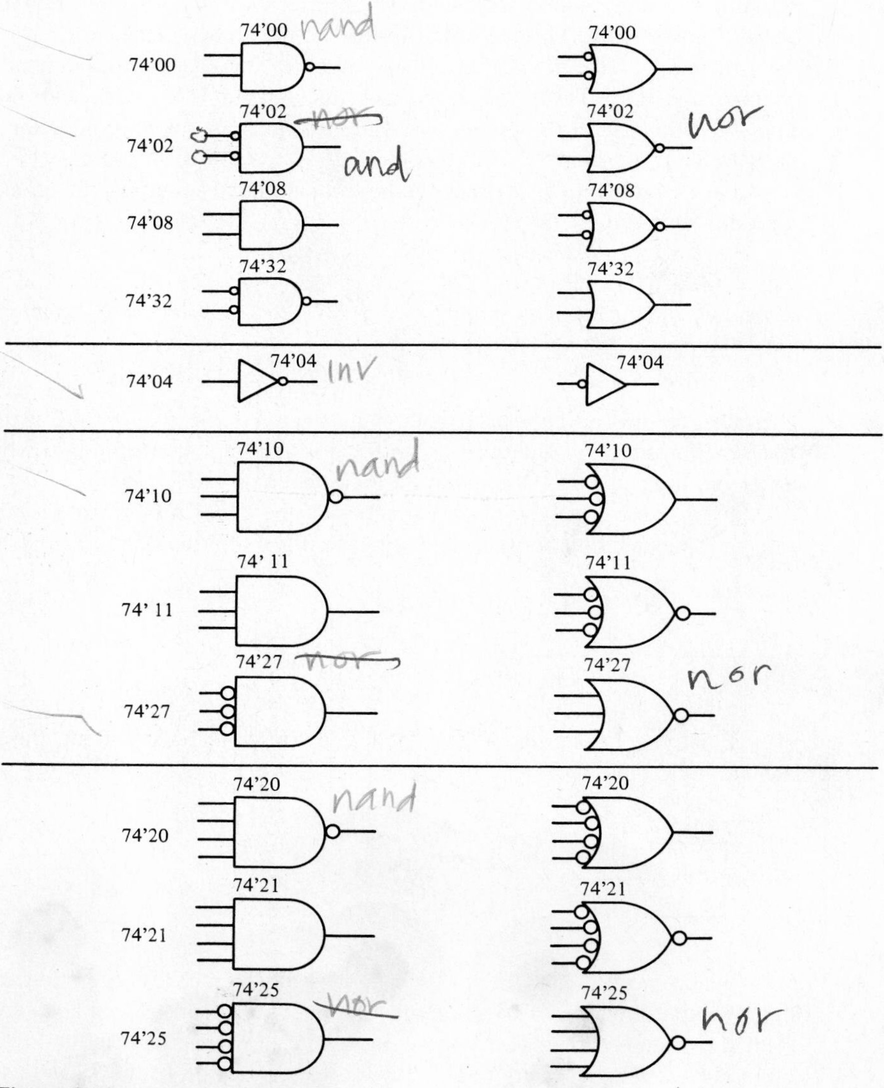

Figure 3.12 TTL basic gate summary.

3.4 SYNTHESIS OF DIGITAL CIRCUITS

Our main goal in this chapter is to learn how to synthesize (design) digital circuits with SSI components, starting with truth table or K-map specifications of the Boolean functions to be implemented. For this design process we must learn how to select the proper SSI components and how to interconnect them such that the resulting digital circuit performs the logic function specified by the logic expression (or truth table). Part of the design process is to make certain that the voltage representations are consistent among the signals. Specifically, we must connect active-high signals to active-high terminals, and connect active-low signals to active-low terminals.

3.4.1 Synthesis Based on the Positive-Logic Convention

One sure way to guarantee that the voltage representations among signals are consistent is to assign them all to be active high. As mentioned, this approach is commonly called the *positive-logic convention*. Currently, it is the most popular approach.

For the positive-logic convention, we can assume in Fig. 3.12 that all the device terminal output circles correspond to logic inversion. (The same is true for the input circles.) In other words, these circles at the outputs have the same effect as NOT gates (logic inverters). This observation follows from the facts that there are no active-low signals in the positive-logic convention and, as we proved in Sec. 3.3, $X.L = \overline{X}.H$. Thus, we see from Fig. 3.12 that the 74'00, 74'10, and 74'20 are positive-logic NAND gates; the 74'02, 74'27, and 74'25 are positive-logic NOR gates; and the 74'04 is a logic inverter. These positive-logic descriptions are the ones that digital-circuit manufacturers specify in their data books along with the 74 labels.

We will study the positive-logic convention approach by way of examples. In them, we will use only MSOPs since the extension to designing with MPOSs should be apparent.

EXAMPLE 3.4

Using the positive-logic convention, design a digital circuit based on the following truth table:

A	B	C	Z
0	0	0	0
0	0	1	1
0	1	0	1
0	1	1	0
1	0	0	1
1	0	1	1
1	1	0	0
1	1	1	0

Solution. Our first step is to obtain an MSOP, as follows:

BC \ A	0	1
00	0	1
01	1	1
11	0	0
10	1	0

$Z = A\bar{B} + \bar{B}C + \bar{A}B\bar{C}$

As should be apparent, to implement this MSOP we need 2 two-input AND gates, 1 three-input AND gate, 1 three-input OR gate and some inverters. Of course, the implementation would be very straightforward if we had the following gates: 74'08 (two-input AND), 74'11 (three-input AND), 74'04 (inverter), and a three-input OR gate. Since there is no three-input TTL OR gate, we could, instead, use two 74'32 gates (two-input OR), as follows:

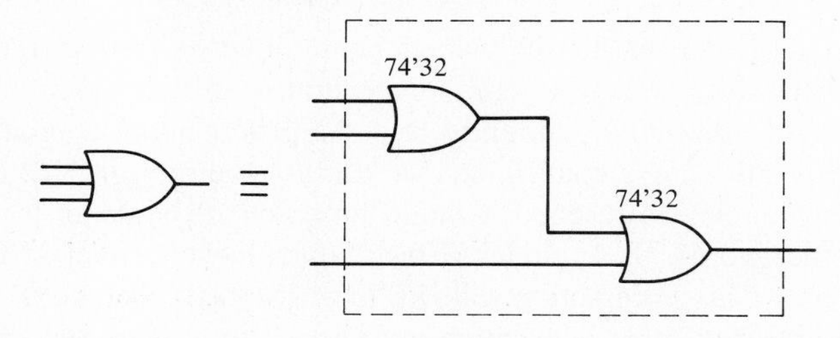

A problem arises, though, if, say, we have only gates 74'00, 74'02, 74'04, 74'10, and 74'27, since in this list there are no AND and OR gates. Because we fixed our voltage representation (all signals are active-high), our interpretations of these gates are also fixed. From Fig. 3.12 they are

74'00—two-input NAND	74'10—three-input NAND
74'02—two-input NOR	74'27—three-input NOR
74'04—NOT	

So all we have available are NAND and NOR gates and some inverters. Consequently, to implement Z we must transform the MSOP (which is in terms of AND, OR, and NOT) into an equivalent expression in terms of NAND, NOR, or NOT. To change an OR expression into a NAND expression, we can double complement (which gives an equivalent expression), and in so doing, use one complement to change the ORs to ANDs as required for NAND, and use the other complement for the inversion part of the NAND operation:

$$
\begin{aligned}
Z &= A\bar{B} + \bar{B}C + \bar{A}B\bar{C} \\
&= \overline{\overline{A\bar{B} + \bar{B}C + \bar{A}B\bar{C}}} \\
&= \overline{\overline{A\bar{B}} \cdot \overline{\bar{B}C} \cdot \overline{\bar{A}B\bar{C}}} \qquad \text{(DeMorgan's law)}
\end{aligned}
$$

By using a pair of two-input NAND gates (74'00) and a three-input NAND gate (74'10) at the input level, we can obtain $\overline{A\bar{B}}$, $\overline{\bar{B}C}$, and $\overline{\bar{A}B\bar{C}}$. Then, with a single three-input NAND gate (74'10) at the second level, we can obtain the NAND of these three terms. Figure 3.13 shows the resulting digital circuit. Note that the literal inputs at the first level are the literals appearing in the original MSOP. This is generally true for a two-level NAND implementation. (Similarly, a two-level NOR realization has as inputs the literals appearing in an MPOS.) ■ ■

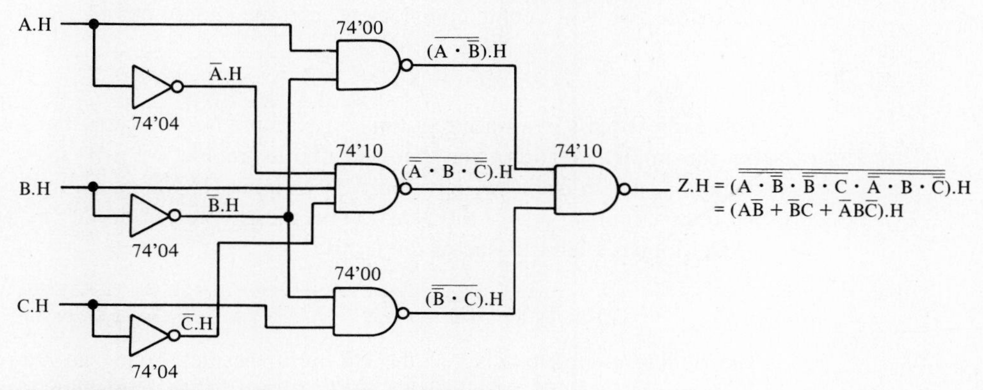

Figure 3.13 Implementation for Example 3.4.

EXAMPLE 3.5

Using positive logic, design a digital circuit for producing the function Z defined in the following truth table.

W	X	Y	Z
0	0	0	0
0	0	1	0
0	1	0	1
0	1	1	1
1	0	0	0
1	0	1	0
1	1	0	1
1	1	1	0

Solution. As before, our first step is to find an MSOP:

XY \ W	0	1
00	0	0
01	0	0
11	1	0
10	1	1

$Z = \overline{W}X + X\overline{Y}$

For our implementation of $Z = \overline{W}X + X\overline{Y}$, again assume that we have only the following gates with the corresponding interpretations (because all signals are active-high):

74'00—two-input NAND 74'10—three-input NAND

74'02—two-input NOR   74'27—three-input NOR
74'04—NOT

Since all we have available are NAND and NOR gates and inverters, we must transform the MSOP expression into an expression in terms of these operations. As before, we will double complement, and then apply one of DeMorgan's laws:

$$Z = \overline{W}X + X\overline{Y} = \overline{\overline{\overline{W}X + X\overline{Y}}} = \overline{\overline{\overline{W}X} \cdot \overline{X\overline{Y}}}$$

This expression is in a form requiring 3 two-input NAND gates and some inverters for the implementation. For the sake of illustration, we will make another assumption, which is that we have only 2 two-input NAND gates. In this case, then, we must eliminate one of the NAND operations. We can do this by applying one of DeMorgan's laws to one of the terms:

$$Z = \overline{\overline{\overline{W}X} \cdot \overline{X\overline{Y}}} = \overline{\overline{\overline{W}X} \cdot (\overline{X} + Y)} = \overline{\overline{\overline{W}X} \cdot (\overline{\overline{\overline{X} + Y}})}$$

The double complementing of the OR term, besides giving an equivalent term, corresponds to NOR followed by NOT. Now we can implement the expression with a pair of two-input NAND gates, 1 two-input NOR gate, and some inverters, as shown in Fig. 3.14. ■ ■

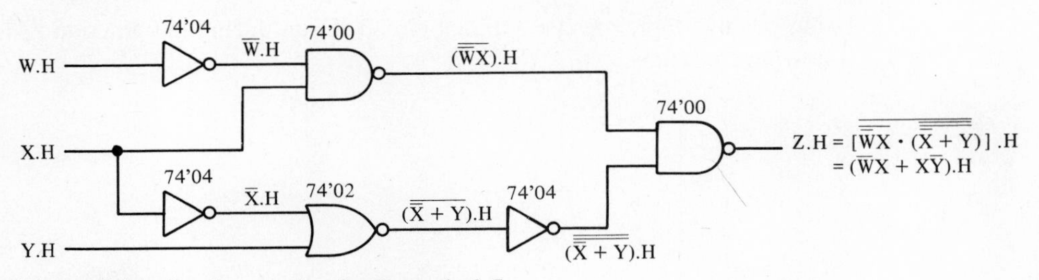

Figure 3.14 Implementation for Example 3.5.

As is evident, a major advantage of the positive-logic approach is the consistency of signal voltage representations since all signals are active-high. A disadvantage is that we sometimes have to transform the original logic expression into some initially obscure expression in order to implement it with the available gates. For this, DeMorgan's laws are very helpful.

3.4.2 Synthesis Based on the Negative-Logic Convention

Another sure way to guarantee that the voltage representations are consistent is to assign them all to be active-low. This approach, commonly called the *negative-logic convention,* is analogous to the approach based on the positive-logic convention. Since the voltage representations are fixed (all signals are active-low), we are, for example, limited to the use of only one of the eight logic/voltage assignments for a two-input, single-output gate, such as the 74'00. Consequently, we must still transform the original logic expression to "fit" that interpretation of the gate. Because of the similarity to the positive-logic convention, we will not consider any examples here.

3.4.3 Synthesis Based on the Mixed-Logic Convention

The third approach to digital circuit synthesis is based on the *mixed-logic convention* in which both active-high and active-low signals are allowed. Since we are not restricted to a particular logic/voltage assignment (where *all* the signals are either active-high or active-low), we can choose the voltage assignment of each signal at implementation time. As a result, we can use all eight logic/voltage assignments of a gate such as the 74'00. Also, we can base our designs on the AND, OR, and NOT operations, and so directly implement an MSOP or an MPOS without any algebraic manipulation. With this approach, we must, of course, take care to ensure that the voltage representations are consistent between signals and terminals. Perhaps the mixed-logic approach can be best introduced by way of an example.

EXAMPLE 3.6

Using the mixed-logic convention, directly implement $Z = \overline{W}X + X\overline{Y}$ from Example 3.5 in Sec. 3.4.1. The available SSI gates are the 74'00, 74'02, 74'04, 74'10, and 74'27. The input signals are W.H, X.H, and Y.H, and the output signal is to be Z.H.

Solution. To implement the logic expression directly for Z, we need a pair of two-input AND gates, a two-input OR gate, and some inverters. (Note that we do *not* make any transformation of the original expression.) We will make a preliminary sketch of the circuit diagram, showing the AND and OR gate placements.

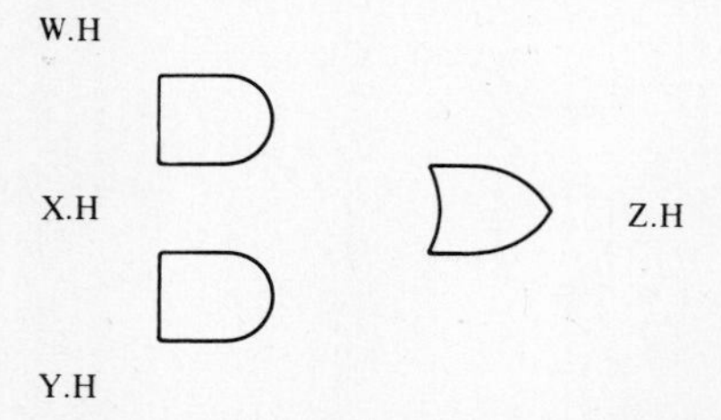

Now, do we have any AND and OR gates among our available gates? Looking at Fig. 3.12, we see that the 74'00 is a two-input AND gate if we use the following voltage assignments:

74'00
A.H
B.H
(AB).L

Furthermore, the 74'02 is also a two-input AND gate if we use the following voltage assignments:

74'02
A.L
B.L
(AB).H

So, in our list of available gates, we have not one, but two types of two-input AND gates! Similarly, we have two types of two-input OR gates:

Both these AND and OR gates are available for our selection. Now, back to the example. Do we use the 74'00 or the 74'02 for our two-input AND gates? Since at this point it really doesn't matter, we will arbitrarily select the 74'00 for the first two-input AND gate.

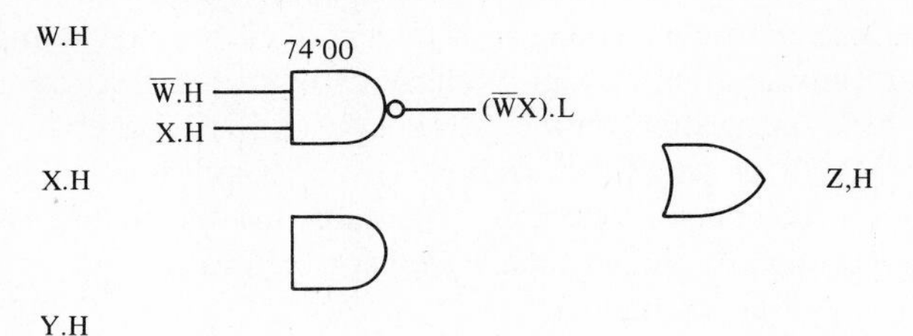

With the operation specified, we have fixed the voltage representations of all the signals of this 74'00. For the AND operation, the inputs are active-high and the output is active-low. Since we want this gate to generate the term $\overline{W}X$, we need inputs of X.H and $\overline{W}$.H. Of course, X.H is available. In addition, we can generate $\overline{W}$.H from the available W.H by using a 74'04 as a logic inverter. This is possible because $\overline{W}$.H = W.L.

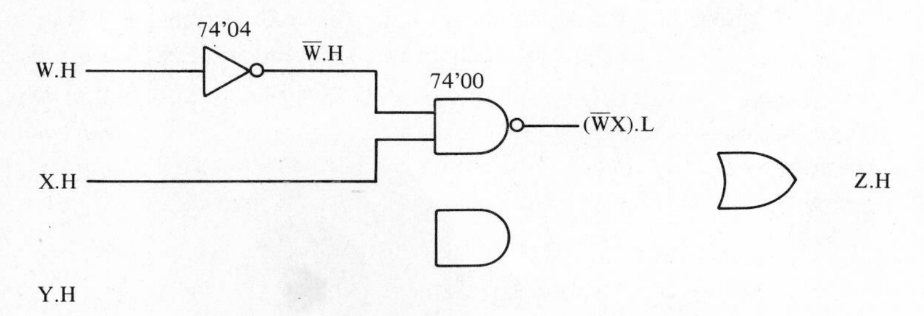

Similarly, we can generate the term $X\overline{Y}$.

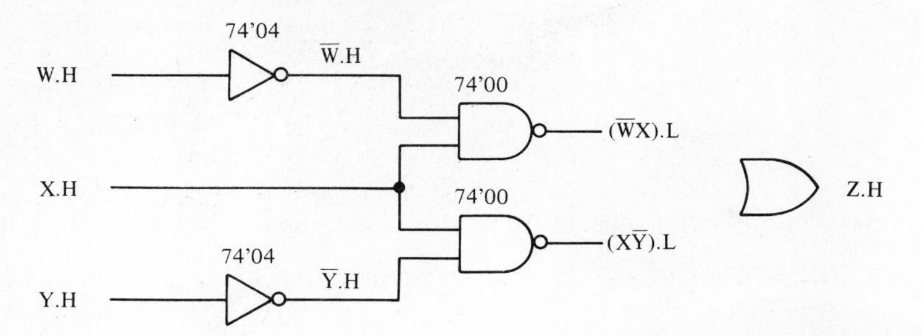

There is no special reason to choose the 74'00 for this second AND gate other than to have the voltage representation (active-low) of the output of this gate the same as the voltage representation of the output of the other AND gate.

Finally, we need to OR $\overline{W}X$ and $X\overline{Y}$. For this, we should use the 74'00 version of the OR gate, rather than the 74'02 version, since these two signals are active-low, and the 74'00 OR gate requires active-low inputs. An added bonus is

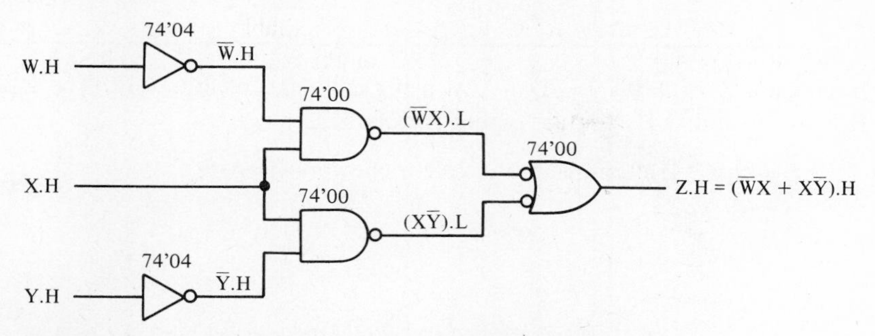

Figure 3.15 Mixed-logic implementation for Example 3.6.

that the output of the 74'00 OR gate is active-high, as is required for the output signal. The complete circuit is shown in Fig. 3.15.

This implementation of $Z = \overline{W}X + X\overline{Y}$ is functionally equivalent to that of Fig. 3.14 which, although for the same function, is based on a different approach—that of positive logic. Even with the same logic convention, we can obtain different implementations, some of which may be better than others. Figure 3.16 shows another mixed-logic implementation of $Z = \overline{W}X + X\overline{Y}$. ■■

The functionally equivalent implementations of Figs. 3.15 and 3.16 have the same number of gates. Usually, though, different implementations have different numbers of gates, primarily of inverters. Optimally, of course, we want an implementation with the least number of inverters, and of gates in general. There are no definitive rules for obtaining such a "best" implementation. It is a skill that can be enhanced through practice. The following guidelines are helpful, however.

1. Determine the framework of the digital circuit by defining all the input and output signals and sketching in the required AND and OR gates.
2. Pick one of the gates, preferably at the input level, and select a specific TTL gate for it. This selection fixes the voltage assignments for the gate input and output signals.
3. Obtain the required input signals for this gate from the available signals, using inverters and/or the "other label" ($A.H = \overline{A}.L$), when necessary.
4. Repeat steps 2 and 3 for the remaining gates until the circuit is complete. To minimize the number of inverters, you may, from time to time, have to back up and reconsider your choice of specific TTL gates when it is obvious that you have made a poor choice.

We will illustrate these guidelines with an example.

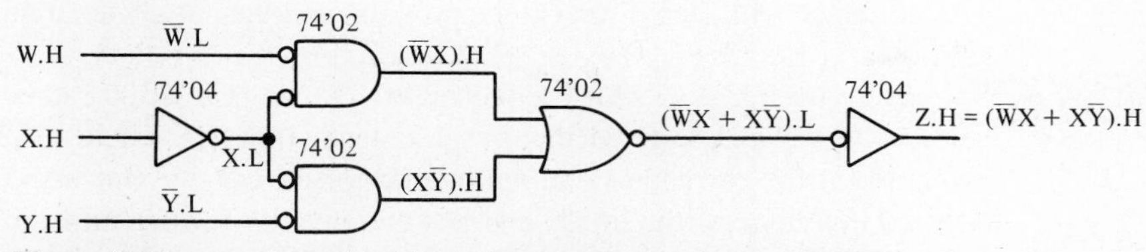

Figure 3.16 Another mixed-logic implementation for Example 3.6.

EXAMPLE 3.7

Implement $Z = \overline{A}BC + CD + A\overline{C}$ using any gates of Fig. 3.12. The inputs are A.H, B.L, C.H, and D.H, and the output is to be Z.L.

Solution. Determine the circuit framework (step 1).

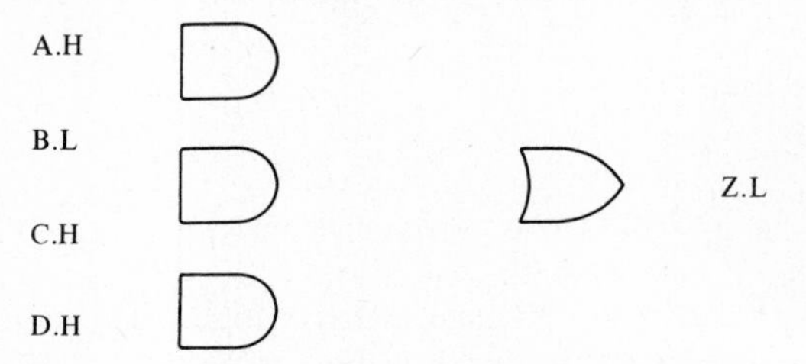

Now (step 2) pick a gate at the first level and select the specific TTL gate for it. We will start with the first three-input AND gate. Which TTL gate should we use?

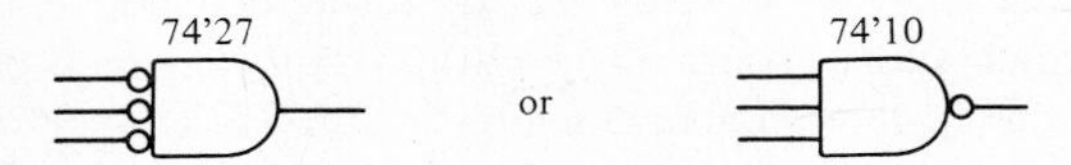

Normally, at this point in the design it would not matter which TTL gate we selected. But we know that at the second level we want a three-input OR gate with an active-low output to generate the output Z.L, and in Fig. 3.12 the only such OR gate is the 74'27, which requires active-high inputs.

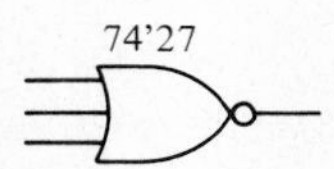

Therefore, for the first AND gate we prefer the 74'27, at least temporarily, since it has an active-high output. Now obtain the required input signals (step 3).

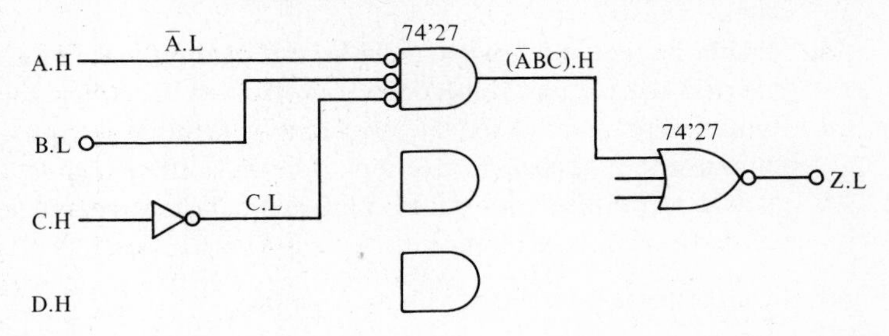

In accordance with step 4, repeat the process until the complete circuit is designed, as shown in Fig. 3.17.

Note in Fig. 3.17 that the second AND gate is a 74'00 rather than a 74'02 version. If we had selected the 74'02 instead, then we would have needed two inverters for the two inputs and no inverter for the output. But with the 74'00, we need no inverters for the inputs and just one inverter for the output. Sometimes by making the appropriate choice of gates, we can reduce the number of inverters

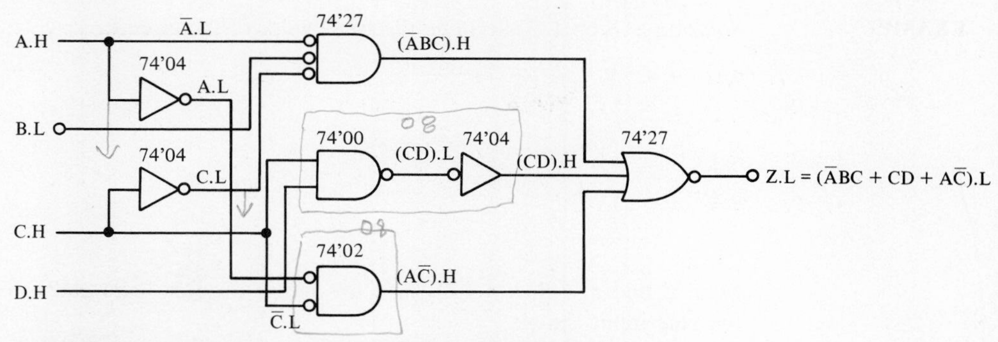

Figure 3.17 Mixed-logic implementation for Example 3.7.

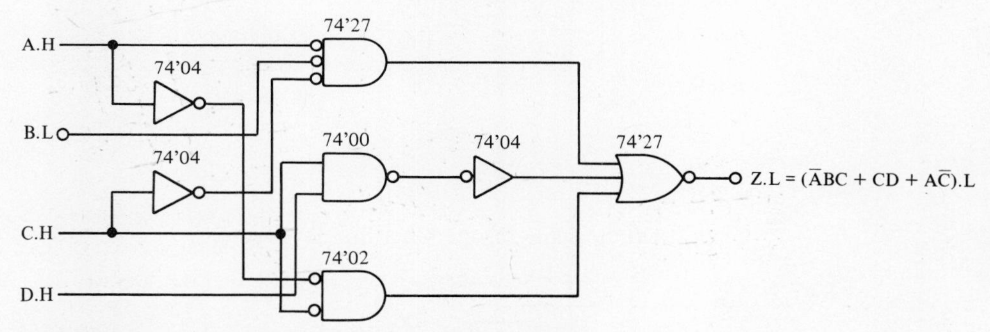

Figure 3.18 Preferred circuit diagram for Example 3.7.

required. Saving inverters is desirable, especially when board space is limited. But that should not be an overriding concern. Having a clear and structured design is more important.

In Fig. 3.17 all the intermediate terms are labeled. Usually, though, this is not desirable, and they can be omitted, as shown in Fig. 3.18.

When we finish our design, we should check our circuit diagram. In analyzing a circuit diagram we should realize that the circle at an input or output terminal can implement a voltage inversion or a logic inversion operation, depending on the context. If there is a mismatched logic/voltage assignment between a signal and a terminal, then it is a logic inversion operation. Otherwise, it is a voltage inversion operation and can be ignored in the analysis of the circuit. With this in mind, we can readily see in Fig. 3.18 that the output of the three-input 74'27 AND gate is $\bar{A}BC$, and the output of the 74'02 AND gate is $A\bar{C}$, even though these intermediate terms are not labeled. ■■

EXAMPLE 3.8

Directly implement $Z = \overline{(A + \bar{B} + C)} \cdot (A + \bar{D})$. The inputs are A.H, A.L, B.L, C.L, and D.L, and the output is to be Z.H.

Solution. Step 1: Determine the framework of the circuit.

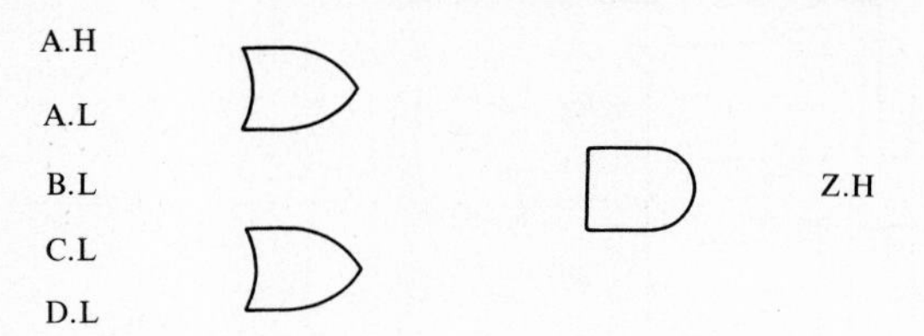

Steps 2 and 3: Select a TTL gate for one of the first-level gates. Then, generate the gate input signals.

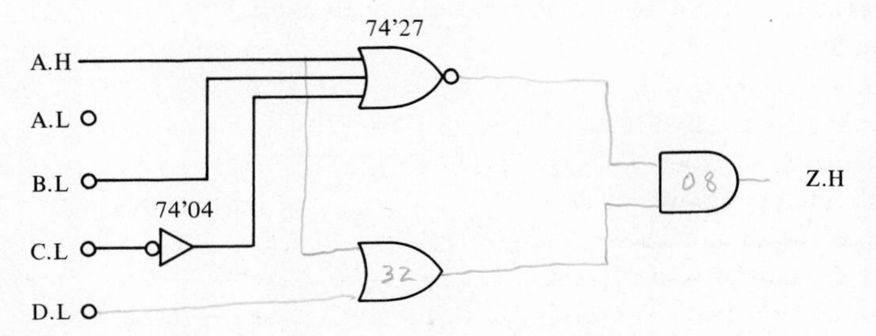

Step 4: Do the same for the other first-level gate.

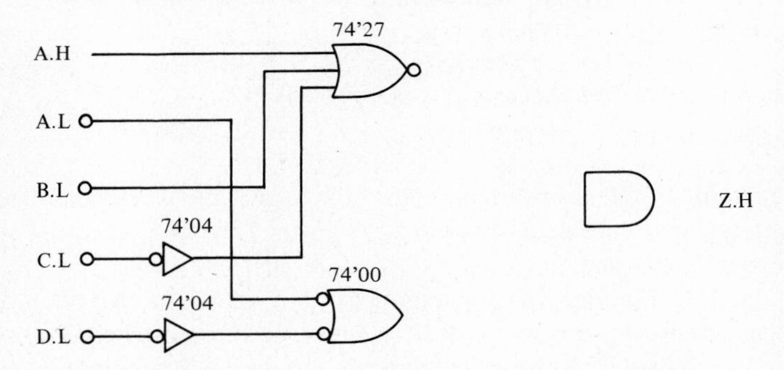

Continuing for the second-level gate, we obtain the circuit of Fig. 3.19. Note how the term $(A + \overline{B} + C)$ is complemented to $\overline{(A + \overline{B} + C)}$.

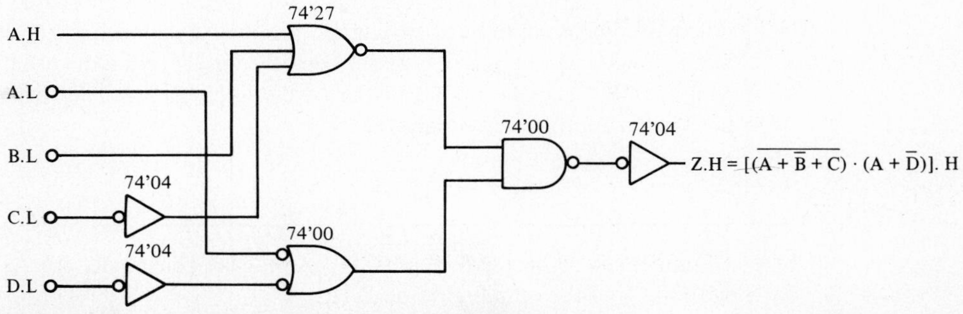

Figure 3.19 Mixed-logic implementation for Example 3.8.

As a check, and also as an exercise, you should analyze the circuit diagram of Fig. 3.19 to determine whether the output is $Z = (A + \overline{B} + C) \cdot (A + \overline{D})$. As mentioned, to do this you should ignore the circles if they are voltage inverters and consider each mismatched logic/voltage assignment as producing a logic NOT.

Incidentally, with different choices for the AND and OR gates, the number of inverters required can be reduced from three to one. We will leave it to the reader to try it. ■ ■

3.5 SUMMARY—LOGIC CONVENTIONS

The positive- and negative-logic conventions are simply special cases of the mixed-logic convention. An advantage of the positive-logic approach (and also of the negative-logic approach) is the certainty of matching between signals and gate terminals, since all are active-high (or active-low). The trade-off is a loss of some flexibility. For example, for a gate such as the 74'00 with three signals, there are $2^3 = 8$ different possible logic/voltage assignments. Yet, with the positive-logic approach we can select only one of the eight assignments—that of all signals being active-high. Consequently, we are forced to transform the original logic expression to "fit" *that* interpretation of the gate. This lack of flexibility is also true of the negative-logic approach.

With the mixed-logic approach we do not assign the logic/voltage assignment until implementation time. Therefore we have the freedom to use all possible assignments for a gate. Consequently, we do not need to transform the original expression to fit the gates. Rather, we change the interpretations of the gates *to fit the logic expression*. The trade-off for the mixed-logic approach is the necessity of making certain that the voltage representations between signals and gate input requirements are consistent (sometimes using inverters, sometimes using the "other label").

For the positive- and negative-logic approaches, circuit synthesis generally requires Boolean algebraic manipulation of logic expressions. For the mixed-logic approach, however, circuit synthesis generally requires graphical manipulation. For the design of simple circuits that are implemented with SSI devices, the selection of the easiest approach is a matter of personal preference. But the advantages of the mixed-logic approach become more apparent in the design of more complex circuits with MSI and LSI components. Also, the mixed-logic approach becomes increasingly valuable as the clarity of the digital design becomes more important.

SUPPLEMENTARY READING (see Bibliography)

[Boole 54], [Fletcher 80], [Kintner 71], [Mano 84], [Motorola], [Prosser 87], [Taub 82], [Texas Instruments]

PROBLEMS

3.1. Explain the differences among the positive-logic, negative-logic, and mixed-logic conventions.

3.2. What is the difference between logic values and voltage levels?

3.3. Complete the solution for Example 3.3 in Sec. 3.3.

3.4. Consider the device of Fig. 3.20.

(a) Complete the following table:

Input voltage values			Output logic values (0 = false, 1 = true)	
S1	S2	S3	SA	SB
L	L	L		
L	L	H		
L	H	L		
L	H	H		
H	L	L		
H	L	H		
H	H	L		
H	H	H		

(b) Find the logic equations for SA and SB as functions of S1, S2, and S3.

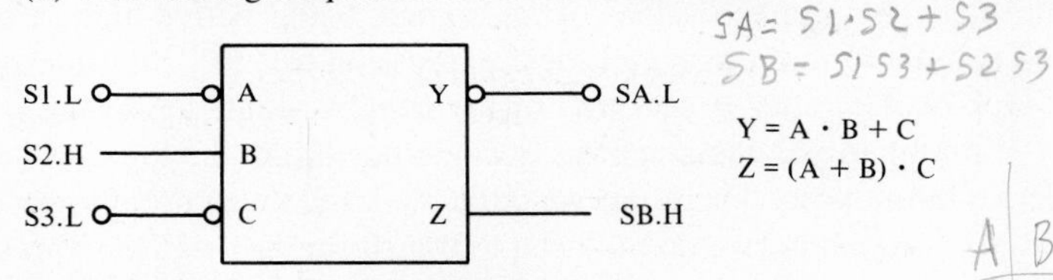

Figure 3.20 Device for Problem 3.4.

3.5. Repeat Problem 3.4 for the device of Fig. 3.21.

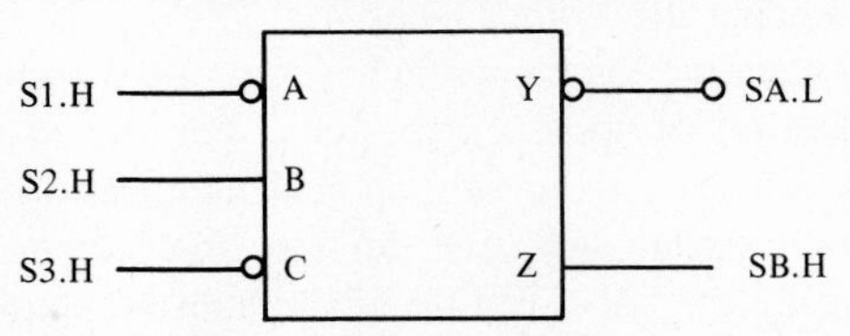

Y = A · B + C
Z = (A + B) · C

Figure 3.21 Device for Problem 3.5.

3.6. Repeat Problem 3.4 for the device of Fig. 3.22.

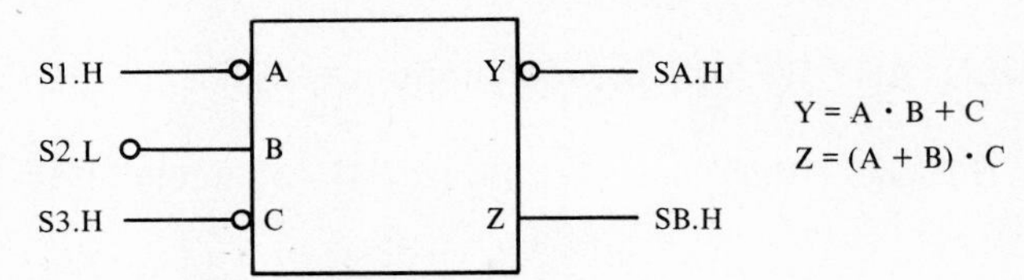

Figure 3.22 Device for Problem 3.6.

3.7. Given the following logic table and that the logic/voltage assignment is A active-low, B active-high, C active-high, and Z active-low, determine the corresponding voltage table. Arrange it in the conventional manner.

A	B	C	Z
F	F	F	F
F	F	T	T
F	T	F	T
F	T	T	F
T	F	F	T
T	F	T	F
T	T	F	T
T	T	T	T

3.8. **(a)** Show how a 74'00 gate can be made into an inverter.
(b) Is it a logic inverter or a voltage inverter? Explain.

3.9. **(a)** By looking into a TTL data book, find the 74'Y component corresponding to the following logic symbol:

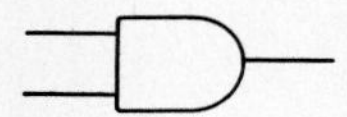

(b) Determine the voltage table for it.
(c) List the eight possible logic/voltage assignments.
(d) For each of the eight assignments, determine (using the voltage and logic tables) the logic function that this component performs.
(e) What logic function does it perform for the positive-logic convention?
(f) What logic function does it perform for the negative-logic convention?

3.10. Repeat Problem 3.9 for the following logic symbol:

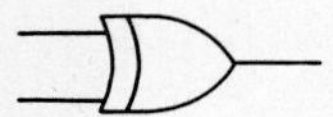

3.11. Find voltage tables for the devices of Fig. 3.23.

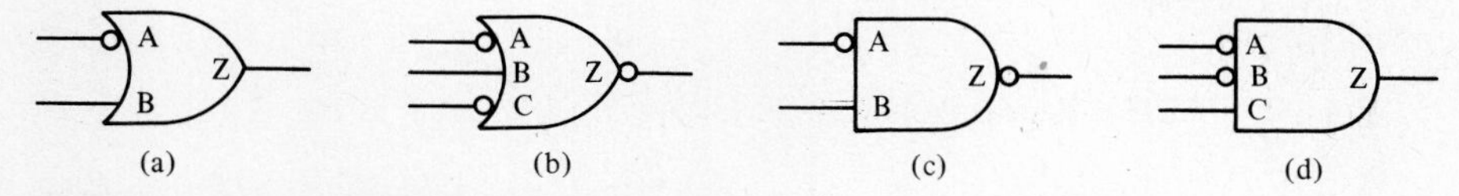

Figure 3.23 Devices for Problem 3.11.

3.12. Find the logic expressions for the Z outputs in Fig. 3.24.

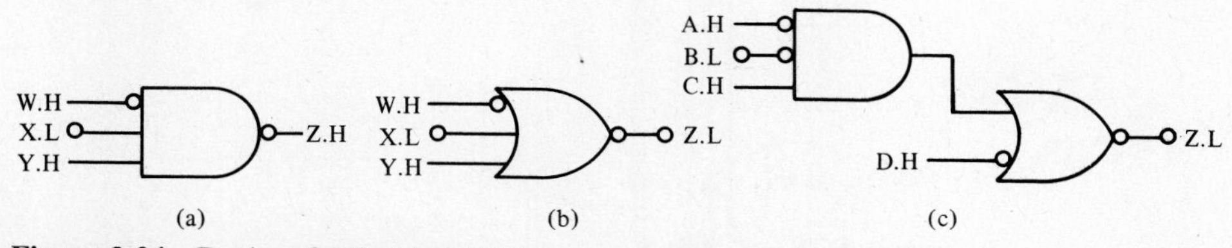

Figure 3.24 Devices for Problem 3.12.

3.13. Fill in all the intermediate signal names for the mixed-logic circuit diagram of Fig. 3.25. Also, find a logic expression for Z.

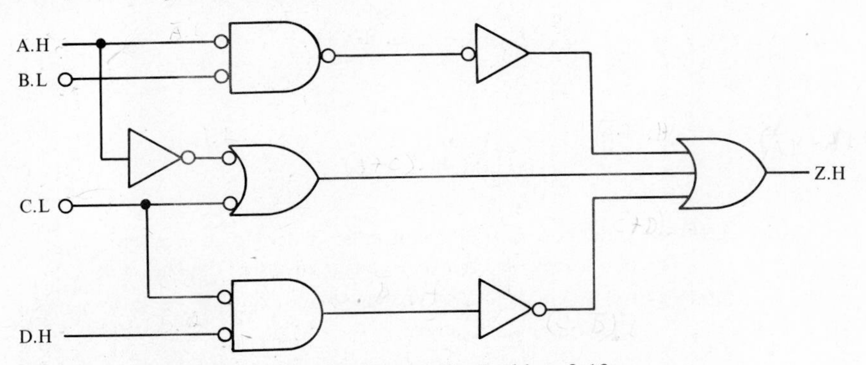

Figure 3.25 Mixed-logic circuit diagram for Problem 3.13.

3.14. Repeat Problem 3.13 for the circuit diagram of Fig. 3.26.

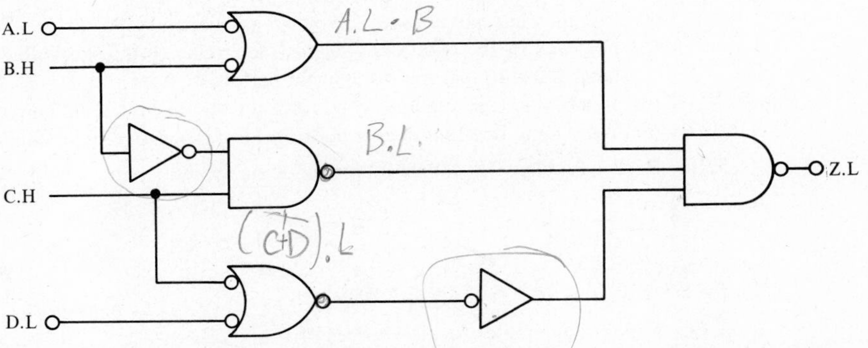

Figure 3.26 Mixed-logic circuit diagram for Problem 3.14.

3.15. Repeat Problem 3.13 for the circuit diagram of Fig. 3.27.

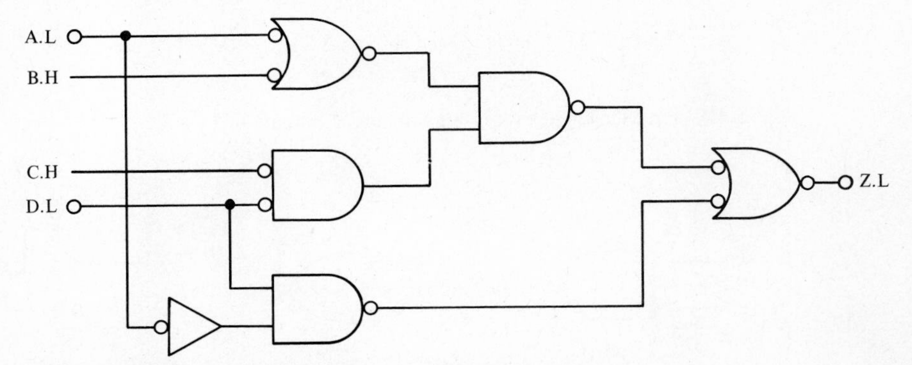

Figure 3.27 Mixed-logic circuit diagram for Problem 3.15.

3.16. Figure 3.28 shows a circuit based on the positive-logic convention. Fill in all the intermediate signal names and find a logic expression for Z. Compare with the answer to Problem 3.13.

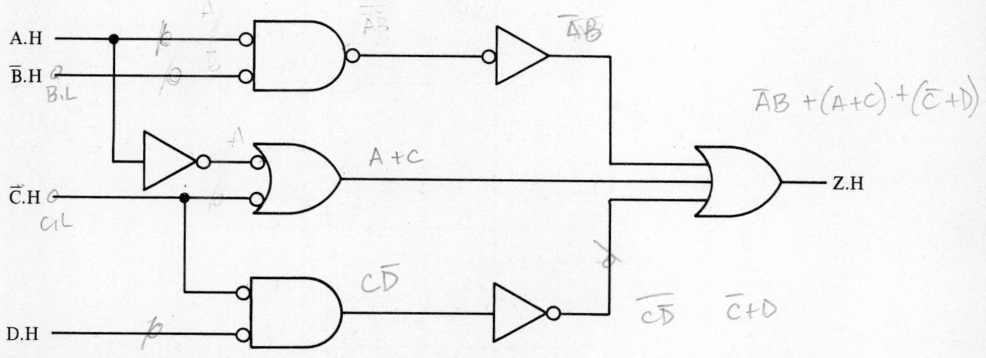

Figure 3.28 Positive-logic circuit diagram for Problem 3.16.

3.17. Repeat Problem 3.16 for the circuit of Fig. 3.29. Compare with the answer to Problem 3.14.

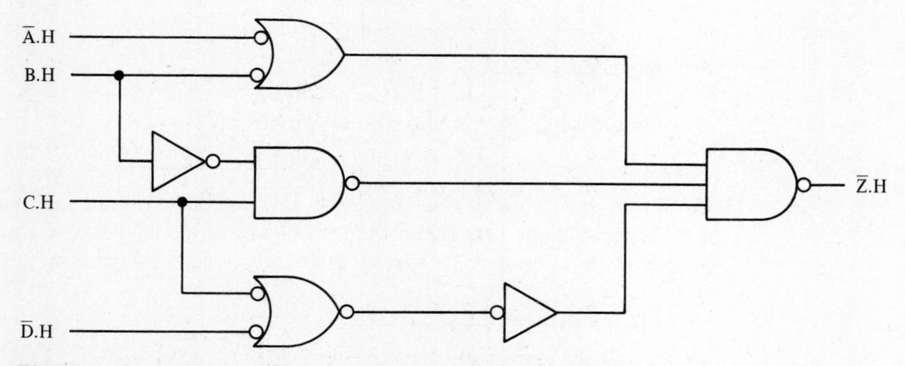

Figure 3.29 Positive-logic circuit diagram for Problem 3.17.

3.18. Repeat Problem 3.16 for the circuit of Fig. 3.30. Compare with the answer to Problem 3.15.

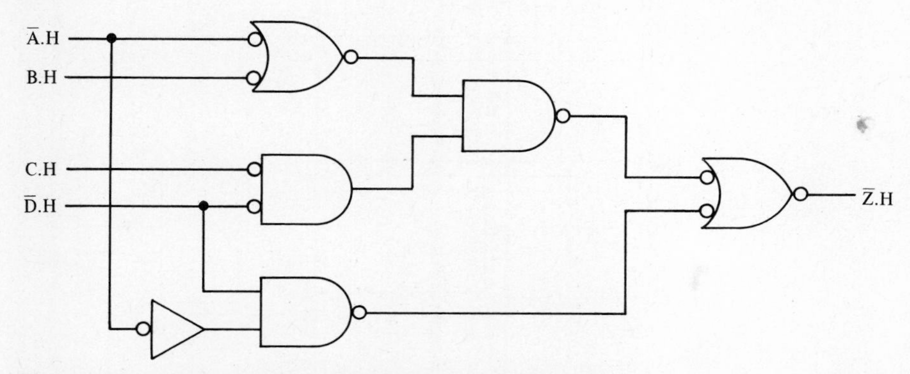

Figure 3.30 Positive-logic circuit diagram for Problem 3.18.

3.19. Analyze each circuit diagram of Fig. 3.31 and determine the logic equation for Z based on the positive-logic convention. Remember that in the positive-logic convention, all inverters and inverting circles perform a logic inversion.

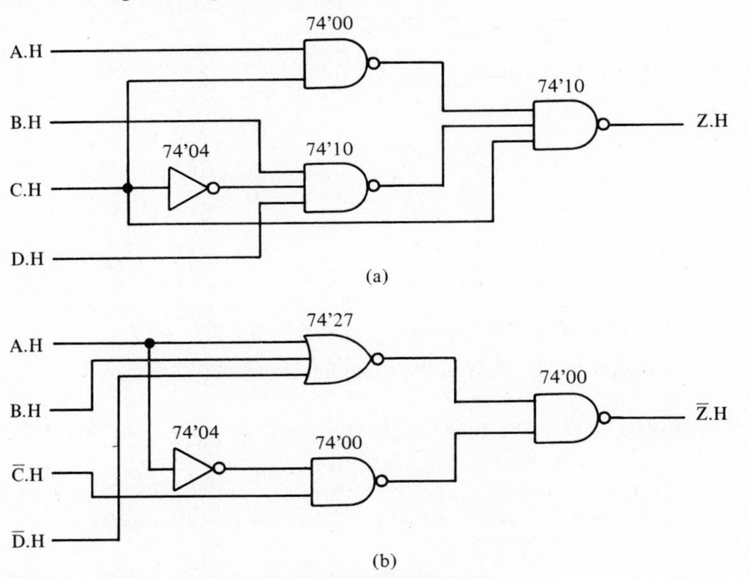

Figure 3.31 Circuit diagrams for Problem 3.19.

3.20. Analyze each circuit diagram of Fig. 3.32 and determine the logic equation for Z based on the mixed-logic convention. Remember that in the mixed-logic convention, a logic NOT occurs as a result of a ''mismatched'' logic/voltage assignment.

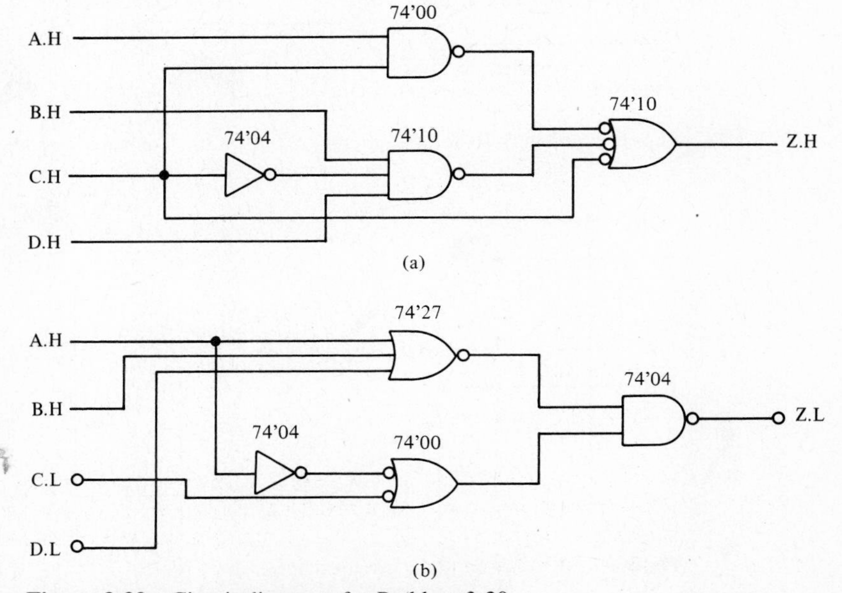

Figure 3.32 Circuit diagrams for Problem 3.20.

3.21. Figure 3.33 shows a two-level NAND logic diagram based on the positive-logic convention. (The number of levels is the maximum number of gates that signals must pass through.) Find an SOP expression for Z.

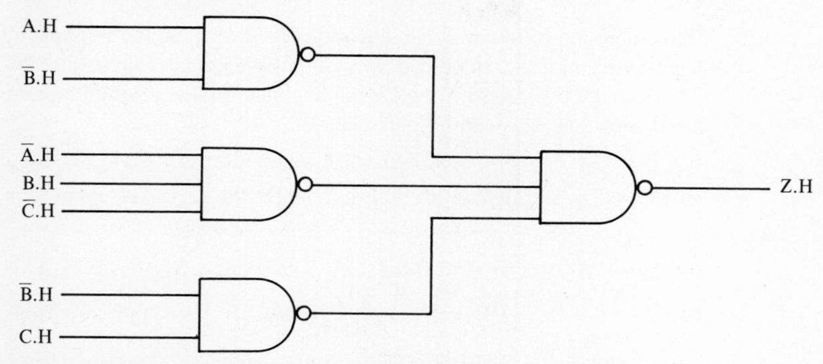

Figure 3.33 Logic diagram for Problem 3.21.

3.22. Implement each of the following using an optimum two-level positive-logic NAND realization. In other words, only NAND gates are available for the implementation, and they are to be used in no more than two levels. Assume that the variables and their complements are both available for inputs.

(a) $Z = \overline{A}\,\overline{B} + \overline{B}C$

(b) $Z = (\overline{A} + B + \overline{C})(\overline{B} + C)$

(c) $Z = \overline{A}CD + AC\overline{D} + AB + B\overline{C}\,\overline{D}$

(d) $Z = (B + \overline{D})(A + \overline{B})(\overline{B} + C + D)$

3.23. Figure 3.34 shows a two-level NOR logic diagram based on the positive-logic convention. Find a POS expression for Z.

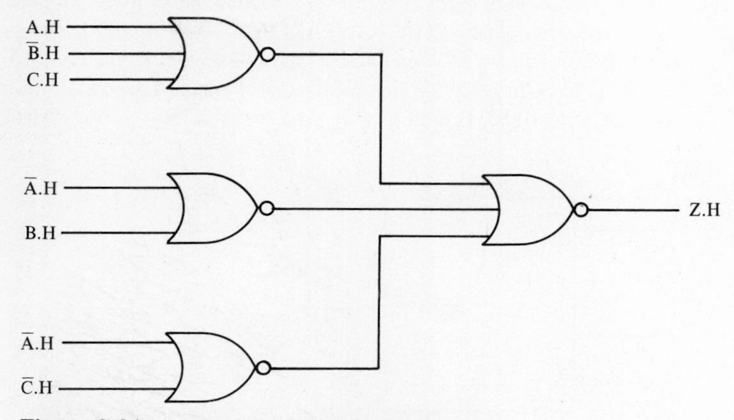

Figure 3.34 Logic diagram for Problem 3.23.

3.24. Implement each of the following using an optimum two-level positive-logic NOR realization. In other words, only NOR gates are available for the implementation, and they are to be used in no more than two levels. Assume that the variables and their complements are both available for inputs.

(a) $Z = (\overline{A} + \overline{B})(\overline{B} + C)$

(b) $Z = A\overline{B}\,\overline{C} + \overline{A}C$

(c) $Z = (\overline{A} + C + D)(A + C + \overline{D})(A + B)(B + \overline{C} + \overline{D})$

(d) $Z = B\overline{D} + A\overline{B} + \overline{B}CD$

3.25. Implement the following logic equations for Z, basing the implementations on the positive-logic convention. Assume that only the following gates are available: 74'00, 74'02, 74'04, 74'10, and 74'27. Also, for each implementation use a minimum number of gates and draw the circuit diagram with all gates labeled.

(a) $Z = AB + \overline{C} + \overline{A}\,\overline{B}\,\overline{D}$ for Z.H. The inputs are A.H, B.H, C.H, and D.L.

(b) $Z = (A + D)(\overline{A} + C + \overline{D})(\overline{C} + D)$ for Z.H. The inputs are A.H, B.L, C.H, and D.H.

(c) $Z = \overline{A\overline{B}} + \overline{C}D + C\overline{D}$ for Z.H. The inputs are A.L, B.H, C.H, and D.L.

(d) $Z = \overline{(A + \overline{D})}(\overline{B} + C)(B + \overline{C})$ for Z.H. The inputs are A.H, B.L, C.H, D.H, and D.L.

(e) $Z = \overline{A\overline{C}} + BD + \overline{B}\,\overline{D}$ for Z.L. The inputs are A.L, B.L, C.H, and D.H.

(f) $Z = A\overline{C} + BD + C\overline{D}$ for Z.L. The inputs are A.L, B.L, C.H, and D.L.

3.26. Repeat Problem 3.25 for implementations based on the mixed-logic convention.

3.27. Implement the following logic equations for Z, basing the implementations on the positive-logic convention. Draw the circuit diagrams with all gates labeled. Select any gates from a TTL data book, and the 74'86 Exclusive OR gate in particular. But, minimize the number of *IC packages* used. In other words, try to use the same types of gates when possible.

(a) $Z = AB + \overline{A \odot C} + \overline{A}C$ for Z.L. The inputs are A.H, A.L, B.L, C.H, and C.L.

(b) $Z = \overline{A \oplus B} + \overline{B} \odot C + \overline{\overline{C}D}$ for Z.H. The inputs are A.H, B.H, C.L, and D.H.

(c) $Z = A\overline{B} + B \oplus D$ for Z.H. The inputs are A.H, B.L, C.L, and D.H.

3.28. Repeat Problem 3.27 for the mixed-logic convention.

3.29. Figure 3.35 shows a parity detector for detecting the parity of an input 4-bit number ABCD. The detector output PARITY is 0 (PARITY = 0) if ABCD has even parity, which means that ABCD contains an even number of 1s. And the output PARITY is 1 (PARITY = 1) if ABCD has odd parity, which means that it contains an odd number of 1s. For example, for ABCD = 0110, PARITY = 0. And for ABCD = 1101, PARITY = 1.

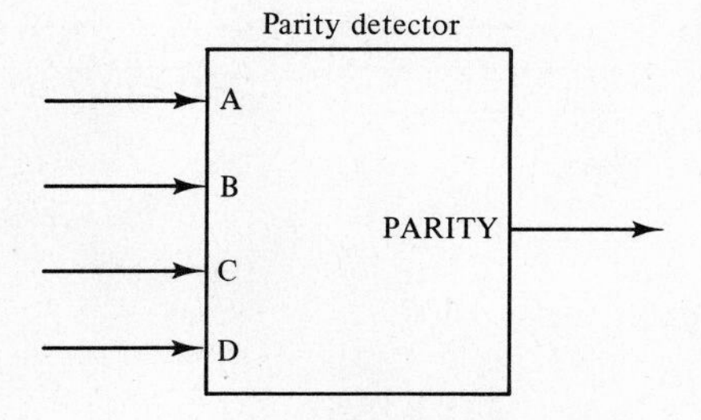

Figure 3.35 Parity detector for Problem 3.29.

(a) Make a truth table for the parity detector, with input columns of ABCD and an output column of PARITY.

(b) Determine an MSOP expression for PARITY.

(c) For inputs of A.H, B.H, C.H, and D.H, implement the MSOP logic equation for PARITY.H using the positive-logic convention.

(d) Repeat part (c) for the mixed-logic convention.

3.30. A combinational circuit with inputs A, B, C, and D and an output Z is to be designed such that Z = 1 if and only if three or more of the inputs are 1.

(a) Make a truth table for the circuit.

(b) Determine an MSOP expression for Z.

(c) For inputs of A.H, B.H, C.H, and D.H, implement the MSOP logic equation for Z.H using the positive-logic convention.

(d) Repeat part (c) for the mixed-logic convention.

3.31. Figure 3.36 shows an excess-3 code generator and table. An excess-3 code is a binary code for decimal numbers in which each *decimal digit* is represented by its binary equivalent *plus 3*. For example, the excess-3 code for the digit 0 is 0011, and for the digit 9 it is 1100. A characteristic of the excess-3 code is that every coded digit has at least one 1, which is important in some applications.

(a) Complete the truth table for the circuit, using don't cares for invalid inputs.

(b) Determine MSOP expressions for the four outputs X_3, X_2, X_1, and X_0.

(c) Implement the MSOP logic equations for the four outputs, basing the implementations on the positive-logic convention. Assume that all inputs and outputs are active-high.

(d) Repeat part (c) for the mixed-logic convention, again assuming that all inputs and outputs are active-high.

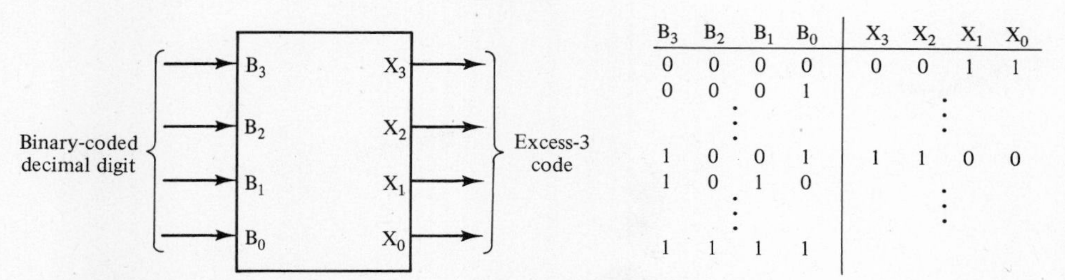

B_3	B_2	B_1	B_0	X_3	X_2	X_1	X_0
0	0	0	0	0	0	1	1
0	0	0	1				
⋮				⋮			
1	0	0	1	1	1	0	0
1	0	1	0				
⋮				⋮			
1	1	1	1				

Figure 3.36 Excess-3 code generator and table for Problem 3.31.

3.32. A large room has three entrances, each with a light switch that controls an overhead light. A combinational circuit is to be designed with inputs A, B, and C from the individual light switches and an output LIGHT for controlling the energization of the overhead light. When all three switches are down (i.e., A = 0, B = 0, and C = 0), then the light is to be off (LIGHT = 0). Also, a change in position of any switch will change the state of the light. Assume that only one switch can be changed at a time.

(a) Draw a block diagram of this circuit

(b) Make a truth table for the circuit.

(c) Determine an MSOP expression for LIGHT.

(d) Implement the MSOP logic equation for $\overline{\text{LIGHT}}$.H, basing the implementation on the positive-logic convention. Assume that the inputs are active-high.

(e) Repeat part (d) for the mixed-logic convention, again assuming that all inputs are active-high.

3.33. The input to the combinational circuit of Fig. 3.37 is a 4-bit binary number $B_3B_2B_1B_0$. The function of this circuit is to convert this binary number into the corresponding negative number $N_3N_2N_1N_0$ in 2s-complement form.

(a) Complete the truth table for the circuit.

(b) Determine MSOP expressions for the four outputs N_3, N_2, N_1, and N_0.

(c) Implement the MSOP logic equations for the four outputs, basing the implementations on the positive-logic convention. Assume that all inputs and outputs are active-high.

(d) Repeat part (c) for the mixed-logic convention, again assuming that all inputs and outputs are active-high.

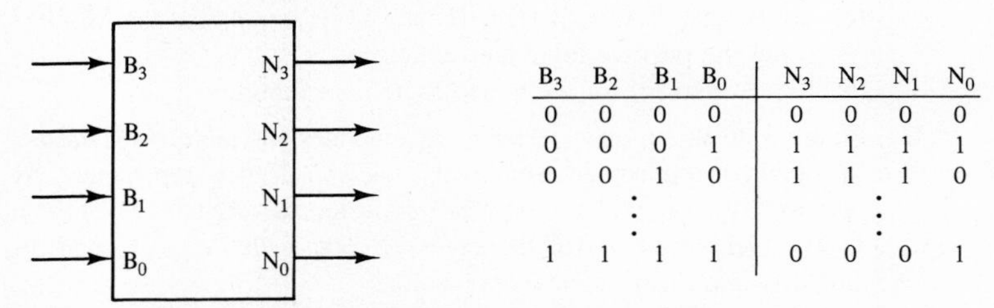

B_3	B_2	B_1	B_0	N_3	N_2	N_1	N_0
0	0	0	0	0	0	0	0
0	0	0	1	1	1	1	1
0	0	1	0	1	1	1	0
		⋮				⋮	
1	1	1	1	0	0	0	1

Figure 3.37 Combinational circuit and table for Problem 3.33.

3.34. What are the advantages and disadvantages of synthesis based on the positive-logic convention?

3.35. What are the advantages and disadvantages of synthesis based on the mixed-logic convention?

Chapter 4

Combinational MSI Circuit Elements

4.1 INTRODUCTION

A number of logic functions are used in the design of a typical digital circuit. The circuit elements that realize these logic functions are classified as one of two types: *combinational* or *sequential*. For a combinational circuit element, the output values at any time are functions only of a combination of the present input values. In contrast, for a sequential circuit element, the output values are functions not only of the present inputs but also of the conditions of some internal states of the circuit element. And the conditions of these states are, in turn, functions of previous inputs. Consequently, for a sequential circuit element the outputs depend on both the present and past values of the inputs. Because past inputs affect present outputs, a sequential circuit element must have some type of memory capability.

In this and the next two chapters we will study the designs and applications of circuit elements that realize some of the commonly used logic functions. These circuit elements form the building blocks used in the design of a digital circuit, as will be discussed in Chapter 7. In the present chapter, combinational MSI (medium-scale integration) circuits will be considered. In Chapter 5 sequential MSI circuit elements will be considered; and in Chapter 6 we complete the study of digital building blocks by considering some of the commonly used LSI (large-scale integration) circuit elements.

In all the following sections of this chapter, the presentation format for each logic function is consistent. First, a functional description is given. Then, there is the design and realization of the logic function based on the SSI design methods of the preceding chapters. Next, commercially available MSI realizations are described. Finally, some applications of these MSI circuit elements are presented.

4.2 BINARY ADDER AND SUBTRACTOR

4.2.1 Half Adder and Full Adder

Binary addition was discussed in Chapter 1. As should be recalled, the binary addition of two bits (A_i and B_i) is represented by the *addition table* shown in Fig. 4.1(a). Here, 0 and 1 represent the binary bits zero and one. If we associate the binary bit 0 with the logic value false, and the binary bit 1 with the logic value true, then the table of Fig. 4.1(a) is *also* the *truth table* for a binary addition circuit element. Using the methods of Chapter 2, we can determine the logic expressions for SUM_i and $CARRY_{i+1}$, and formulate the corresponding circuit, as shown in Fig. 4.1(b). This circuit is called a *half adder*. (The upper graphic symbol represents Exclusive OR.)

The half-adder circuit does not suffice for general additions. To see this, consider the following addition of two multibit binary numbers:

carry	0	1	1	0	
A		0	1	1	0
B	+	0	0	1	1
sum		1	0	0	1

As is evident, when the binary numbers to be added are multibit, then we need to consider the carry that is generated from the preceding stage of addition. Consequently, except

A_i	B_i	SUM_i	$CARRY_{i+1}$
0	0	0	0
0	1	1	0
1	0	1	0
1	1	0	1

(a) Addition table

$$SUM_i = \overline{A}_i B_i + A_i \overline{B}_i = A_i \oplus B_i$$

$$CARRY_{i+1} = A_i B_i$$

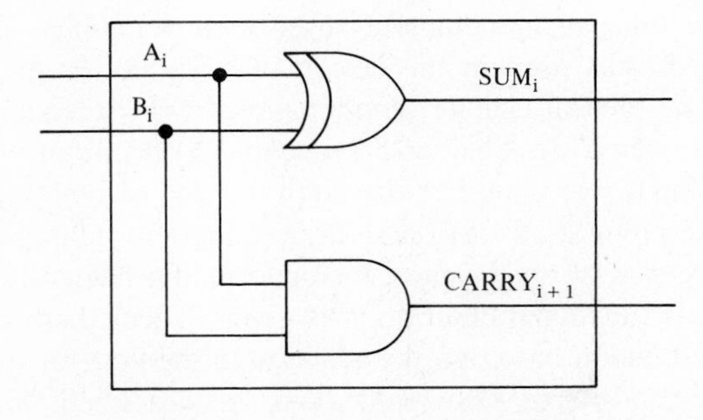

(b) Half-adder circuit

Figure 4.1 Half adder.

for the addition of the least significant bits, a half adder is not adequate. What is needed is a *full adder*.

The functional block diagram and the truth table describing a full adder (FA) are given in Figs. 4.2(a) and (b), respectively. The inputs to the full adder are the current bits of the numbers to be added (A_i and B_i) and the carry-in (C_i) from the preceding addition stage. The outputs are the sum (S_i) and the carry-out (C_{i+1}) generated from the current stage of addition. This carry-out is the carry-in for the next stage.

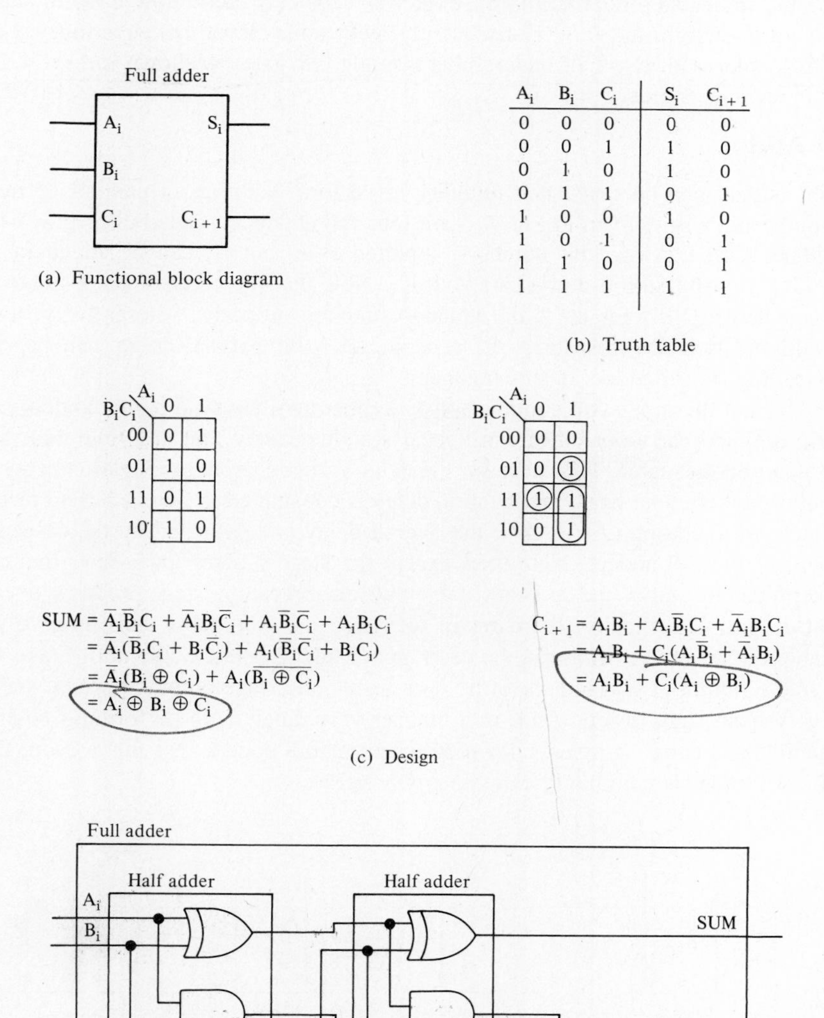

(a) Functional block diagram

A_i	B_i	C_i	S_i	C_{i+1}
0	0	0	0	0
0	0	1	1	0
0	1	0	1	0
0	1	1	0	1
1	0	0	1	0
1	0	1	0	1
1	1	0	0	1
1	1	1	1	1

(b) Truth table

B_iC_i \ A_i	0	1
00	0	1
01	1	0
11	0	1
10	1	0

B_iC_i \ A_i	0	1
00	0	0
01	0	1
11	1	1
10	0	1

(c) Design

(d) Realization

Figure 4.2 Full adder.

Using the methods of Chapters 2 and 3 we can design and realize a full adder, as shown in Figs. 4.2(c) and (d). Note that the groupings of the 1s in the K-map for C_{i+1} do *not* produce the minimum SOP expression for C_{i+1}, which is

$$C_{i+1} = A_iB_i + A_iC_i + B_iC_i$$

This expression would result in the minimum amount of hardware used if C_{i+1} was the only output. However, since S_i is also an output, less *overall* hardware is required if we use the indicated "nonminimum" expression for C_{i+1} and allow C_{i+1} to share with S_i the common term $A_i \oplus B_i$. Consequently, with some clever maneuvering, we can realize a full adder with two half adders plus a single OR gate, as shown in Fig. 4.2(d).

4.2.2 Parallel Adder

Full adders can be connected together to perform addition in parallel of two multibit binary numbers. Shown in Fig. 4.3 are four full adders cascaded to form a 4-bit parallel adder. With it, two 4-bit numbers, inputted at A and B, can be added in parallel to produce a 4-bit sum S and a carry-out C_4. Note that the carry-in C_0 of Stage 0 must be connected to "0" for the 4-bit adder to function properly. Alternatively, a half adder could be used for this stage. In general, an *N*-bit parallel adder can be realized by cascading *N* full adders in this manner.

If all the inputs are simultaneously applied to a physical parallel adder, the correct sum bits and carry-out bit do not appear simultaneously, but at a time that may be tens of nanoseconds later. The cause for the delay is the inherent propagation delay that every real digital element has. (Propagation delay is considered in Sec. 4.8.3.) For the parallel adder configuration of Fig. 4.3, the overall delay is aggravated by the cascade arrangement of the full adders. Note that, except for Stage 0, each carry-in is the carry-out of the preceding stage, and so is not stable until the preceding stage produces a stable carry-out output. Specifically, the carry-in for Stage 1 is not stable until Stage 0 produces a stable output at C_1. Similarly the carry-in at Stage 2 is not stable until Stage 1 produces a stable output at C_2, and so forth. As a result, the stage outputs become stable successively from right to left. And, in a manner very much as in performing binary addition manually, a carry "ripples" down the chain of full adders. For this reason, this type of parallel adder is commonly called a *ripple adder*.

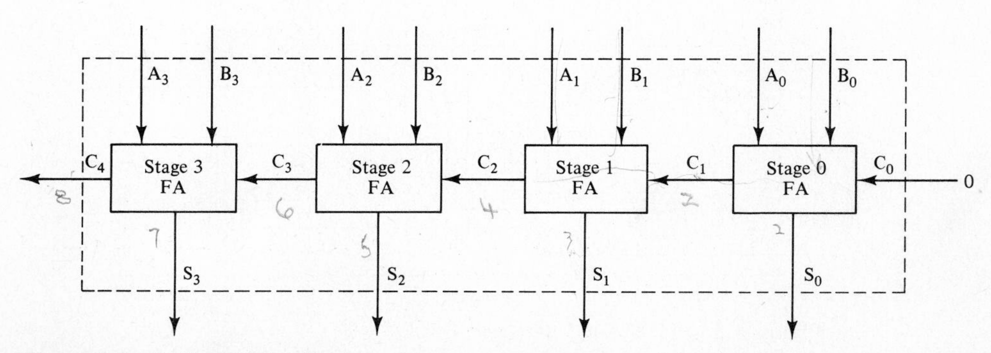

Figure 4.3 4-bit parallel adder.

4.2.3 MSI Parallel Adders

Since binary addition is an important function in digital design, integrated-circuit manufacturers produce multibit parallel adders in the form of MSI chips. Two examples are the 74'83 and the 74'283. Functionally, both perform the same function as the 4-bit parallel adder described in the last section. In other words, each adds two 4-bit binary numbers with a carry-in, and produces a 4-bit sum and a carry-out. Additionally, both of these adders feature *look-ahead* circuitry to eliminate the relatively slow rippling effect of the carry bits of the ripple adder. Carry look-ahead circuitry is discussed in more detail in Chapter 6.

The voltage table for the 74'283 is shown in Fig. 4.4(a). Recall that a voltage table defines the physical behavior of a digital device. It shows how the device really

A_i	B_i	C_i	S_i	C_{i+1}
L	L	L	L	L
L	L	H	H	L
L	H	L	H	L
L	H	H	L	H
H	L	L	H	L
H	L	H	L	H
H	H	L	L	H
H	H	H	H	H

(a) Voltage table

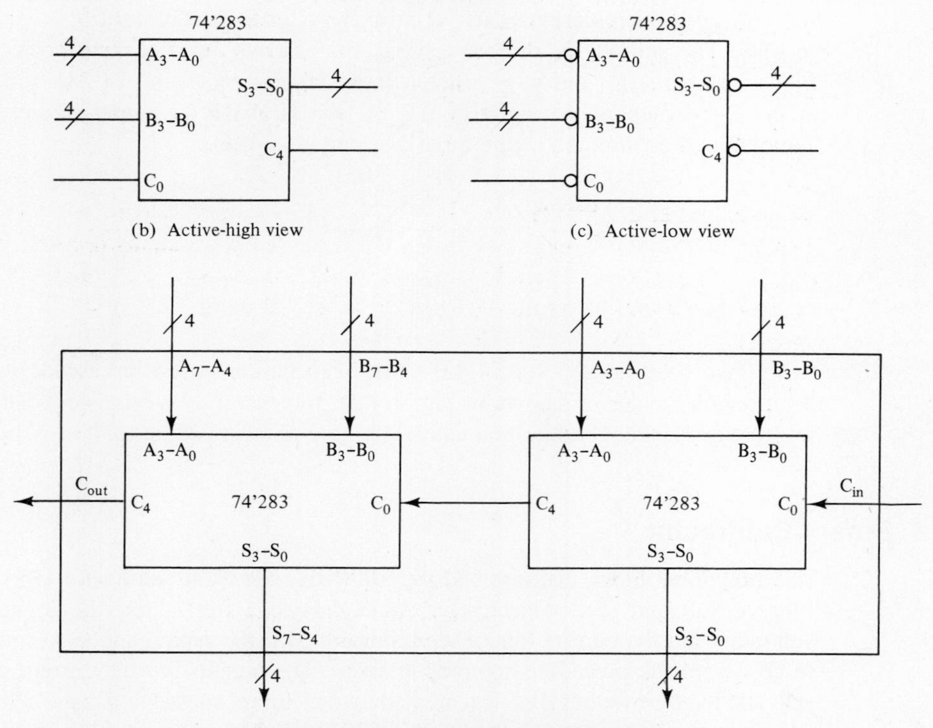

(b) Active-high view

(c) Active-low view

(d) An 8-bit adder constructed from two 74'283 adders

Figure 4.4 74'283 4-bit parallel adder.

works. In contrast, the logic operations that the device performs depend on the assignment of the voltage representation to the logic values. Due to the symmetry of the binary add function, the 74'283 can function as a 4-bit binary adder for two different voltage assignments. The functional block diagram for the active-high view is shown in Fig. 4.4(b), and that for the active-low view is shown in Fig. 4.4(c).

Figure 4.4 introduces a commonly used shorthand notation:

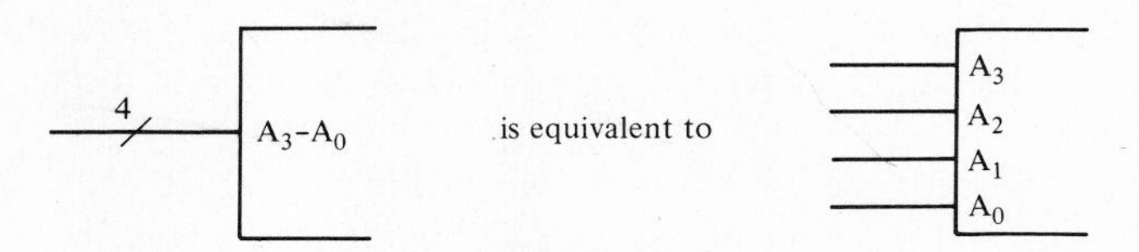

Often, it is convenient to group a set of related signals. Notationally, a slash indicates that a line represents more than one signal. And, the number associated with the slash indicates the number of signals.

The most common view of the 74'283 is the active-high view shown in Fig. 4.4(b). If, however, most of the signals that require the use of the adder are active-low, then the active-low view should be used. Forcing the active-high view in this case requires additional inverters. By mapping the voltage table of Fig. 4.4(a) into the respective logic tables, as done in Chapter 3, we can see that both views result in the binary addition function. (See Problem 4.1.)

For an illustration, we will use the active-low view of a 74'283 adder and show the voltage-level representations for all signals for the addition of 9 + 3 with a carry-in of 1. The sum is 13, of course. And, the carry-out is 0 because no more than four bits are required for the sum. So, for the active-low view, the 74'283 will have LLHL at the SUM outputs, representing the 13, and H at the C_4 output, representing the 0. Following is a summary of the input and output signals.

	A	B	C_0	SUM	C_4
decimal	9	3	1	13	0
binary	1001	0011	1	1101	0
74'283 voltage level	LHHL	HHLL	L	LLHL	H

Two 4-bit 74'283 (or 74'83) adders can be connected in cascade to produce an 8-bit parallel adder, as shown in Fig. 4.4(d). In general, N 4-bit adders can be cascaded to produce a parallel adder that can add binary numbers of up to $4 \times N$ bits each.

4.2.4 Binary Subtractor

The functional block diagram and the truth table for a full subtractor (FS) are given in Figs. 4.5(a) and (b), respectively. Analogous to a full adder, the inputs to the full subtractor are the current bits of the minuend (M_i), the subtrahend (S_i), and the borrow-in (B_i) from the preceding subtraction stage. The outputs are the current difference bit (D_i) and the borrow-out (B_{i+1}) generated by the current subtraction stage. Full subtractors can also be cascaded for the subtraction in parallel of two multibit binary numbers, as shown in Fig. 4.5(c).

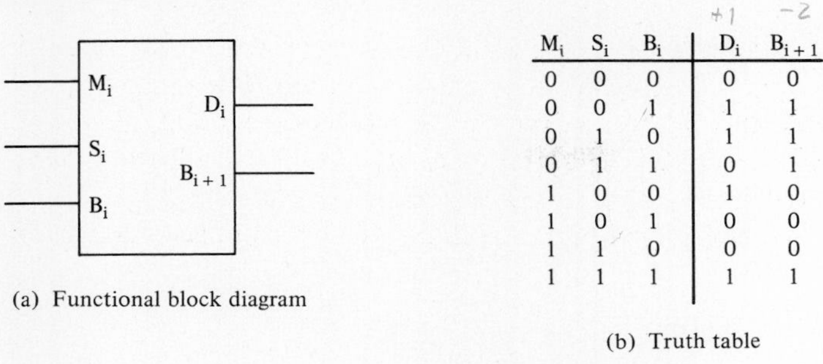

M_i	S_i	B_i	D_i	B_{i+1}
0	0	0	0	0
0	0	1	1	1
0	1	0	1	1
0	1	1	0	1
1	0	0	1	0
1	0	1	0	0
1	1	0	0	0
1	1	1	1	1

(a) Functional block diagram

(b) Truth table

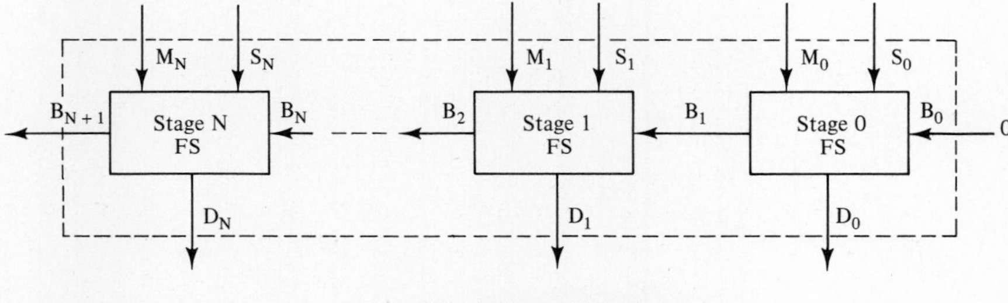

(c) (N + 1)-bit parallel subtractor

Figure 4.5 Binary subtractor.

Binary subtractors are commercially available in the form of MSI circuit elements. One example is, interestingly enough, the 74'283. With a change in voltage assignment from that of Fig. 4.4(b) or (c), it can be made to subtract instead of add. In fact, as shown in Figs. 4.6(a) and (b), two additional views, both realizing 4-bit binary subtractors, for the 74'283 are possible. Consider the 74'283 for the voltage assignment of Fig. 4.6(a). For the voltage table for stage *i* of the 74'283 shown in Fig. 4.4(a), we obtain the following logic table:

VOLTAGE TABLE

M_i	S_i	B_i	D_i	B_{i+1}
L	L	L	L	L
L	L	H	H	L
L	H	L	H	L
L	H	H	L	H
H	L	L	H	L
H	L	H	L	H
H	H	L	L	H
H	H	H	H	H

→

LOGIC TABLE

M_i	S_i	B_i	D_i	B_{i+1}
0	1	1	0	1
0	1	0	1	1
0	0	1	1	1
0	0	0	0	0
1	1	1	1	1
1	1	0	0	0
1	0	1	0	0
1	0	0	1	0

rearrange →

LOGIC TABLE

M_i	S_i	B_i	D_i	B_{i+1}
0	0	0	0	0
0	0	1	1	1
0	1	0	1	1
0	1	1	0	1
1	0	0	1	0
1	0	1	0	0
1	1	0	0	0
1	1	1	1	1

We see from the second logic table that the 74'283 realizes a 4-bit binary subtractor for an assignment of an active-high minuend and difference, and an active-low subtrahend, borrow-in, and borrow-out. Consider what needs to be added for the 4-bit subtractor to have an active-high minuend and subtrahend. (See Problem 4.3.) Similarly, a proof can be derived for the voltage assignment shown in Fig. 4.6(b). (See Problem 4.2.)

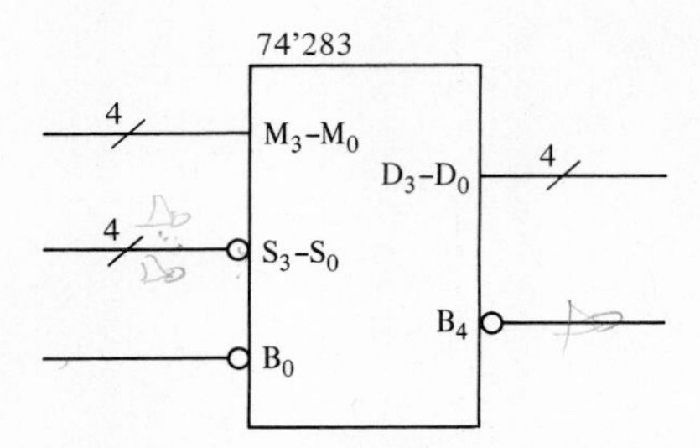

(a) Active-high minuend and difference

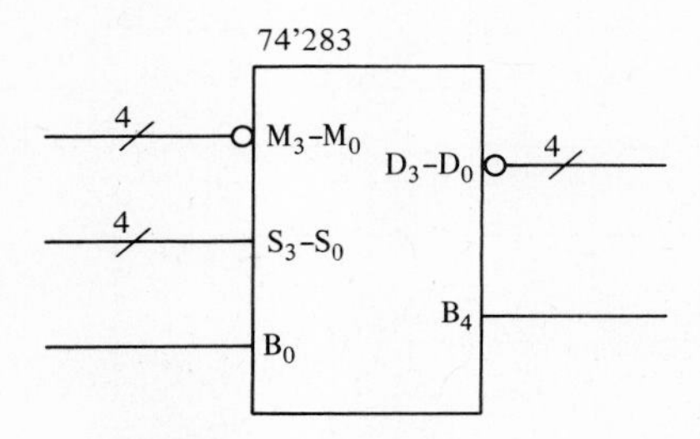

(b) Active-low minuend and difference

Figure 4.6 74'283 binary subtractor.

4.3 MAGNITUDE COMPARATOR

A magnitude comparator is a combinational circuit element that produces outputs that are functions of the relative magnitudes of two inputted binary numbers. A functional block diagram for a 2-bit magnitude comparator is shown in Fig. 4.7(a). For this comparator the inputs are two 2-bit numbers A and B. The outputs are signals indicating whether the binary number A is greater than B, equal to B, or less than B. The functional description of the 2-bit comparator, in the form of a truth table, is given in Fig. 4.7(b). As shown, the signal $(A = B)$ is true when A and B are both equal to 00, 01, 10, or 11. Otherwise, A is either less than B, as shown in rows 2, 3, 4, 7, 8, and 12 of the truth table; or A is greater than B, as shown in rows 5, 9, 10, 13, 14, and 15.

With this truth table, we can use the methods of Chapters 2 and 3 to design and realize the 2-bit comparator, as shown in Figs. 4.7(c) and (d). Note that we need to implement only two out of the three outputs directly since we can generate the third output more economically by using the fact that it is true only when both other outputs are false. So, we can generate one output indirectly by using one of the following equations:

$$(A < B) = \overline{(A > B)} \cdot \overline{(A = B)}$$
$$(A = B) = \overline{(A > B)} \cdot \overline{(A < B)}$$

or

$$(A > B) = \overline{(A = B)} \cdot \overline{(A < B)}$$

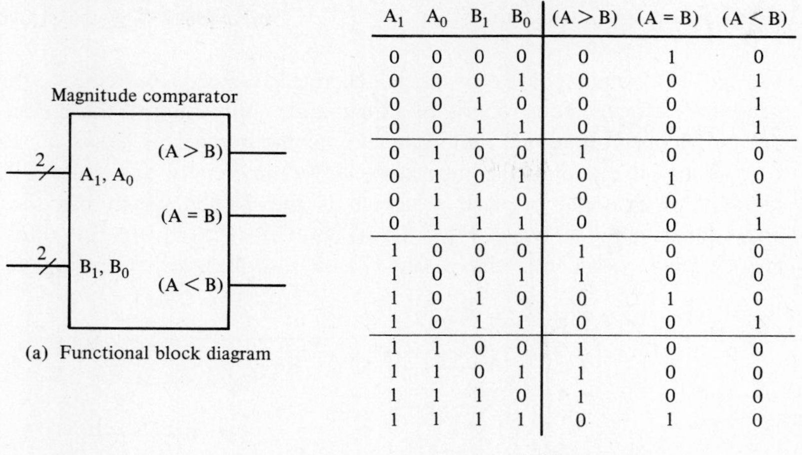

(a) Functional block diagram

A_1	A_0	B_1	B_0	(A > B)	(A = B)	(A < B)
0	0	0	0	0	1	0
0	0	0	1	0	0	1
0	0	1	0	0	0	1
0	0	1	1	0	0	1
0	1	0	0	1	0	0
0	1	0	1	0	1	0
0	1	1	0	0	0	1
0	1	1	1	0	0	1
1	0	0	0	1	0	0
1	0	0	1	1	0	0
1	0	1	0	0	1	0
1	0	1	1	0	0	1
1	1	0	0	1	0	0
1	1	0	1	1	0	0
1	1	1	0	1	0	0
1	1	1	1	0	1	0

(b) Truth table.

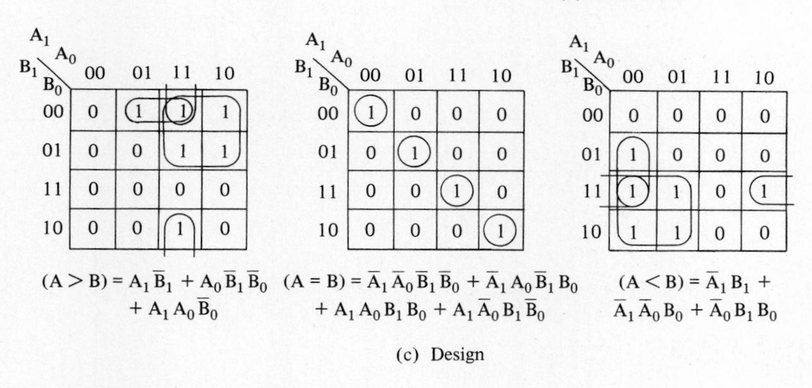

$$(A > B) = A_1 \overline{B}_1 + A_0 \overline{B}_1 \overline{B}_0 + A_1 A_0 \overline{B}_0$$

$$(A = B) = \overline{A}_1 \overline{A}_0 \overline{B}_1 \overline{B}_0 + \overline{A}_1 A_0 \overline{B}_1 B_0 + A_1 A_0 B_1 B_0 + A_1 \overline{A}_0 B_1 \overline{B}_0$$

$$(A < B) = \overline{A}_1 B_1 + \overline{A}_1 \overline{A}_0 B_0 + \overline{A}_0 B_1 B_0$$

(c) Design

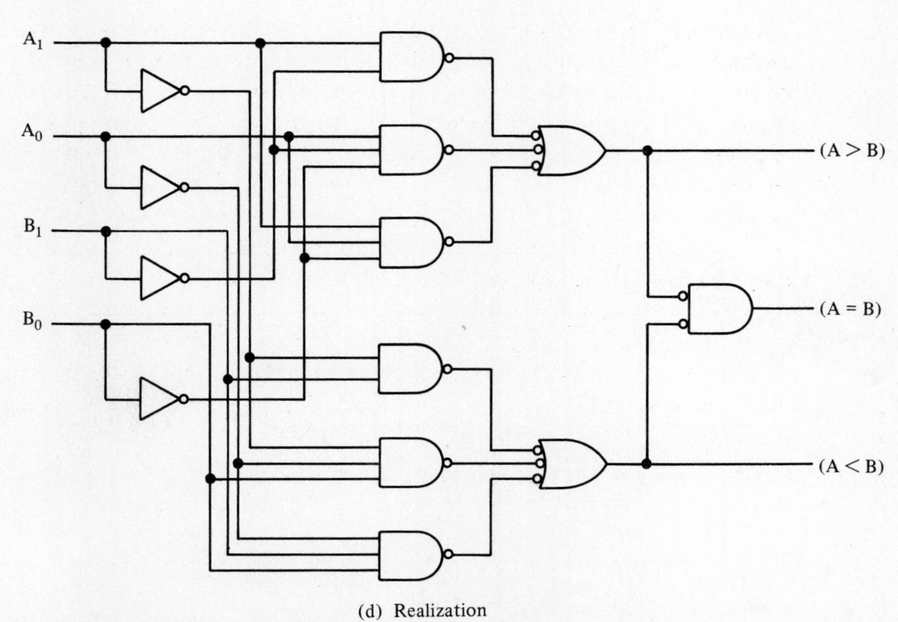

(d) Realization

Figure 4.7 2-bit magnitude comparator.

In Fig. 4.7(d) the signal (A = B) is generated indirectly. Of course, this design can be generalized for the construction of a magnitude comparator of more than 2 bits. But, the design becomes more and more unwieldy as the number of inputs increases.

Four-bit magnitude comparators are commercially available as MSI circuit elements. An example of such a device is the 74'85, which has the block diagram and voltage table shown in Figs. 4.8(a) and (b), respectively. In addition to the inputs and outputs considered above, the 74'85 also has the following cascading inputs:

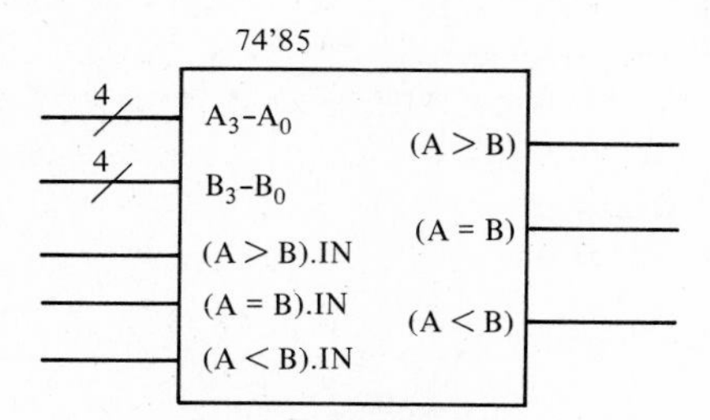

(a) Functional block diagram

Comparing inputs				Cascading inputs			Outputs		
A_3, B_3	A_2, B_2	A_1, B_1	A_0, B_0	(A > B).IN	(A < B).IN	(A = B).IN	(A > B)	(A < B)	(A = B)
$A_3 > B_3$	X	X	X	X	X	X	H	L	L
$A_3 < B_3$	X	X	X	X	X	X	L	H	L
$A_3 = B_3$	$A_2 > B_2$	X	X	X	X	X	H	L	L
$A_3 = B_3$	$A_2 < B_2$	X	X	X	X	X	L	H	L
$A_3 = B_3$	$A_2 = B_2$	$A_1 > B_1$	X	X	X	X	H	L	L
$A_3 = B_3$	$A_2 = B_2$	$A_1 < B_1$	X	X	X	X	L	H	L
$A_3 = B_3$	$A_2 = B_2$	$A_1 = B_1$	$A_0 > B_0$	X	X	X	H	L	L
$A_3 = B_3$	$A_2 = B_2$	$A_1 = B_1$	$A_0 < B_0$	X	X	X	L	H	L
$A_3 = B_3$	$A_2 = B_2$	$A_1 = B_1$	$A_0 = B_0$	H	L	L	H	L	L
$A_3 = B_3$	$A_2 = B_2$	$A_1 = B_1$	$A_0 = B_0$	L	H	L	L	H	L
$A_3 = B_3$	$A_2 = B_2$	$A_1 = B_1$	$A_0 = B_0$	X	X	H	L	L	H
$A_3 = B_3$	$A_2 = B_2$	$A_1 = B_1$	$A_0 = B_0$	H	H	L	L	L	L
$A_3 = B_3$	$A_2 = B_2$	$A_1 = B_1$	$A_0 = B_0$	L	L	L	H	H	L

(b) Voltage table

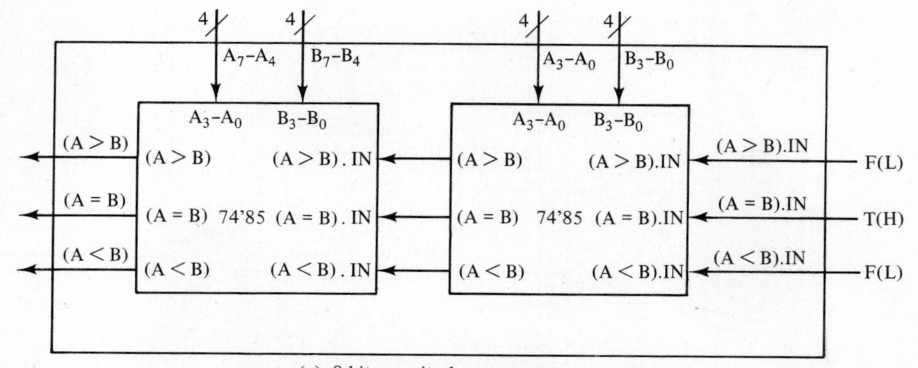

(c) 8-bit magnitude comparator

Figure 4.8 74'85 magnitude comparator.

(A > B).IN, (A = B).IN, and (A < B).IN. These inputs make it easy to cascade 74'85s to produce an N-bit magnitude comparator, where N is a multiple of 4. An 8-bit comparator is shown in Fig. 4.8(c). Note for this connection that if the most significant 4 bits of A have a value greater than those of B, then the signal (A > B) is true regardless of the cascading inputs from the comparator of the preceding stage. Conversely, if the value of the most significant 4 bits of A is less than that of B, then the signal (A < B) is true regardless of the inputs from the preceding stage. When, however, these two values are equal, then the comparator outputs depend on the cascading inputs from the comparator of the preceding stage. Again, in general, 4-bit magnitude comparators can be cascaded in this manner to produce a magnitude comparator of any reasonable length. Note that for the least significant stage of a chain of 74'85 comparators, the (A = B).IN input must be connected to true, and the (A > B).IN and (A < B).IN inputs must be connected to false.

4.4 DECODER

A binary code of N bits can encode up to 2^N different elements of information. So, a 2-bit code can encode up to four elements. A 3-bit code can encode up to eight elements, and so forth. In the design of a digital circuit, it is often necessary to use a decoding function. A *decoder* is a combinational circuit element that will decode an N-bit code. It has up to 2^N output lines, and activates the output signals as a function of the N-bit code applied at the inputs.

Figure 4.9 shows a 3-to-8 decoder. Its functional block diagram is shown in Fig. 4.9(a), and its functional description, in the form of a truth table, is given in Fig. 4.9(b). The input to the decoder is a 3-bit code at the A_2, A_1, and A_0 inputs. Since a 3-bit code can encode up to eight different elements of information, this decoder has eight outputs, each of which represents one of the eight different elements. For example, if the input code is 000 (i.e., A_2 = false, A_1 = false, A_0 = false), then the Z_0 output is activated and the other outputs are all false. If the input code is 001, then only the Z_1 output is activated, and so forth. From the truth table of Fig. 4.9(b), the logic equations for the eight outputs can be determined in a straightforward manner. The result is the circuit shown in Fig. 4.9(c).

Decoders are commercially available as MSI circuit elements in the form of 2-to-4, 3-to-8, and 4-to-10 decoders. An example of a commercially available MSI 3-to-8 decoder is the 74'138 shown in Fig. 4.10. As shown in the functional block diagram of Fig. 4.10(a), the three inputs A_2, A_1, and A_0 are active-high and the eight outputs are active-low. Furthermore, there are two active-low enable inputs E_1 and E_2, and an active-high enable input E_3. As shown in the voltage table of Fig. 4.10(b), the 74'138 functions as a 3-to-8 decoder only if all three enable inputs are true: E_1 = L, E_2 = L, and E_3 = H. Otherwise all eight outputs are false (H).

The function of the enable inputs is to permit convenient expansion. Figures 4.10(c) and (d) show a 4-to-16 decoder, with an active-low enable, constructed from two 74'138 decoders without any additional logic. Note that the 74'138 decoder on the top is enabled only if A_3 is 0, and the bottom 74'138 decoder is enabled only if A_3 is 1. Consequently, each different element of the 4-bit code activates a unique output. In general, a decoder

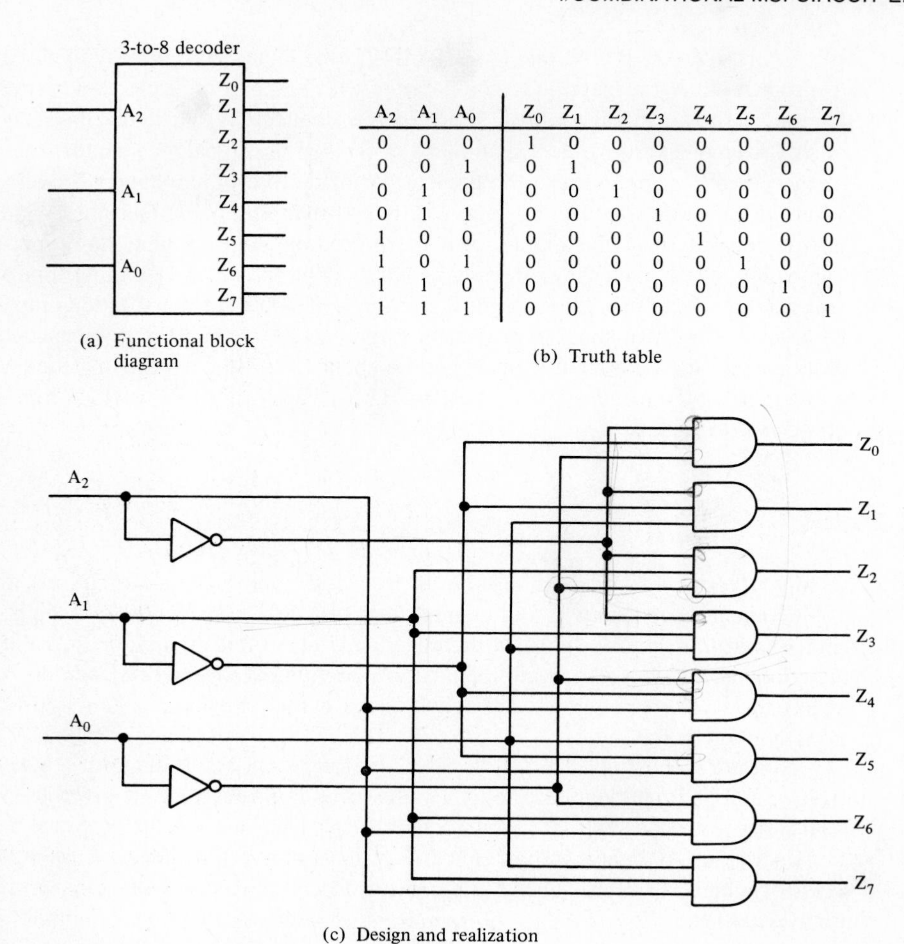

(a) Functional block diagram

A_2	A_1	A_0	Z_0	Z_1	Z_2	Z_3	Z_4	Z_5	Z_6	Z_7
0	0	0	1	0	0	0	0	0	0	0
0	0	1	0	1	0	0	0	0	0	0
0	1	0	0	0	1	0	0	0	0	0
0	1	1	0	0	0	1	0	0	0	0
1	0	0	0	0	0	0	1	0	0	0
1	0	1	0	0	0	0	0	1	0	0
1	1	0	0	0	0	0	0	0	1	0
1	1	1	0	0	0	0	0	0	0	1

(b) Truth table

(c) Design and realization

Figure 4.9 3-to-8 decoder.

74'138 decoder

A2, A1, A0, E1, E2, E3 / Z0, Z1, Z2, Z3, Z4, Z5, Z6, Z7

(a) Functional block diagram

Inputs						Outputs							
E_1	E_2	E_3	A_2	A_1	A_0	Z_0	Z_1	Z_2	Z_3	Z_4	Z_5	Z_6	Z_7
H	X	X	X	X	X	H	H	H	H	H	H	H	H
X	H	X	X	X	X	H	H	H	H	H	H	H	H
X	X	L	X	X	X	H	H	H	H	H	H	H	H
L	L	H	L	L	L	L	H	H	H	H	H	H	H
L	L	H	L	L	H	H	L	H	H	H	H	H	H
L	L	H	L	H	L	H	H	L	H	H	H	H	H
L	L	H	L	H	H	H	H	H	L	H	H	H	H
L	L	H	H	L	L	H	H	H	H	L	H	H	H
L	L	H	H	L	H	H	H	H	H	H	L	H	H
L	L	H	H	H	L	H	H	H	H	H	H	L	H
L	L	H	H	H	H	H	H	H	H	H	H	H	L

(b) Voltage table

Figure 4.10 74'138 decoder.

4-to-16 decoder

4

A_3–A_0

Z_0–Z_{15}

16

EN

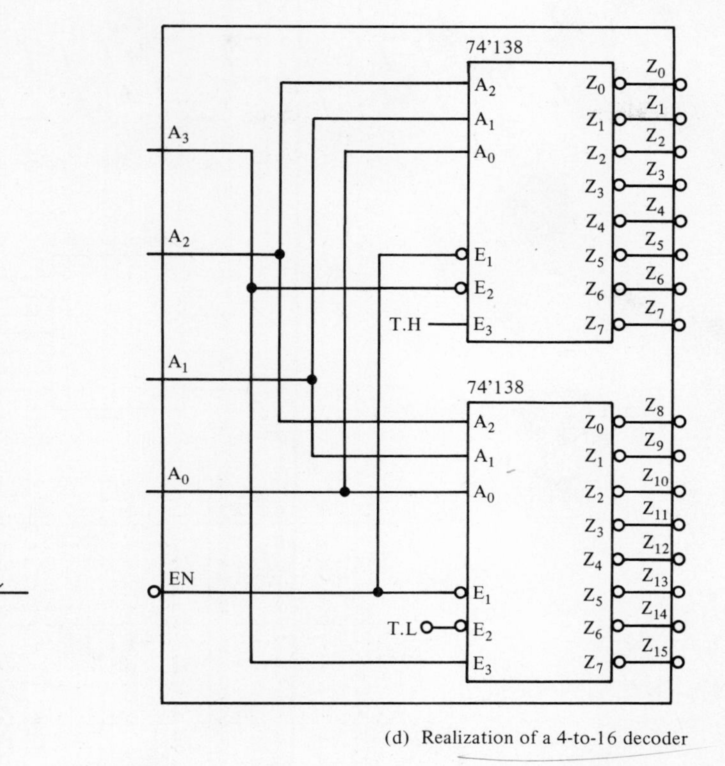

(c) Functional block diagram

(d) Realization of a 4-to-16 decoder

Figure 4.10 *(cont.)*

of a larger size can be constructed by using decoders of a smaller size along with some additional circuitry. Is such circuitry needed to construct a 5-to-32 decoder from four 74'138 decoders? (See Problem 4.8.)

4.4.1 BCD-to-7-Segment Decoder

Outputs of a digital circuit are often displayed as decimal digits. The most common and simplest device for displaying a decimal digit is a 7-segment display, as shown in Fig. 4.11(a). Each of the segments is an LED (light-emitting diode) that will glow when a true signal is applied to it. By a proper selection of the segments to be lit, the decimal digits can be displayed as shown in Fig. 4.11(a).

Seven-segment displays are commercially available in two forms: common anode and common cathode. The common-anode display, with all the LED anodes connected together, is active-low. And, the common-cathode display, with all the LED cathodes connected together, is active-high.

A BCD-to-7-segment decoder is a combinational circuit element that converts a BCD number into the signals required for the display of the value of that number on a 7-segment display. The functional block diagram for such a decoder is shown in Fig. 4.11(b). The seven decoder outputs (a, b, c, d, e, f, g) correspond to the seven segments with the same labels of the 7-segment display. The functional description of the decoder,

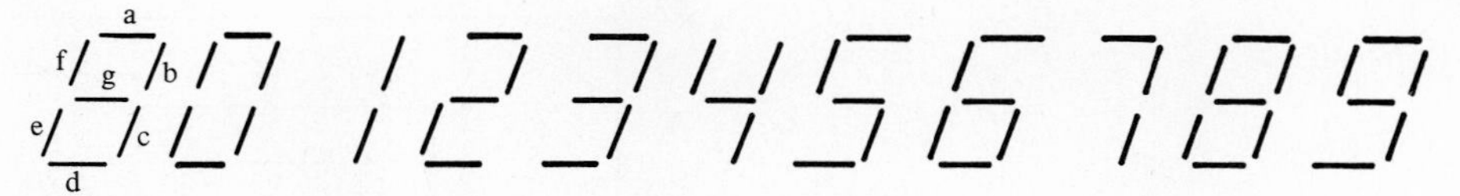

(a) A 7-segment display

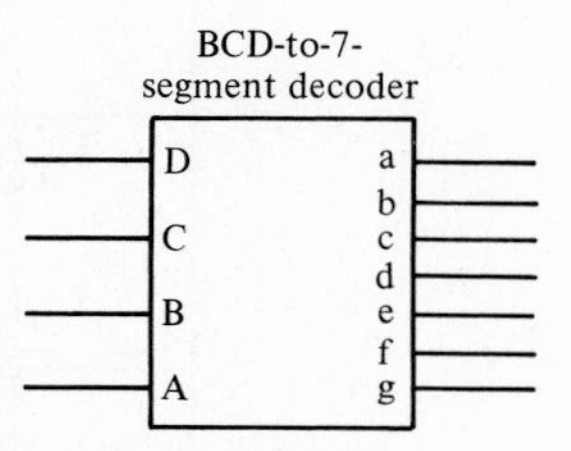

(b) Functional block diagram

D	C	B	A	a	b	c	d	e	f	g
0	0	0	0	1	1	1	1	1	1	0
0	0	0	1	0	1	1	0	0	0	0
0	0	1	0	1	1	0	1	1	0	1
0	0	1	1	1	1	1	1	0	0	1
0	1	0	0	0	1	1	0	0	1	1
0	1	0	1	1	0	1	1	0	1	1
0	1	1	0	1	0	1	1	1	1	1
0	1	1	1	1	1	1	0	0	0	0
1	0	0	0	1	1	1	1	1	1	1
1	0	0	1	1	1	1	1	0	1	1
1	0	1	0	X	X	X	X	X	X	X
1	0	1	1	X	X	X	X	X	X	X
1	1	0	0	X	X	X	X	X	X	X
1	1	0	1	X	X	X	X	X	X	X
1	1	1	0	X	X	X	X	X	X	X
1	1	1	1	X	X	X	X	X	X	X

(c) Truth table

Figure 4.11 BCD-to-7-segment decoder.

in the form of a truth table, is shown in Fig. 4.11(c). Observe from the first row of the truth table that for the display of the digit 0, segments a, b, c, d, e, and f have to be lit, as is evident from Fig. 4.11(a). For the display of the digit 1, the second row specifies that segments b and c have to be lit, and so forth. From another point of view, this truth table specifies that the output corresponding to segment a has to be true for the displays of digits 0, 2, 3, 5, 6, 7, 8, and 9. Also, the output corresponding to segment b has to be true for digits 0, 1, 2, 3, 4, 7, 8, and 9, and so forth. Note that since binary numbers 1010 through 1111 are not valid BCD representations, the outputs for these inputs are designated as don't cares in the truth table. The logic equations for the seven outputs can be determined in a straightforward manner. (See Problem 4.10.)

BCD-to-7-segment decoders are commercially available with either active-low or active-high outputs. Examples are the 74'47 (active-low) and the 74'48 (active-high). The functional block diagrams and voltage tables of both are shown in Figs. 4.12(a) and (b). Note from the voltage tables of Fig. 4.12(b) that the inverse in the output levels is

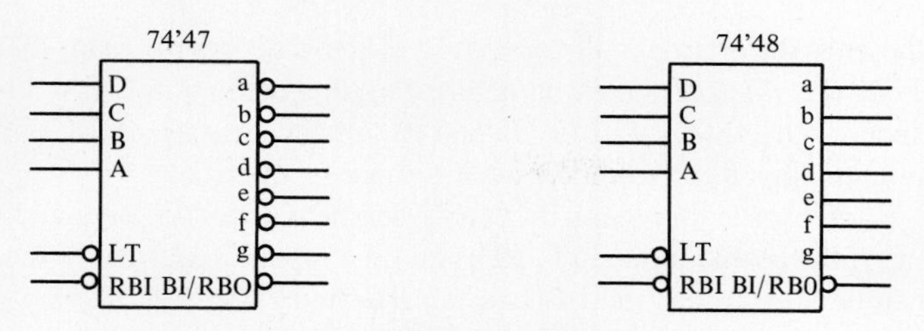

(a) Functional block diagrams for the74'47 and 74'48

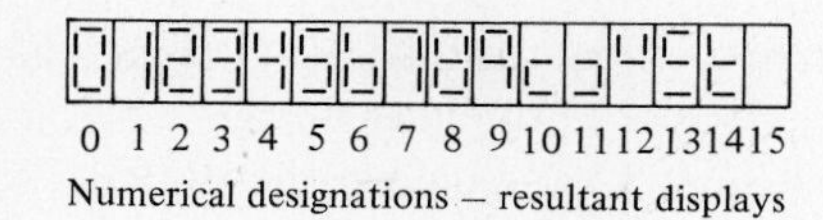

Numerical designations – resultant displays

74'47

Decimal or function	Inputs							Outputs						
	LT	RBI	D	C	B	A	BI/RBO	a	b	c	d	e	f	g
0	H	H	L	L	L	L	H	L	L	L	L	L	L	H
1	H	X	L	L	L	H	H	H	L	L	H	H	H	H
2	H	X	L	L	H	L	H	L	L	H	L	L	H	L
3	H	X	L	L	H	H	H	L	L	L	L	H	H	L
4	H	X	L	H	L	L	H	H	L	L	H	H	L	L
5	H	X	L	H	L	H	H	L	H	L	L	H	L	L
6	H	X	L	H	H	L	H	H	H	L	L	L	L	L
7	H	X	L	H	H	H	H	L	L	L	H	H	H	H
8	H	X	H	L	L	L	H	L	L	L	L	L	L	L
9	H	X	H	L	L	H	H	L	L	L	H	H	L	L
10	H	X	H	L	H	L	H	H	H	H	L	L	H	L
11	H	X	H	L	H	H	H	H	H	L	L	H	H	L
12	H	X	H	H	L	L	H	H	L	H	H	H	L	L
13	H	X	H	H	L	H	H	L	H	H	L	H	L	L
14	H	X	H	H	H	L	H	H	H	H	L	L	L	L
15	H	X	H	H	H	H	H	H	H	H	H	H	H	H
BI	X	X	X	X	X	X	L	H	H	H	H	H	H	H
RBI	H	L	L	L	L	L	L	H	H	H	H	H	H	H
LT	L	X	X	X	X	X	H	L	L	L	L	L	L	L

74'48

Decimal or function	Inputs							Outputs						
	LT	RBI	D	C	B	A	BI/RBO	a	b	c	d	e	f	g
0	H	H	L	L	L	L	H	H	H	H	H	H	H	L
1	H	X	L	L	L	H	H	L	H	H	L	L	L	L
2	H	X	L	L	H	L	H	H	H	L	H	H	L	H
3	H	X	L	L	H	H	H	H	H	H	H	L	L	H
4	H	X	L	H	L	L	H	L	H	H	L	L	H	H
5	H	X	L	H	L	H	H	H	L	H	H	L	H	H
6	H	X	L	H	H	L	H	L	L	H	H	H	H	H
7	H	X	L	H	H	H	H	H	H	H	L	L	L	L
8	H	X	H	L	L	L	H	H	H	H	H	H	H	H
9	H	X	H	L	L	H	H	H	H	H	L	L	H	H
10	H	X	H	L	H	L	H	L	L	L	H	H	L	H
11	H	X	H	L	H	H	H	L	L	H	H	L	L	H
12	H	X	H	H	L	L	H	L	H	L	L	L	H	H
13	H	X	H	H	L	H	H	H	L	L	H	L	H	H
14	H	X	H	H	H	L	H	L	L	L	H	H	H	H
15	H	X	H	H	H	H	H	L	L	L	L	L	L	L
BI	X	X	X	X	X	X	L	L	L	L	L	L	L	L
RBI	H	L	L	L	L	L	L	L	L	L	L	L	L	L
LT	L	X	X	X	X	X	H	H	H	H	H	H	H	H

(b) Displays and voltage tables for the 74'47 and 74'48

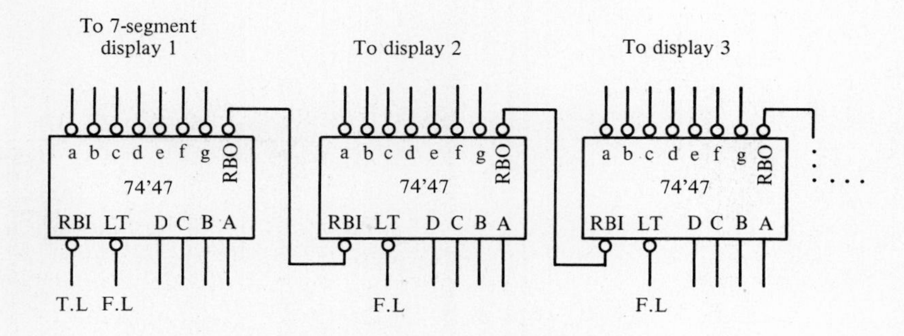

(c) Use of RBI and RBO in a cascade of 7-segment displays

Figure 4.12 74'47 and 74'48 BCD-to-7-segment decoders.

the only difference in the operation of these decoders. Also, the outputs for 1010 through 1111 are *not* don't cares, in contrast to those shown in Fig. 4.11(c). Instead, the specified special characters will be displayed. In the display of BCD numbers, however, these outputs should never occur unless there is an error.

Observe that these BCD-to-7-segment decoders have additional features, which are provided by the inputs LT, RBI, and the input/output BI/RBO. The lamp test (LT) input can be used to test the LEDs of an attached 7-segment display. As shown in the last row of the voltage tables of Fig. 4.12(b), all segments of the display can be lit by making LT true (L) and BI/RBO false (H) for either the 74'47 or the 74'48. Now consider the BI/RBO input/output for which the BI is an abbreviation for blanking input, and RBO for ripple-blanking output. The BI/RBO input/output serves two functions. Used as an input, as shown in row 17 of either voltage table, a true (L) applied to the BI/RBO input will blank the display. As an output, the BI/RBO signal is used in conjunction with RBI (ripple-blanking input) to suppress the display of leading zeros in a cascade of 7-segment displays. As illustrated in Fig. 4.12(c), for decoder 1 the RBI input is connected to true (L). Then, if the digit to be displayed is 0, the output is a blank (all outputs false) and the RBO output is true (L), as indicated by row 18 (RBI) of the voltage tables. As a result, the RBI input of decoder 2 is true, and if the digit to be displayed on display 2 is 0, then its output is a blank and its RBO output is true (L), and so forth. In this manner, all the leading zeros of the 7-segment displays will be displayed as blanks. The first nonzero digit will cause its RBO to be false (H). Therefore, any subsequent embedded zeros (e.g., as in 430103) will be displayed as zeros and not blanks, as indicated by the first row of either voltage table. How should RBI and RBO be connected if leading zeros are desired? (See Problem 4.11.)

4.5 ENCODER

The inverse of the decoding function is the encoding function. For N different inputs, only one of which is activated, an *encoder* is a combinational circuit element that generates an M-bit binary code that uniquely identifies the activated input. Here, $2^M \geq N$. An example of an 8-to-3 encoder is shown in Fig. 4.13. Note that in this definition of an encoder, one and only one input can be activated at a time. Otherwise the circuit has

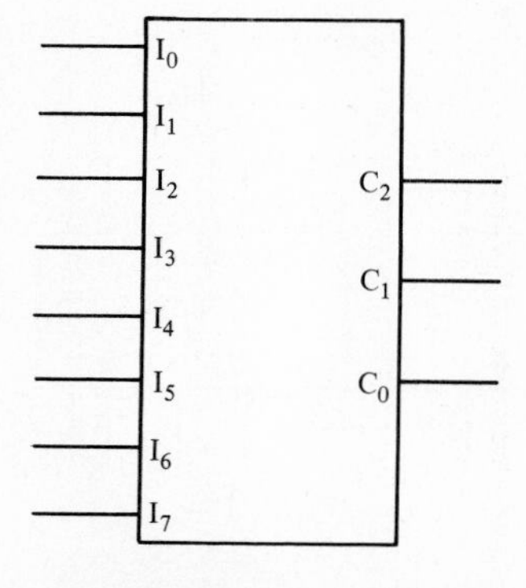

(a) Functional block diagram

I_0	I_1	I_2	I_3	I_4	I_5	I_6	I_7	C_2	C_1	C_0
1	0	0	0	0	0	0	0	0	0	0
0	1	0	0	0	0	0	0	0	0	1
0	0	1	0	0	0	0	0	0	1	0
0	0	0	1	0	0	0	0	0	1	1
0	0	0	0	1	0	0	0	1	0	0
0	0	0	0	0	1	0	0	1	0	1
0	0	0	0	0	0	1	0	1	1	0
0	0	0	0	0	0	0	1	1	1	1

(b) Truth table

Figure 4.13 8-to-3 encoder.

no meaning. In other words, this circuit does not allow the case where several inputs are activated simultaneously or where no input is activated.

4.5.1 Priority Encoder

Encoders are available as MSI circuit elements in the form of *priority* encoders, which allow the activation of several inputs simultaneously, or no input at all. An example of an 8-to-3 priority encoder is the 74'148 shown in Fig. 4.14. As illustrated in the functional

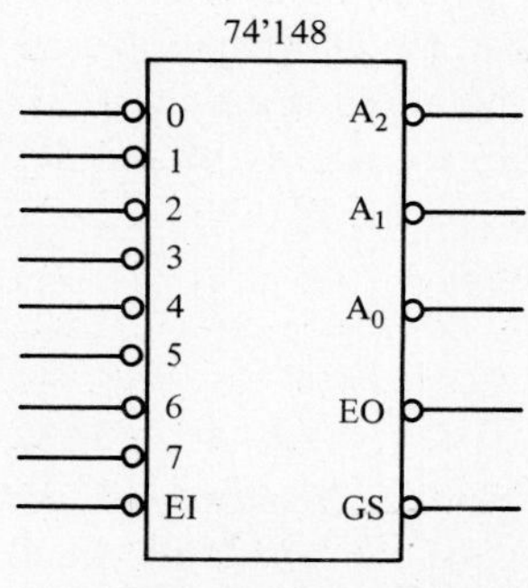

(a) Functional block diagram

Inputs									Outputs				
EI	0	1	2	3	4	5	6	7	A_2	A_1	A_0	GS	EO
H	X	X	X	X	X	X	X	X	H	H	H	H	H
L	H	H	H	H	H	H	H	H	H	H	H	H	L
L	X	X	X	X	X	X	X	L	L	L	L	L	H
L	X	X	X	X	X	X	L	H	L	L	H	L	H
L	X	X	X	X	X	L	H	H	L	H	L	L	H
L	X	X	X	X	L	H	H	H	L	H	H	L	H
L	X	X	X	L	H	H	H	H	H	L	L	L	H
L	X	X	L	H	H	H	H	H	H	L	H	L	H
L	X	L	H	H	H	H	H	H	H	H	L	L	H
L	L	H	H	H	H	H	H	H	H	H	H	L	H

(b) Voltage table

Inputs									Outputs				
EI	0	1	2	3	4	5	6	7	A_2	A_1	A_0	GS	EO
0	X	X	X	X	X	X	X	X	0	0	0	0	0
1	0	0	0	0	0	0	0	0	0	0	0	0	1
1	X	X	X	X	X	X	X	1	1	1	1	1	0
1	X	X	X	X	X	X	1	0	1	1	0	1	0
1	X	X	X	X	X	1	0	0	1	0	1	1	0
1	X	X	X	X	1	0	0	0	1	0	0	1	0
1	X	X	X	1	0	0	0	0	0	1	1	1	0
1	X	X	1	0	0	0	0	0	0	1	0	1	0
1	X	1	0	0	0	0	0	0	0	0	1	1	0
1	1	0	0	0	0	0	0	0	0	0	0	1	0

(c) Truth table

Figure 4.14 74'148 priority encoder.

block diagram of Fig. 4.14(a), the eight inputs (0, 1, . . . , 7) and the three outputs (A_2, A_1, A_0) are active-low. Furthermore, there is an active-low enable input (EI) and two active-low outputs, GS and enable output (EO).

The 74'148 functions as an encoder only when the enable input EI is true. Otherwise all outputs are false, as shown in row 1 of the truth table of Fig. 4.14(c). Note that when more than one input is activated, the 74'148 encodes the input with the highest priority, with input 7 having priority over input 6, which has priority over input 5, and so forth. When input 0 is the only input activated (the last row), then the output code generated is $A_2A_1A_0 = 000$. When no input is activated (row 2), the code generated is also $A_2A_1A_0 = 000$. In this case, the GS output is used to make the distinction. GS is true only when at least one of the inputs is activated. The enable output EO and enable input EI can be used to readily cascade 74'148 priority encoders for octal expansion. (See Problem 4.13.)

4.6 MULTIPLEXER

In the design of a digital circuit, another frequently required logic function is the selection function. A *multiplexer* (MUX) is a combinational circuit element that selects data from one of many inputs and directs it to a single output. Conceptually, the function of a multiplexer can be illustrated by the multipositional switch of Fig. 4.15, which has a number of input lines, but a single output line. Depending on the selection control, the switch will connect a specific one of the inputs to the output Z. This electrical circuit connection is the popular way of considering multiplexer action. But, as will be seen, with a digital circuit multiplexer, there is no direct connection between any input and the output. Despite this, though, the state of the output is the same as that of the selected input.

The block diagram of a four-input MUX is shown in Fig. 4.16(a). As specified in the truth table of Fig. 4.16(b), if for the selection signals $S_1S_0 = 00$, then the output Z is electrically connected to I_0. If $S_1S_0 = 01$, then Z is electrically connected to I_1, and so forth. Note that the truth table in Fig. 4.16(b) is a condensed version of the actual truth table, which would have six input variables (I_0, I_1, I_2, I_3, S_1, and S_0) and 2^6 rows. (See Problem 4.14.) The design and realization of the four-input MUX is shown in Fig. 4.16(c).

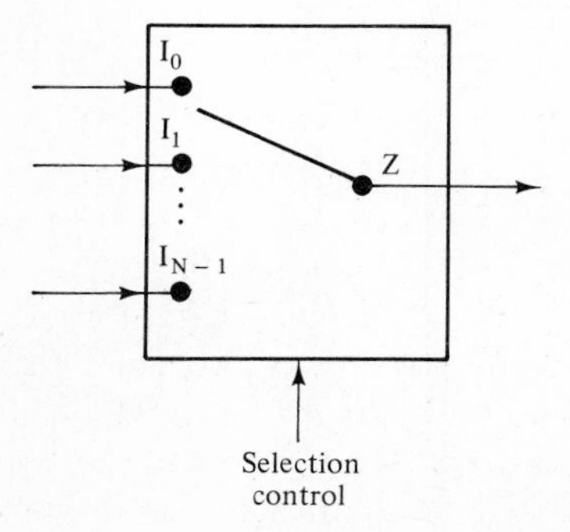

Figure 4.15 *N*-input switch.

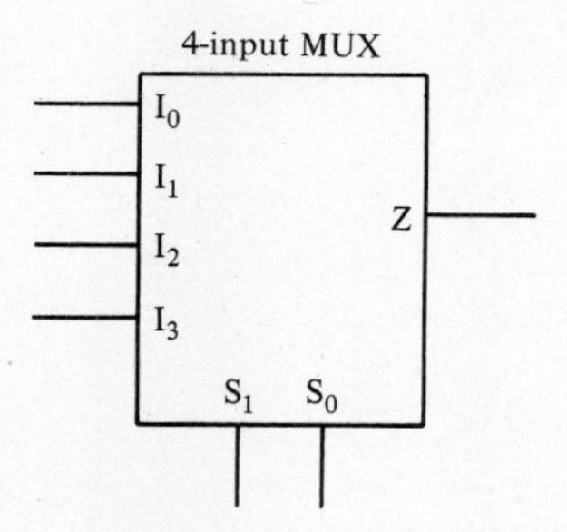

(a) Functional block diagram

S_1	S_0	Z
0	0	I_0
0	1	I_1
1	0	I_2
1	1	I_3

(b) Condensed truth table

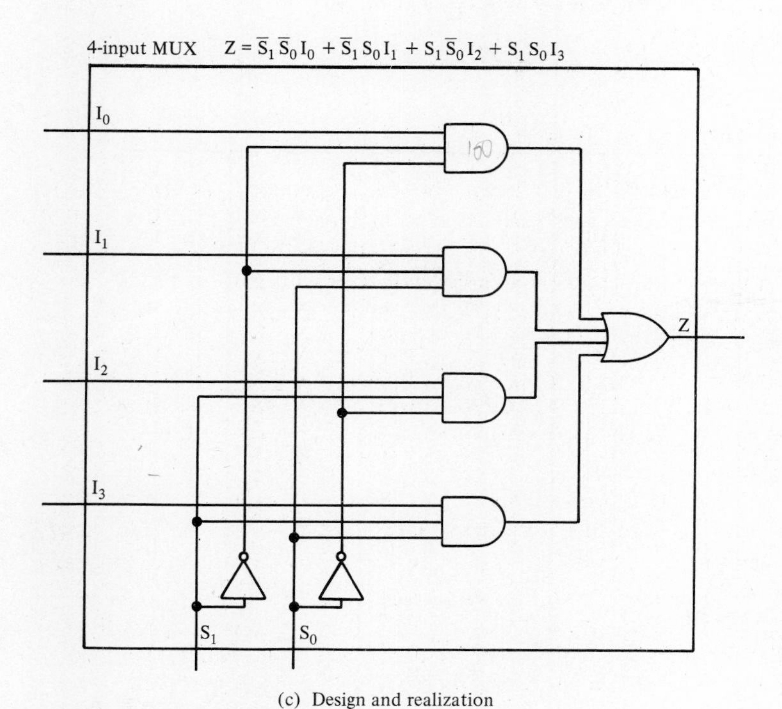

(c) Design and realization

Figure 4.16 Four-input multiplexer.

Multiplexers are commercially available as MSI circuit elements with 2, 4, 8, and 16 inputs and with inverting and/or noninverting outputs (or active-low and active-high outputs). An example of a commercially available eight-input MUX with inverting and noninverting outputs is the 74'151 shown in Fig. 4.17. As specified in the voltage table and the logic equation, the 74'151 functions as a multiplexer only if the enable input E is true (L). Otherwise, the output Z is false. When the enable input E is true, then one of the inputs (selected by the selection inputs) is electrically connected to the output Z.

Multiplexers are useful in the design of a digital circuit where the data for an input of a device is from several sources. Using a multiplexer, we can easily control the source

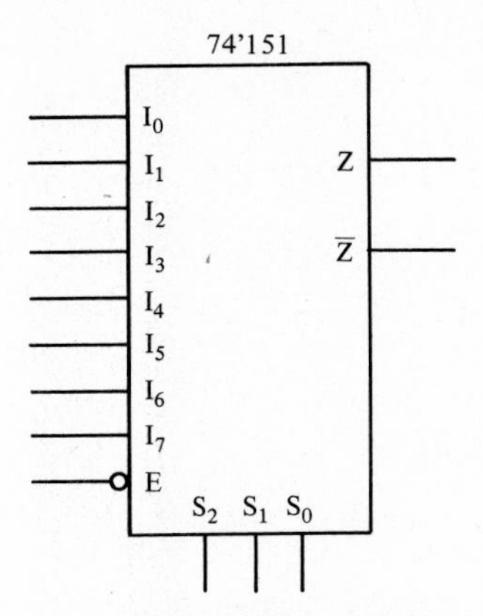

(a) Functional block diagram

E	S_2	S_1	S_0	I_0	I_1	I_2	I_3	I_4	I_5	I_6	I_7	$\overline{Z}$	Z
H	X	X	X	X	X	X	X	X	X	X	X	H	L
L	L	L	L	L	X	X	X	X	X	X	X	H	L
L	L	L	L	H	X	X	X	X	X	X	X	L	H
L	L	L	H	X	L	X	X	X	X	X	X	H	L
L	L	L	H	X	H	X	X	X	X	X	X	L	H
L	L	H	L	X	X	L	X	X	X	X	X	H	L
L	L	H	L	X	X	H	X	X	X	X	X	L	H
L	L	H	H	X	X	X	L	X	X	X	X	H	L
L	L	H	H	X	X	X	H	X	X	X	X	L	H
L	H	L	L	X	X	X	X	L	X	X	X	H	L
L	H	L	L	X	X	X	X	H	X	X	X	L	H
L	H	L	H	X	X	X	X	X	L	X	X	H	L
L	H	L	H	X	X	X	X	X	H	X	X	L	H
L	H	H	L	X	X	X	X	X	X	L	X	H	L
L	H	H	L	X	X	X	X	X	X	H	X	L	H
L	H	H	H	X	X	X	X	X	X	X	L	H	L
L	H	H	H	X	X	X	X	X	X	X	H	L	H

(b) Voltage table

$$Z = E(\overline{S}_2\overline{S}_1\overline{S}_0 I_0 + \overline{S}_2\overline{S}_1 S_0 I_1 + \overline{S}_2 S_1\overline{S}_0 I_2 + \overline{S}_2 S_1 S_0 I_3 + S_2\overline{S}_1\overline{S}_0 I_4 + S_2\overline{S}_1 S_0 I_5 + S_2 S_1\overline{S}_0 I_6 + S_2 S_1 S_0 I_7)$$

(c) Logic equation

Figure 4.17 74'151 MUX.

of the input. Figure 4.18 shows how four 4-input MUXs can be used to select one of four 4-bit data to be processed by the device. For S_1S_0 inputs of 00, 01, 10, or 11, either the 4-bit data A, B, C, or D is connected to input X of the device.

4.6.1 Three-State Logic Element

Multiplexing can also be realized with a *three-state* (sometimes called tristate) logic element. Ordinarily, the voltage level of an output of a device can only be in one of two

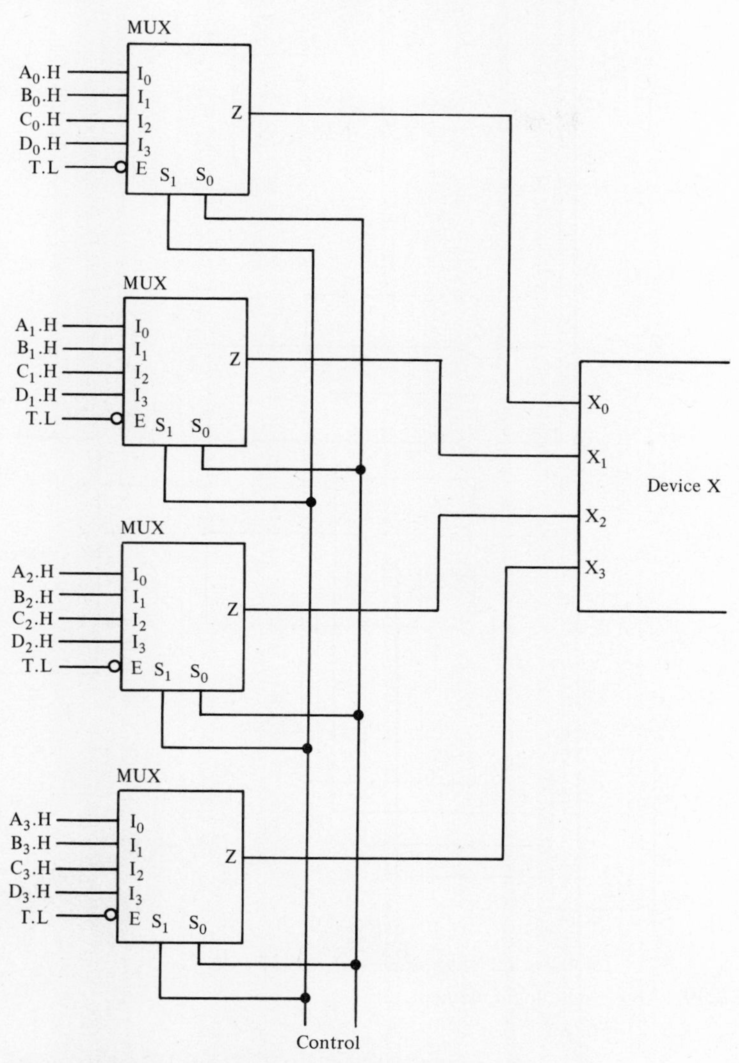

Figure 4.18 Using MUXs for data selection.

states: high or low. However, for a three-state device, such as the 74'125 shown in Fig. 4.19, a third high-impedance state is possible. As shown by the voltage table of Fig. 4.19(b), if the enable input E is true (L), then the device behaves normally with the two states of high and low. If, however, the enable input E is false (H), then the output is in the high-impedance state.

In the high-impedance state, the output does not drive or load any circuit connected to it. In other words, the output behaves as if it were electrically disconnected. As a result, we can use this three-state logic element to realize the multiplexing function. An example of this application is shown to Fig. 4.19(c). Note that the circuit shown here is equivalent to the part of the circuit that is connected to X_0 of the device in Fig. 4.18.

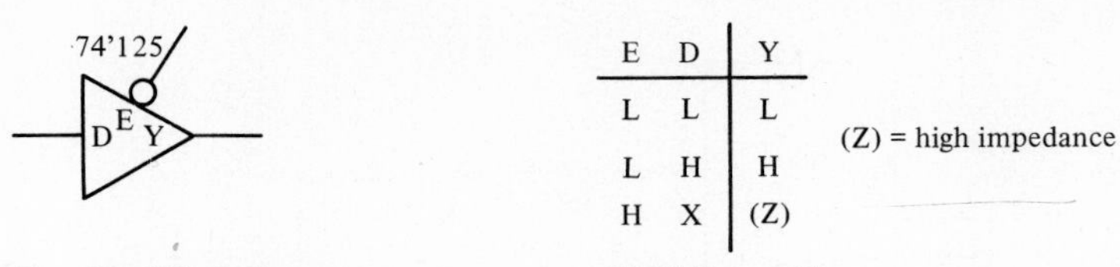

E	D	Y
L	L	L
L	H	H
H	X	(Z)

(Z) = high impedance

(a) Functional block diagram (b) Voltage table

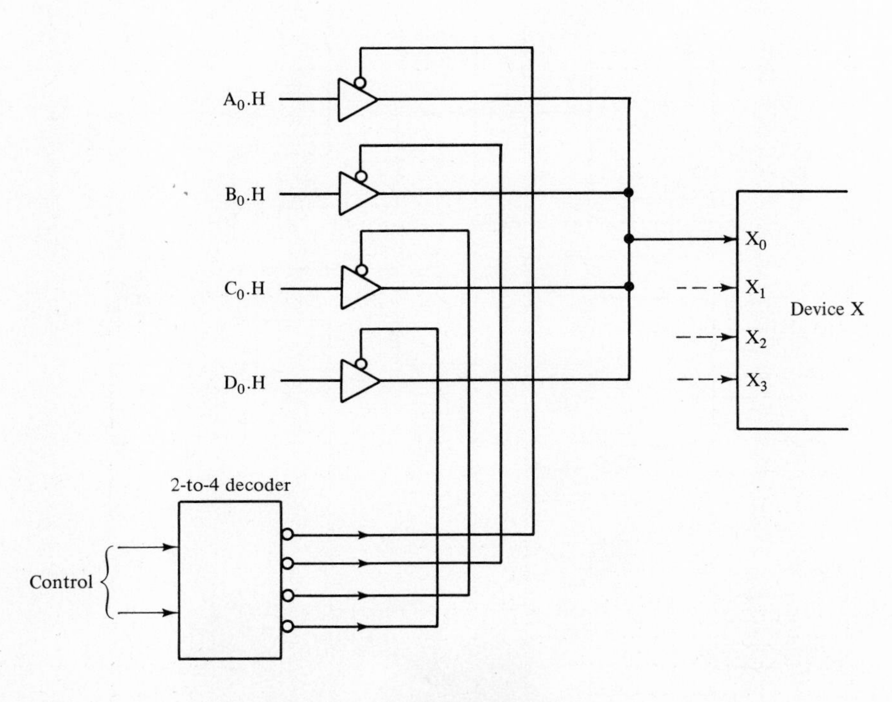

(c) Use of a three-state device for data selection

Figure 4.19 Three-state logic element.

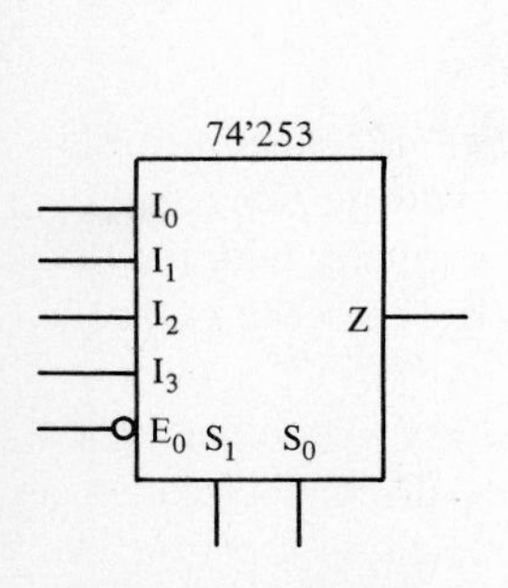

Select inputs		Data inputs				Output enable	Output
S_1	S_0	I_0	I_1	I_2	I_3	E_0	Z
X	X	X	X	X	X	H	(Z)
L	L	L	X	X	X	L	L
L	L	H	X	X	X	L	H
L	H	X	L	X	X	L	L
L	H	X	H	X	X	L	H
H	L	X	X	L	X	L	L
H	L	X	X	H	X	L	H
H	H	X	X	X	L	L	L
H	H	X	X	X	H	L	H

Where (Z) is a high impedance

(a) Functional block (b) Voltage table

Figure 4.20 MUX with three-state outputs.

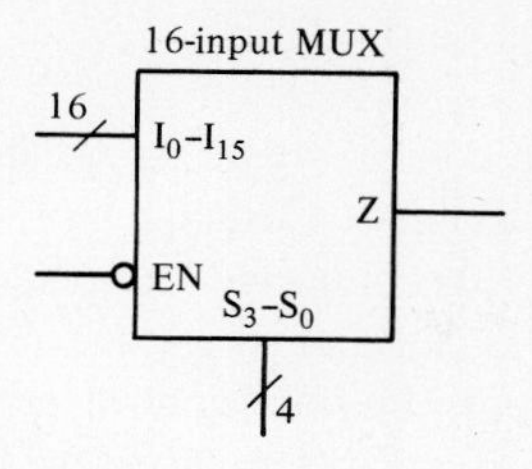

(c) Functional block diagram of a 16-input MUX

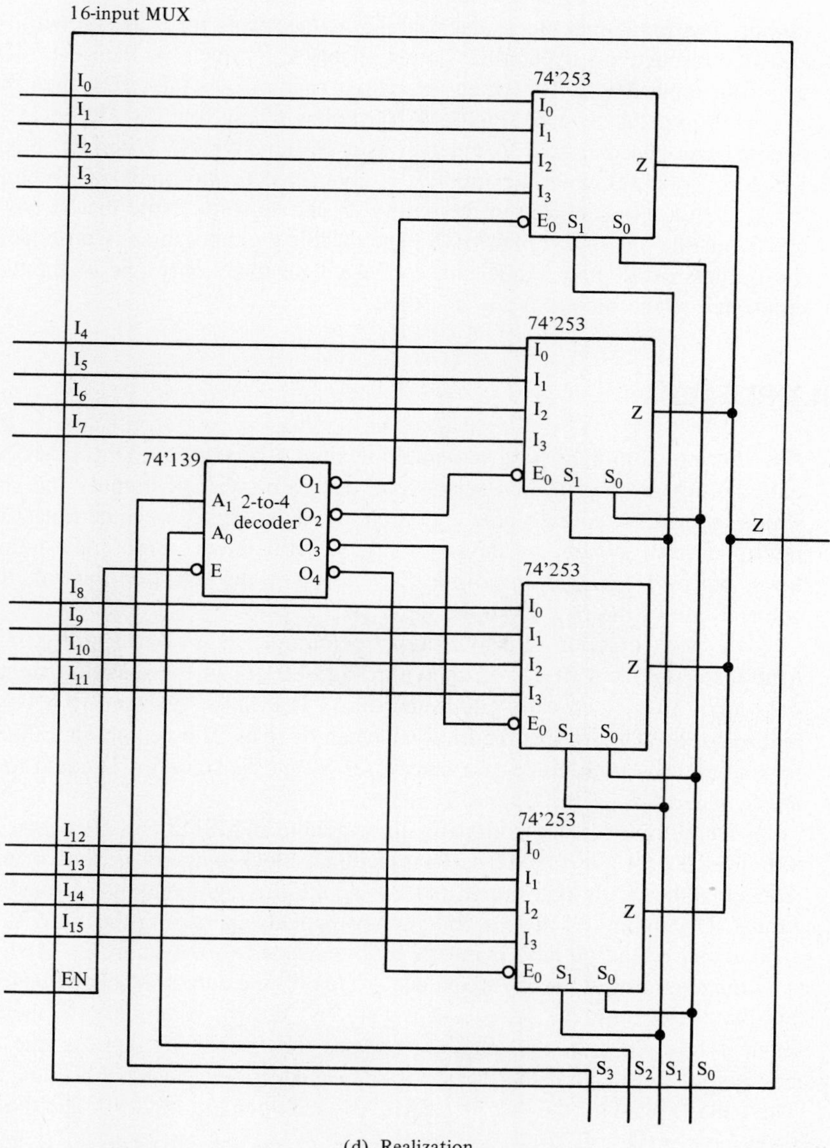

(d) Realization

Figure 4.20 (*cont.*)

Specifically, when the control signal is 00, then the three-state device for A_0 is enabled while the others are disabled. Consequently, only A_0 is connected to X_0 since B_0, C_0, and D_0 are electrically disconnected from X_0. Similarly, when the control signal is 01, then B_0 is connected to X_0, and so forth.

Three-state outputs for bus control are often required in the designs of digital circuits. In other words, it is often necessary to connect together a number of three-state outputs from different devices to form a bus for the transfer of data among devices. Consequently, many of the commercially available circuit elements have built-in three-state outputs. For example, the 74'253 is a four-input multiplexer with a three-state output. The functional block diagram and voltage table for it are shown in Figs. 4.20(a) and (b), respectively. When the output enable E_0 is true (L), then the 74'253 functions as a four-input MUX. But when the output enable E_0 is false (H), then the output Z is in a high-impedance state. Figure 4.20(d) shows how four 74'253 MUXs and a 2-to-4 decoder can be connected to function as a 16-input MUX. When its selection signal is $S_3S_2S_1S_0 = 0000$, then the input I_0 is connected to the output Z. When $S_3S_2S_1S_0 = 0001$, then I_1 is connected to the output Z, and so forth. Note that at any one time, S_3 and S_2 enable one of the four MUXs and disable the other three. Consequently, although the outputs of all four MUXs are connected together, only one of them is electrically connected to the output at any one time.

4.7 DEMULTIPLEXER

The inverse of multiplexing is demultiplexing. A *demultiplexer* (DEMUX) is a combinational circuit element that selects one from a number of outputs and connects it to a single input. Conceptually, the operation of a demultiplexer can be illustrated by another multipositional switch, as shown in Fig. 4.21(a). Here, there are a number of output lines, but only a single input line. Depending on the selection control, the switch will connect one of the outputs to the input D.

A block diagram of a four-output demultiplexer is shown in Fig. 4.21(b), and its truth table in Fig. 4.21(c). As shown in Fig. 4.21(c), if the selection signal $S_1S_0 = 00$, then input D is electrically connected to Z_0. If $S_1S_0 = 01$, then D is connected to Z_1, and so forth. Observe that the truth table can be reduced by combining the first four rows to a single row with don't-care entries for S_1 and S_0 since for D equal to 0, all outputs are 0, regardless of the S_1 and S_0 values.

Demultiplexers are commercially available as MSI circuit elements, an example of which is the 74'138 DEMUX. Its functional block diagram is shown in Fig. 4.22(a). The input E_1 is the data input and Z_7–Z_0 are the eight outputs. E_2 and E_3 are enable inputs. Depending on the selection control signals energizing the $A_2A_1A_0$ inputs, a particular one of the outputs is electrically connected to the input E_1. At least this is the way the operation is usually explained. Actually, the output is not connected to the input, but the operation is the same as if it were. As shown, what really happens is that if the input is false (H), then all outputs are false (H). But if the input is true (L), then only one output is true (L), the particular output depending on the selection control signal. The voltage table for the 74'138 DEMUX is shown in Fig. 4.22(b), and the corresponding truth table in Fig. 4.22(c).

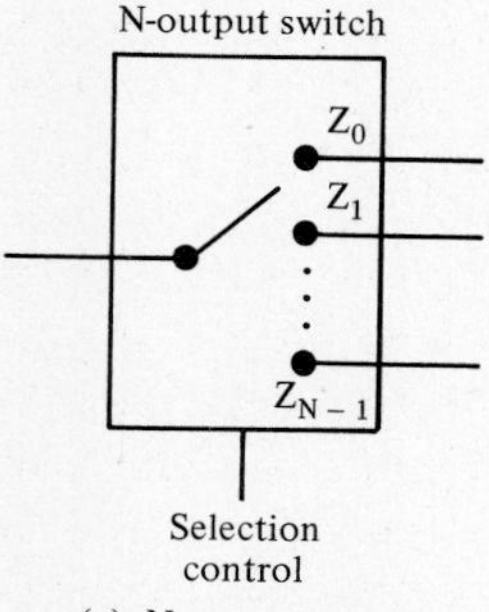

(a) N-output switch

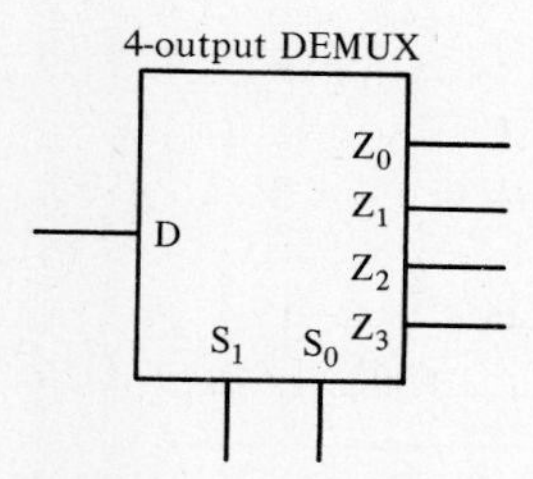

(b) Functional block diagram

D	S_1	S_0	Z_0	Z_1	Z_2	Z_3
0	0	0	0	0	0	0
0	0	1	0	0	0	0
0	1	0	0	0	0	0
0	1	1	0	0	0	0
1	0	0	1	0	0	0
1	0	1	0	1	0	0
1	1	0	0	0	1	0
1	1	1	0	0	0	1

(c) Truth table

Figure 4.21 Demultiplexer.

Recall that this 74'138 DEMUX was used as a decoder in Sec. 4.4 (Fig. 4.10). This double usage is possible because the truth table for a demultiplexer is identical to that of a decoder that has one or more enable inputs. Consequently, any commercially available decoder with one or more enable inputs can also function as a demultiplexer.

4.8 DESIGN CONSIDERATIONS FOR INTEGRATED CIRCUIT (IC) ELEMENTS

In this chapter we have studied some of the commonly used logic functions and the designs and applications of the circuit elements that realize these logic functions. These combinational MSI circuit elements are a part of the building blocks that are required in

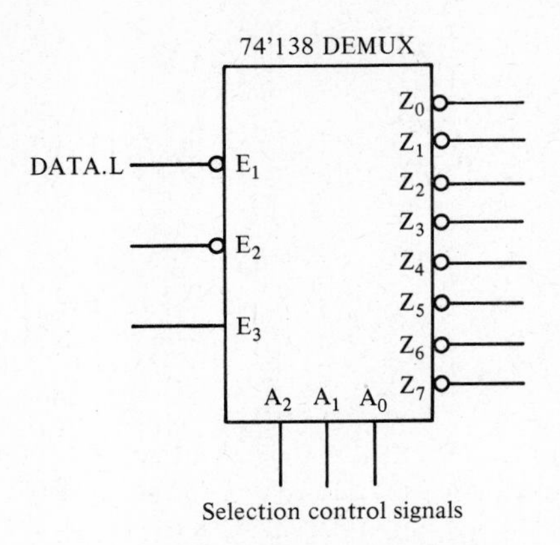

(a) Functional block diagram

Inputs						Outputs							
E_1	E_2	E_3	A_2	A_1	A_0	Z_0	Z_1	Z_2	Z_3	Z_4	Z_5	Z_6	Z_7
H	X	X	X	X	X	H	H	H	H	H	H	H	H
X	H	X	X	X	X	H	H	H	H	H	H	H	H
X	X	L	X	X	X	H	H	H	H	H	H	H	H
L	L	H	L	L	L	L	H	H	H	H	H	H	H
L	L	H	L	L	H	H	L	H	H	H	H	H	H
L	L	H	L	H	L	H	H	L	H	H	H	H	H
L	L	H	L	H	H	H	H	H	L	H	H	H	H
L	L	H	H	L	L	H	H	H	H	L	H	H	H
L	L	H	H	L	H	H	H	H	H	H	L	H	H
L	L	H	H	H	L	H	H	H	H	H	H	L	H
L	L	H	H	H	H	H	H	H	H	H	H	H	L

(b) Voltage table

Enable		Data	Selection			Outputs							
E_2	E_3	E_1	A_2	A_1	A_0	Z_0	Z_1	Z_2	Z_3	Z_4	Z_5	Z_6	Z_7
X	X	0	X	X	X	0	0	0	0	0	0	0	0
0	X	X	X	X	X	0	0	0	0	0	0	0	0
X	0	X	X	X	X	0	0	0	0	0	0	0	0
1	1	1	0	0	0	1	0	0	0	0	0	0	0
1	1	1	0	0	1	0	1	0	0	0	0	0	0
1	1	1	0	1	0	0	0	1	0	0	0	0	0
1	1	1	0	1	1	0	0	0	1	0	0	0	0
1	1	1	1	0	0	0	0	0	0	1	0	0	0
1	1	1	1	0	1	0	0	0	0	0	1	0	0
1	1	1	1	1	0	0	0	0	0	0	0	1	0
1	1	1	1	1	1	0	0	0	0	0	0	0	1

(c) Truth table

Figure 4.22 74'138 DEMUX.

the designs of digital circuits, as will be discussed in Chapter 7. Before we proceed with the study of the sequential MSI circuit elements in the next chapter, let us consider some of the design considerations for these integrated circuit elements.

4.8.1 TTL Digital Logic Family

TTL (Transistor-Transistor Logic), which was briefly considered in Sec. 3.2, is currently the dominant digital logic family for SSI and MSI ICs. The popularity of the TTL family results from its ease of use, low cost, low power consumption, relatively high speed of operation, low noise susceptibility, and good output drive capability. Since its introduction in the early 1960s, it has been expanded to include hundreds of logic functions, and has evolved into the following series:

1. Standard TTL
2. H-TTL (high-speed TTL)
3. L-TTL (low-power TTL)
4. S-TTL (Schottky TTL)
5. LS-TTL (low-power Schottky TTL)
6. ALS-TTL (advanced LS-TTL)
7. AS-TTL (advanced S-TTL)

As mentioned in Sec. 3.2, a TTL element is identified by a label, the first two digits of which are 74 for commercial-grade products and 54 for military-grade products. If the element is other than standard TTL, then letters follow the 74 to identify the series. No letter corresponds to standard TTL, the letter H to high-speed TTL, L to low-power, S to Schottky, LS to low-power Schottky, ALS to advanced low-power Schottky, and AS to advanced Schottky. The remaining part of the label is a two- or three-digit number that identifies the type of element and perhaps some other features, such as the number of elements per chip. For example, 00 identifies a quadruple two-input positive-logic NAND gate, which means that there are four NAND gates on a chip, and each of the NAND gates has two inputs. Manufacturers identify the elements by their positive-logic functions.

The corresponding circuit elements of each of the TTL series are functionally identical. For example, a 7400, a 74LS00, and a 74ALS00 are all two-input NAND gates. The differences among them are physical characteristics such as speed of operation, power dissipation, input loading, output drive capacity, noise margin, and so forth.

A comparison of the various series in terms of speed and power is given in Table 4.1. Perhaps a couple of the column headings need some explanation. As will be described in more detail in Sec. 4.8.3, propagation delay is approximately the time required for a signal to pass through an element. So, the *smaller* the propagation delay, the greater the maximum speed of operation. The power-delay product (PDP) is the product of the propagation delay and power dissipation. This product is one measure of the desirability of an element. Often, but not always, the smaller this product, the better the element.

As shown in Table 4.1, the standard TTL, which was the first series, is relatively fast, but has a fairly high power dissipation. The H-TTL is a high-speed series. Its power dissipation, however, is enormous. For high-speed applications, the H-TTL ICs have

TABLE 4.1 TYPICAL PERFORMANCE COMPARISON OF THE TTL LOGIC FAMILY

Series	Gate propagation delay (ns)	Power dissipation (mW)	Power-delay product (pJ)
Standard TTL	10	10	100
H-TTL	6	22	132
L-TTL	33	1	33
S-TTL	3	19	57
LS-TTL	9.5	2	19
ALS-TTL	4	1	4
AS-TTL	1.5	10	15

been replaced by those of the newer S-TTL Schottky TTL series. The S-TTL series is faster than the H-TTL series. Also, the power dissipation is less, but still substantial.

The L-TTL is a low-power series. Unfortunately, it also has very slow speed. The L-TTL series has been replaced by the newer LS-TTL low-power Schottky series. The LS-TTL ICs consume slightly more power than those of the L-TTL series, but are significantly faster. In fact, as is evident from Table 4.1, the LS-TTL series has one of the best power-delay products. Additionally, there are an extensive number of logic functions available in this series, and the chips are reasonably priced. Consequently, the LS-TTL series is currently the most popular series in the TTL family. The ALS-TTL (advanced LS-TTL) and the AS-TTL (advanced S-TTL) are high-performance series that are improvements over the LS-TTL and the S-TTL series, respectively. The ALS-TTL series has the best power-delay product, but the AS-TTL series is the fastest. However, product offerings in these two series have yet to match those of the LS-TTL and the S-TTL series.

Ordinarily, circuit elements from the same TTL series are used throughout a digital circuit. Sometimes, however, circuit elements from different series are used together to obtain optimum performance. High-performance AS-TTL circuit elements can be used, for example, in speed-critical portions of a digital circuit. And, for portions of the digital circuit where speed is not crucial, slower circuit elements with lower power dissipation can be used. The actual mix of ICs from the different TTL series is, of course, a function of the overall circuit specifications and requirements. For an optimum selection of a TTL series or a mix of circuit elements from different TTL series, it is important to understand the digital IC physical characteristics presented in the following sections.

4.8.2 Parameters for Static Characteristics

The various series within the TTL logic family are characterized by static (dc) parameters and switching (ac) parameters. Static parameters describe the input and output behavior of the IC elements under stable operating conditions. The major static parameters are illustrated in Fig. 4.23 and defined as follows:

I_{IH} High-level input current; the current that flows into an input when a high voltage is applied. This represents the current requirement of an input that is in a high state.

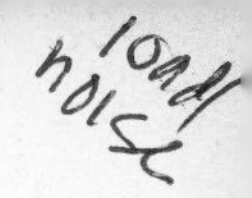

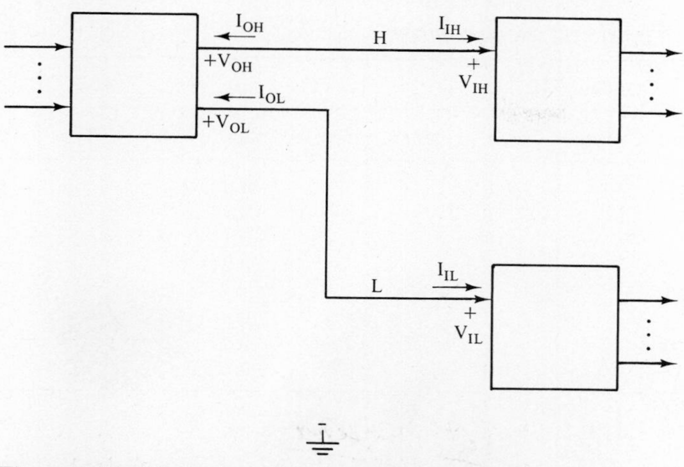

Figure 4.23 Illustration of static parameters.

I_{IL} Low-level input current; the current that flows into an input when a low voltage is applied. This represents the current requirement of an input that is in a low state.

I_{OH} High-level output current; the current that flows into an output that is in a high state.

I_{OL} Low-level output current; the current that flows into an output that is in a low state.

V_{IH} High-level input voltage; the input voltage that is recognized by the circuit element as a high level.

V_{IL} Low-level input voltage; the input voltage that is recognized by the circuit element as a low level.

V_{OH} High-level output voltage; the output voltage that the circuit element will provide when the output is at a high level.

V_{OL} Low-level output voltage; the output voltage that the circuit element will provide when the output is at a low level.

I_{CC} Supply current; the current flowing into the V_{CC} supply terminal of an IC. The product of this current and the supply voltage is the power dissipation of the IC.

Note that both input and output current parameters have references into the TTL device. Consequently, if an input or output current is specified as being negative, the actual current flow is out of the TTL device input or output.

The worst-case and typical values for these parameters for three TTL series are given in Table 4.2. Applications for these parameters are presented in the next two sections.

TABLE 4.2 TTL STATIC PARAMETERS

	Standard TTL			LS-TTL			ALS-TTL			
	Min.	Typ.	Max.	Min.	Typ.	Max.	Min.	Typ.	Max.	Units
I_{IH}			40.0			20.0			20.0	μA
I_{IL}			−1.6			−0.4			−0.1	mA
I_{OH}			−0.4			−0.4			−0.4	mA
I_{OL}			16.0			8.0			8.0	mA
V_{IH}	2.0			2.0			2.0			V
V_{IL}			0.8			0.8			0.8	V
V_{OH}	2.4	3.4		2.7	3.5		2.5	3.0		V
V_{OL}		0.2	0.4		0.35	0.5		0.35	0.5	V

Input Loading and Output Drive

In a digital circuit an output of a circuit element is usually connected to inputs of other circuit elements. Each additional input presents an additional load to the output because it requires a certain amount of current. But an output can supply only a limited amount of current. Exceeding the specified maximum amount will cause the corresponding circuit element to function improperly and possibly be damaged. The *output drive* capability of an output (called the fan-out) is a measure of the number of inputs that the output of a circuit element can drive without impairing its operation. The output drive capability of a TTL element can be calculated from the following formulas:

For high-voltage level,

$$\text{Output drive} = \left|\frac{I_{OH}(\max)}{I_{IH}(\max)}\right|$$

For low-voltage level,

$$\text{Output drive} = \left|\frac{I_{OL}(\max)}{I_{IL}(\max)}\right|$$

In other words, the output drive is measured by the maximum amount of current that an output can supply compared to the maximum amount of current that an input will require. For example, from Table 4.2 we see that for the LS-TTL series, $I_{OH}(\max) = -0.4$ mA and $I_{IH}(\max) = 20$ μA. Therefore, an LS-TTL series output in the high state can drive up to 20 LS-TTL inputs. For the low-voltage level, $I_{OL}(\max) = 8$ mA and $I_{IL}(\max) = -0.4$ mA. Therefore, an LS-TTL series output can also drive up to 20 LS-TTL inputs in the low state. So, the fan-out is 20. Similarly, we can determine that an LS-TTL series output in the high state can drive up to 20 ALS-TTL inputs, but up to 80 ALS-TTL inputs in the low state. When the two numbers differ, as they do here, the smaller of the two numbers must be used to ensure proper operation. In other words, an LS-TTL output can safely drive up to 20 ALS-TTL inputs.

Noise Margin

In the operation of a digital circuit, "noise" voltages often occur. These are nonsignal voltage pulses or spikes caused by electrical disturbances such as lightning, automobile ignitions, or sudden changes in supply voltage levels. It is, of course, not desirable for noise voltages to affect the circuit operation.

Noise margin is a measure of the ability of a digital IC to withstand these noise voltages. Noise margin is defined as follows:

$$\text{High-level noise margin} = V_{OH}(\text{min}) - V_{IH}(\text{min})$$
$$\text{Low-level noise margin} = V_{IL}(\text{max}) - V_{OL}(\text{max})$$

For the LS-TTL series, for example, the high-level noise margin is 2.7 − 2.0 = 0.7 V. So, in the worst case, a high-level signal can drop 0.7 V in going from an LS-TTL output to an LS-TTL input and still be recognized as a high level. For the LS-TTL series, the low-level noise margin is 0.8 − 0.5 = 0.3 V.

4.8.3 Parameters for Switching Characteristics

For combinational circuit elements, the most important switching (ac) parameters are the *propagation delays*. A signal takes a finite amount of time in propagating through a circuit element from an input to an output. This amount of time is the propagation delay. There are two types of propagation delay:

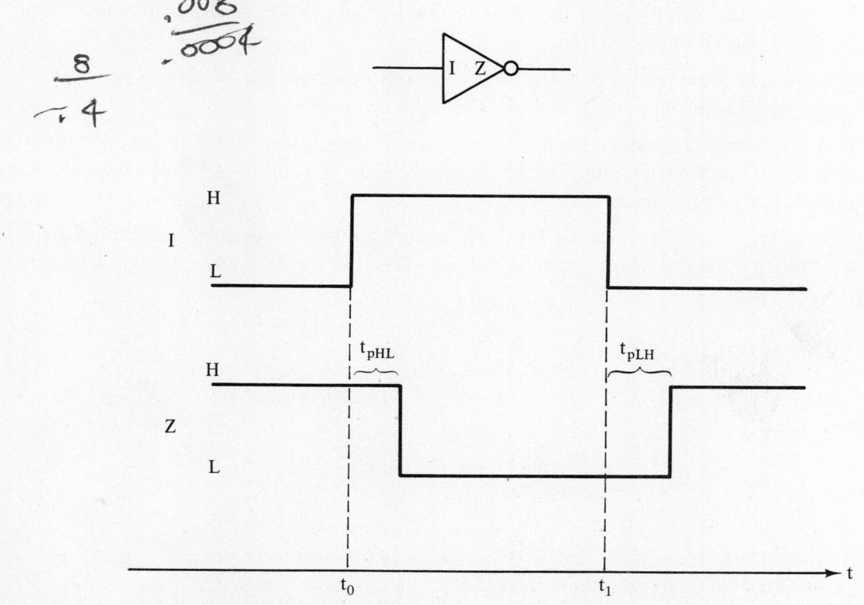

Figure 4.24 Propagation delays.

t_{pHL} The delay from the time the input changes to the time the output switches from H to L.

t_{pLH} The delay from the time the input changes to the time the output switches from L to H.

The propagation delays t_{pHL} and t_{pLH} of an inverter are illustrated by the *timing diagrams* of Fig. 4.24, which are plots of the voltage levels versus time. In Fig. 4.24, the voltage level at input I changes from L to H at time t_0. But, the output Z does not change from H to L until the time $t_0 + t_{pHL}$. Similarly, the voltage level at I changes from H to L at t_1, but the output Z does not change from L to H until $t_1 + t_{pLH}$. Timing diagrams are very useful in the analysis of digital circuits, and will be encountered frequently throughout the remainder of this text.

SUPPLEMENTARY READING (see Bibliography)

[Bartee 85], [Blakeslee 79], [Fletcher 80], [Hill 81], [Mano 79], [McCluskey 75], [Motorola], [Peatman 80], [Prosser 87], [Texas Instruments]

PROBLEMS

4.1. The active-high view and the active-low view for the 74'283 adder are shown in Figs. 4.4(b) and 4.4(c), respectively. Using the voltage table given in Fig. 4.4(a), prove that both of these views perform the binary addition function.

4.2. Prove that the 74'283 adder performs the binary subtraction function for the voltage assignment shown in Fig. 4.6(b).

4.3. Design a 4-bit subtractor that has an active-high minuend and an active-high subtrahend. Use a 74'283 plus any additional gates that are needed.

4.4. If all the inputs are applied simultaneously to the ripple adder shown in Fig. 4.3, how long does it take before the sum and C_4 become valid? Assume that the delay of each *gate* (within each adder stage) is t_p.

4.5. Design the 4-bit BCD adder of Fig. 4.25 using two 74'283s plus any additional gates that are needed. The following examples may be helpful in clarifying the problem specification:

If	A =	B =	C_{in} =	then	C_{out} =	BCDSUM =
	0101	0011	0		0	1000
	0101	0100	0		0	1001
	0101	0101	0		1	0000
	0101	0110	0		1	0001
	0101	0111	0		1	0010
	0101	1000	0		1	0011
		.				.
		.				.
		.				.
	1001	1001	1		1	1001

Note that A, B, or BCDSUM > 1001 is not allowed in the BCD notation.

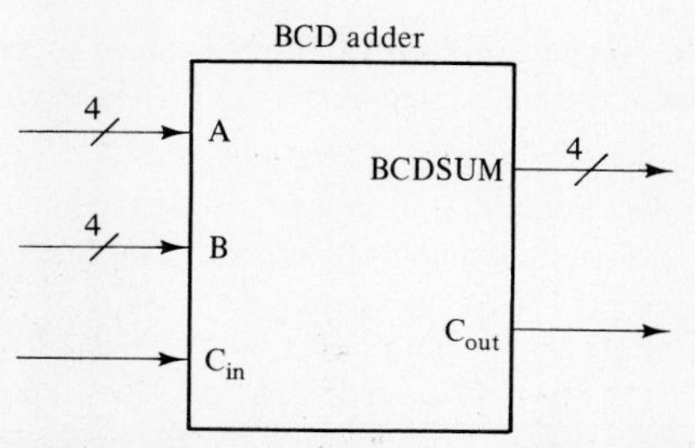

Figure 4.25 BCD adder for Problem 4.5.

4.6. Using 75'85 comparators, design a 16-bit magnitude comparator.

4.7. Design the 4-bit magnitude comparator of Fig. 4.26, using a 74'85 comparator plus any additional gates that are needed. Note that this comparator has three more than the usual number of outputs. These are $<=$, $>=$, and $<>$, which represent less than or equal to, greater than or equal to, and not equal to, respectively.

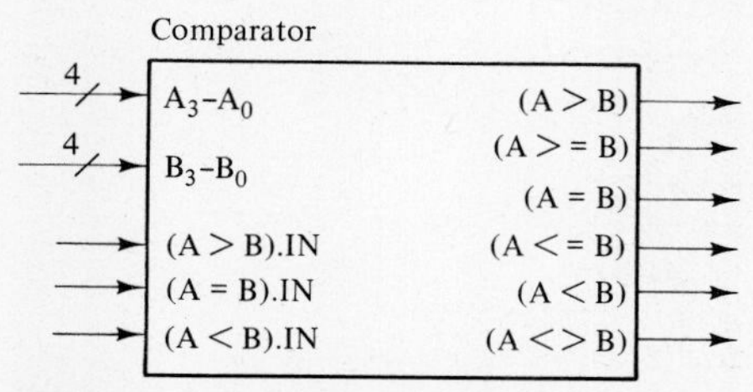

Figure 4.26 Comparator for Problem 4.7.

4.8. Design a 5-to-32 decoder using 74'138 decoders and any additional gates that are required.

4.9. Given an 8-bit address A_7–A_0, what are the addresses that will enable the modules M_0, M_1, M_2, and M_3 shown in Fig. 4.27. For convenience, use X for a don't-care address bit.

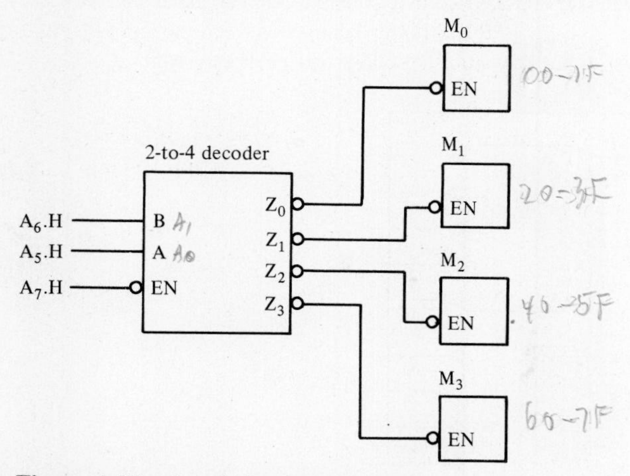

Figure 4.27 Circuit for Problem 4.9.

4.10. Using the truth table for a BCD-to-7-segment decoder shown in Fig. 4.11(c), derive the logic equations for the seven outputs a, b, c, d, e, f, and g.

4.11. A chain of 74'47 BCD-to-7-segment decoders can be connected together as shown in Fig. 4.12(c) to display leading zeros as blanks. Reconnect the 74'47s in such a way that leading zeros are displayed as zeros.

4.12. Design Module M in Fig. 4.28 to obtain a 16-to-4 priority encoder. (*Hint:* Module M is a combinational circuit with eight inputs and five outputs. You are to determine the logic equations for the five outputs.)

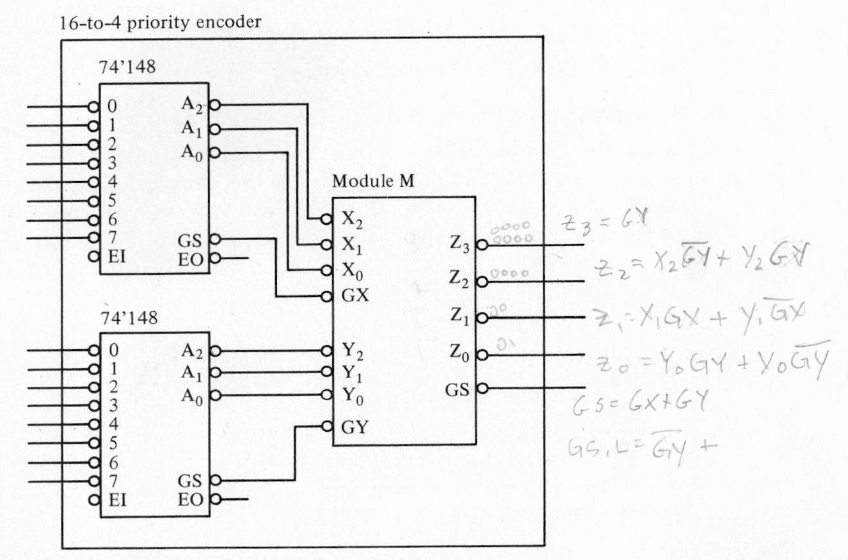

Figure 4.28 Encoder for Problem 4.12.

4.13. The enable output EO and enable input EI of a 74'148 can be used to cascade 74'148 priority encoders for easy octal expansion. Such an expansion for the 74'148s of Fig. 4.29(a) is shown in Fig. 4.29(b). Show connections for the 74'148s of Fig. 4.29(a) that will accomplish this expansion. No additional gates are required.

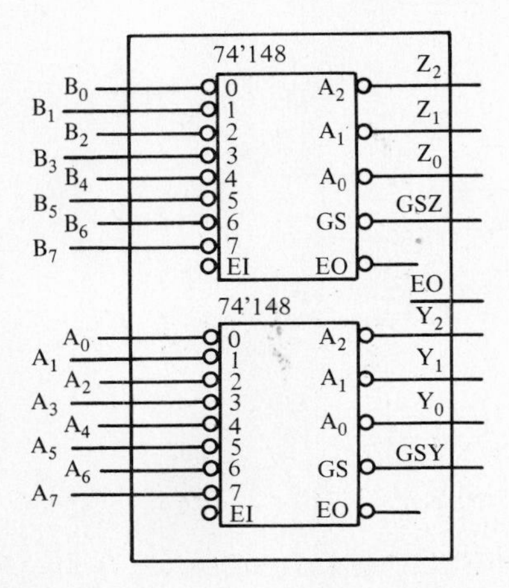

(a)

EI	A_0	A_1	A_2	A_3	A_4	A_5	A_6	A_7	B_0	B_1	B_2	B_3	B_4	B_5	B_6	B_7	Z_2	Z_1	Z_0	Y_2	Y_1	Y_0	GSZ	GSY	EO
0	X	X	X	X	X	X	X	X	X	X	X	X	X	X	X	X	0	0	0	0	0	0	0	0	0
1	0	0	0	0	0	0	0	0	0	0	0	0	0	0	0	0	0	0	0	0	0	0	0	0	1
1	1	0	0	0	0	0	0	0	0	0	0	0	0	0	0	0	0	0	0	0	0	0	0	1	0
1	X	1	0	0	0	0	0	0	0	0	0	0	0	0	0	0	0	0	0	0	0	1	0	1	0
1	X	X	1	0	0	0	0	0	0	0	0	0	0	0	0	0	0	0	0	0	1	0	0	1	0
1	X	X	X	1	0	0	0	0	0	0	0	0	0	0	0	0	0	0	0	0	1	1	0	1	0
1	X	X	X	X	1	0	0	0	0	0	0	0	0	0	0	0	0	0	0	1	0	0	0	1	0
1	X	X	X	X	X	1	0	0	0	0	0	0	0	0	0	0	0	0	0	1	0	1	0	1	0
1	X	X	X	X	X	X	1	0	0	0	0	0	0	0	0	0	0	0	0	1	1	0	0	1	0
1	X	X	X	X	X	X	X	1	0	0	0	0	0	0	0	0	0	0	0	1	1	1	0	1	0
1	X	X	X	X	X	X	X	X	1	0	0	0	0	0	0	0	0	0	0	0	0	0	1	0	0
1	X	X	X	X	X	X	X	X	X	1	0	0	0	0	0	0	0	0	1	0	0	0	1	0	0
1	X	X	X	X	X	X	X	X	X	X	1	0	0	0	0	0	0	1	0	0	0	0	1	0	0
1	X	X	X	X	X	X	X	X	X	X	X	1	0	0	0	0	0	1	1	0	0	0	1	0	0
1	X	X	X	X	X	X	X	X	X	X	X	X	1	0	0	0	1	0	0	0	0	0	1	0	0
1	X	X	X	X	X	X	X	X	X	X	X	X	X	1	0	0	1	0	1	0	0	0	1	0	0
1	X	X	X	X	X	X	X	X	X	X	X	X	X	X	1	0	1	1	0	0	0	0	1	0	0
1	X	X	X	X	X	X	X	X	X	X	X	X	X	X	X	1	1	1	1	0	0	0	1	0	0

(b)

Figure 4.29 Table and components for Problem 4.13.

4.14. **(a)** Determine the complete truth table for the four-input MUX shown in Fig. 4.16.
(b) Derive the logic equation for Z.

4.15. In Fig. 4.20, a 16-input MUX is realized with 4 four-input MUXs and a decoder. But, suppose no decoder is available. Design a 16-input MUX using any number of 4-input MUXs (74'253), but no decoders.

4.16. Using a 74'148 priority encoder and any additional gates that are required, design the circuit of Fig. 4.30 for generating the S_0 and S_1 control inputs for the circuit of Fig. 4.18. Each of the requesting devices (A, B, C, and D) can request a connection to device X through the signals REQA, REQB, REQC, and REQD, respectively. If there are competing requests, then the order of priority is as follows: D, C, B, and A, with D having the highest priority. If no request is made, then device X is connected to A by default.

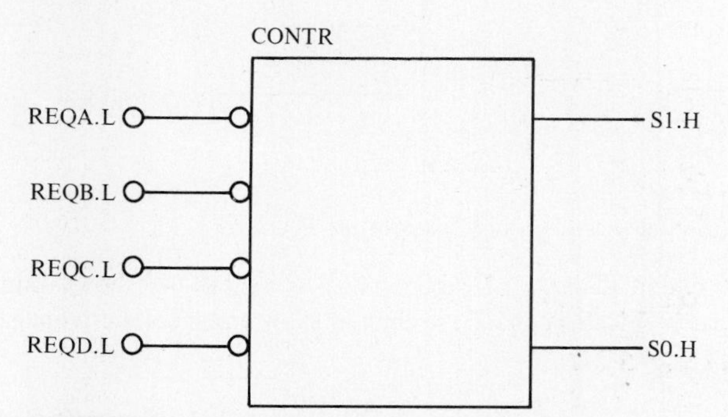

Figure 4.30 Circuit for Problem 4.16.

4.17. Design the circuit BIDIR of Fig. 4.31 such that the signal DATA is bidirectional. Specifically, when IOCTR is 1 (H), then DATA is connected to INPUT and the direction of the data flow is "in." But when IOCTR is 0 (L), then DATA is connected to OUTPUT and the direction of the data flow is "out." (*Hint:* Use 74'125 three-state logic elements.)

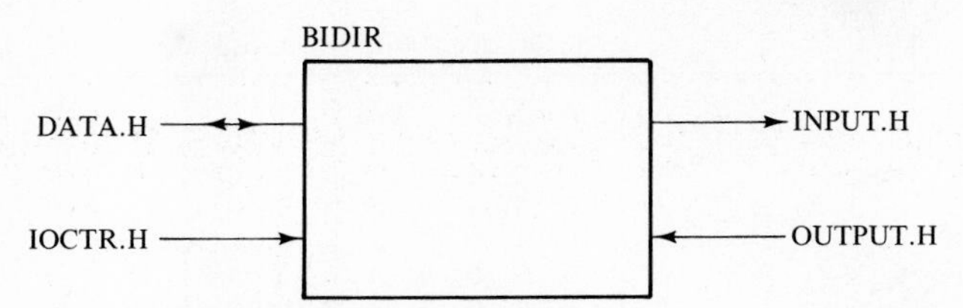

Figure 4.31 Circuit for Problem 4.17.

4.18. **(a)** Discuss the similarities and the differences between a decoder and a demultiplexer.
(b) An encoder performs the inverse function of a decoder, and a multiplexer performs the inverse function of a demultiplexer. Then, is there any relationship between an encoder and a multiplexer? Explain.

4.19. What is the maximum number of standard-TTL inputs that an ALS-TTL output can drive?

4.20. If the signal MEMCS shown in Fig. 4.32 activates the CS inputs of a bank of memory chips with the specified characteristics, how many CS inputs can it safely drive?

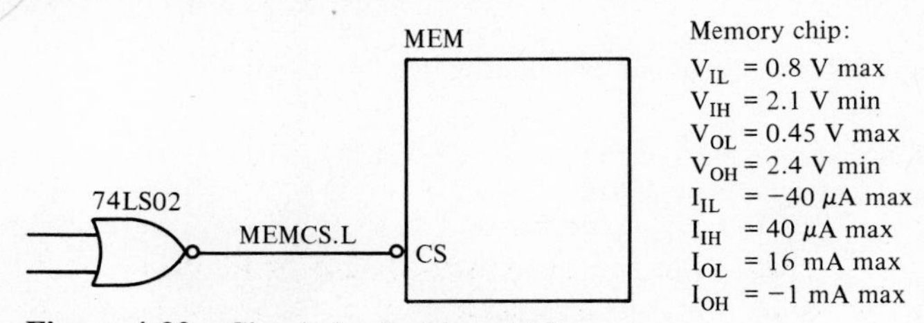

Figure 4.32 Circuit for Problem 4.20.

4.21. What is the noise margin for the ALS-TTL series components?

4.22. If an ALS-TTL output drives a number of LS-TTL inputs, what is the resultant noise margin?

4.23. Given the Exclusive OR gate of Fig. 4.33(a), complete the shown timing diagram for Z in Fig. 4.33(b). Be sure to show and label the propagation delays t_{pHL} and t_{pLH}.

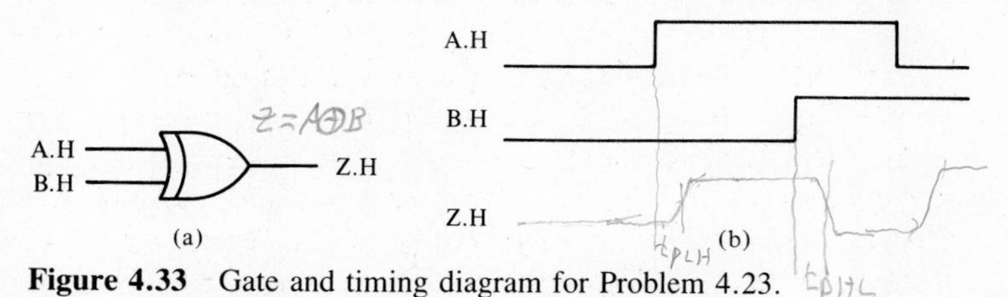

Figure 4.33 Gate and timing diagram for Problem 4.23.

4.24. Given the circuit diagram of Fig. 4.34(a), complete the shown timing diagram in Fig. 4.34(b) for the signals $\overline{A}$ and Z. Be sure to show and label the propagation delays t_{pHL} and t_{pLH}.

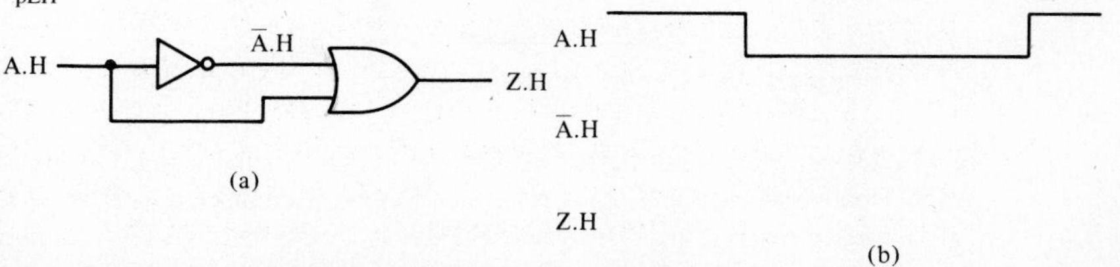

Figure 4.34 Gate and timing diagram for Problem 4.24.

Chapter 5

Sequential MSI Circuit Elements

5.1 INTRODUCTION

In Chapter 4 we studied the designs and applications of some combinational MSI circuit elements that realize many of the commonly used logic functions. In the present chapter we will study the designs and applications of *sequential* MSI circuit elements. Unlike combinational circuit element output values, which are functions only of the present input values, the outputs of a sequential circuit element depend on both the present and past input values. In effect, a sequential circuit element has a *memory* in which the effects of past input values can be stored.

Sequential circuit elements are classified as either *clocked* (synchronous) or *unclocked* (asynchronous). A clocked sequential circuit element responds to a change of input signals only at discrete instants of time, as determined by a *clock* input. Most sequential circuit elements that are available and used in the designs of digital circuits are of the clocked type. On the other hand, an unclocked circuit element is not regulated by a clock input and so can respond to a change in the inputs at any instant of time. The most important examples of unclocked sequential elements are the random access memory (RAM) and the read-only memory (ROM), which are presented in Chapter 6. The circuit elements considered in the present chapter are mainly clocked circuit elements.

5.2 THE CLOCK SIGNAL

As stated, the *clock* input is the controlling input to a clocked sequential circuit element. A clock input signal is generally a periodic waveform consisting of equally spaced pulses. Two examples of clock signals are shown in the timing diagrams of Figs. 5.1(a) and (b). The *pulse width* of each clock signal is T_p and the *period* is T, as shown. The *duty cycle* of a clock signal is the percentage of the time in which the signal is true, and is equal to $(T_p/T) \times 100$ percent. For example, the clock signal of Fig. 5.1(b) has a 50 percent duty cycle.

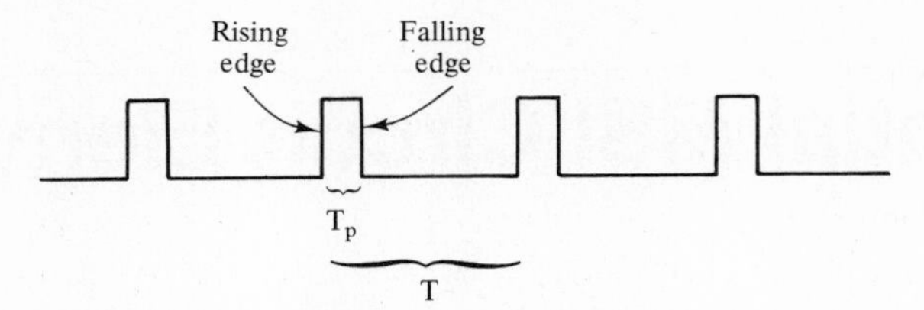

(a) A clock signal with a pulse width of T_p and a period of T

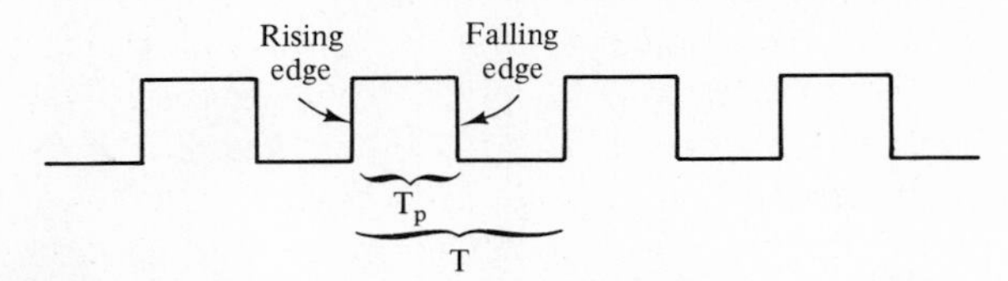

(b) A clock signal with a 50% duty cycle

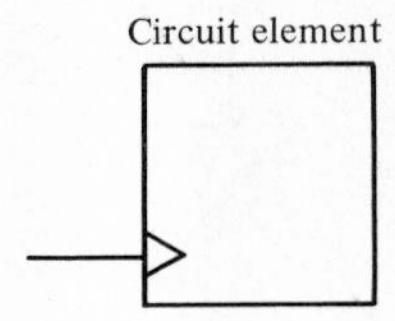

(c) Symbol for an active-high clock input

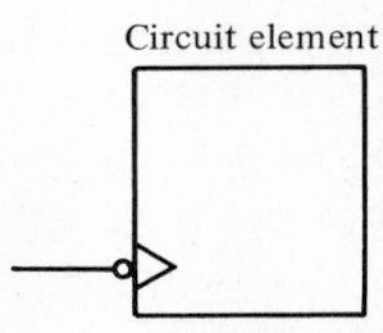

(d) Symbol for an active-low clock input

Figure 5.1 Clock signals and clock inputs.

As shown in Figs. 5.1(a) and (b), the *rising edge* of a clock pulse is the positive-going (from low to high) transition, and the *falling edge* of a clock pulse is the negative-going (from high to low) transition. A clocked sequential circuit element responds to the input signals only at the *active edges* of the clock signals. For an *active-high* clock input, the active edge is the rising edge of the clock pulse. In other words, during a clock cycle, a circuit element with an active-high clock input will respond to the values of the inputs only "at the moment" that the clock signal goes from low to high. For the rest of the clock cycle, any change in the input values will have no effect on the state and outputs of the circuit element. For an *active-low* clock input, the active edge is the falling edge of the clock pulse. The symbols for an active-high and active-low clock input are shown in Figs. 5.1(c) and (d), respectively.

Now that we have defined the relevant parameters of a clock signal, we are ready to see how the clock input is used to control the functioning of clocked sequential circuit elements.

5.3 FLIP-FLOPS

Flip-flops are fundamental memory devices that can assume one of two stable states: 0 (false) or 1 (true). Consequently, a flip-flop is capable of storing 1 bit of information. The state of a flip-flop will remain stable (''remembering'' a 0 or a 1), as long as there is no change in the inputs. A change in the flip-flop inputs, however, can produce a change in the flip-flop state, causing it to go from the *present state* to the *next state*. The next state of a flip-flop is a function of the flip-flop inputs *and* the present state of the flip-flop. In other words, the same values applied at the flip-flop inputs may produce different next states (and consequently different outputs), depending on whether the present state of the flip-flop is a 0 or a 1. A flip-flop is characterized by the types of inputs and the manner in which the inputs affect the operation. In the following sections, we will study the types of flip-flops that are commonly used in the designs of digital circuits. We will study flip-flops in a top-down manner, considering first the function of each type of flip-flop. Then, after the functions are well understood, we will consider the realization details of the flip-flops. This material is presented in Sec. 5.5.

5.3.1 The J-K Flip-Flop

Figure 5.2(a) shows the functional block diagram of a clocked J-K flip-flop. It has two inputs, J and K, in addition to an active-high clock input. It also has two outputs: Q, which represents the state of the flip-flop, and $\overline{Q}$, which is simply the inverted value of the flip-flop state.

The truth table for a J-K flip-flop, commonly called the *characteristic table,* is shown in Fig. 5.2(b). Observe that Q^+, which is the next state of the flip-flop, is a function of the flip-flop inputs J and K, and the present state Q of the flip-flop. Note that Q and Q^+ in the truth table represent the value of the *same* flip-flop output Q, but *at different moments in time*. More specifically, let Q represent the value of the output Q at some moment in time, then Q^+ represents the value of the output Q at a time ''just after'' the *next* active clock edge. For this flip-flop, the active clock edge is the rising edge, as indicated in the truth table by the upward arrow ($\uparrow$).

The function of the J-K flip-flop can be best understood from a timing diagram, as shown in Fig. 5.2(c). Recall that a timing diagram illustrates the behavior of a device over time. For the timing diagram of Fig. 5.2(c), time is divided into intervals (*T*'s), each corresponding to a period of the clock signal CLK. Since the J-K flip-flop shown in Fig. 5.2(a) has an active-high clock input, the diagram has, for emphasis, dashed lines at the rising edges of the clock signal.

For an explanation of this timing diagram, assume that the present interval is T_i, and that the present state of the J-K flip-flop, represented by Q, is 1, as shown. At the next active clock transition (the dashed line between T_i and T_{i+1}), we see that J = 0, K = 1, and the present state Q = 1. For this condition, the fourth row of the truth table of Fig. 5.2(b) specifies that the next state Q^+ is 0. This is graphically illustrated in the timing diagram by the transition of Q from 1 to 0 ''immediately'' after the active clock transition between T_i and T_{i+1}. Actually, there is no ''immediate'' transition of Q, but rather one that occurs later after a time equal to the propagation delay t_{pHL}, as shown.

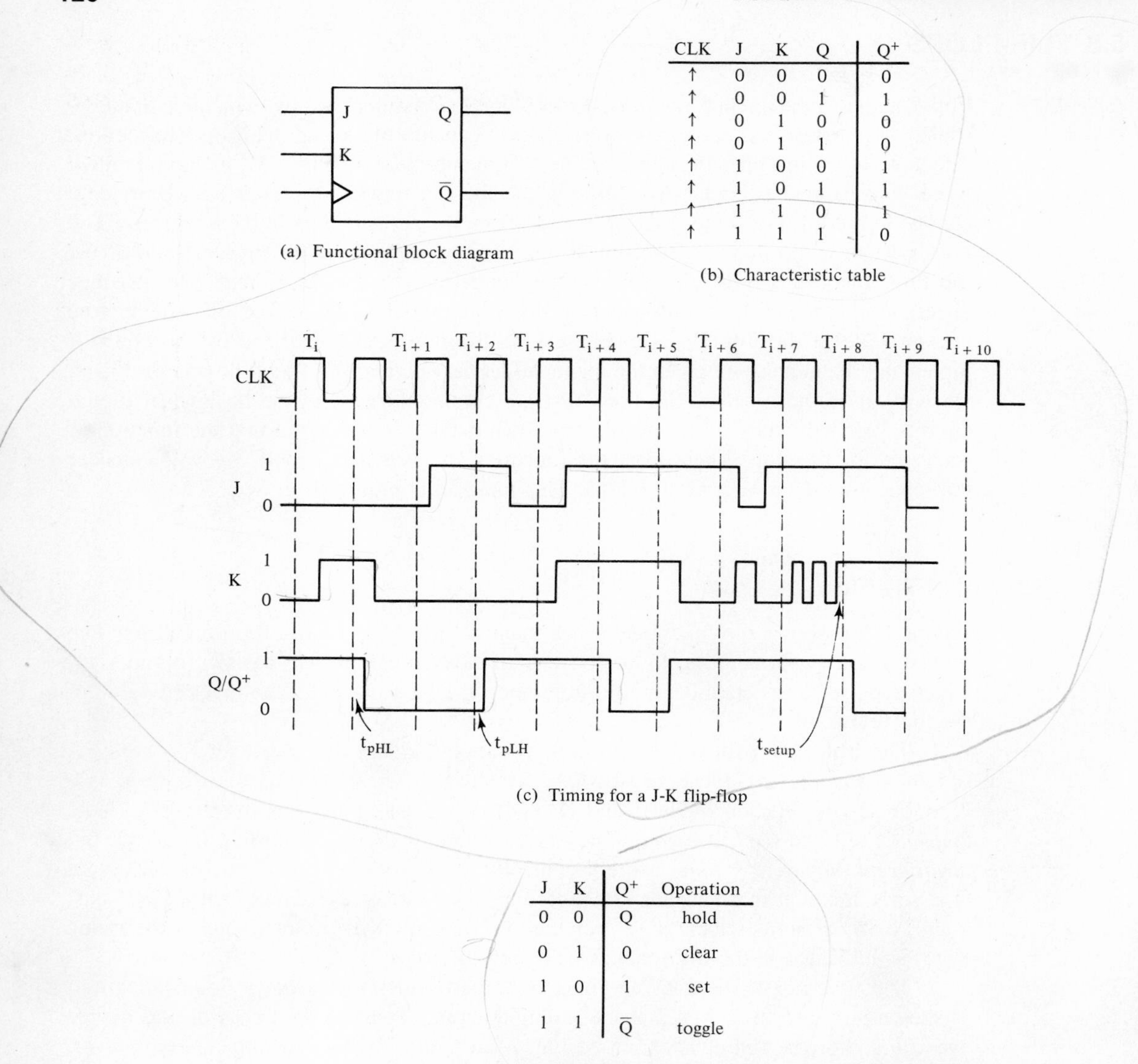

(a) Functional block diagram

CLK	J	K	Q	Q+
↑	0	0	0	0
↑	0	0	1	1
↑	0	1	0	0
↑	0	1	1	0
↑	1	0	0	1
↑	1	0	1	1
↑	1	1	0	1
↑	1	1	1	0

(b) Characteristic table

(c) Timing for a J-K flip-flop

J	K	Q+	Operation
0	0	Q	hold
0	1	0	clear
1	0	1	set
1	1	$\overline{Q}$	toggle

(d) Condensed characteristic table

Figure 5.2 The J-K flip-flop.

For the next interval T_{i+1}, the present flip-flop state Q is equal to 0. And, at the next active clock transition, we see that J = 0, K = 0, and Q = 0. For this condition, the first row of the truth table of Fig. 5.2(b) specifies that the next state Q^+ is 0. Again, this is graphically illustrated in the timing diagram by the fact that Q remains 0 after the next active clock transition (the dashed line between T_{i+1} and T_{i+2}).

In this manner, the timing diagram of Fig. 5.2(c) illustrates the effect of the eight possible combinations of inputs and present state of a J-K flip-flop. Note that the flip-flop responds only to the input values at the next active clock transition, and not at any other time. Consequently, no matter how many times the inputs change during a clock cycle, only the values at the time of the next active clock transition will affect the next state of the flip-flop. This is clearly illustrated in intervals T_{i+7} and T_{i+8} of the timing diagram.

This brings up an interesting point. What happens if the inputs are changed "right at" the active clock transition, as shown in the timing diagram between T_{i+9} and T_{i+10}? In this case, the input J is ambiguous. It may be "seen" by the flip-flop as either a 0 or a 1. This is, of course, an undesirable situation. Care must be taken to ensure that the desired value is placed at the input at some specified setup time, t_{setup}, before the next active clock edge. This t_{setup} is a switching (ac) parameter that is included in the manufacturer's data sheet on the sequential circuit element, and is the time prior to an active clock transition that an input must be stable to assure reliable operation. Later in this

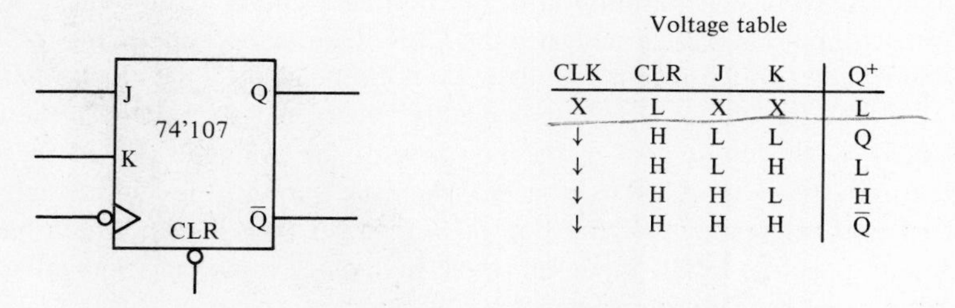

Voltage table

CLK	CLR	J	K	Q^+
X	L	X	X	L
↓	H	L	L	Q
↓	H	L	H	L
↓	H	H	L	H
↓	H	H	H	$\overline{Q}$

(a) 74'107 with a direct clear

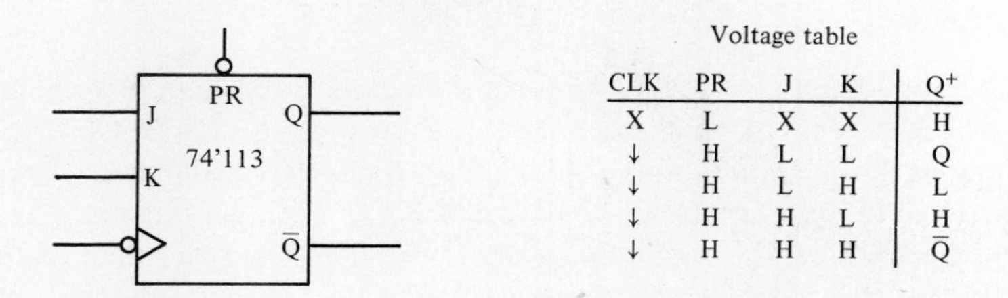

Voltage table

CLK	PR	J	K	Q^+
X	L	X	X	H
↓	H	L	L	Q
↓	H	L	H	L
↓	H	H	L	H
↓	H	H	H	$\overline{Q}$

(b) 74'113 with a direct set

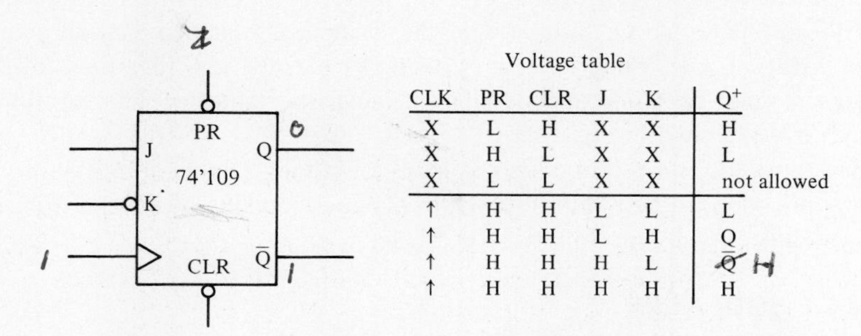

Voltage table

CLK	PR	CLR	J	K	Q^+
X	L	H	X	X	H
X	H	L	X	X	L
X	L	L	X	X	not allowed
↑	H	H	L	L	L
↑	H	H	L	H	Q
↑	H	H	H	L	$\overline{Q}$
↑	H	H	H	H	H

(c) 74'109 with direct set and direct clear

Figure 5.3 Commercially available J-K flip-flops.

book, we will study design methods for ensuring that inputs are synchronized to change and become stable in the early part of the clock cycle so that the setup times of the inputs of the sequential circuit elements are never violated.

The truth table of Fig. 5.2(b) can be condensed into the one shown in Fig. 5.2(d) by specifying the present state as simply Q instead of a 0 or a 1. From this condensed truth table, we can clearly see the four operations that can be performed by a J-K flip-flop. If J and K are both 0 at an active clock transition, then the next state Q^+ is the same as the present state Q. If J = 0 and K = 1, then the next state is 0, regardless of the value of the present state. On the other hand, if J = 1 and K = 0, then the next state is 1, regardless of the value of the present state. Finally, if J and K are both 1, then the next state is the complement of the present state, a behavior called *toggling*.

J-K flip-flops are commercially available with a variety of active-high and active-low inputs and outputs. Several common ones are shown in Fig. 5.3, along with their voltage tables. Note that in addition to the normal J and K inputs, these J-K flip-flops have a preset input PR and/or a clear input CLR. Unlike the J and K inputs, these inputs are *asynchronous,* which means that they are not regulated by the clock input. When PR is true (L), then the output Q is ''immediately'' reset to 1 regardless of the clock input, as indicated by the don't care in the first row of the voltage table of the 74'113 and 74'109. Similarly, when CLR is true (L), then the output Q is ''immediately'' cleared to 0. The most interesting J-K flip-flop of Fig. 5.3 is the 74'109, which has an active-high J input, an active-low K input, an active-high clock input, and both an asynchronous preset PR and an asynchronous clear CLR.

5.3.2 The D Flip-Flop

Another type of commonly used flip-flop is the clocked D flip-flop, also called the delay flip-flop. The functional block diagram for it is shown in Fig. 5.4(a). Observe that it has only one input, D, in addition to the clock input. It also has two outputs, Q and $\overline{Q}$. The characteristic table for a D flip-flop is shown in Fig. 5.4(b). As illustrated in the timing diagram of Fig. 5.4(c), the operation of the D flip-flop is simpler than that of the J-K flip-flop. In fact, the next state Q^+ is a function of the D input only and, unlike the J-K flip-flop, is independent of the present state Q. If the D input is equal to 1 at an active clock transition, as at T_i/T_{i+1} and T_{i+3}/T_{i+4}, then Q^+ is equal to 1 regardless of the value of the present state Q. Similarly, if the D input is equal to 0 at an active clock transition, as at T_{i+1}/T_{i+2} and T_{i+2}/T_{i+3}, then Q^+ is equal to 0 regardless of the value of the present state Q. Consequently, the characteristic table for a D flip-flop can be condensed to that shown in Fig. 5.4(d).

The D flip-flop is a useful circuit element for storing 1 bit of information. Another common application of the D flip-flop is for delaying the value of a signal for one clock cycle, as is illustrated by the following example.

Example 5.1 Serial Adder

The parallel adder described in Chapter 4 can add two *N*-bit numbers simultaneously by using *N* full adders. But, with only one full adder, a serial adder can add two *N*-bit numbers. The serial adder, however, will require *N* clock cycles to perform the addition.

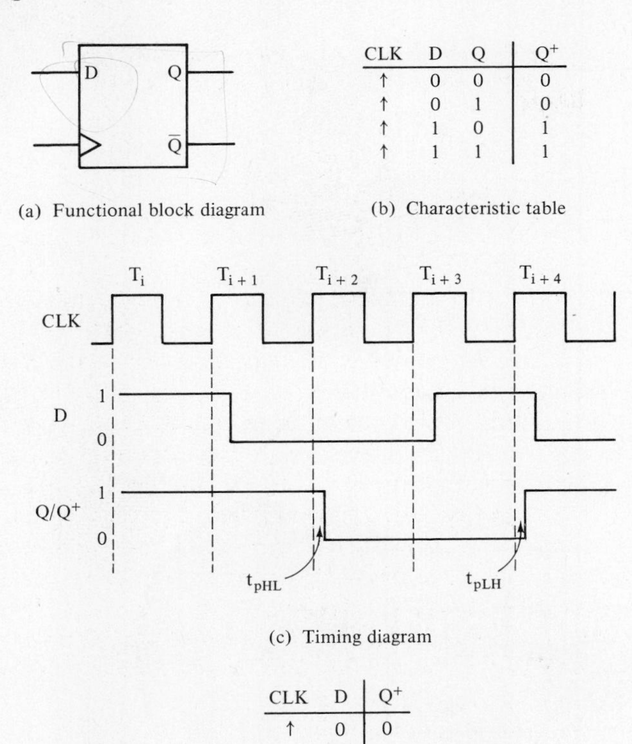

(a) Functional block diagram

CLK	D	Q	Q^+
↑	0	0	0
↑	0	1	0
↑	1	0	1
↑	1	1	1

(b) Characteristic table

(c) Timing diagram

CLK	D	Q^+
↑	0	0
↑	1	1

(d) Condensed characteristic table

Figure 5.4 The D flip-flop.

A circuit diagram of a serial adder is shown in Fig. 5.5(a). It consists of a full adder and a D flip-flop. The operation of the serial adder is described below. We will consider this example in some detail since it illustrates some important characteristics of combinational and sequential circuit elements.

In the serial adder of Fig. 5.5(a), the N-bit numbers X and Y are fed in serially, one pair of bits at a time, at inputs A and B, respectively. First, the least significant bits X_0 and Y_0 are fed in, then X_1 and Y_1, and so forth until finally the most significant bits X_{N-1} and Y_{N-1} are fed in. The carry-in CY_i for each stage of the addition is provided by the output of the D flip-flop. Since the input to the D flip-flop is CY_{i+1}, the carry-in CY_i of the current stage of addition is the CY_{i+1} of the preceding stage of addition. In other words, the D flip-flop delays the value of the carry-out of the preceding stage of addition by one clock cycle so that this carry-out can be used as the carry-in for the current addition stage, as is required for serial addition. Initially, the contents of the D flip-flop should be 0.

For proper operation, the inputs X and Y must be synchronized with the clock input of the D flip-flop so that a new pair of bits of X and Y is present at each clock cycle. This can be done by storing the bits for X and Y in sequential circuit elements called *shift registers*, and then by shifting them out one at a time to be fed into the adder

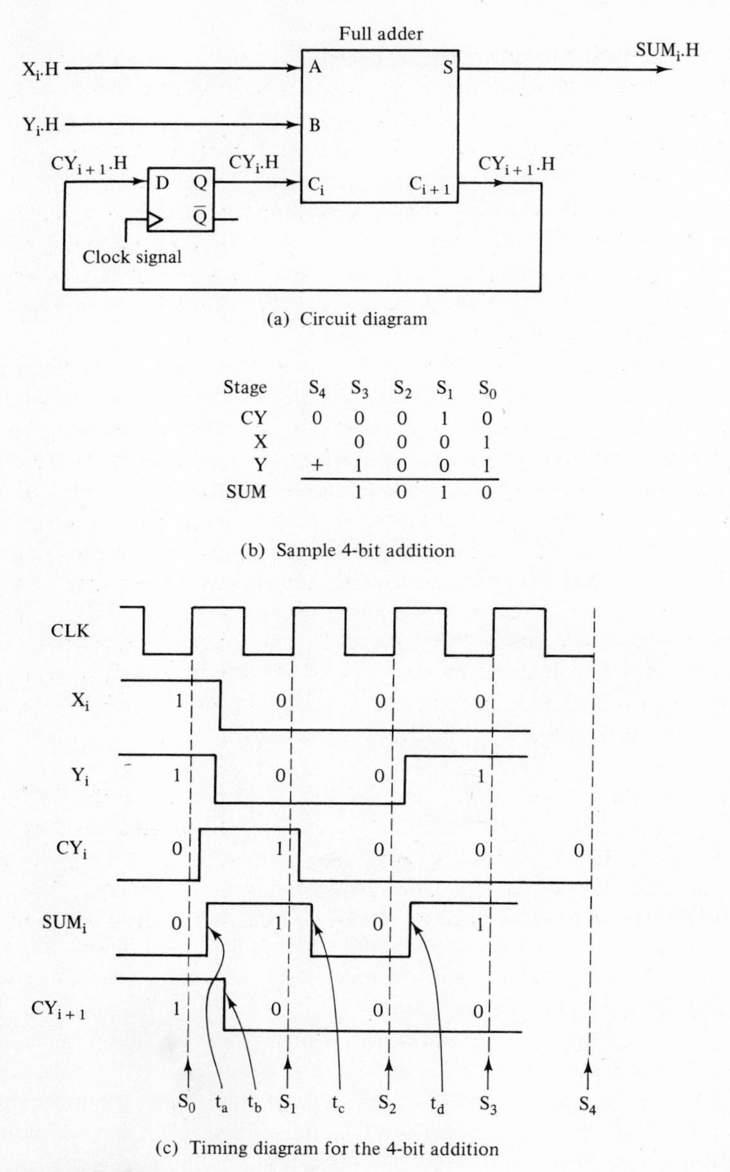

(a) Circuit diagram

Stage	S_4	S_3	S_2	S_1	S_0
CY	0	0	0	1	0
X		0	0	0	1
Y	+	1	0	0	1
SUM		1	0	1	0

(b) Sample 4-bit addition

(c) Timing diagram for the 4-bit addition

Figure 5.5 A serial adder.

inputs at each clock cycle. Similarly, the output SUM can be stored by shifting each SUM_i bit into a shift register. The operation of a shift register is explained in Sec. 5.7.2.

Perhaps the operations of the serial adder can be best understood by tracing through a timing diagram such as the one shown in Fig. 5.5(c). As shown, each stage of addition in the sample 4-bit addition of Fig. 5.5(b) is represented by the values of each of the active clock transitions in Fig. 5.5(c). In other words, the values of the inputs and outputs are valid only at the active clock transitions. In this timing diagram, the active clock transitions are labeled S_0, S_1, S_2, S_3, and S_4. At S_0, the least significant bits X_0, Y_0, and

CY_0 are added together to produce SUM_0 and CY_1. At S_1, the inputs X_1, Y_1, and CY_1 are added together to produce SUM_1 and CY_2. The process continues until at S_3, the most significant bits X_3, Y_3, and CY_3 are added together to produce SUM_3 and CY_4.

Note that the D flip-flop is a clocked sequential circuit element. Consequently, its output CY_i changes only at each active clock transition, as can be seen graphically in the timing diagram. The value of CY_{i+1} at each clock transition is loaded into the D flip-flop to be used as CY_i at the next active clock transition. For example, the value of CY_{i+1} is 1 at S_0, at which time it is loaded into the D flip-flop. It then becomes available at input C_i at time $S_0 + t_{pLH}$, but it will not be used as the new value of CY_i until the next active clock transition S_1.

The full adder, on the other hand, is a combinational circuit element. Consequently, its outputs respond to the inputs "immediately" after (actually a propagation delay after) the change in the inputs. This can be seen graphically in the timing diagram at t_a where SUM_i responds to the change in CY_i, at t_b where CY_{i+1} responds to the changes in X_i and Y_i, at t_c where SUM_i responds to the change in CY_i, and at t_d where SUM_i responds to the change in Y_i. Due to this behavior, the inputs and outputs of the serial adder are guaranteed to be valid and stable only at the active clock transitions. For example, between S_0 and S_1, the inputs and outputs are changing at various times. In fact, at time t_a, the values are $X_i = 1$, $Y_i = 1$, $CY_i = 1$, $SUM_i = 1$, and $CY_{i+1} = 1$, which do not correspond to any of the addition stages for this example. However, all the inputs and outputs have become stable by the time t_b, which is well before the required setup time for the next active clock transition at S_1. At this time, the values of the inputs and outputs correspond to Stage 1 of this addition example. ■■

D flip-flops are commercially available in a form such as the 74'74 D flip-flop shown in Fig. 5.6. Like some J-K flip-flops, this D flip-flop also has a preset input PR and a clear input CLR. Again, these inputs are asynchronous, and so are not regulated by the clock input. Consequently, they can be used to "immediately" set or clear the state of the flip-flop during any part of the clock cycle, if either is required.

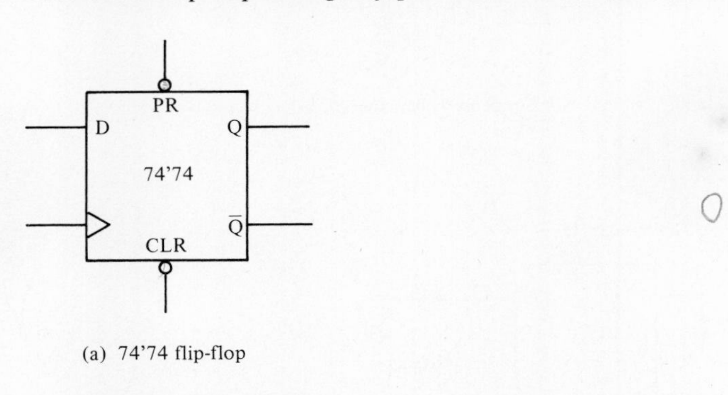

(a) 74'74 flip-flop

CLK	PR	CLR	D	Q^+
X	L	H	X	H
X	H	L	X	L
↑	H	H	L	L
↑	H	H	H	H

(b) Voltage table

Figure 5.6 Commercially available D flip-flop.

5.3.3 The T Flip-Flop

A less common, but still useful, type of flip-flop is the clocked T flip-flop, also called the toggle flip-flop. The functional block diagram for it is shown in Fig. 5.7(a). This flip-flop has one input T, in addition to the clock input. It also has two outputs Q and $\overline{Q}$. The characteristic table for the T flip-flop is shown in Fig. 5.7(b). A condensed version of this table is shown in Fig. 5.7(c).

The function of the T flip-flop is quite simple. If the T input is false (0), then at the next active clock transition nothing happens. So the previous state of the flip-flop is retained. But if the T input is true (1), then at the next active clock transition the output of the flip-flop is complemented. Note that the T flip-flop performs the "hold" and "toggle" functions of the J-K flip-flop.

T flip-flops are not commercially available in IC form since they can be easily derived from other commercially available flip-flops. For example, a T flip-flop can be obtained from a J-K flip-flop by simply connecting the J and K inputs together for the T input, as shown in Fig. 5.7(d).

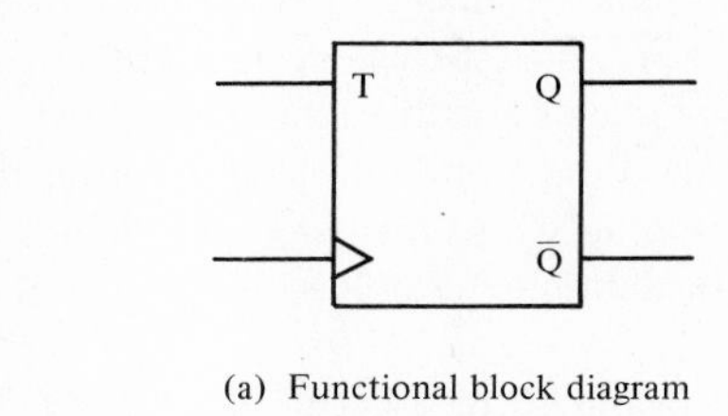

(a) Functional block diagram

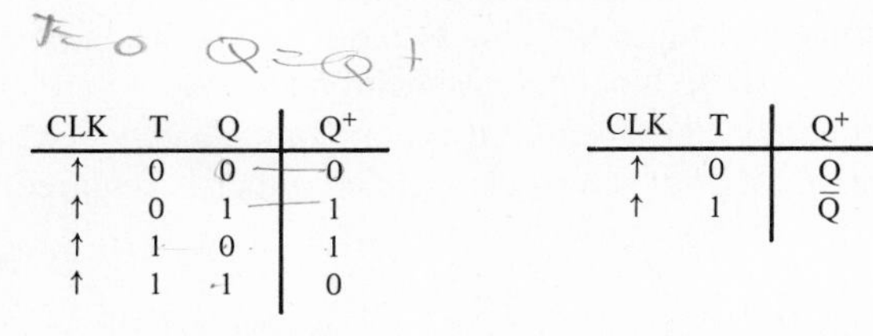

CLK	T	Q	Q^+
↑	0	0	0
↑	0	1	1
↑	1	0	1
↑	1	1	0

(b) Characteristic table

CLK	T	Q^+
↑	0	Q
↑	1	$\overline{Q}$

(c) Condensed characteristic table

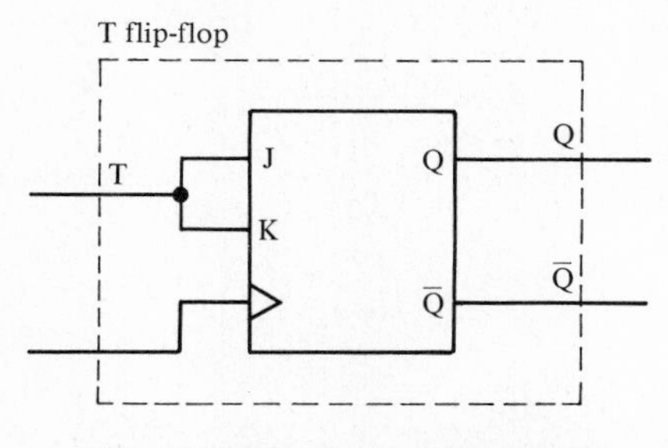

(d) A T flip-flop from a J-K flip-flop

Figure 5.7 The T flip-flop.

5.3.4 Flip-Flop Conversion

As seen from the preceding section, the conversion from a J-K flip-flop to a T flip-flop can be done ''intuitively'' since the conversion is so simple. When the conversion becomes more complex, however, a systematic procedure is required.

Example 5.2 A Gated D Flip-Flop Using a J-K Flip-Flop
In this example we will design a modified D flip-flop that has a controlling load input LD. The functional block diagram and the truth table for it are shown in Figs. 5.8(a) and (b), respectively. This flip-flop differs from the ordinary D flip-flop which, at each active clock transition, loads in the value at its D input indiscriminantly. With this modified D flip-flop, however, the LD input can be used to load in the value of the D

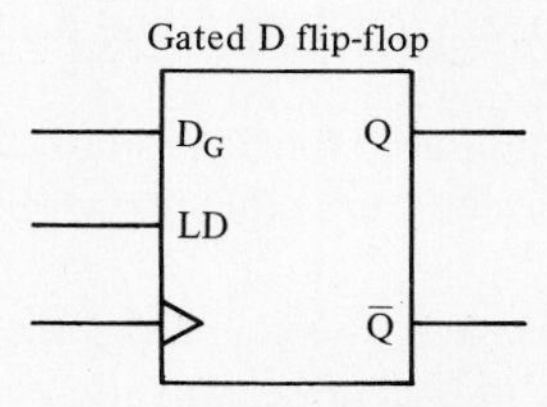

(a) Functional block diagram

CLK	LD	D_G	Q	Q^+
↑	0	0	0	0
↑	0	0	1	1
↑	0	1	0	0
↑	0	1	1	1
↑	1	0	0	0
↑	1	0	1	0
↑	1	1	0	1
↑	1	1	1	1

(b) Truth table

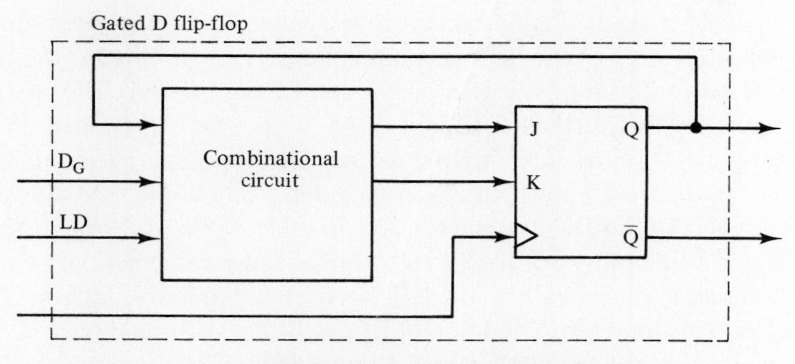

(c) Block diagram of the flip-flop conversion

Figure 5.8 Gated D flip-flop of Example 5.2.

input only at selected active clock transitions. Specifically, if the LD input is false (0), then at the next active clock transition, nothing happens. The previous state of the flip-flop is retained. But if the LD input is true (1), then at the next active clock transition, the gated D flip-flop loads in the value at the D_G input.

In this design our approach will be to convert a J-K flip-flop into a gated D flip-flop, as shown in the block diagram of Fig. 5.8(c). With the flip-flop selected, the problem reduces to the design of the combinational circuit that will transform the inputs D_G, LD, and Q, the present flip-flop state, into the corresponding J and K inputs for the J-K flip-flop so that the correct next state is outputted at the next active clock transition. A systematic procedure will now be outlined to facilitate this conversion process.

Step 1: Determine the functional block diagram of the combinational circuit. For our example, this functional block diagram is shown in Fig. 5.9(a). The inputs to it are LD, D_G, and the present state Q, which is fed back from the output of the J-K flip-flop. The outputs of the combinational circuit are J and K, which correspond to the J and K inputs of the J-K flip-flop.

Step 2: Determine the truth table for the combinational circuit. In other words, for our circuit we need to complete the truth table shown in Fig. 5.9(b). This step is divided into two substeps.

Step 2(a): Transform the characteristic table of the source flip-flop into its *excitation table*. In this case, the source flip-flop to be converted is a J-K flip-flop, the characteristic table for which is shown in Fig. 5.9(c). This characteristic table specifies the value of the next-state output Q^+ as a function of the inputs J, K, and the present state Q. On the other hand, the excitation table, also shown in Fig. 5.9(c), specifies the values that the inputs J and K must be for the output to change from the specified present state Q to the specified next state Q^+. For example, the first row of the excitation table specifies that for the J-K flip-flop to change from the present state Q of 0 to the next state Q^+ of 0, the value of the J input must be 0, but the value of the K input can be either 0 or 1 (i.e., a don't care). This information is obtained from rows 1 and 3 of the original characteristic table. Similarly, the second row of the excitation table specifies that for the J-K flip-flop to change from a present state of 0 to a next state of 1, the value of the J input must be 1, but the value of the K input can be either 0 or 1. This information is obtained from rows 5 and 7 of the characteristic table. The remainder of the excitation table can be determined in the same manner. This excitation table for a J-K flip-flop will be used in the next step to obtain the truth table for the combinational circuit.

Step 2(b): Use the excitation table for the source flip-flop to determine the output values for the truth table of the combinational circuit. For our circuit, the truth table of the desired gated D flip-flop is shown on the left of Fig. 5.9(d). The first row of that truth table specifies that for inputs LD $= 0$, $D_G = 0$, and a present Q $= 0$, the next state Q^+ is to be 0. Now consider the source J-K flip-flop. For it to change from $Q = 0$ to $Q^+ = 0$, J must be 0 but K can be a don't care at the next active clock transition, as specified by the J-K excitation table. In other words, looking at Fig. 5.8(c), if LD $= 0$, $D_G = 0$, and $Q = 0$, then for Q^+ to be 0, the combinational circuit must be designed such that it generates $J = 0$ and $K = 0$ or 1. Similarly, for row 2 of the truth table of Fig. 5.9(d), under the condition of LD $= 0$, $D_G = 0$, and $Q = 1$, the quantity J can be a don't care but K must be a 0 to obtain $Q^+ = 1$. Continuing in this manner we can determine the rest of the values for J and K. Note, as shown by the arrow in

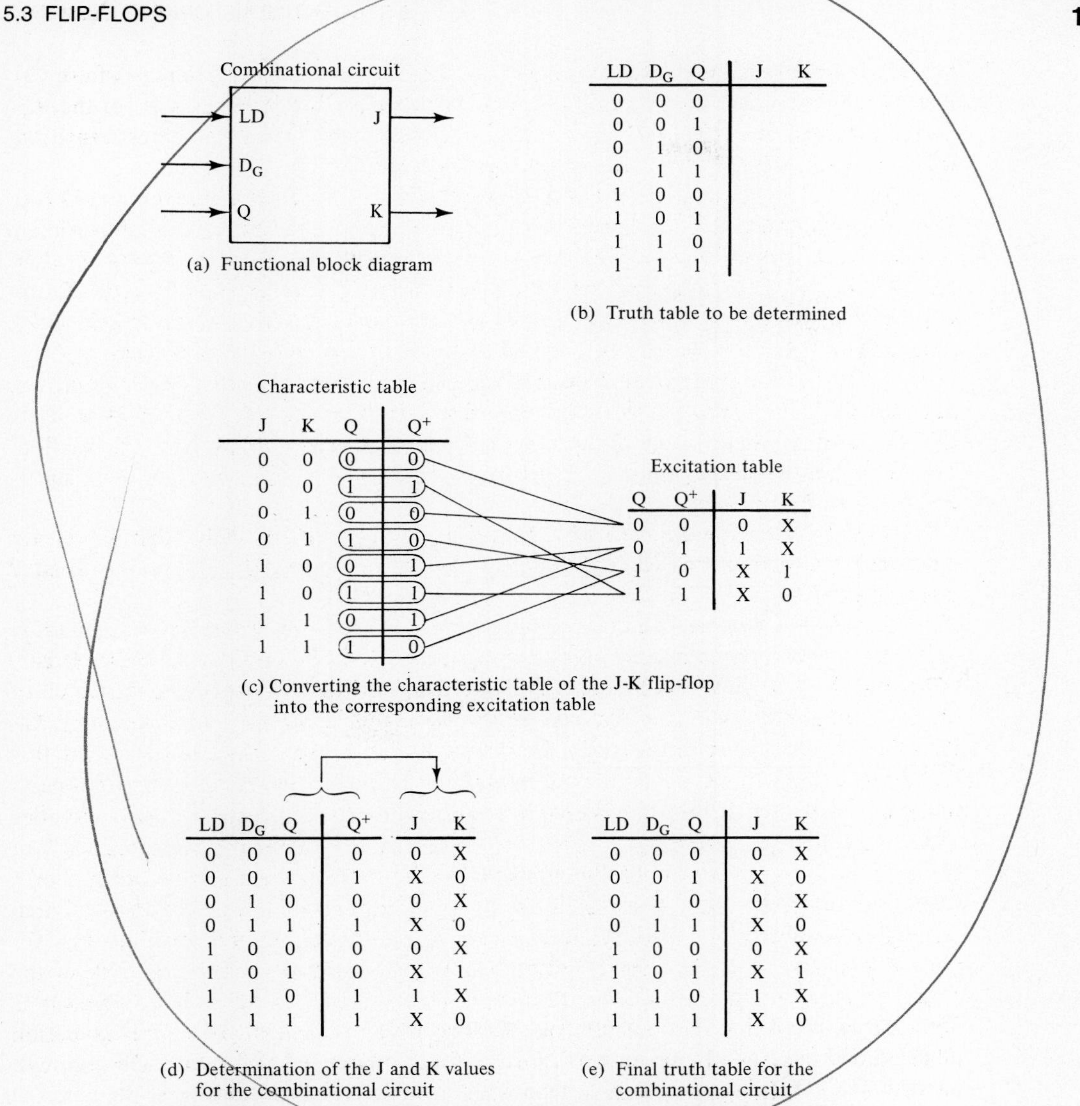

LD	D_G	Q	J	K
0	0	0		
0	0	1		
0	1	0		
0	1	1		
1	0	0		
1	0	1		
1	1	0		
1	1	1		

(b) Truth table to be determined

Characteristic table

J	K	Q	Q^+
0	0	0	0
0	0	1	1
0	1	0	0
0	1	1	0
1	0	0	1
1	0	1	1
1	1	0	1
1	1	1	0

Excitation table

Q	Q^+	J	K
0	0	0	X
0	1	1	X
1	0	X	1
1	1	X	0

(c) Converting the characteristic table of the J-K flip-flop into the corresponding excitation table

LD	D_G	Q	Q^+	J	K
0	0	0	0	0	X
0	0	1	1	X	0
0	1	0	0	0	X
0	1	1	1	X	0
1	0	0	0	0	X
1	0	1	0	X	1
1	1	0	1	1	X
1	1	1	1	X	0

(d) Determination of the J and K values for the combinational circuit

LD	D_G	Q	J	K
0	0	0	0	X
0	0	1	X	0
0	1	0	0	X
0	1	1	X	0
1	0	0	0	X
1	0	1	X	1
1	1	0	1	X
1	1	1	X	0

(e) Final truth table for the combinational circuit

Figure 5.9 Design and realization of the gated D flip-flop using a J-K flip-flop.

Fig. 5.9(d), that the values of J and K for each row are determined by the Q to Q^+ transition of that row, based on the information specified in the J-K excitation table of Fig. 5.9(c). Since the J and K inputs of the J-K flip-flop are the J and K outputs of the combinational circuit, the truth table for the combinational circuit can be determined, as shown in Fig. 5.9(e).

Step 3: Realize the combinational circuit. Once the truth table for the combinational circuit is determined from step 2, the realization of this circuit is straightforward using the techniques presented in Chapters 2 and 3. The resultant circuit diagram is shown in

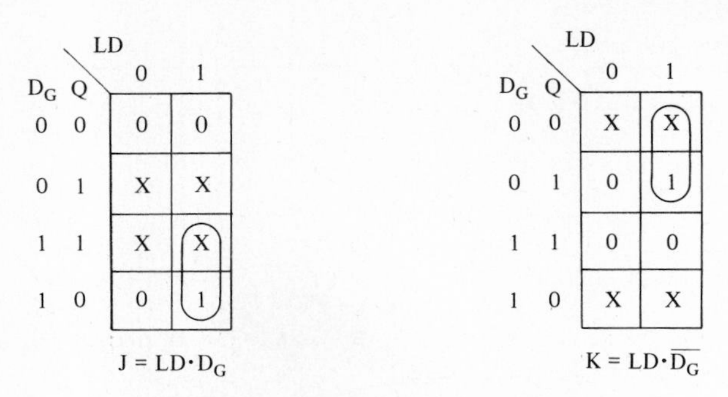

(f) K-maps for the J and K outputs

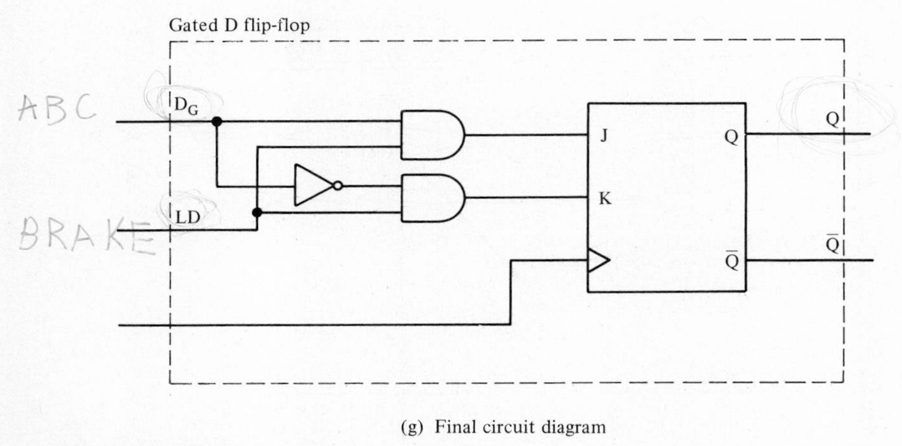

(g) Final circuit diagram

Figure 5.9 *(cont.)*

Fig. 5.9(g). Verifying its operation, we see that if LD is 0, then both AND gates are disabled and the J and K inputs are 0. Consequently, the gated D flip-flop will retain its present state. If, however, LD is 1, then at the next active clock transition, the gated D flip-flop will load in the value of D_G. ■■

Example 5.3 A Gated D Flip-Flop Using a D Flip-Flop

In this example, we will redesign the gated D flip-flop, using an ordinary D flip-flop. The procedure outlined in Example 5.2 still applies. Step 2(a), however, will be different since the source flip-flop is now a D flip-flop.

Step 1: Determine the functional block diagram of the combinational circuit. The functional block diagrams of the flip-flop conversion and of the combinational circuit are shown in Figs. 5.10(a) and (b), respectively. Observe that the inputs to the combinational circuit are LD, D_G, and the present state Q, just as before. The output, however, is now D, corresponding to the D input of the D flip-flop.

Step 2: Determine the truth table for the combinational circuit.

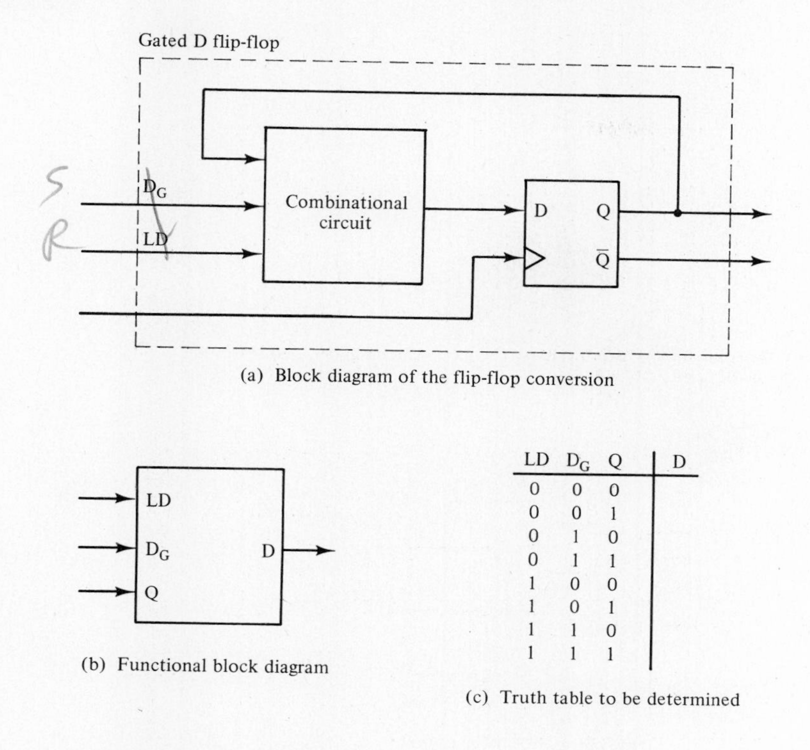

(a) Block diagram of the flip-flop conversion

(b) Functional block diagram

LD	D_G	Q	D
0	0	0	
0	0	1	
0	1	0	
0	1	1	
1	0	0	
1	0	1	
1	1	0	
1	1	1	

(c) Truth table to be determined

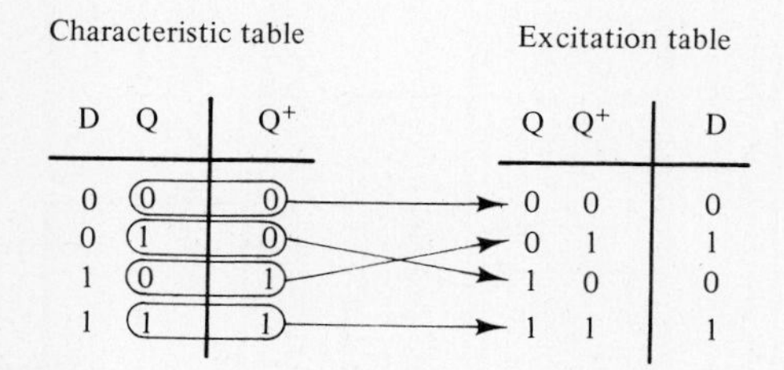

Characteristic table

D	Q	Q^+
0	0	0
0	1	0
1	0	1
1	1	1

Excitation table

Q	Q^+	D
0	0	0
0	1	1
1	0	0
1	1	1

(d) Determining the excitation table for a D flip-flop

Figure 5.10 Design and realization of the gated D flip-flop using a D flip-flop.

Step 2(a): Transform the characteristic table of the source flip-flop into its excitation table. For our source D flip-flop, the transformation process, which is shown in Fig. 5.10(d), is analogous to that for the J-K flip-flop in Example 5.2. For example, row 2 of the excitation table specifies that for the D flip-flop to change from the present state (Q) of 0 to the next state (Q^+) of 1, the value of the D input must be 1 at the next active clock transition. This information is obtained from row 3 of the characteristic table.

Step 2(b): Use the excitation table for the source flip-flop to determine the output values for the truth table of the combinational circuit. The process, which is shown in Fig. 5.10(e), is again analogous to that for the J-K flip-flop of Example 5.2. For example, in row 2 of the truth table in Fig. 5.10(e), under the condition of LD $= 0$, $D_G = 0$,

LD	D_G	Q	Q^+	D
0	0	0	0	0
0	0	1	1	1
0	1	0	0	0
0	1	1	1	1
1	0	0	0	0
1	0	1	0	0
1	1	0	1	1
1	1	1	1	1

(e) Determination of the D values for the combinational circuit

LD	D_G	Q	D
0	0	0	0
0	0	1	1
0	1	0	0
0	1	1	1
1	0	0	0
1	0	1	0
1	1	0	1
1	1	1	1

(f) Final truth table for the combinational circuit

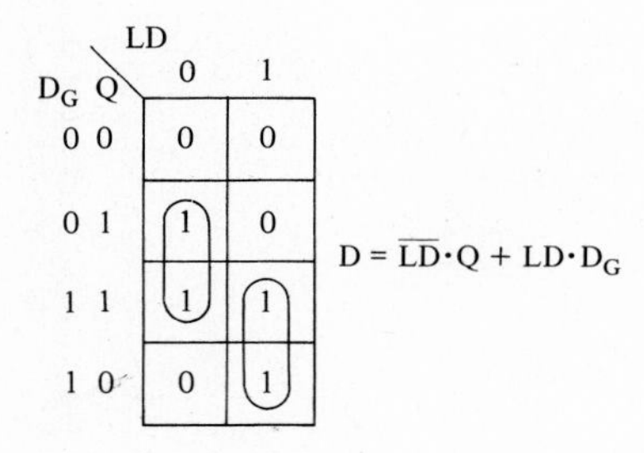

(g) K-map for the D output

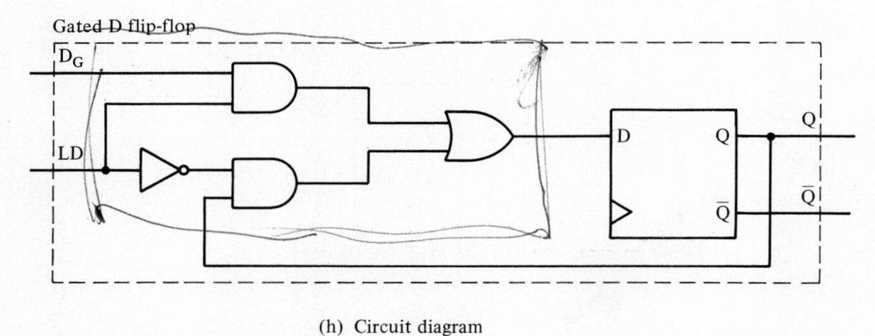

(h) Circuit diagram

Figure 5.10 (*cont.*)

and Q = 1, the D input must be a 1 to obtain $Q^+ = 1$ at the next active clock transition. Again note, as shown by the arrow in Fig. 5.10(e), that the value of D for each row is determined by the Q to Q^+ transition of that row, based on the information given in the D flip-flop excitation table of Fig. 5.10(d). The final truth table for the combinational circuit is shown in Fig. 5.10(f).

Step 3: Realize the combinational circuit. The realization of the combinational circuit is shown in Figs. 5.10(g) and (h). Verifying the operation of the circuit, we see that if LD is 0, then the top AND gate is disabled. However, the bottom AND gate is activated, thereby enabling the present state Q to be passed through and loaded back into the flip-flop at the next active clock transition. If LD is 1, though, the bottom AND gate is disabled and the top AND gate is activated, thereby enabling the new value at D to be loaded into the flip-flop at the next active clock transition. ■ ■

In summary, the procedure outlined in this section can be used to convert any type of flip-flop to any other type of flip-flop. In fact, this procedure can be modified to be used for the design of more complex sequential circuits, as we will see later in this chapter.

5.4 THE UNCLOCKED S-R FLIP-FLOP

The functional block diagram for an *unclocked* S-R flip-flop is shown in Fig. 5.11(a). It has two inputs S (set) and R (reset), but has no clock input. The S-R flip-flop also has two outputs Q and $\overline{Q}$. As shown in the truth table in Fig. 5.11(b), the operation of an S-R flip-flop is similar to that of a J-K flip-flop, except that both inputs of 1 (S = R =

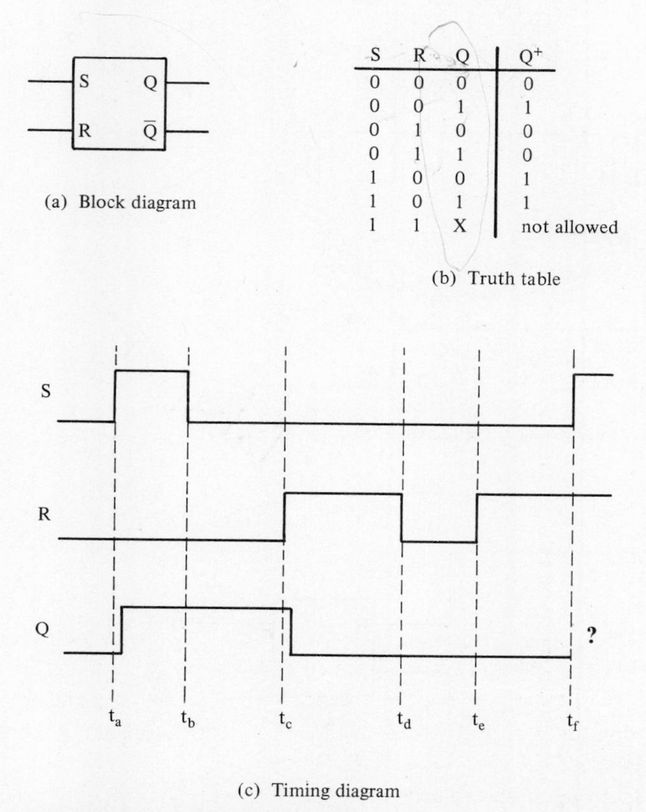

(a) Block diagram

S	R	Q	Q^+
0	0	0	0
0	0	1	1
0	1	0	0
0	1	1	0
1	0	0	1
1	0	1	1
1	1	X	not allowed

(b) Truth table

(c) Timing diagram

Figure 5.11 The S-R flip-flop.

1) are not allowed. Furthermore, unlike the clocked flip-flops that have been presented, the unclocked S-R flip-flop is not controlled by a clock input, but rather responds asynchronously to changes in the inputs.

The operation of the S-R flip-flop is illustrated in the timing diagram of Fig. 5.11(c). Note the asynchronous nature of this flip-flop. The transition to the next state Q^+ occurs ''immediately'' after a change in one of the inputs, and is not synchronized by an active transition of a clock input. For example, at $t = t_a$, the input S change from 0 to 1 causes Q, at $t = t_a + t_{pLH}$, to change from 0 to 1. The t_{pLH} here is the low-to-high propagation delay of the S-R flip-flop. Also note that at $t = t_f$, S and R are both 1, which is not allowed for an S-R flip-flop. We will see the reason for this in Sec. 5.5, when the realization details of flip-flops are discussed.

Example 5.4 Using an S-R Flip-Flop for Switch Debouncing

A common application of the S-R flip-flop is for the debouncing of a mechanical switch. When a mechanical switch is first closed, it does not make contact cleanly, but rather makes and breaks contact several times before remaining closed. Similarly, when a mechanical switch is first opened, it will ''bounce'' several times before remaining open.

The problem with switch bouncing is illustrated by the waveform of the switch signal SW shown in Fig. 5.12(a). If this signal is used as an input to other circuit

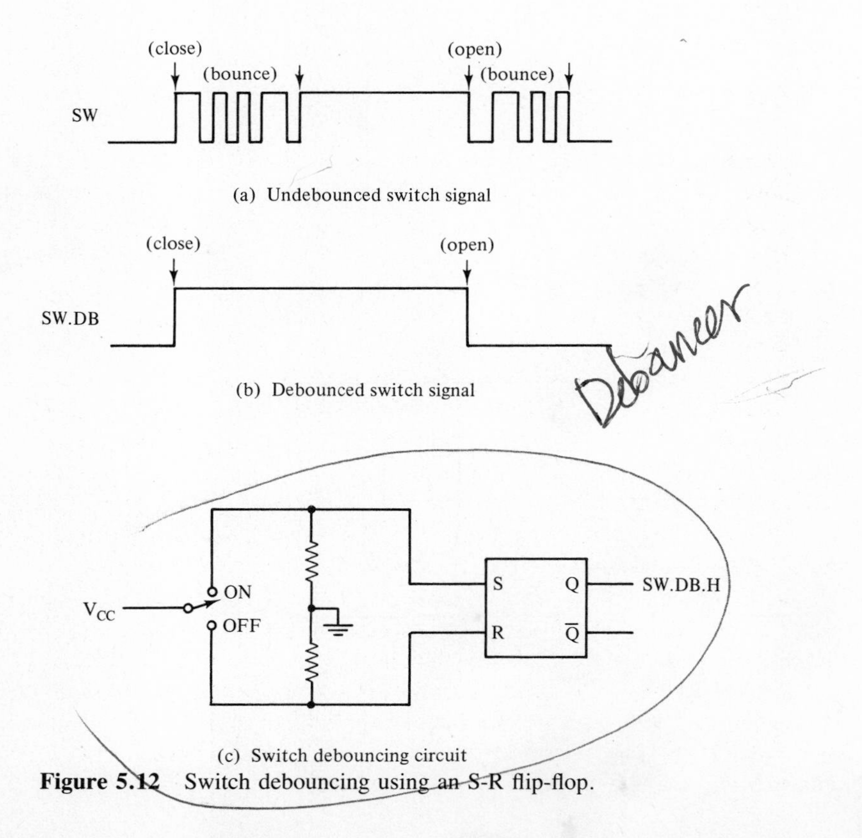

Figure 5.12 Switch debouncing using an S-R flip-flop.

elements, undesirable circuit behavior may result because each bounce may be interpreted as a signal change. What is desired is a debounced switch signal, as shown in Fig. 5.12(b), that has a single low-to-high transition for each switch closure, and a single high-to-low transition for each switch opening.

Figure 5.12(c) shows the use of an S-R flip-flop in a switch debouncing circuit for a single-pole, double-throw switch. For an understanding of the operation let us assume that initially the switch arm is between OFF and ON and that SW.DB is low. Then, when the switch arm first makes contact with the ON terminal, the S input of the S-R flip-flop becomes 1, causing SW.DB to make a low-to-high transition. During the time that the switch bounces after this initial closing, the switch arm makes and breaks contact with the ON terminal. Each time the contact is broken, the values at the S and R inputs of the S-R flip-flop are both 0. And, each time the contact is made, the values are S = 1 and R = 0. In either case, SW.DB remains 1. Consequently, there is only a single low-to-high transition of SW.DB, as shown.

Now if the switch arm is moved from the ON terminal to the OFF terminal, the value of SW.DB remains 1 until the switch arm first makes contact with the OFF terminal. At that moment, the inputs to the S-R flip-flop become R = 1 and S = 0, with the result that SW.DB makes a high-to-low transition. Again, the switch bounces, thereby causing the value of R to alternate between 0 and 1 while the value of S stays at 0. In either case, SW.DB remains 0. Consequently, there is only one high-to-low transition of SW.DB, as shown. ■ ■

5.5 REALIZATION OF FLIP-FLOPS

Our primary concern with flip-flops is the *use* of them as circuit elements in the design of digital systems. It is nevertheless important to have some insight into the internal structures of these devices. Consequently, we will now digress to study some realization details of flip-flops. We do this now rather than at the beginning of our consideration of flip-flops because of the top-down manner of our study. In the preceding sections we learned how the common types of flip-flop function. Now that we understand this, we can better appreciate the realization details of flip-flops.

The unclocked S-R flip-flop of the preceding section is a basic flip-flop from which other flip-flops can be derived. So, we will consider it first. A realization of the S-R flip-flop is shown in Fig. 5.13(a). It is a simple logic circuit consisting of two interconnected NOR gates. The feature of this circuit that distinguishes it from the combinational circuits presented in Chapter 4 is the *feedback* connections from Q to NOR gate 1 and from $\overline{Q}$ to NOR gate 2. It is these feedback paths that provide the memory property, allowing the flip-flop to assume one of two stable states: 0 (false) and 1 (true).

The operation of this flip-flop circuit is illustrated by the timing diagram shown in Fig. 5.13(b). As a starting point, let us assume that initially (before $t = t_a$), the flip-flop circuit is in a stable state with the following values: S = 0, R = 0, Q = 0, and $\overline{Q}$ = 1. At $t = t_a$, though, S changes from 0 to 1. Consequently, the output of NOR gate 1, $\overline{Q}$, goes from 1 to 0, as shown in the timing diagram at $t_a + t_{pHL}$. And since $\overline{Q}$ is fed back to the input of NOR gate 2, it causes Q to change from 0 to 1 at $t_a + t_{pHL} + t_{pLH}$. Similarly, Q is fed back to the input of NOR gate 1. Since Q is now 1 and S is 1, then

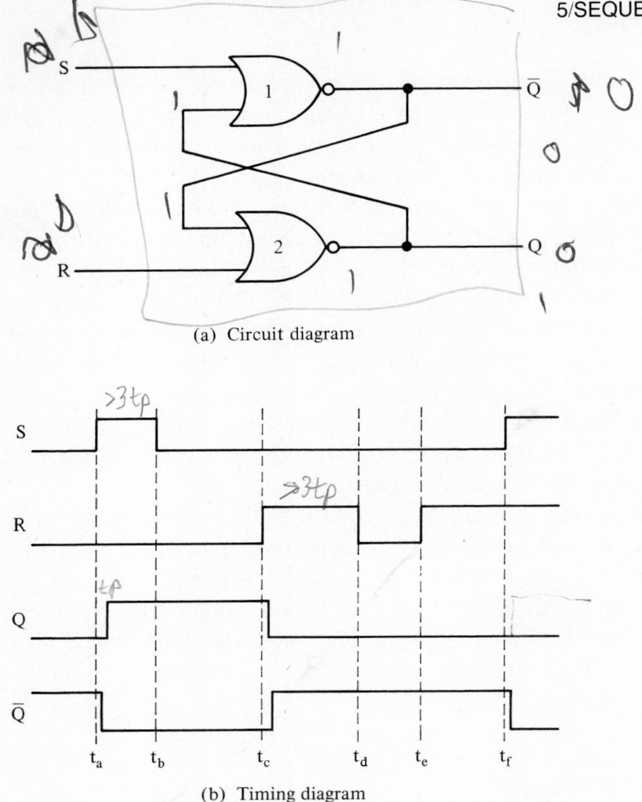

(a) Circuit diagram

(b) Timing diagram

Figure 5.13 Realization of an S-R flip-flop.

$\overline{Q}$ should be 0. However, $\overline{Q}$ is already 0. Therefore, the circuit has become stable at $t_a + t_{pHL} + t_{pLH}$, with the following values: S = 1, R = 0, Q = 1, and $\overline{Q}$ = 0.

The state of the flip-flop circuit remains in that stable state until time t_b, when S changes from 1 to 0. Although the S input to NOR gate 1 goes to 0, there is still a 1 input from Q. Consequently, the output $\overline{Q}$ of NOR gate 1 remains 0, and again the flip-flop circuit is in a stable state with the following values: S = 0, R = 0, Q = 1, and $\overline{Q}$ = 0. Note that even though the S input has returned to 0, the flip-flop "remembers" the last set command by remaining in a Q = 1 state.

At time $t = t_c$, the input R changes from 0 to 1, causing the output Q of NOR gate 2 to change from 1 to 0 at $t_c + t_{pHL}$. Since Q was providing the only 1 input to NOR gate 1, this change in Q causes $\overline{Q}$ to change from 0 to 1 at $t_c + t_{pHL} + t_{pLH}$. This $\overline{Q}$ is fed back to NOR gate 2. However, the circuit is already stable with the following values: S = 0, R = 1, Q = 0, and $\overline{Q}$ = 1.

At $t = t_d$ and $t = t_e$, the changes at the R input have no effect on the state of the flip-flop circuit, and the outputs Q and $\overline{Q}$ remain at 0 and 1, respectively. At $t = t_f$, however, S changes from 0 to 1 to make both the S and R inputs equal to 1. Then, $\overline{Q}$ becomes 0, but Q also remains 0, and the flip-flop circuit is in a stable state with S = 1, R = 1, Q = 0, and $\overline{Q}$ = 0. Since it is not meaningful to have both Q and $\overline{Q}$

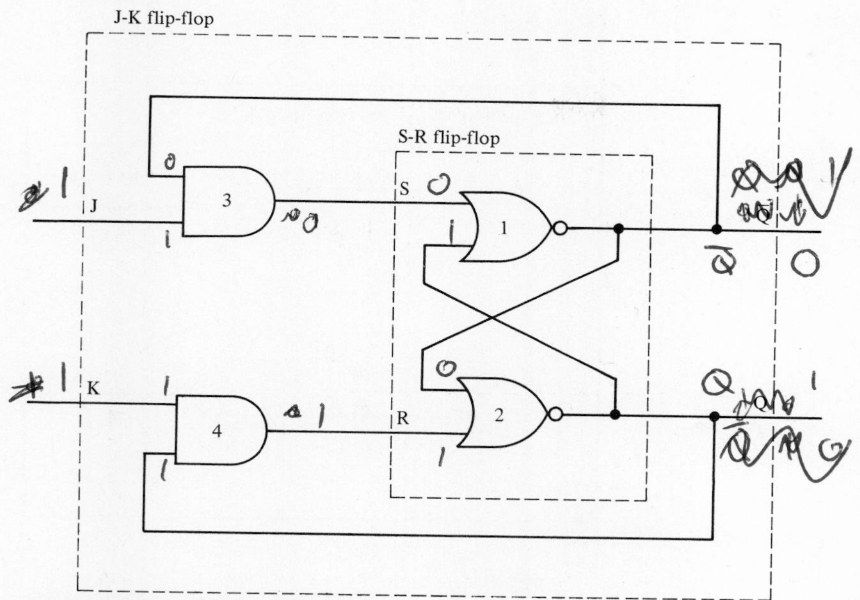

Figure 5.14 Unclocked J-K flip-flop.

equal to 0, this combination of inputs (S = R = 1) is not allowed for the unclocked S-R flip-flop.

The circuit for the unclocked S-R flip-flop has been considered in some detail here because it is the basic flip-flop circuit upon which the more complex flip-flops are built.

5.5.1 J-K Flip-Flops

The flip-flop circuit of Fig. 5.13(a) can be modified to allow both flip-flop inputs to be equal to 1. The result is the unclocked J-K flip-flop circuit shown in Fig. 5.14. As shown, two AND gates are added along with feedback connections from $\overline{Q}$ to AND gate 3 and from Q to AND gate 4. As can be verified with an analysis similar to that of the last section, these additions change only the operation for two 1 inputs. Specifically, for J = 1 and K = 1 applied, the output of either AND gate 3 or AND gate 4 becomes 1, the particular one depending on the values of Q and $\overline{Q}$. If Q = 1 (and $\overline{Q}$ = 0), then the output of AND gate 4 becomes 1 and so the next state Q^+ becomes 0. But, if $\overline{Q}$ = 1 (and Q = 0), then the output of AND gate 3 becomes 1 and so the next state Q^+ becomes 1. In other words, for the inputs J = K = 1, the state of the flip-flop is toggled.

This J-K flip-flop has two serious operational problems. First, to toggle the state of the flip-flop, the J and K inputs must be made 1 simultaneously. Otherwise, a *race condition* exists in which the flip-flop may be set to 1 if J is made 1 slightly earlier, or it may be cleared to 0 if K is made 1 slightly earlier, and then the flip-flop will toggle and return to its original state. Second, to toggle the state of the flip-flop just once, the J and K inputs have to be *pulsed* to 1, with the duration of the pulse being less than the propagation delay of the flip-flop. Otherwise, if J and K are 1 too long, the change in

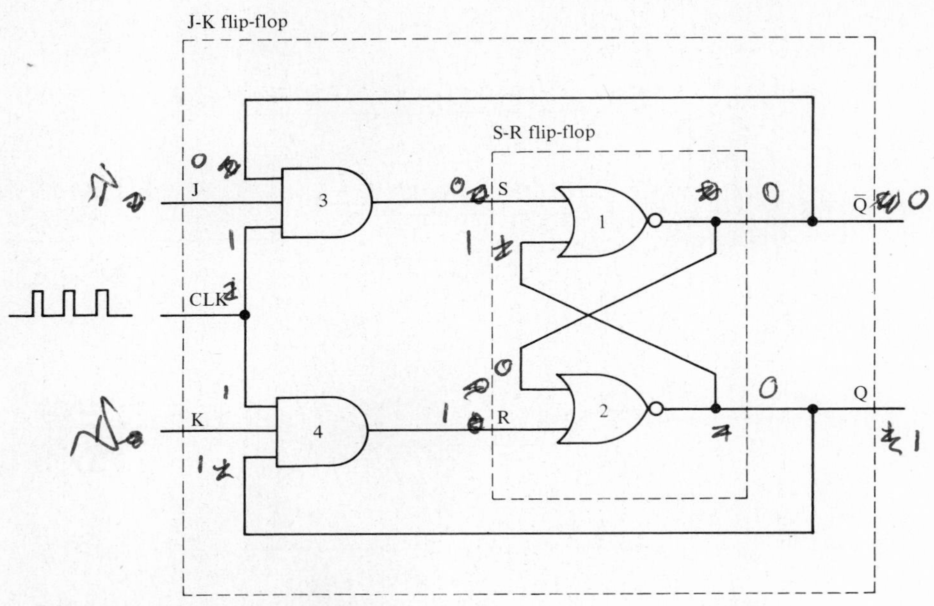

Figure 5.15 Clocked J-K flip-flop.

Q will propagate back to the AND gates while J and K are still 1, thereby causing another change of state.

The race condition problem can be eliminated by adding a clock input (CLK), as shown in Fig. 5.15. For this clocked J-K flip-flop, the changes in the J and K inputs will not affect the state of the flip-flop unless the clock input is 1. Consequently, provided that the changes in the J and K inputs are made before the clock pulse becomes 1, the timing of these inputs is not crucial as it was for the unclocked J-K flip-flop of Fig. 5.14. However, unlike the edge-triggered flip-flops that were presented earlier in this chapter, the duration of the clock pulse is still undesirably restricted to being less than the propagation delay of the flip-flop. To obtain an edge-triggered flip-flop, additional modifications to the basic flip-flop circuit are required. An example of such modifications is shown in Fig. 5.16, which is a circuit diagram for a 74'114 negative edge-triggered J-K flip-flop, with asynchronous preset and clear inputs. A detailed analysis of this flip-flop is quite cumbersome and so will not be presented here.

At this point of our study, we have gained enough of an understanding of the internal structure of flip-flops to use them effectively as circuit elements in digital systems. Therefore, we will not belabor the realization details, but instead will proceed to consider the applications of these devices.

5.6 COUNTERS

A digital *counter* is a sequential circuit element that counts through a prescribed sequence of numbers repeatedly. A counter may, for example, count through a 3-bit binary se-

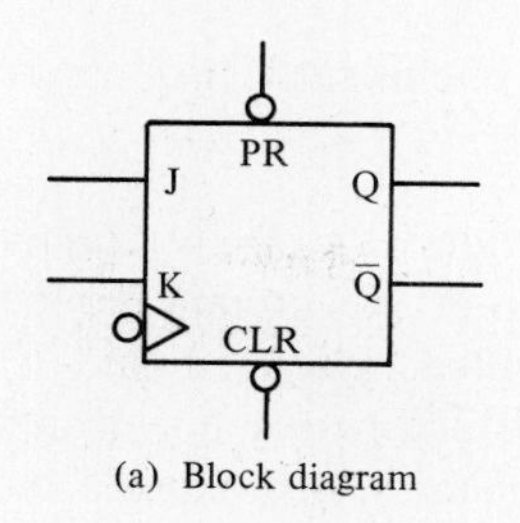

(a) Block diagram

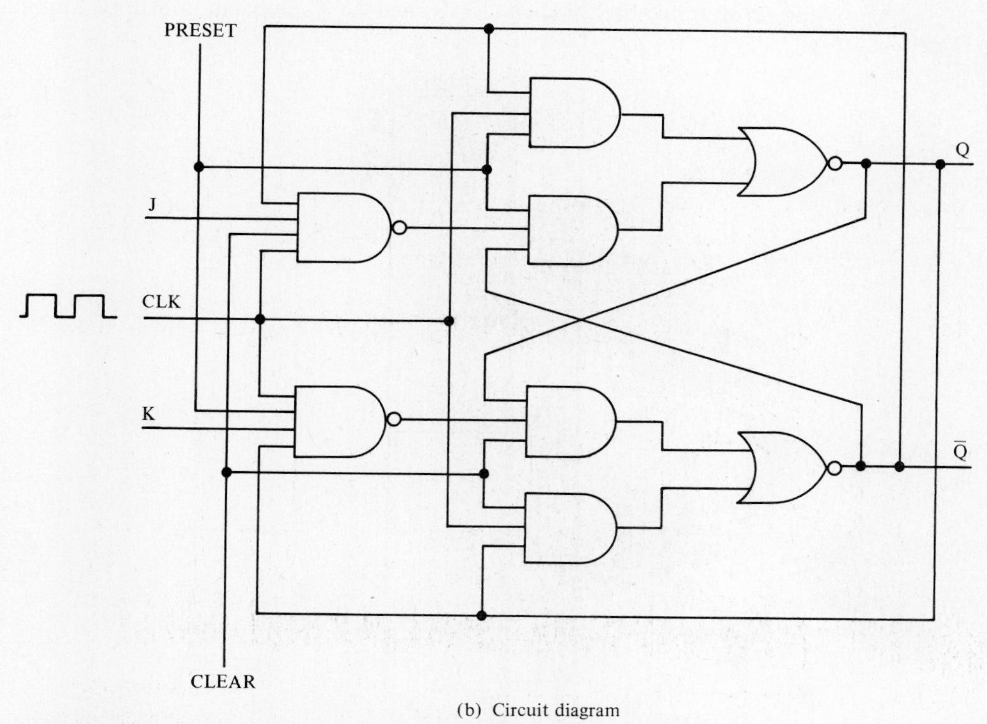

(b) Circuit diagram

Figure 5.16 74'114 negative edge-triggered clocked J-K flip-flop.

quence such as

. . . , 101, 110, 111, 000, 001, 010, 011, 100, 101, 110, 111, . . .

or a 4-bit arbitrary sequence such as

. . . , 0111, 1010, 1001, 1111, 0000, 0110, 0011, 0111, 1010, 1001, . . .

Counters are commonly used building blocks in the designs of digital systems.

This section begins with methods for the design and realization of counters using flip-flops. Then some commercially available realizations of counters, integrated as MSI circuit elements, will be described. Finally, the applications of MSI counters will be considered.

5.6.1 The Design and Realization of Synchronous Counters

In this subsection we will use examples to outline a systematic procedure for the design and realization of synchronous counters using flip-flops. We will design a specific synchronous 3-bit binary counter using D flip-flops first, and then using J-K flip-flops. Last, we will design and realize a synchronous counter with an arbitrary sequence.

Consider the synchronous 3-bit binary counter described in Fig. 5.17. As shown in the functional block diagram of Fig. 5.17(a), it has a single active-high clock input and three active-high outputs: C, B, and A. The operation of this 3-bit counter can be

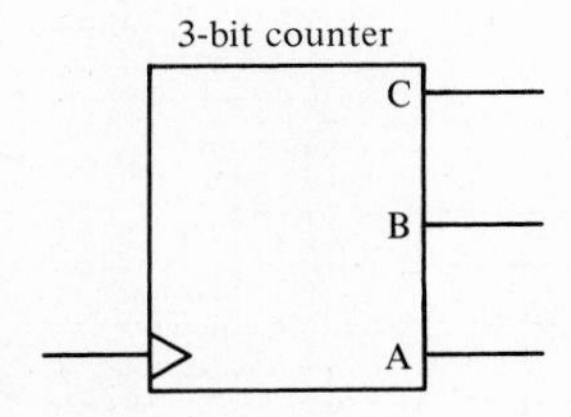

(a) Functional block diagram

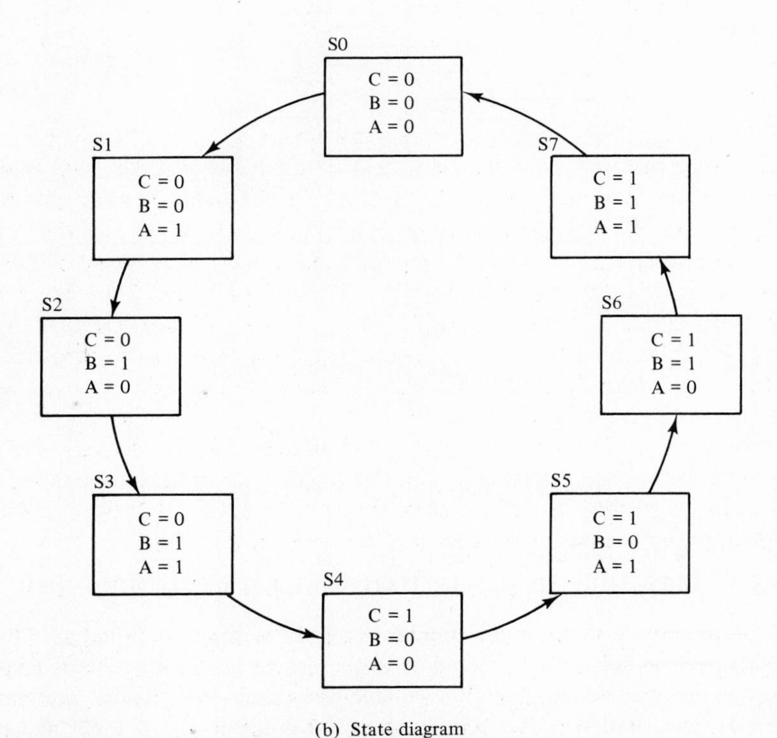

(b) State diagram

Figure 5.17 Definition of a 3-bit binary counter.

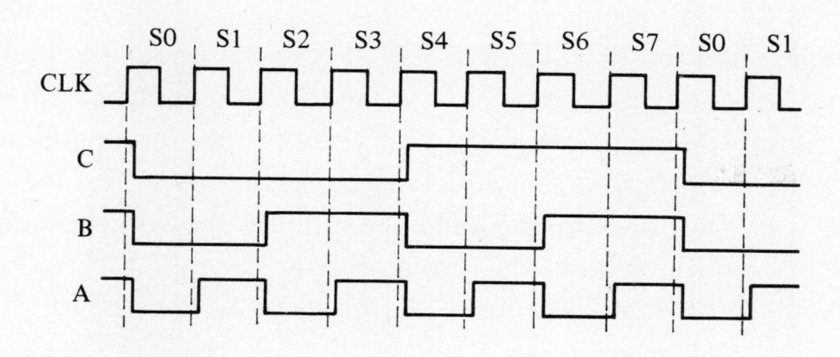

(c) Timing diagram with explicit propagation delays

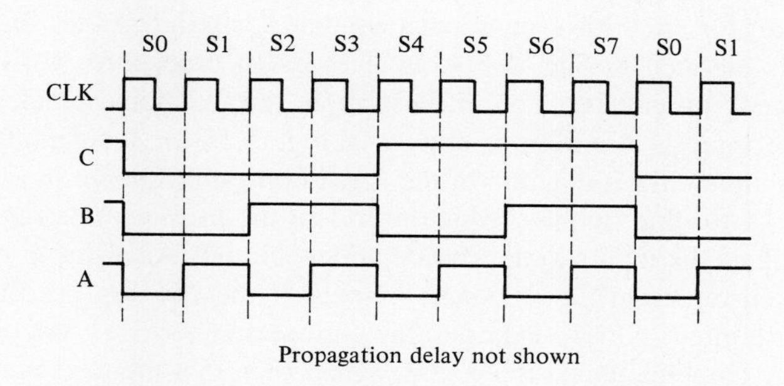

(d) Timing diagram without explicit propagation delays

Figure 5.17 (*cont.*)

described through the use of the *state diagram* shown in Fig. 5.17(b). In a state diagram, a *state* is represented by a rectangular box with the name of the state labeled on the outside. For the state diagram of Fig. 5.17(b), there are eight states: S0, S1, S2, S3, S4, S5, S6, and S7. The duration of each state is one clock cycle of the counter clock input. Conceptually, a state diagram describes the outputs of a sequential circuit element over time. Specifically, the state diagram of Fig. 5.17(b) specifies that if the present state outputs of the counter are C = 0, B = 0, A = 0, then the next state outputs should be C = 0, B = 0, A = 1. Continuing, if the present state outputs are C = 0, B = 0, A = 1, then the next state outputs should be C = 0, B = 1, A = 0. And if the present state outputs are C = 1, B = 0, A = 1, then the next state outputs should be C = 1, B = 1, A = 0, and so forth. Essentially, the state diagram of Fig. 5.17(b) specifies that the 3-bit binary counter is a modulo 8 counter that counts the number of active clock transitions. In other words, it counts repeatedly through the following sequence at a rate equal to the frequency of the counter clock input.

. . . , 101, 110, 111, 000, 001, 010, 011, 100, 101, 110, 111, 000, . . .

The operation of this 3-bit counter can also be described through the use of the timing diagram shown in Fig. 5.17(c). This timing diagram contains the same information as the state diagram of Fig. 5.17(b), but being more familiar, can be used to clarify and confirm the explanation of the state diagram. Note from Fig. 5.17(c) that the counter outputs change synchronously a propagation delay after the next active clock transition.

This is the reason that this type of counter is called a synchronous counter. Note also that the propagation delays are shown explicitly in the timing diagram, as they have been in all the previous timing diagrams. However, in practice, the propagation delays are usually not explicitly shown. Therefore, unless they are necessary to resolve ambiguities, the propagation delays will just be assumed and not explicitly shown in subsequent timing diagrams in this text. As an illustration, the timing diagram of Fig. 5.17(c) is reproduced in Fig. 5.17(d) without an explicit showing of the propagation delays.

Example 5.5 Design and Realization of a Synchronous 3-Bit Counter Using D Flip-Flops

The design procedure for an N-bit synchronous counter is similar to that for a gated D flip-flop, which was outlined in Sec. 5.3.4. In this design procedure, N flip-flops are required to realize an N-bit counter. For an illustration, the functional block diagram of a 3-bit counter with three D flip-flops is shown in Fig. 5.18(a). Incidentally, note its similarity to the functional block diagram of the gated D flip-flop shown in Fig. 5.10(a). In Fig. 5.18(a) the D flip-flop outputs, which represent the *present*-state outputs of the counter, are fed back as the inputs to the combinational circuit. Also, the combinational circuit outputs, which represent the *next*-state outputs of the counter, are the inputs to the D flip-flops. With inputs corresponding to the current number (i.e., the present-state counter outputs), the combinational circuit is designed such that the next number in the

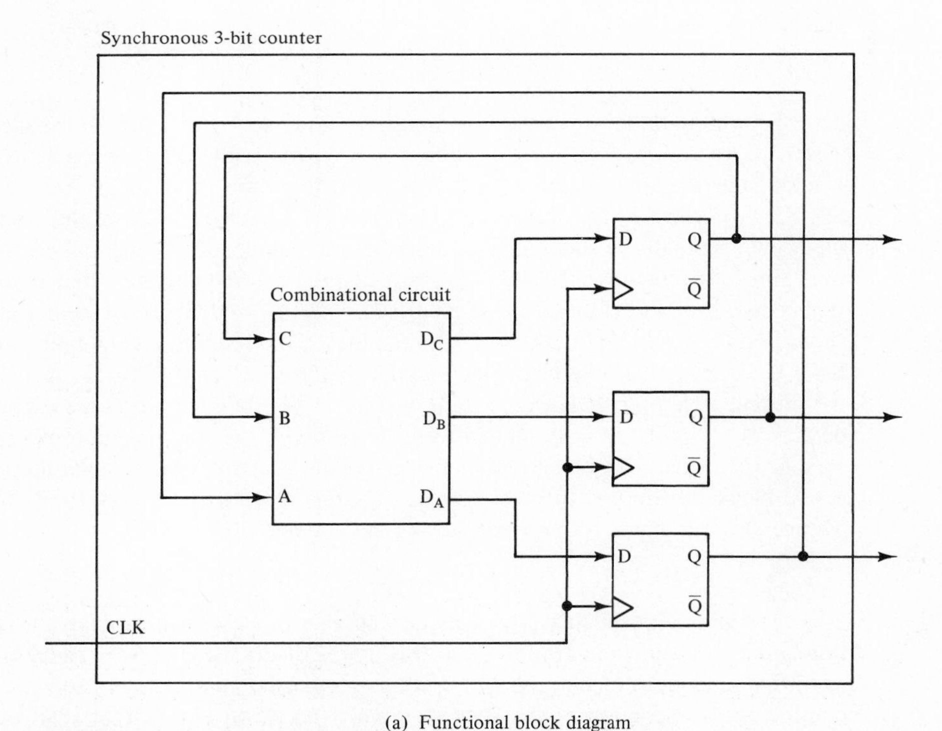

(a) Functional block diagram

Figure 5.18 Design and realization of a synchronous 3-bit counter.

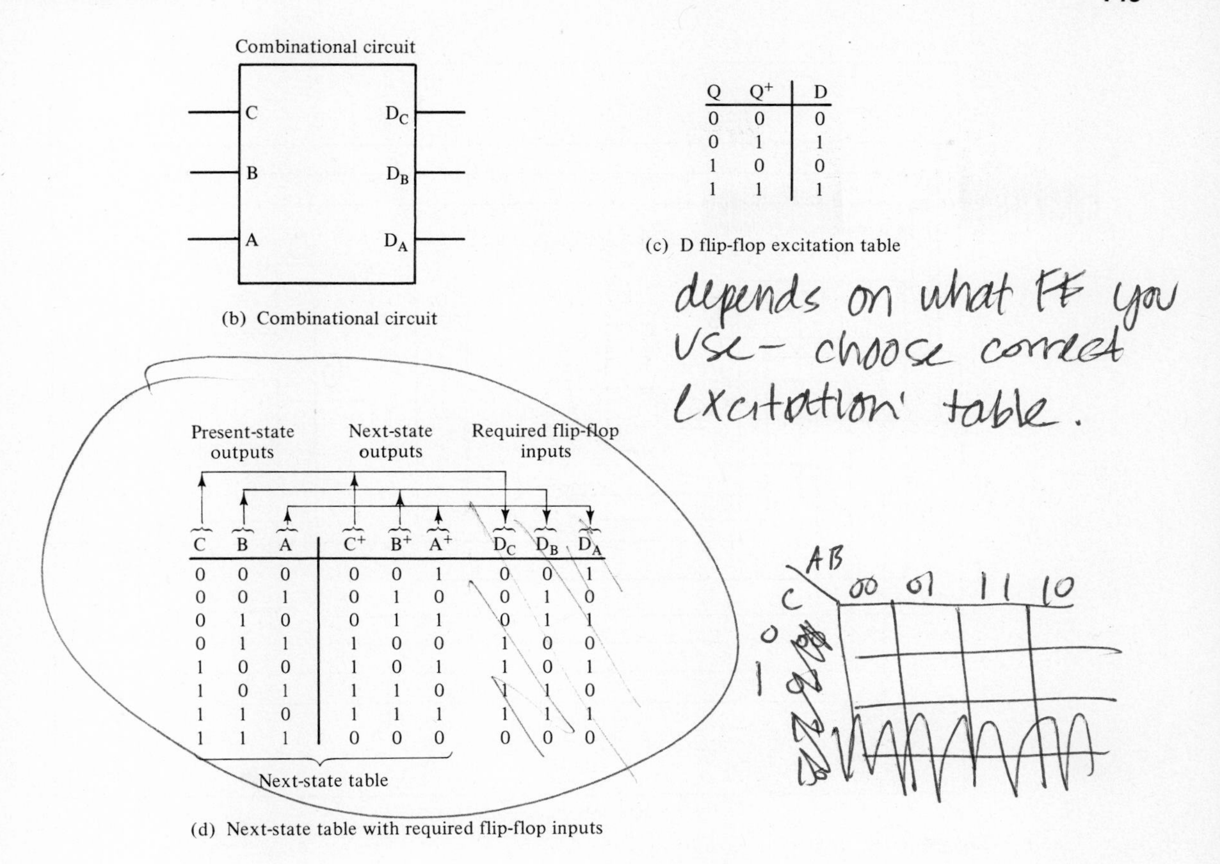

Q	Q^+	D
0	0	0
0	1	1
1	0	0
1	1	1

C	B	A	C^+	B^+	A^+	D_C	D_B	D_A
0	0	0	0	0	1	0	0	1
0	0	1	0	1	0	0	1	0
0	1	0	0	1	1	0	1	1
0	1	1	1	0	0	1	0	0
1	0	0	1	0	1	1	0	1
1	0	1	1	1	0	1	1	0
1	1	0	1	1	1	1	1	1
1	1	1	0	0	0	0	0	0

(b) Combinational circuit

(c) D flip-flop excitation table

(d) Next-state table with required flip-flop inputs

BA \ C	0	1
00	0	1
01	0	1
11	1	0
10	0	1

$D_C = AB\overline{C} + \overline{B}C + \overline{A}C$

BA \ C	0	1
00	0	0
01	1	1
11	0	0
10	1	1

$D_B = A\overline{B} + \overline{A}B = A \oplus B$

BA \ C	0	1
00	1	1
01	0	0
11	0	0
10	1	1

$D_A = \overline{A}$

(e) K-maps for flip-flop inputs

Figure 5.18 (*cont.*)

sequence (i.e., the next-state counter outputs) is outputted. Consequently, at the next active clock transition, this next number in the sequence is loaded into the D flip-flops and outputted. It is apparent that the problem of counter design is reduced to the design of the combinational circuit that will transform the present-state counter outputs into the appropriate inputs for the flip-flops. A systematic procedure for the design and realization of a synchronous counter will now be outlined.

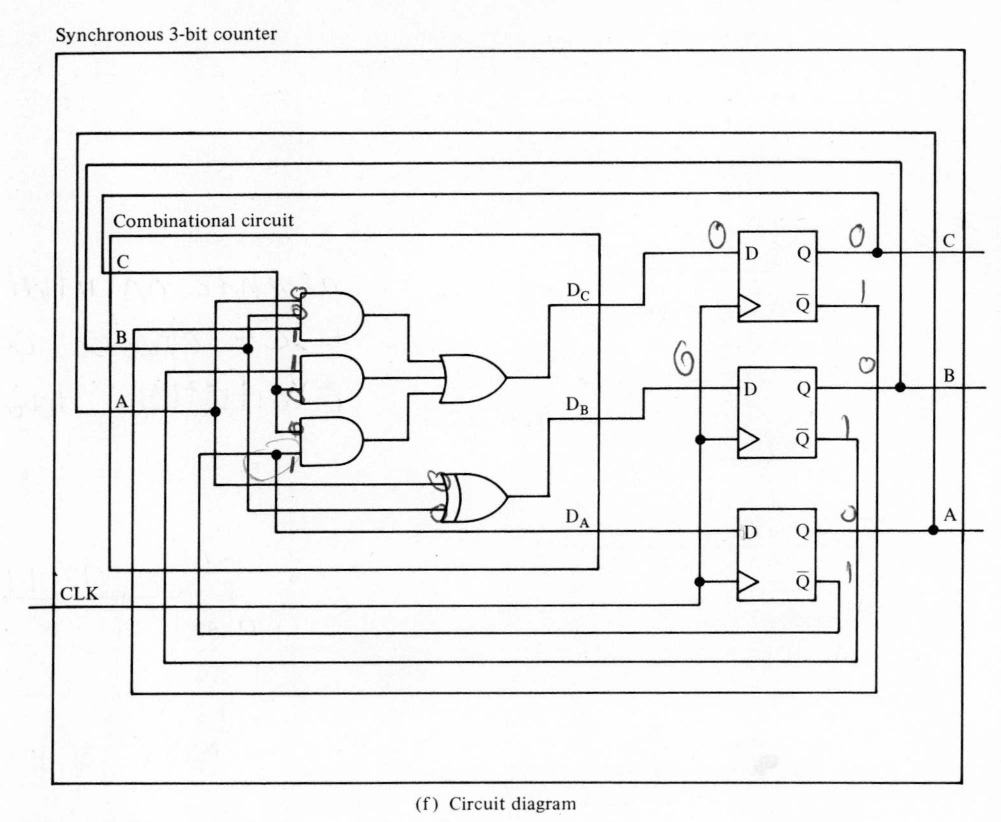

(f) Circuit diagram

Figure 5.18 (*cont.*)

Step 1: Using a state diagram, define the count sequence of the counter to be designed. For this example, the count sequence for the 3-bit counter is defined by the state diagram shown in Fig. 5.17(b).

Step 2: Determine the functional block diagram of the *N*-bit counter to be designed. It consists of *N* flip-flops and a combinational circuit for generating the valid inputs for the flip-flops. For this example, the functional block diagram of the 3-bit counter is shown in Fig. 5.18(a).

Step 3: Determine the functional block diagram of the combinational circuit. It can be readily extracted from the functional block diagram of the counter obtained in step 2. For this example, the functional block diagram for the combinational circuit is shown in Fig. 5.18(b). The inputs are the present-state counter outputs C, B, A, which are fed back from the outputs of the flip-flops. The outputs of the combinational circuit are D_C, D_B, D_A, which correspond to the respective D flip-flop inputs.

Step 4: Using the excitation table for the selected flip-flop type, determine the truth table for the combinational circuit. For this example, the D flip-flop excitation table of Fig. 5.18(c) is used.

Step 4(a): Determine the next-state table for the counter. The next-state table specifies, in a tabular form, the next-state outputs of the counter corresponding to the present-state outputs. Since the next-state table contains the same information as the state diagram, it can be easily derived from this diagram. For the 3-bit counter, the next-state table, which is shown in Fig. 5.18(d), is derived from the state diagram of Fig. 5.17(b).

Step 4(b): Determine the required flip-flop inputs. Once the next-state table is determined, we can readily obtain the required flip-flop inputs by using the excitation table of the selected flip-flop type. For this example, the pertinent excitation table is that of Fig. 5.18(c) for the D flip-flop. With it we can determine the D inputs required to produce the desired D flip-flop output transitions. As indicated at the top of the table in Fig. 5.18(d), the values of D_C are derived from those of C and C^+, the values of D_B are derived from those of B and B^+, and the values of D_A from those of A and A^+.

Step 5: Realize the combinational circuit. Once the truth table for the combinational circuit is determined from step 4, then the realization of this circuit is straightforward, using the techniques presented in Chapters 2 and 3. The resultant circuit diagram is shown in Fig. 5.18(f).

The operation of this counter circuit is verifiable from an exhaustive analysis of the circuit. Specifically, for each of the eight possible present-state outputs, the next-state counter output (i.e., the next number in the sequence) should agree with the state diagram of Fig. 5.17(b). ■ ■

Example 5.6 Design and Realization of a Synchronous 3-Bit Counter Using J-K Flip-Flops

In this example, we will redesign the 3-bit counter, using J-K instead of D flip-flops. The procedure outlined in Example 5.5 still applies, but in step 4 the excitation table of a J-K flip-flop must be used.

Step 1: Using a state diagram, define the count sequence of the counter to be designed. Since we are designing the same 3-bit counter, the state diagram of Fig. 5.17(b) still applies.

Step 2: Determine the functional block diagram of the 3-bit counter. As shown in Fig. 5.19(a), it consists of three J-K flip-flops and a combinational circuit for generating the valid inputs for the J-K flip-flops.

Step 3: Determine the functional block diagram for the combinational circuit. The functional block diagram is shown in Fig. 5.19(b). The inputs are the present-state counter inputs C, B, A, and the outputs are J_C, K_C, J_B, K_B, J_A, K_A corresponding to the J and K inputs of the three flip-flops.

Step 4: Using the excitation table for the selected flip-flop, determine the truth table for the combinational circuit. In this case, we use the excitation table for the J-K flip-flop shown in Fig. 5.19(c).

Step 4(a): Determine the next-state table. Since we are designing the same 3-bit counter, the next-state table of Fig. 5.18(d) again applies. It is reproduced in Fig. 5.19(d).

Step 4(b): Determine the required flip-flop inputs. Once the next-state table is defined, we can readily determine the required flip-flop inputs by using the J-K flip-flop excitation table of Fig. 5.19(c). As indicated at the top of the table in Fig. 5.19(d), the

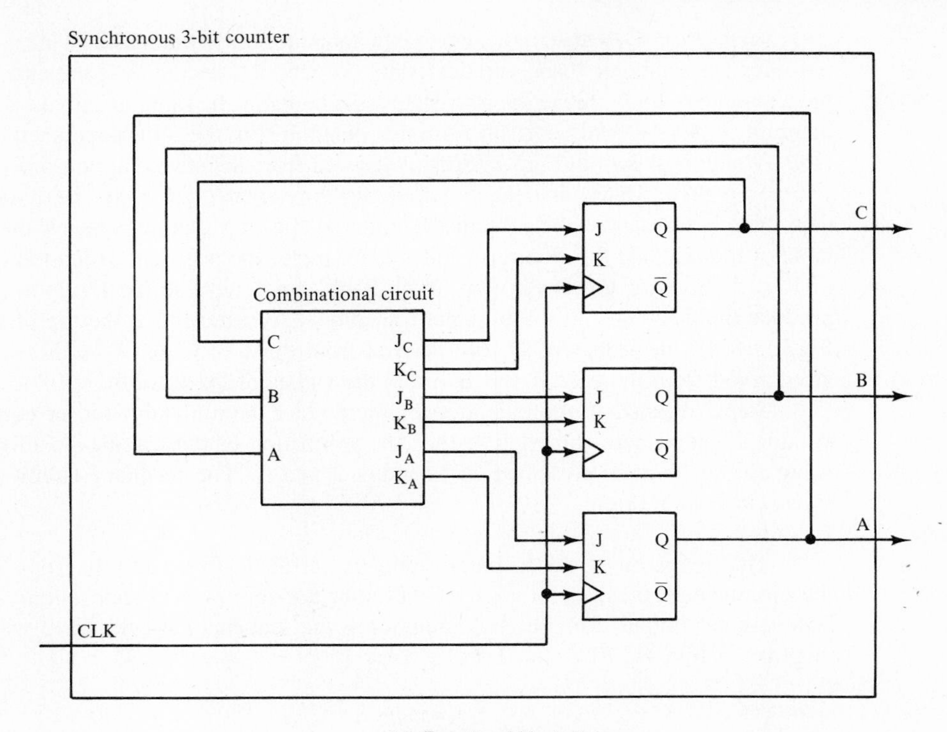

(a) Functional block diagram

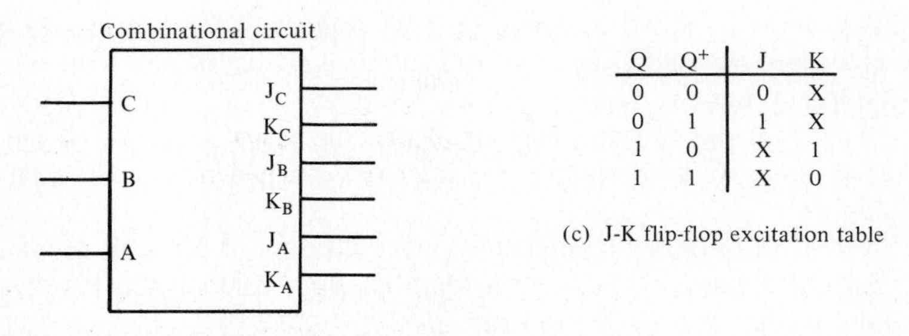

(b) Combinational circuit

Q	Q^+	J	K
0	0	0	X
0	1	1	X
1	0	X	1
1	1	X	0

(c) J-K flip-flop excitation table

Figure 5.19 Design and realization of a synchronous 3-bit counter using J-K flip-flops.

values for J_C and K_C are derived from those of C and C^+, the values of J_B and K_B are derived from those of B and B^+, and the values of J_A and K_A from those of A and A^+.

Step 5: Realize the combinational circuit. Once the truth table for the combinational circuit is determined from step 4, the realization is easy to obtain with the techniques of Chapters 2 and 3. The resultant logic equations for the J-K flip-flops are given in Fig. 5.19(e). We will leave it to the reader to complete the drawing and verify the circuit diagram. ■ ■

Present-state outputs			Next-state outputs			Required flip-flop inputs					
C	B	A	C^+	B^+	A^+	J_C	K_C	J_B	K_B	J_A	K_A
0	0	0	0	0	1	0	X	0	X	1	X
0	0	1	0	1	0	0	X	1	X	X	1
0	1	0	0	1	1	0	X	X	0	1	X
0	1	1	1	0	0	1	X	X	1	X	1
1	0	0	1	0	1	X	0	0	X	1	X
1	0	1	1	1	0	X	0	1	X	X	1
1	1	0	1	1	1	X	0	X	0	1	X
1	1	1	0	0	0	X	1	X	1	X	1

Next-state table

(d) Next-state table with required flip-flop inputs

$$J_C = K_C = AB$$
$$J_B = K_B = A$$
$$J_A = K_A = 1$$

(e) Logic equations for flip-flop inputs

Figure 5.19 (*cont.*)

Example 5.7 The Design and Realization of a Synchronous Counter with an Arbitrary Sequence

Counters with an arbitrary counting sequence can also be designed with the above procedure. In this example, a counter is to be designed to count through the following sequence:

. . . , 101, 001, 000, 010, 110, 100, 101, 001, 000, 010, . . .

Additionally, there is an external input CLEAR that when true will synchronously clear the counter. In other words, when CLEAR is made true, the normal count sequence is interrupted. And, at the next active clock transition, the counter outputs are cleared to 000. Then, normal counting will continue from this 000. An example of this clearing is shown in the following sequence:

. . . , 101, 001, 000, 010, 110, 100, 000, 010, 110, 100, 101, 001, 000, . . .
↑
CLEAR becomes true here

Step 1: Using a state diagram, define the count sequence of the counter that is to be designed. For this example, the count sequence for the counter is defined by the state diagram shown in Fig. 5.20(a). Note that when the CLEAR signal is false, the counter counts through the normal sequence. At any time, however, that CLEAR becomes true, the next state is S0.

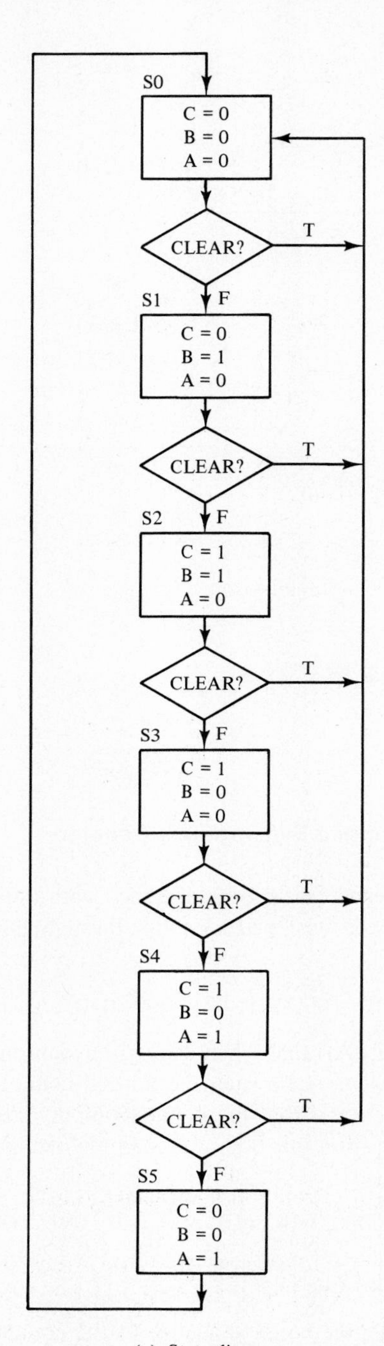

(a) State diagram

Figure 5.20 Counter for Example 5.7.

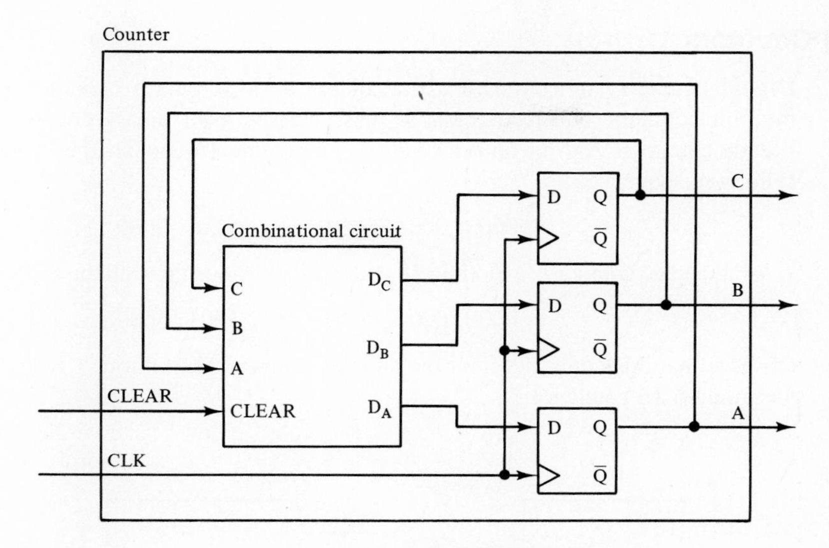

(b) Functional block diagram of counter

CLEAR	C	B	A	C^+	B^+	A^+	D_C	D_B	D_A
0	0	0	0	0	1	0	0	1	0
0	0	0	1	0	0	0	0	0	0
0	0	1	0	1	1	0	1	1	0
0	0	1	1	X	X	X	X	X	X
0	1	0	0	1	0	1	1	0	1
0	1	0	1	0	0	1	0	0	1
0	1	1	0	1	0	0	1	0	0
0	1	1	1	X	X	X	X	X	X
1	X	X	X	0	0	0	0	0	0

(c) Next-state table with required flip-flop inputs

$$D_C = \overline{\text{CLEAR}} \cdot B + \overline{\text{CLEAR}} \cdot C \cdot \overline{A}$$
$$D_B = \overline{\text{CLEAR}} \cdot \overline{C} \cdot \overline{A}$$
$$D_A = \overline{\text{CLEAR}} \cdot C \cdot \overline{B}$$

(d) Logic equations for flip-flop inputs

Figure 5.20 (*cont.*)

Step 2: Determine the functional block diagram of the counter to be designed. In this example, we will arbitrarily decide to use D flip-flops. Since it is a 3-bit counter, three D flip-flops are required, as is shown in Fig. 5.20(b).

Step 3: Determine the functional block diagram for the combinational circuit. This circuit is also shown in Fig. 5.20(b).

Step 4: Using the excitation table for the selected flip-flop, determine the truth table for the combinational circuit. For the selected D flip-flop, the result is shown in Fig. 5.20(c). Note that the count sequence does not include the binary numbers 011 and 111. Consequently, in the next-state table shown in Fig. 5.20(c), the entries corresponding to those two numbers are don't cares.

Step 5: Realize the combinational circuit. The resulting logic equations for the flip-flop inputs are shown in Fig. 5.20(d). ■■

5.6.2 MSI Counters

Digital counters are commercially available as MSI circuit elements in the form of multibit counters. Most common of these are the 4-bit binary counters and the 4-bit decade counters. A 4-bit binary counter is a modulo 16 counter that counts through the following sequence:

$$\ldots, 0000, 0001, 0010, \ldots, 1111, 0000, \ldots$$

A 4-bit decade counter is a modulo 10 counter that counts through the following sequence:

$$\ldots, 0000, 0001, 0010, \ldots, 1001, 0000, \ldots$$

Observe that this count does not include the numbers 1010 through 1111 of the count of the modulo 16 counter.

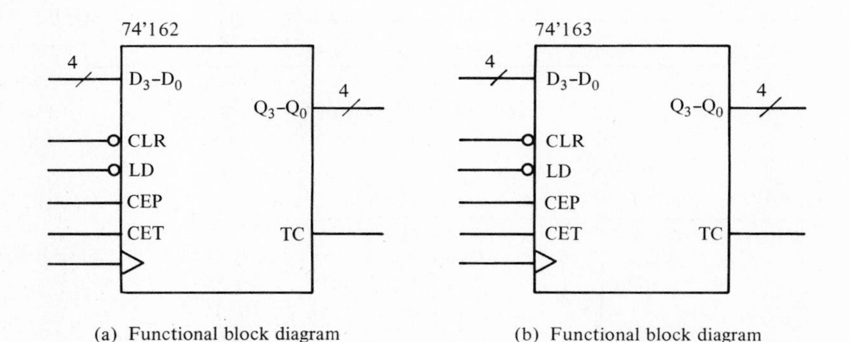

(a) Functional block diagram for the 74'162

(b) Functional block diagram for the 74'163

CLR	LD	CEP	CET	Function
0	0	1	1	Normal count
0	1	X	X	Parallel load
1	X	X	X	Clear

For the 74'162

$$TC = CET \cdot Q_3 \cdot \overline{Q}_2 \cdot \overline{Q}_1 \cdot Q_0$$

For the 74'163

$$TC = CET \cdot Q_3 \cdot Q_2 \cdot Q_1 \cdot Q_0$$

(c) Functional descriptions

$$\cdots, 1111, 0000, 0001, 0010, 1011, 1100, 1101, 1110, 1111, 0000, \cdots$$

↑ during this clock cycle $LD = 1$ and $D_3\text{-}D_0 = 1011$

(d) Example of the LD input function

$$\cdots, 1111, 0000, 0001, 0010, 0000, 0001, 0010, 0011, 0100, \cdots$$

↑ during this clock cycle $CLR = 1$

(e) Example of the CLR input function

Figure 5.21 MSI counters.

The 74'163 is an example of a synchronous 4-bit binary counter, and the 74'162 is an example of a synchronous 4-bit decade counter. The functional block diagrams and the functional descriptions of both counters are shown in Fig. 5.21. Except for the count sequence, both counters operate in an identical manner. Specifically, for a count in a normal manner, both the count enable parallel input (CEP) and the count enable trickle input (CET) have to be true (H). Also, the parallel load input (LD) and the clear input (CLR) must be false (H).

There are other modes of operation. If the LD input is true (L) and the CLR input is false (H), then any 4-bit number applied at the D_3–D_0 inputs is synchronously loaded into the counter at the next active clock transition. So, the normal count sequence is interrupted. Further, when the normal counting resumes, it will continue from the loaded number, as shown in the example in Fig. 5.21(d). Finally, if the CLR input is true (L), then the counter is synchronously cleared to 0000 at the next active clock transition. In other words, the normal count sequence is again interrupted. And when normal counting resumes, it will continue from 0000, as shown in the example in Fig. 5.21(e). Incidentally, since in this example, as well as that of Fig. 5.21(d), the count is shown as exceeding 1001, the specific counts shown must be from the 74'163 binary counter.

Both the 74'162 and the 74'163 have a terminal count (TC) output. This output becomes true (H) at the active clock transition at which the terminal count (1001 for the 74'162 and 1111 for the 74'163) of the counter is reached, resulting in a positive-going pulse that lasts for one clock period. This pulse can be applied as a control signal to other circuit elements. Most often this TC output is used in the cascading of 4-bit counters to obtain a counter of greater length. As an illustration, in Fig. 5.22 an 8-bit binary counter is shown as constructed from two 74'163 4-bit counters without any additional circuitry. Note that the TC output from the low-order 4-bit counter controls the operation of the high-order 4-bit counter so that this high-order counter counts only once for every 16 clock cycles. In general, a number of 4-bit counters can be cascaded in this manner to produce a binary counter of any reasonable length.

5.7 REGISTERS

A register is a group of flip-flops that are used for data storage and perhaps also for performing some function on the stored data. Strictly speaking, the counters of the last section are also registers in which the function performed is counting through a prescribed sequence. In this section, we will consider two common types of registers: storage registers and shift registers.

5.7.1 Storage Registers

A storage register is a group of flip-flops that are used simply for data storage. The simplest storage register consists of several flip-flops with common clock inputs. An example of a 4-bit storage register is shown in Fig. 5.23. It consists of four D flip-flops in which 4 bits of data (D_3–D_0) are loaded into the storage register at each active clock

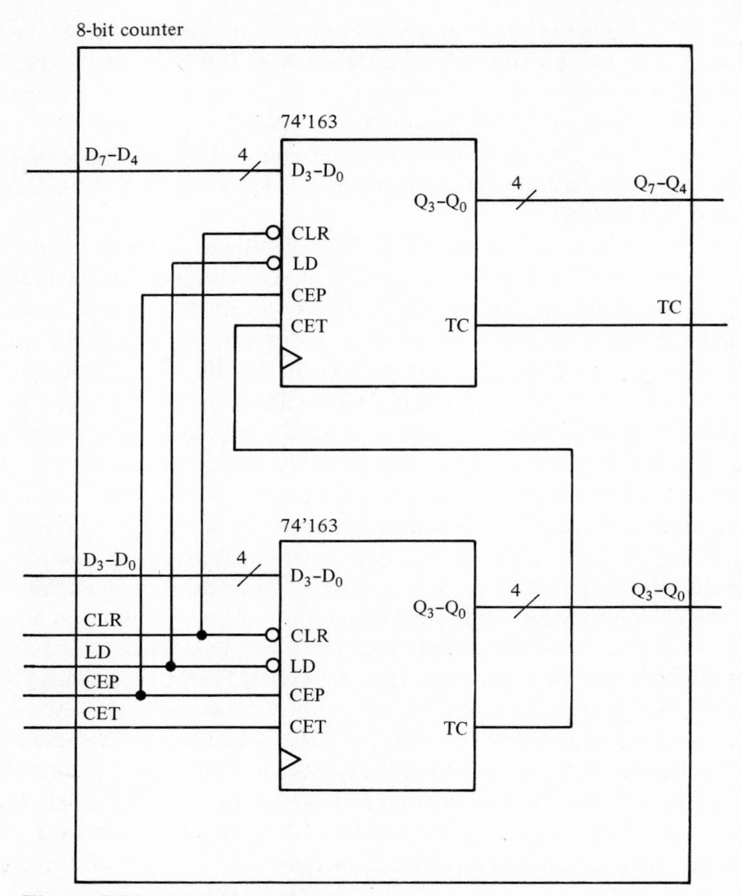

Figure 5.22 8-bit binary counter.

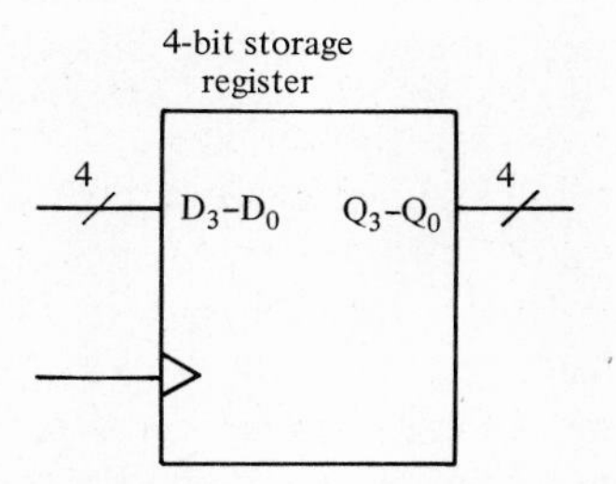

(a) Functional block diagram

Figure 5.23 4-bit storage register.

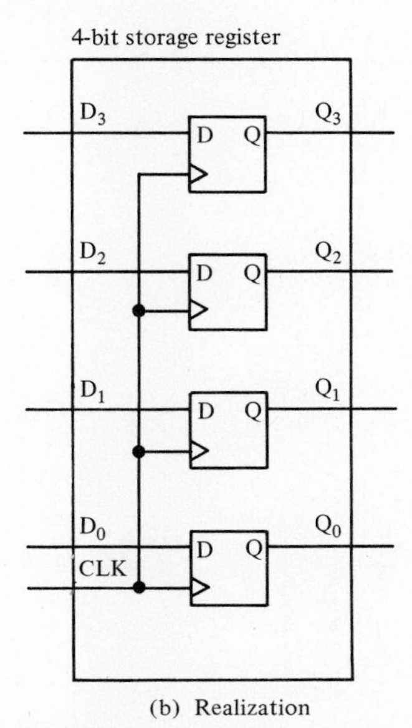

(b) Realization

Figure 5.23 (*cont.*)

transition. With this type of storage register the data is stored in the register until the next active clock transition.

Storage registers are commercially available as MSI circuit elements in various forms, with various features. Three examples are shown in Fig. 5.24. The 74'273 storage register of Fig. 5.24(a) has an asynchronous master reset input MR. As shown in the accompanying voltage table, if the MR input is false (H), then the 74'273 functions as a normal storage register. If, however, the MR input is true (L), then the 8-bit register is *asynchronously* cleared to 0.

The difference between an asynchronous operation and a synchronous operation is illustrated in Fig. 5.25. An asynchronous operation is performed ''immediately'' after the control input is applied, as shown in Figs. 5.25(a) and (b). In contrast, a synchronous operation is synchronized by a clock input, and so the operation is performed only if the control signal is true at the active clock transition, as is illustrated in Figs. 5.25(c) and (d). In Fig. 5.25(d) note that the pulse at the MR input has no effect on the synchronous clear operation.

In Fig. 5.24(b) is illustrated a 74'378 storage register with an enable input E. As shown in the accompanying voltage table, the 6-bit storage register is loaded with new data (D_5–D_0) only if the enable input is true (L). Otherwise, the existing data will be stored in the register indefinitely. A storage register is most useful if it has an enable input, because then we can connect the system clock signal directly to the clock input of the register and a control signal to the enable input.

Shown in Fig. 5.24(c) is a 74'163 counter, arranged to be a 4-bit storage register with a *synchronous* master reset input (CLR) and an enable input (LD). This storage feature is obtained by disabling the count enable (CEP = CET = false) and not using the terminal count (TC) output.

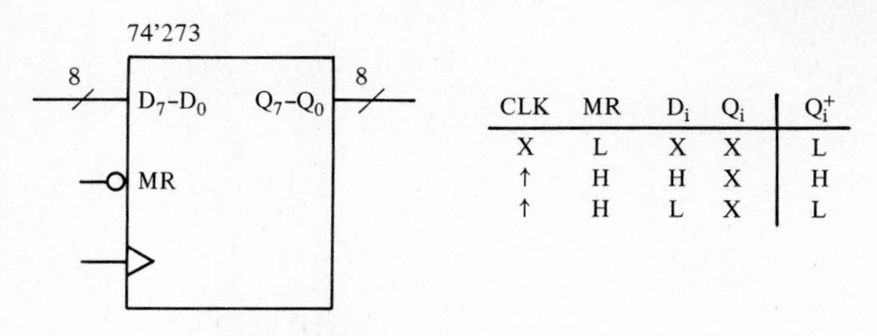

CLK	MR	D_i	Q_i	Q_i^+
X	L	X	X	L
↑	H	H	X	H
↑	H	L	X	L

(a) 74'273 8-bit storage register

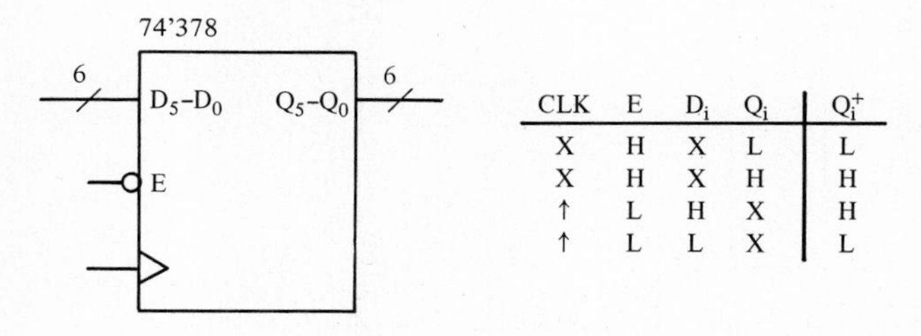

CLK	E	D_i	Q_i	Q_i^+
X	H	X	L	L
X	H	X	H	H
↑	L	H	X	H
↑	L	L	X	L

(b) 74'378 6-bit storage register

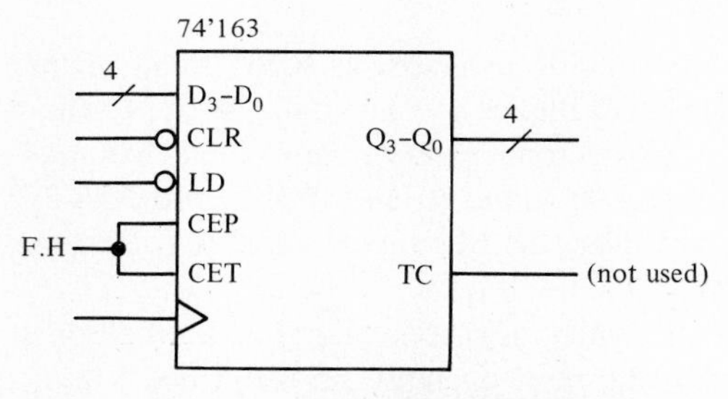

(c) 74'163 4-bit storage register

Figure 5.24 Commercially available storage registers.

5.7.2 Shift Registers

A shift register is a register in which the stored data can be shifted to the left or to the right. The simplest shift register consists of flip-flops connected as shown in Fig. 5.26(b). For this shift register, the contents of each flip-flop are loaded into the adjacent one on the right at each active clock transition. In other words, the stored contents are shifted to the right by one flip-flop. In this process, the old contents of the rightmost flip-flop are lost, and the new contents of the leftmost flip-flop are the data applied at DIN.

Shift registers are commercially available as MSI circuit elements in various forms and with various features. An example of one with many features is the 74'194 4-bit

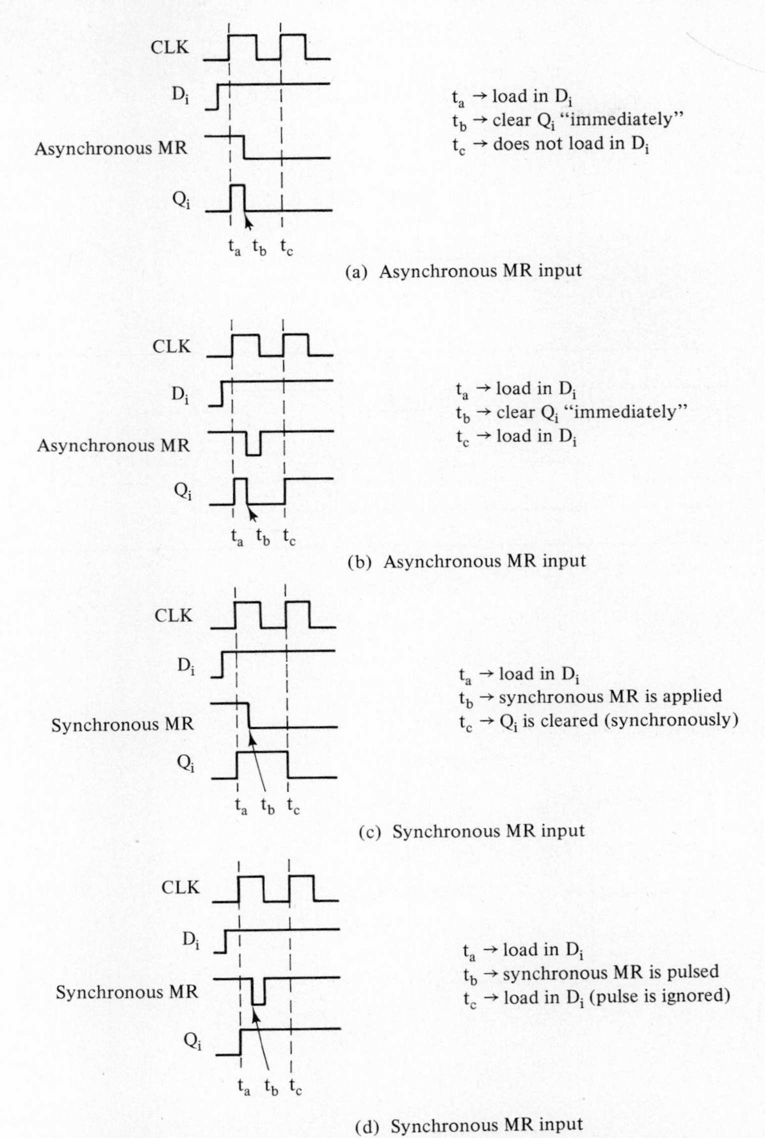

Figure 5.25 Asynchronous and synchronous control inputs.

bidirectional universal shift register, which is illustrated in Fig. 5.27. As shown in the voltage table, it performs five major operations, all synchronously. If the CLR input is true (L), then regardless of the other inputs, the shift register is synchronously cleared. But if the CLR input is false (H), the operation of the shift register depends on the mode control inputs S_1 and S_0. If S_1S_0 = LL, then the shift register retains the existing data and functions essentially as a storage register. If S_1S_0 = LH, then the contents of the shift register are shifted to the right by 1 bit at the next active clock transition, with the

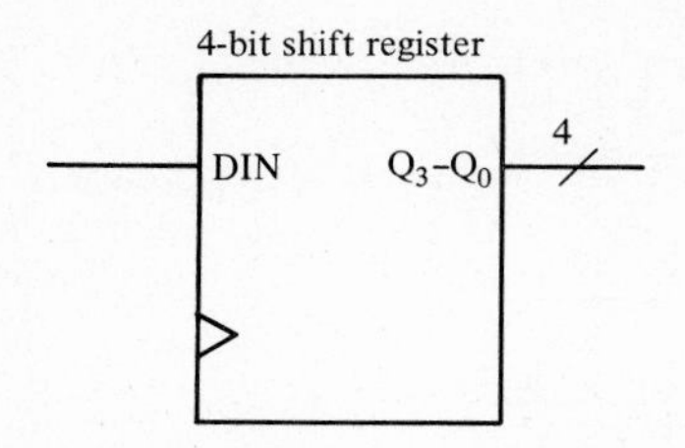

(a) Functional block diagram

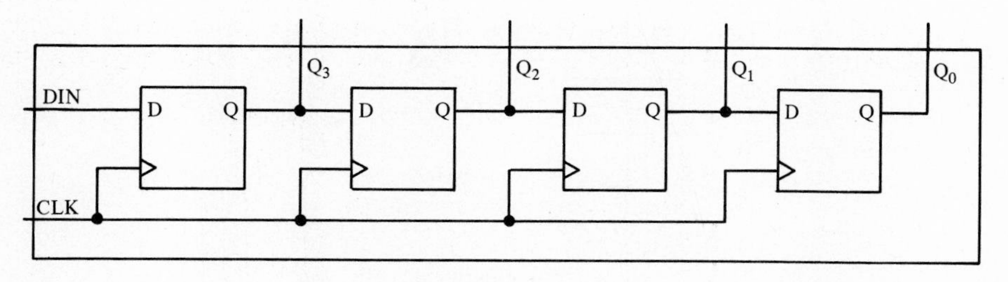

(b) Realization

Figure 5.26 4-bit shift register.

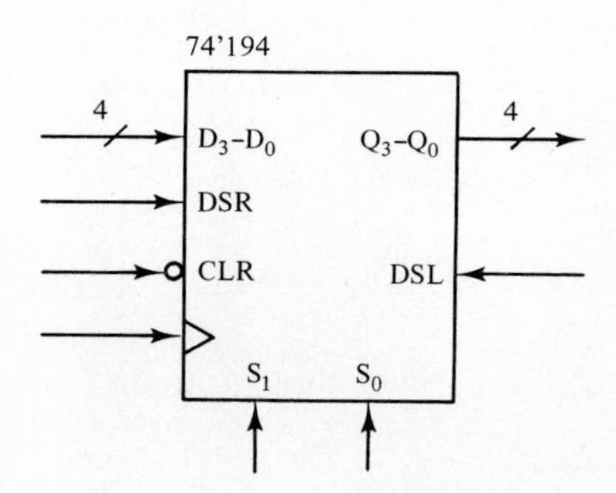

(a) Functional block diagram

Operation	CLK	CLR	S_1	S_0	DSR	DSL	Q_3^+	Q_2^+	Q_1^+	Q_0^+
Clear	↑	L	X	X	X	X	L	L	L	L
Hold	↑	H	L	L	X	X	Q_3	Q_2	Q_1	Q_0
Shift right	↑	H	L	H	L	X	L	Q_3	Q_2	Q_1
	↑	H	L	H	H	X	H	Q_3	Q_2	Q_1
Shift left	↑	H	H	L	X	L	Q_2	Q_1	Q_0	L
	↑	H	H	L	X	H	Q_2	Q_1	Q_0	H
Parallel load	↑	H	H	H	X	X	D_3	D_2	D_1	D_0

(b) Voltage table

Figure 5.27 74'194 bidirectional universal shift register.

leftmost flip-flop (Q_3) receiving the data applied at DSR. If S_1S_0 = HL, then the contents are shifted to the left by 1 bit, with the rightmost flip-flop (Q_0) receiving the data applied at DSL. Finally, if S_1S_0 = HH, then the data applied at D_3–D_0 is parallel loaded into the shift register, and there is no shifting.

5.8 SYNCHRONOUS VERSUS ASYNCHRONOUS DESIGNS

The difference between asynchronous and synchronous operations was illustrated by the examples in Fig. 5.25. Generally, in the design of digital circuits, synchronous operations using synchronous sequential circuit elements are preferred. To understand why, consider the simple circuit shown in Fig. 5.28(a), and assume that the initial values for the inputs

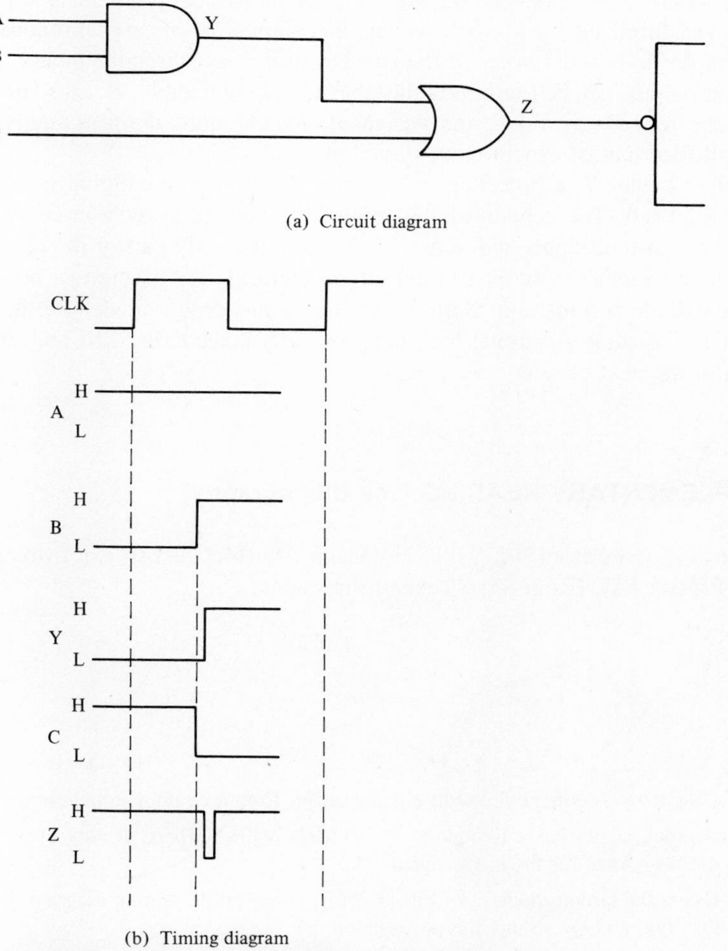

Figure 5.28 Glitch resulting from unequal path delays.

are A = H, B = L, and C = H. For these inputs, the output Z, of course, should be H, as shown in Fig. 5.28(b). Now if B is changed from L to H and at the same time C is changed from H to L, then logically the output Z should remain H. Unfortunately, as shown in the timing diagram of Fig. 5.28(b), these changes of inputs cause a momentarily unwanted glitch to occur at the output Z. This is a result of the unequal path delays through the combinational circuit between B to Z and C to Z. In other words, a *race condition* exists between the two signals. The change in C arrives at Z first, after one gate delay, causing Z to become L momentarily. Finally, after two gate delays, the change in B arrives and the Z output settles down to its correct H value. The circuit becomes stable at two gate delays after the changes. But if the Z output is used as an input to an *asynchronous* control input such as a clear input, then the circuit element would be cleared erroneously by the glitch.

In general, asynchronous operations and asynchronous circuit elements are to be avoided in the design of a digital circuit. In addition to erroneous operations such as the one just illustrated, there can be other serious operational problems and restrictions. Not being synchronized by a clock signal, the outputs of an asynchronous digital circuit element depend on the *order* of the changes in the asynchronous inputs. As a result, the element outputs can be transiently unstable and unpredictable. Because of these and other problems related to timing, the design of asynchronous digital circuits is much more difficult than that of synchronous digital circuits.

In Chapter 7, a procedure for designing synchronous digital circuits is described and formalized. The general idea is to avoid the use of asynchronous operations and to have the transient inputs and outputs isolated in the early part of the clock cycle. In this manner, all inputs to the sequential circuit elements are assured of being stable at the next active clock transition. Before considering this design procedure, however, we will complete the study of digital building blocks by considering LSI and other circuit elements in the next chapter.

SUPPLEMENTARY READING (see Bibliography)

[Bartee 85], [Blakeslee 79], [Hill 81], [Mano 79], [McCluskey 75], [Motorola], [Peatman 80], [Prosser 87], [Roth 85], [Texas Instruments]

PROBLEMS

5.1. How does a sequential circuit element differ from a combinational circuit element?

5.2. A clock signal has a frequency of 5.0 MHz and a 30 percent duty cycle. Draw and label the waveform for the clock signal.

5.3. Given the circuit diagram of Fig. 5.29(a), complete the timing diagram of Fig. 5.29(b) for Q_A, Q_B, and Q_C. Show the propagation delays t_{pHL} and t_{pLH}.

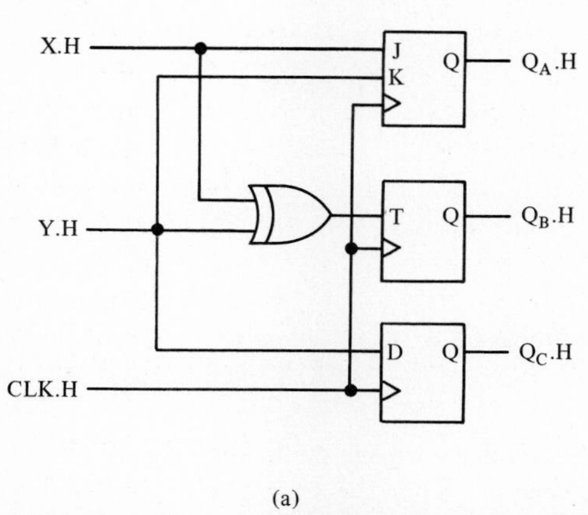

(a)

(b)

Figure 5.29 Circuit diagram and timing diagram for Problem 5.3.

5.4. Figure 5.30(a) shows a J-K flip-flop with an active-low clock input.

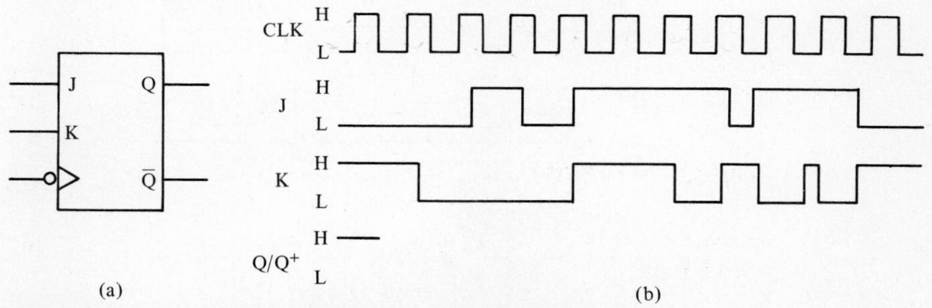

(a) (b)

Figure 5.30 J-K flip-flop and timing diagram for Problem 5.4.

(a) Complete the timing diagram of Fig. 5.30(b) and compare it with the one shown in Fig. 5.2(c). Show the propagation delays t_{pHL} and t_{pLH}.

(b) Redraw the timing diagram without showing the propagation delays.

5.5. Recall that binary subtraction can be performed by adding the 2s-complement form of the subtrahend to the minuend. In other words,

$$A - B = A + (-B)$$

With this in mind, convert the serial adder shown in Fig. 5.5(a) into a serial subtractor. What must be the initial value of the D flip-flop output? Assume that a 74'74 D flip-flop, as shown in Fig. 5.6(a), is to be used. Be sure to initialize the contents of the D flip-flop with an active-low signal, INIT_PULSE (‾‾|_|‾‾).

5.6. Convert the serial adder of Fig. 5.5(a) into a serial adder/subtractor, the block diagram of which is shown in Fig. 5.31. The operation is as follows: When $\overline{A}/S$ is false (L), the addition operation is performed. But when $\overline{A}/S$ is true (H), the subtraction operation is performed. The active-low signal INIT_PULSE is used to initialize the D flip-flop contents. Assume that a 74'74 D flip-flop is to be used. (*Hint:* Also use an XOR gate and an inverter.)

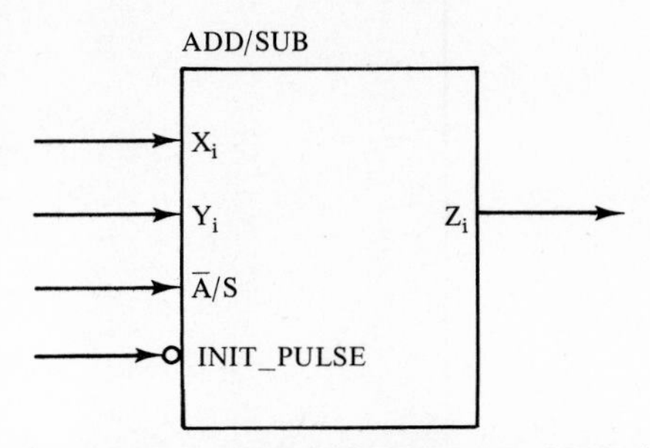

Figure 5.31 Serial adder/subtractor for Problem 5.6.

5.7. **(a)** Analyze the circuit diagram of Fig. 5.32(a) and complete the timing diagram of Fig. 5.32(b). Do not show the propagation delays.

(b) Assuming that the circuit of Fig. 5.32(a) represents a type of flip-flop, derive its characteristic table and its condensed characteristic table.

(c) Derive its excitation table.

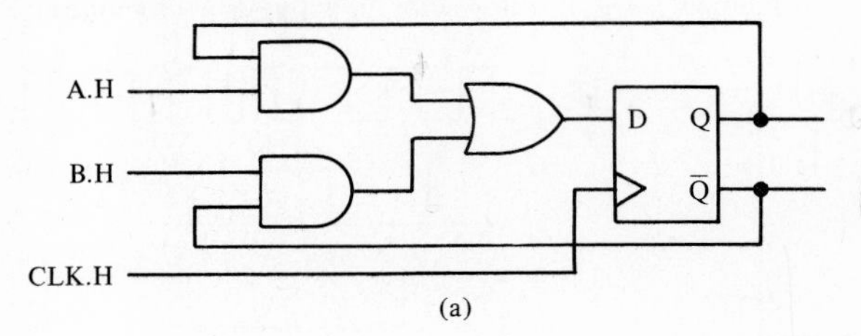

(a)

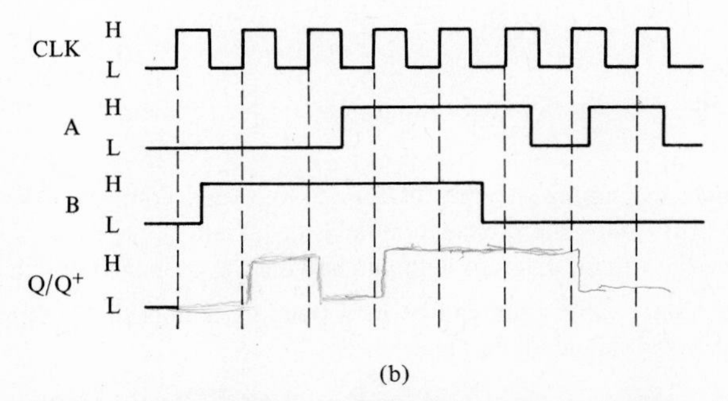

(b)

Figure 5.32 Circuit and timing diagrams for Problem 5.7.

5.8. Repeat Problem 5.7 for the circuit and timing diagrams of Fig. 5.33. Note that the clock input is active-low.

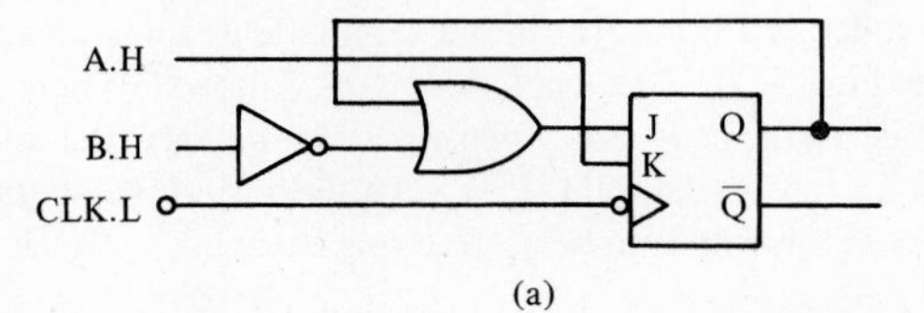

(a)

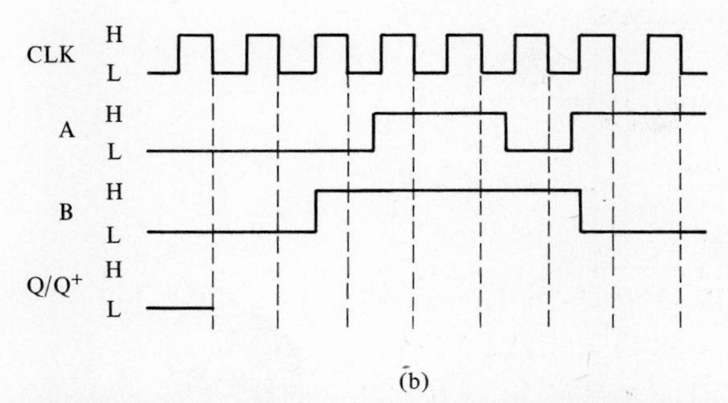

Figure 5.33 Circuit and timing diagrams for Problem 5.8.

5.9. Implement a J-K flip-flop using a 74'74 D flip-flop and any gates that are needed. Label all gates.

5.10. Implement a T flip-flop using a 74'74 D flip-flop and any gates that are needed. Label all gates.

5.11. Implement the unclocked S*-R* flip-flop of Fig. 5.34 with a normal unclocked S-R flip-flop and any gates that are needed. An unclocked S*-R* flip-flop functions exactly like a normal unclocked S-R flip-flop except that S* = R* = 1 is allowed. For this input the S*-R* flip-flop retains its previous Q value.

S*	R*	Q^+
0	0	Q
0	1	0
1	0	1
1	1	Q

Figure 5.34 Flip-flop for Problem 5.11.

5.12. The A-B flip-flop of Fig. 5.35 is to be implemented. Note that this flip-flop functions as a J-K flip-flop except for the inputs A = B = 1, for which the "set" input A dominates.

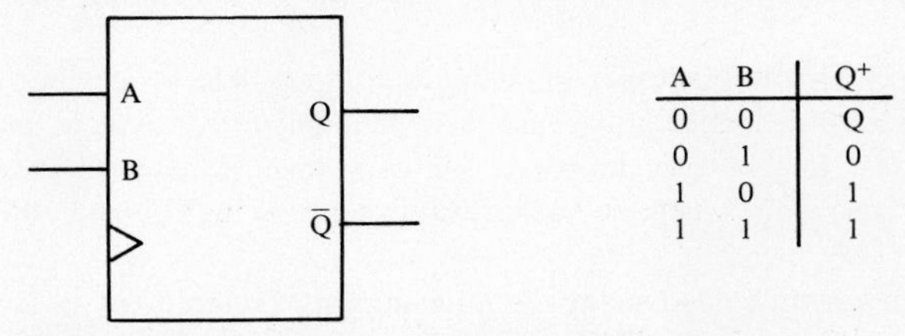

A	B	Q^+
0	0	Q
0	1	0
1	0	1
1	1	1

Figure 5.35 Flip-flop for Problem 5.12.

(a) Implement it using a 74'109 J-K flip-flop plus any gates that are needed.
(b) Implement it using a 74'74 D flip-flop plus any gates that are needed.
(c) Implement it using the T flip-flop of Fig. 5.7 plus any gates that are needed.

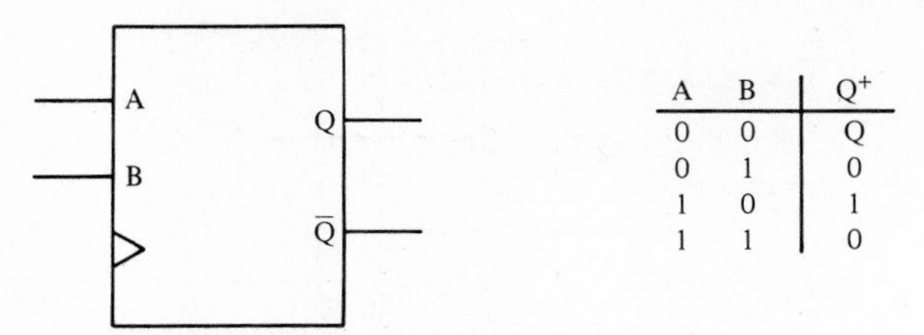

A	B	Q^+
0	0	Q
0	1	0
1	0	1
1	1	0

Figure 5.36 Flip-flop for Problem 5.13.

5.13. Repeat Problem 5.12 for the A-B flip-flop of Fig. 5.36. Note that this flip-flop functions as a J-K flip-flop except for the inputs A = B = 1, for which the "clear" input B dominates.

5.14. **(a)** Given the truth table of Fig. 5.11(b) for an unclocked S-R flip-flop, determine the logic equation for Q^+ as a function of S, R, and Q.
(b) The circuit diagram of Fig. 5.13(a) is the most popular gate implementation for an S-R flip-flop. Algebraically show that this implementation is consistent with your answer to part (a).
(c) Draw a mixed-logic circuit diagram corresponding to your answer to part (a).

5.15. Using commercially available gates, implement the unclocked S-R flip-flop of Fig. 5.37. Note the active-low inputs. Specify and label all components and signals.

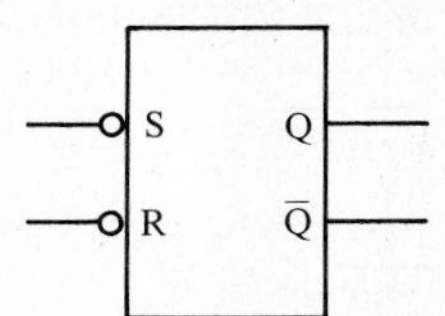

Figure 5.37 S-R flip-flop for Problem 5.15.

5.16. The operation of the switch debouncing circuit of Fig. 5.12(c) depends upon the voltage levels at the S and R inputs being both L when the switch arm is not in contact with either the ON or OFF terminal. For this L voltage level to occur, the resistor resistances must be small enough that the voltages at the S and R inputs are less than or equal to V_{IL}, which is 0.8 V for LS-TTL. With this in mind, determine the maximum resistance, R_{max}, allowable for the resistors. Use LS-TTL values.

5.17. The unclocked S-R flip-flop with active-low inputs that is specified in Problem 5.15 can also be used to debounce a mechanical switch. The circuit will be similar to that shown in Fig. 5.12(c). In this case, however, a *high* voltage is required at the S and R inputs to present a false value when the switch arm is between the ON and OFF terminals. With this in mind,
(a) Design a switch debouncing circuit using this "active-low" S-R flip-flop.
(b) Determine the minimum resistance, R_{min}, of the resistors that will give proper operation when the switch arm is between the ON and OFF terminals. Use LS-TTL values.

5.18. Complete the timing diagram of Fig. 5.38, which is for the unclocked J-K flip-flop shown in Fig. 5.14. Show the propagation delays t_{pHL} and t_{pLH}. Also, what happens after t_f?

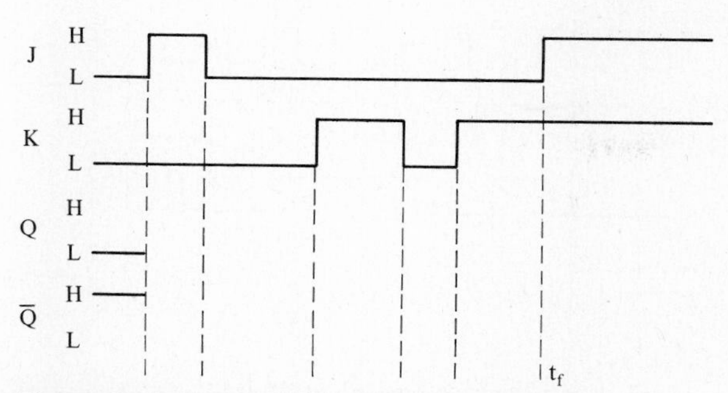

Figure 5.38 Timing diagram for Problem 5.18.

5.19. The circuit diagram of Fig. 5.15 is for a clocked J-K flip-flop.
(a) How does this flip-flop differ from the clocked J-K flip-flop discussed in Sec. 5.3.1?
(b) What is the restriction on the clock signal for the flip-flop shown in Fig. 5.15? Why?

5.20. Complete the timing diagram of Fig. 5.39 for the 74'107 J-K flip-flop of Fig. 5.3(a) to demonstrate the difference between a synchronous and an asynchronous clear operation. Do not show the propagation delays.

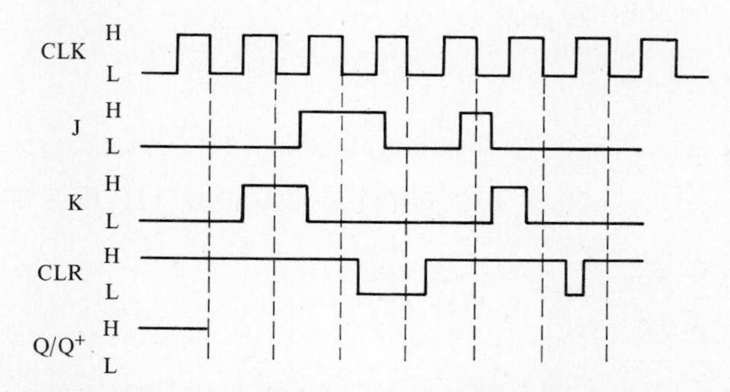

Figure 5.39 Timing diagram for Problem 5.20.

5.21. Given the circuit diagram of Fig. 5.40(a) consisting of a normal D flip-flop and a gated D flip-flop (see Example 5.2 and Fig. 5.8 for details), complete the timing diagram of Fig. 5.40(b). Do not show the propagation delays.

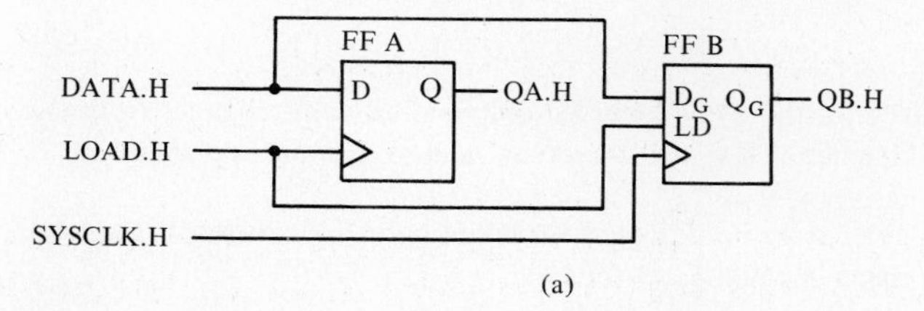

(a)

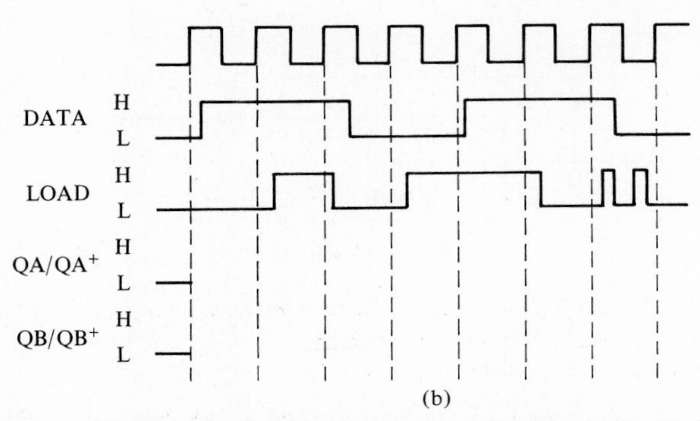

Figure 5.40 Circuit diagram and timing diagram for Problem 5.21.

5.22. A 4-bit binary counter is to be designed and realized using D flip-flops. The count is to be as follows:

. . . , 0000, 0001, 0010, 0011, 0100, . . . , 1110, 1111, 0000, 0001, . . .

(a) Draw the state diagram for the count sequence.
(b) Draw the functional block diagram for the counter, including the D flip-flops and the corresponding combinational circuit, in the manner shown in Fig. 5.18(a).
(c) Determine the required logic equations and draw the circuit diagram for the counter.

5.23. Repeat Problem 5.22 using J-K flip-flops.

5.24. Repeat Problem 5.22 using T flip-flops.

5.25. Design and realize a 3-bit counter that counts in the following sequence:

. . . , 111, 010, 001, 110, 100, 000, 111, 010, 001, . . .

(a) Use D flip-flops.
(b) Use J-K flip-flops.
(c) Use T flip-flops.

5.26. Design and realize a 4-bit decade counter with a synchronous CLEAR input. Use D flip-flops.

5.27. Design and realize a 3-bit Gray-code counter with an enable (EN) input. Use J-K flip-flops. The counter is to count in the prescribed Gray-code sequence if EN is true at the next active clock transition. But if EN is false at this transition, then the counter does not count and, instead, retains its current value. The Gray-code sequence is as follows:

. . . , 000, 001, 011, 010, 110, 111, 101, 100, 000, 001, . . .

Observe for the Gray code that only one bit value changes from one number to the next in the sequence. This is an important feature in some applications.

5.28. A counter is to be designed for counting in four different sequences under the control of two inputs X_1 and X_2 as follows:

For inputs		
X_1	X_2	The sequence is
0	0	. . . , 00, 01, 10, 11, 00, . . .
0	1	. . . , 11, 10, 01, 00, 11, . . .
1	0	. . . , 10, 11, 01, 00, 10, . . .
1	1	. . . , 01, 11, 10, 00, 01, . . .

The inputs X_1 and X_2 can affect the count sequence at any point during the sequence.

(a) Draw a state diagram for this counter.

(b) Design and implement this counter. Use J-K flip-flops.

5.29. Determine the count sequence for the counter of Fig. 5.41. Also, draw a state diagram for it.

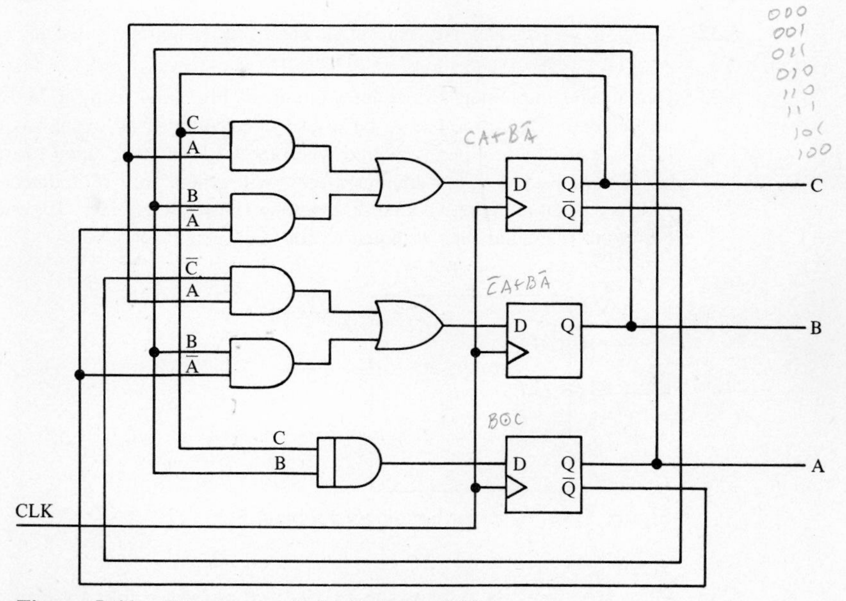

Figure 5.41 Counter circuit for Problem 5.29.

5.30. Design the 4-bit binary down-counter shown in Fig. 5.42 using a 74'163 plus any gates that are necessary. This counter is to have the same features as the 74'163 except for the count sequence, which is as follows:

. . . , 0000, 1111, 1110, 1101, 1100, 1011, . . . , 0010, 0001, 0000, 1111, . . .

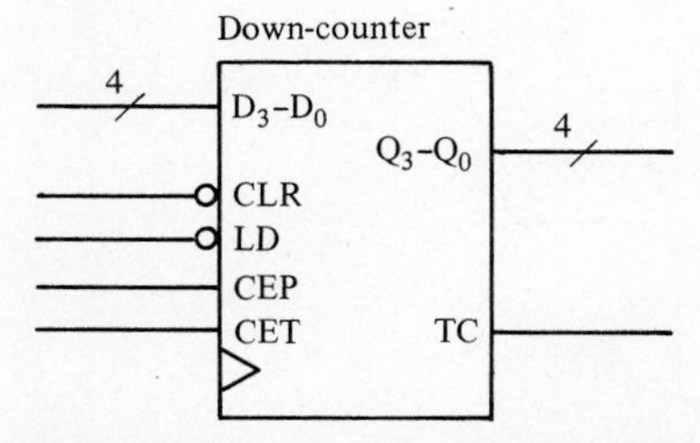

Figure 5.42 Down-counter for Problem 5.30.

5.31. Show that the 4-bit binary down-counter of Fig. 5.43 can be realized with only a 74'163. No additional components are needed.

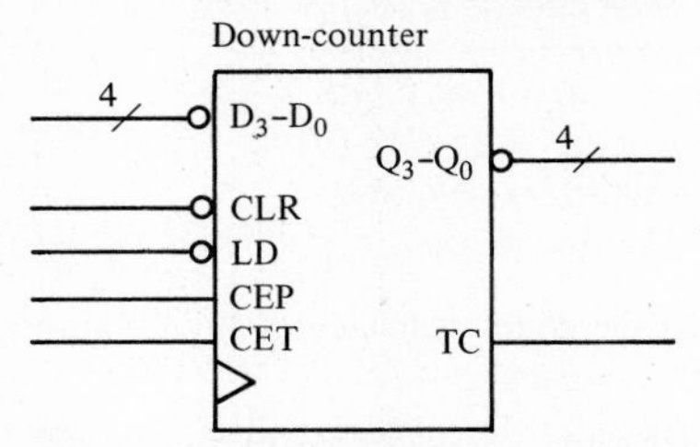

Figure 5.43 Down-counter for Problem 5.31.

5.32. Using a 74'163 and any gates that are needed, realize a decade counter similar to the 74'162.

5.33. Design and implement the counter circuit of Fig. 5.44 using a 74'163 and any gates that are needed. This circuit is to be used to determine if an event has occurred ten or more times. It is synchronously cleared when the CLR input is equal to true or when the count has reached 1111. This counter circuit will count only if it detects a true value at the EVENT input at an active clock transition. The (COUNT >= 10) output is true only when the count is greater than or equal to 1001.

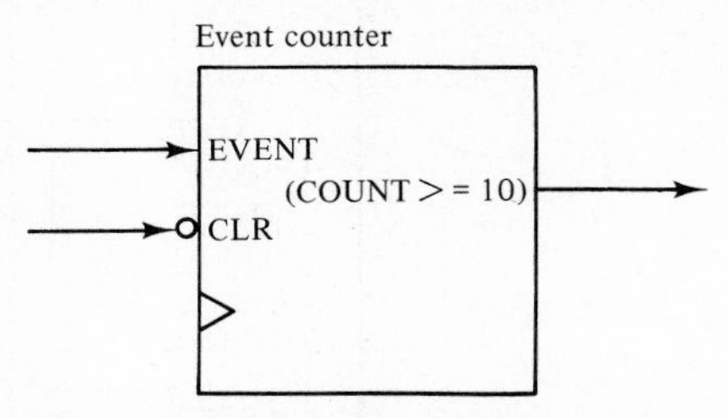

Figure 5.44 Counter circuit for Problem 5.33.

5.34. Show what must be done to the 74'194 to transform it into a 4-bit storage register.

5.35. Using a 74'194 and any gates that are needed, design and implement the sequence detector circuit shown in Fig. 5.45(a). The circuit input is a sequence of 1-bit values (either a 0 or a 1) detected at each active transition of the clock signal. When the circuit detects the sequence 1101, the output FOUND.SEQ becomes true for one clock cycle, as shown in the timing diagram of Fig. 5.45(b).

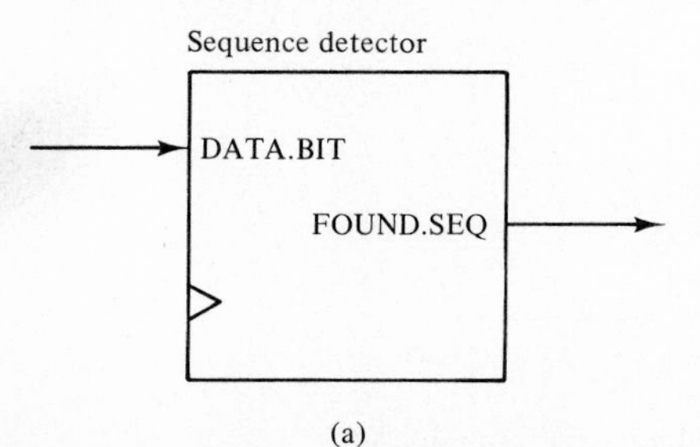

(a)

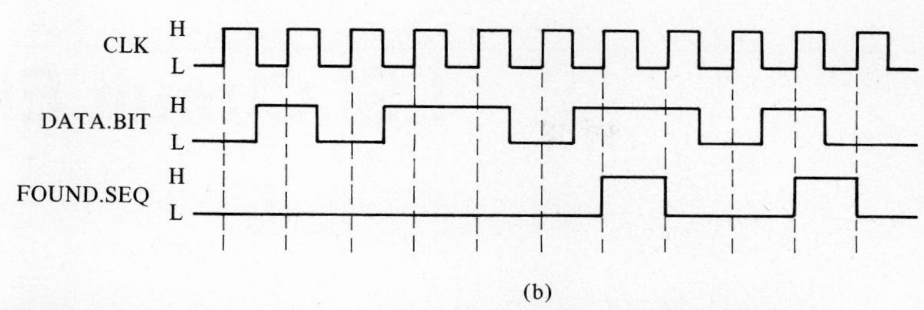

(b)

Figure 5.45 Sequence detector and timing diagram for Problem 5.35.

5.36. By cascading two 74'194 shift registers, construct an 8-bit universal bidirectional shift register.

Chapter 6

LSI Circuit Elements

6.1 INTRODUCTION

Recent rapid advances in large-scale integration technology have resulted in a larger and larger number of standard logic functions being integrated on a single chip. A single LSI (large-scale integrated) chip can now perform a number of logic functions that formerly required an entire circuit board of MSI (medium-scale integrated) and SSI (small-scale integrated) circuit elements.

There are, of course, significant advantages in the use of the LSI circuit elements instead of SSI or MSI. For a given number of logic functions, the use of the LSI circuit elements results in a reduction in the IC package count. This translates into a reduction of power consumption as well as of the PC (printed circuit) board space that is required. A reduction in the number of IC packages also results in an increase in reliability since the number of interconnections, a major source of failure, is reduced. The end result is a reduction in the total cost of the digital product. Less than half the cost of a digital product is in the actual purchase price of the ICs. Most of the cost relates to PC board area, assembly, and the testing associated with the product, all of which are reduced with the use of LSI circuit elements.

A less obvious but more important advantage of using higher-density circuit elements is the decrease in design complexity. In the top-down design approach to be presented in Chapter 7, the design process naturally leads into the use of high-level logic functions that are realized as LSI and MSI circuit elements. This design, being systematic and less complex, results in an increase in the ease of realization, testability, and maintainability of the digital circuits.

In the preceding two chapters common MSI circuit elements were presented. In this chapter we will conclude the presentation of the digital building blocks with the study of the commonly used LSI circuit elements, including the arithmetic logic unit (ALU), the look-ahead carry adder, the programmable logic array (PLA), and the programmable array logic (PAL). Also presented in this chapter are the various types of random access memories, including the read-write memories (static and dynamic RAMs) and the read-only memories (ROM, PROM, and EPROM). The currently most important LSI circuit element, the microprocessor, is omitted from this chapter, but it is the major topic, along with microprocessor-based design, of the second half of this book.

6.2 ARITHMETIC LOGIC UNIT

An arithmetic logic unit (ALU) is a combinational circuit element that performs a set of commonly used arithmetic and logic operations. A representative example of a commercially available ALU is the 74'181, a block diagram of which is shown in Fig. 6.1(a). The corresponding functional description is given by the table of Fig. 6.1(b).

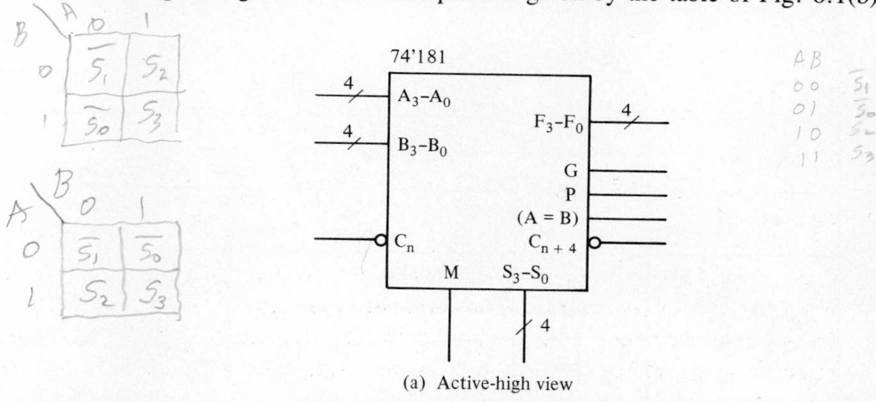

(a) Active-high view

Selection				Active-high data		
				M = H logic functions	M = L; Arithmetic operations	
S3	S2	S1	S0		C_n = H (no carry)	C_n = L (with carry)
L	L	L	L	F = $\overline{A}$	F = A	F = A PLUS 1
L	L	L	H	F = $\overline{A + B}$	F = A + B	F = (A + B) PLUS 1
L	L	H	L	F = $\overline{A}B$	F = A + $\overline{B}$	F = (A + $\overline{B}$) PLUS 1
L	L	H	H	F = 0	F = MINUS 1 (2's COMP)	F = ZERO
L	H	L	L	F = $\overline{AB}$	F = A PLUS A$\overline{B}$	F = A PLUS A$\overline{B}$ PLUS 1
L	H	L	H	F = $\overline{B}$	F = (A + B) PLUS A$\overline{B}$	F = (A + B) PLUS A$\overline{B}$ PLUS 1
L	H	H	L	F = A⊕B	F = A MINUS B MINUS 1	F = A MINUS B
L	H	H	H	F = A$\overline{B}$	F = A$\overline{B}$ MINUS 1	F = A$\overline{B}$
H	L	L	L	F = $\overline{A}$ + B	F = A PLUS AB	F = A PLUS AB PLUS 1
H	L	L	H	F = $\overline{A \oplus B}$	F = A PLUS B	F = A PLUS B PLUS 1
H	L	H	L	F = B	F = (A + $\overline{B}$) PLUS AB	F = (A + $\overline{B}$) PLUS AB PLUS 1
H	L	H	H	F = AB	F = AB MINUS 1	F = AB
H	H	L	L	F = 1	F = A PLUS A	F = A PLUS A PLUS 1
H	H	L	H	F = A + $\overline{B}$	F = (A + B) PLUS A	A = (A + B) PLUS A PLUS 1
H	H	H	L	F = A + B	F = (A + $\overline{B}$) PLUS A	F = (A + $\overline{B}$) PLUS A PLUS 1
H	H	H	H	F = A	F = A MINUS 1	F = A

(b) Functional description of the active-high view

Figure 6.1 The 74'181 ALU.

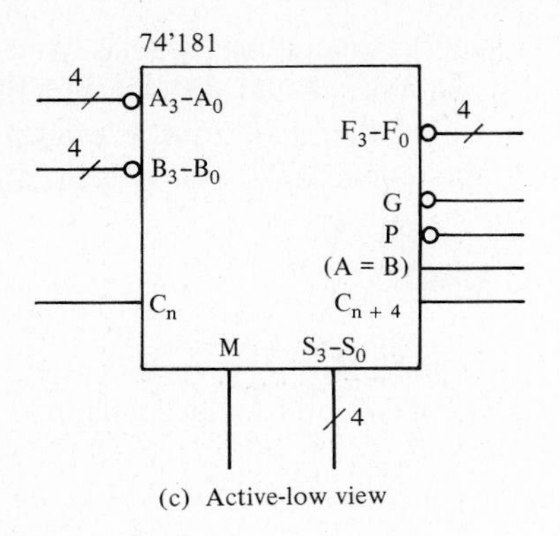

(c) Active-low view

Selection				Active-low data		
				M = H logic functions	M = L; Arithmetic operations	
S3	S2	S1	S0		C_n = L (no carry)	C_n = H (with carry)
L	L	L	L	$F = \overline{A}$	F = A MINUS 1	F = A
L	L	L	H	$F = \overline{AB}$	F = AB MINUS 1	F = AB
L	L	H	L	$F = \overline{A} + B$	F = $A\overline{B}$ MINUS 1	$F = A\overline{B}$
L	L	H	H	F = 1	F = MINUS 1 (2's COMP)	F = ZERO
L	H	L	L	$F = \overline{A + B}$	F = A PLUS $(A + \overline{B})$	F = A PLUS $(A + \overline{B})$ PLUS 1
L	H	L	H	F = B	F = AB PLUS $(A + \overline{B})$	F = AB PLUS $(A + \overline{B})$ PLUS 1
L	H	H	L	$F = \overline{A \oplus B}$	F = A MINUS B MINUS 1	F = A MINUS B
L	H	H	H	$F = A + \overline{B}$	$F = A + \overline{B}$	F = $(A + \overline{B})$ PLUS 1
H	L	L	L	$F = \overline{A}B$	F = A PLUS (A + B)	F = A PLUS (A + B) PLUS 1
H	L	L	H	$F = A \oplus B$	F = A PLUS B	F = A PLUS B PLUS 1
H	L	H	L	F = B	F = $A\overline{B}$ PLUS (A + B)	F = $A\overline{B}$ PLUS (A + B) PLUS 1
H	L	H	H	F = A + B	F = (A + B)	F = (A + B) PLUS 1
H	H	L	L	F = 0	F = A PLUS A	F = A PLUS A PLUS 1
H	H	L	H	$F = A\overline{B}$	F = AB PLUS A	F = AB PLUS A PLUS 1
H	H	H	L	F = AB	F = $A\overline{B}$ PLUS A	F = $A\overline{B}$ PLUS A PLUS 1
H	H	H	H	F = A	F = A	F = A PLUS 1

(d) Functional description of the active-low view

Figure 6.1 *(cont.)*

The primary inputs to this ALU are two 4-bit operands: A_3–A_0 and B_3–B_0. The ALU performs some operation on these operands to produce a 4-bit output F_3–F_0. The specific operation that is performed depends on the function selection inputs, S_3–S_0, as well as on the mode control input M. As shown in the table of Fig. 6.1(b), if M = 1 (H), then one of the 16 logic operations is performed, the particular one depending on the values of S_3–S_0. If M = 0 (L), however, then one of the 32 predominantly arithmetic operations is performed, the particular one depending on the values of S_3–S_0 and C_n.

The input C_n is the carry-in input for arithmetic operations. And C_{n+4} is the carry-out output. The output (A = B) is for magnitude comparison operations. It is true (H) when A_3–A_0 is equal to B_3–B_0, and is false (L) otherwise. The group-generate output G and the group-propagate output P will be discussed in following sections.

With mixed logic, a second view of the 74'181 ALU, with active-low operands, is possible. The block diagram of it is shown in Fig. 6.1(c). As specified by the corresponding functional description of Fig. 6.1(d), this view gives rise to a new set of arithmetic and logic operations.

From Figs. 6.1(b) and (d), it is evident that the 74'181 performs all the commonly used arithmetic and logic operations, along with some not so common ones. The 74'181 provides a powerful building block and an economic means for the design of digital circuits that require a variety of these operations.

6.3 LOOK-AHEAD CARRY CIRCUITS FOR ADDERS AND ALUs

As was explained in Sec. 4.2.2 of Chapter 4, parallel adders may require excessive amounts of time for adding operations. Consider the 4-bit parallel adder of Fig. 4.3, which is reproduced in Fig. 6.2 for convenience. Since the carry-out of each full-adder stage is connected to the carry-in of the next stage, the carry-in for each stage is not stable until the preceding stage produces a stable carry-out output. For example, the carry-in for Stage 1 is not stable until Stage 0 produces a stable output at C_1. Similarly, the carry-in to Stage 2 is not stable until Stage 1 produces a stable output at C_2, and so forth. In this manner, the carry "ripples" down the chain of full adders. Consequently, after the inputs are applied to an N-bit ripple adder, the outputs do not become stable until a time equal to $N \times t_p(\text{FA})$, in which $t_p(\text{FA})$ is the propagation delay of a full-adder stage. If N is large, say 64 bits, then the time for the carry to propagate to the last stage can be substantial.

Since the addition operation is a fundamental arithmetic operation upon which other arithmetic operations are frequently based, it is extremely desirable to optimize its performance and reduce the time for the carry-in of each adder stage to become stable. The most common technique to accomplish this is with *look-ahead carry circuits*. Recall from Sec. 4.2.1 that the equations for the outputs for the full adder at each adder stage are

$$S_i = A_i \oplus B_i \oplus C_i \quad \text{and} \quad C_{i+1} = A_iB_i + (A_i \oplus B_i)C_i$$

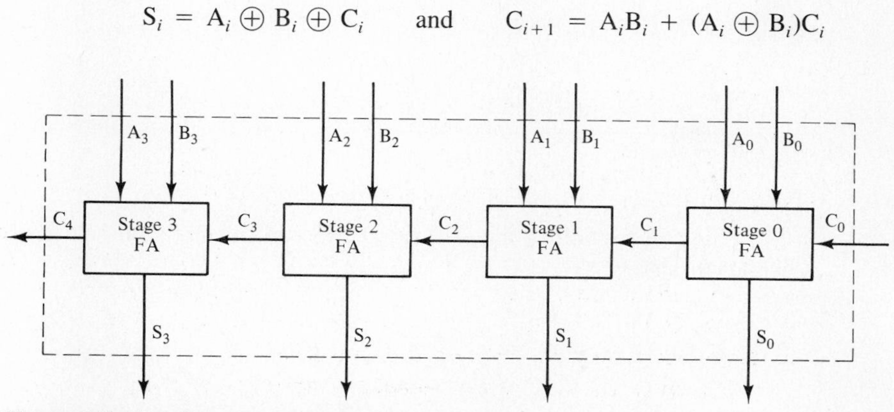

Figure 6.2 Parallel adder circuit diagram.

Let us now define the following variables:

$$G_i = A_iB_i \quad \text{and} \quad P_i = A_i \oplus B_i$$

With the substitution of these variables, the C_{i+1} equation becomes

$$C_{i+1} = G_i + P_iC_i$$

The variable G_i is called the *carry-generate* for the ith adder stage. It is the logic AND of the two input bits to that stage, A_i and B_i. Its significance is that a carry-out is *generated* by Stage i if $G_i = 1$ (i.e., $A_i = 1$ and $B_i = 1$), regardless of what transpires in the adder stages preceding Stage i. In other words, if $G_i = 1$, then $C_{i+1} = 1$ regardless of the value of C_i. For example,

Stage	···	$(i+1)$	(i)	$(i-1)$	···	(0)
C		1	X	X	···	X
A			1	X	···	X
B			1	X	···	X
Sum			X	X	···	X

where X = don't care

The variable P_i is called the *carry-propagate* for the ith adder stage. It is the Exclusive OR of the two input bits A_i and B_i to that stage. Consequently, if $P_i = 1$ (i.e., $A_i = 1$ and $B_i = 0$, or $A_i = 0$ and $B_i = 1$), then the carry-in C_i for this stage will be *propagated* to the carry-out C_{i+1}. In other words, if $P_i = 1$, then $C_{i+1} = C_i$, as is illustrated by the following examples:

Stage	···	$(i+1)$	(i)	$(i-1)$	···	(0)
C		0	0	X	···	X
A			1	X	···	X
B			0	X	···	X
Sum			X	X	···	X

Stage	···	$(i+1)$	(i)	$(i-1)$	···	(0)
C		1	1	X	···	X
A			1	X	···	X
B			0	X	···	X
Sum			X	X	···	X

Using this modified C_{i+1} equation, we can calculate the values of the carries at each adder stage:

$$C_1 = G_0 + P_0C_0$$
$$C_2 = G_1 + P_1C_1 = G_1 + P_1G_0 + P_1P_0C_0$$
$$C_3 = G_2 + P_2C_2 = G_2 + P_2G_1 + P_2P_1G_0 + P_2P_1P_0C_0$$
$$C_4 = G_3 + P_3C_3 = G_3 + P_3G_2 + P_3P_2G_1 + P_3P_2P_1G_0 + P_3P_2P_1P_0C_0$$

In general,

$$\begin{aligned} C_n &= G_{n-1} + P_{n-1}C_{n-1} \\ &= G_{n-1} + P_{n-1}G_{n-2} + P_{n-1}P_{n-2}G_{n-3} + \cdots + P_{n-1}P_{n-2}P_{n-3} \cdots P_0C_0 \end{aligned}$$

Note that C_0 is the only carry that appears in this equation. Consequently, the carry-in of any adder stage in an N-bit parallel adder can be determined from only the A_i and B_i inputs and C_0. Thus, to calculate the value of any C_n we do not need the value of C_{n-1}, C_{n-2}, or of any of the preceding carries except C_0. Clearly, then, it is possible to design a parallel adder in which the carry-in of any adder stage does not have to wait for the carry to be propagated down the adder chain.

In terms of hardware, each C_i can be realized as a combinational circuit consisting of two levels of AND-OR gates, as shown in Fig. 6.3. This circuit is called a *look-ahead carry circuit*. After the inputs have been applied, the outputs of an N-bit adder with look-ahead carry circuitry will be stable at a time equal to $t_p(\text{CLA}) + t_p(\text{FA})$, *independently* of the value of N. Here, $t_p(\text{FA})$ is again the propagation delay of a full adder, and $t_p(\text{CLA})$ is the propagation delay of the two-level AND-OR look-ahead carry circuit.

Look-ahead carry circuits based on the above discussion are frequently used in commercially available ICs. For example, both the 74'83 and 74'283 MSI 4-bit adders, described in Sec. 4.2.2, have look-ahead carry circuitry for faster addition. The 74'181 ALU, described in the preceding section, also has the look-ahead carry feature. Fur-

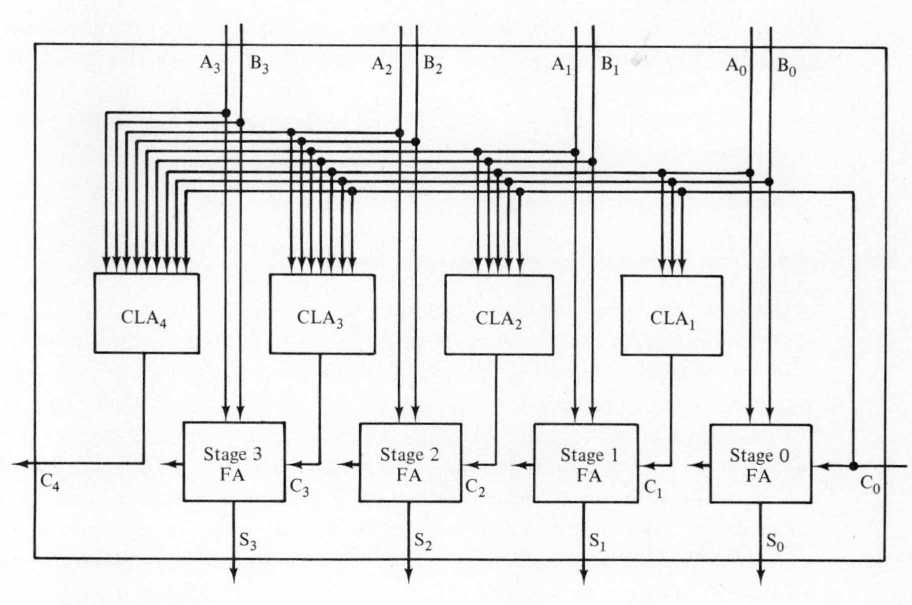

$C_1 = G_0 + P_0C_0$
$C_2 = G_1 + P_1C_1 = G_1 + P_1G_0 + P_1P_0C_0$
$C_3 = G_2 + P_2C_2 = G_2 + P_2G_1 + P_2P_1G_0 + P_2P_1P_0C_0$
$C_4 = G_3 + P_3C_3 = G_3 + P_3G_2 + P_3P_2G_1 + P_3P_2P_1G_0 + P_3P_2P_1P_0C_0$
in which $G_i = A_iB_i$ and $P_i = A_i \oplus B_i$

Figure 6.3 4-bit adder with carry look-ahead circuitry.

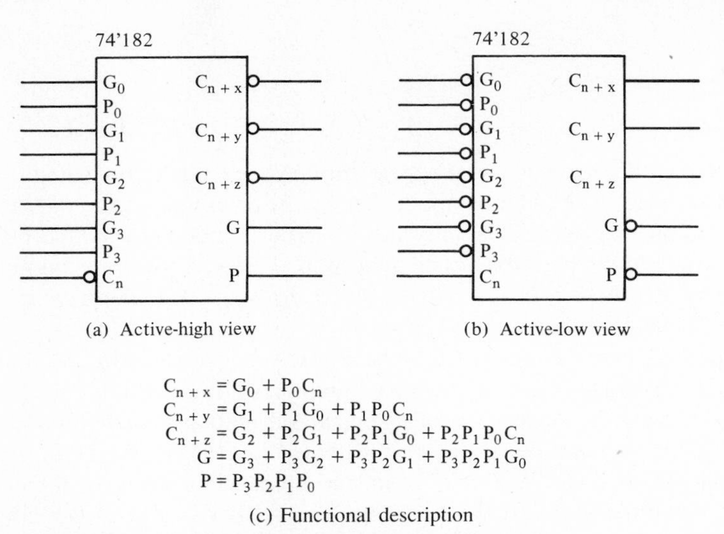

$$
\begin{aligned}
C_{n+x} &= G_0 + P_0 C_n \\
C_{n+y} &= G_1 + P_1 G_0 + P_1 P_0 C_n \\
C_{n+z} &= G_2 + P_2 G_1 + P_2 P_1 G_0 + P_2 P_1 P_0 C_n \\
G &= G_3 + P_3 G_2 + P_3 P_2 G_1 + P_3 P_2 P_1 G_0 \\
P &= P_3 P_2 P_1 P_0
\end{aligned}
$$

(c) Functional description

Figure 6.4 The 74'182 look-ahead carry generator.

thermore, look-ahead carry circuits themselves are commercially available as ICs. An example is the 74'182 look-ahead carry generator. Two functional block diagrams for it are shown in Figs. 6.4(a) and (b), and the corresponding functional description is given in Fig. 6.4(c). The inputs to the look-ahead carry generator are the various carry-generates and carry-propagates (G_i's and P_i's) and the carry-in (C_n) for the group of adders. The outputs from it are the carries (C_{n+x}, C_{n+y}, and C_{n+z}), the group-generate (G), and the group-propagate (P). The use of the 74'182 look-ahead carry generator will be illustrated in the next section.

6.3.1 Modified Look-Ahead Carry Approaches

In the look-ahead carry scheme, the price that is paid for the performance improvement is the additional hardware, which consists of the look-ahead carry circuit for each adder stage. As is evident from Fig. 6.3, for each successive adder stage, the look-ahead carry circuit becomes more complex and can quickly become unmanageable. As an illustration, the look-ahead carry circuit for Stage 15 requires 33 inputs. Clearly, to handle look-ahead carry for larger multibit adders, the look-ahead carry scheme needs to be modified.

One approach is to have a scheme that is a combination of look-ahead carry and ripple carry. This approach is illustrated by the 16-bit adder shown in Fig. 6.5. In this approach, look-ahead circuitry is used *within* each of the 4-bit adder groups, but carries are rippled *between* each group. The propagation delay for this 16-bit adder is equal to $4 \times t_p$(adder group), in which t_p(adder group) is the time required for an adder group to produce the sum and the carry-out for that group. Although this delay time is greater than that of a 16-bit adder that employs full look-ahead carry across all 16 bits, the hardware requirement is far less. Also, this delay time is still substantially less than that of a 16-bit ripple adder without any look-ahead carry circuitry.

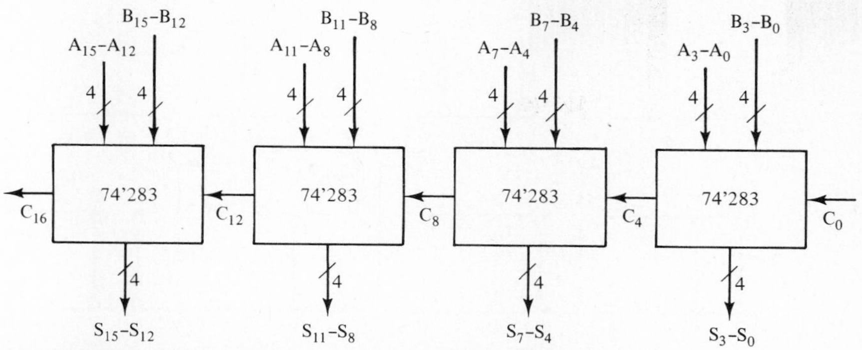

Figure 6.5 16-bit adder with look-ahead carry and ripple carry.

Another approach is to have a multilevel look-ahead carry scheme. Let us define the following variables for a multibit adder group (here they are defined only for a 4-bit adder group for simplicity of explanation):

$$\text{G(group)} = G_3 + P_3G_2 + P_3P_2G_1 + P_3P_2P_1G_0$$
$$\text{P(group)} = P_3P_2P_1P_0$$

The variable G(group) is the *group-generate* for a multibit adder group. In general, it is some logic function of the individual carry-generates (G_i's) and carry-propagates (P_i's) within that group. This definition is such that a carry-out, C_{out}(group), is *generated* out of this multibit adder group if G(group) = 1, regardless of what transpires in the adder groups preceding this group. So if G(group) = 1, then C_{out}(group) = 1. The variable P(group) is the group-propagate for this multibit adder group. It is the logic AND of the individual carry propagates within that group. If P(group) is equal to 1, then the carry-in for this adder group is propagated to the carry-out of the adder group. In other words, if P(group) = 1, then C_{out}(group) = C_{in}(group). From what has been stated, it is evident that the G(group) and P(group) variables are defined such that

$$C_{out}\text{(group)} = \text{G(group)} + \text{P(group)}C_{in}\text{(group)}$$

For the 74'181 ALU of Fig. 6.1, the group-generate G and the group-propagate P outputs are produced in this manner. Using these group-generate and group-propagate outputs with look-ahead carry generators, such as the 74'182 shown in Fig. 6.4, we can connect a number of 74'181 ALUs in a multilevel look-ahead scheme. Shown in Fig. 6.6 is a 16-bit ALU constructed in this manner. Note that in performing the addition operation each of the 74'181 ALUs cannot produce stable outputs until its respective C_n input is stable. This 16-bit ALU functions as follows: Given that the two 16-bit operands and the C_{in} are applied, look-ahead circuitry within each 74'181 produces the group-generate and group-propagate outputs after a delay of t_p(181PG). Then, it takes the 74'182 look-ahead carry generator a delay of t_p(182) to produce C_{n+x}, C_{n+y}, and C_{n+z}. At this point in time, the C_n for each of the 74'181s is stable. Therefore, the outputs for the 16-bit ALU become stable after another delay of t_p(181ADD), which is the time required for the 74'181 to produce the sum. So, the total propagation delay for the 16-bit ALU for the addition operation is t_p(181PG) + t_p(182) + t_p(ADD).

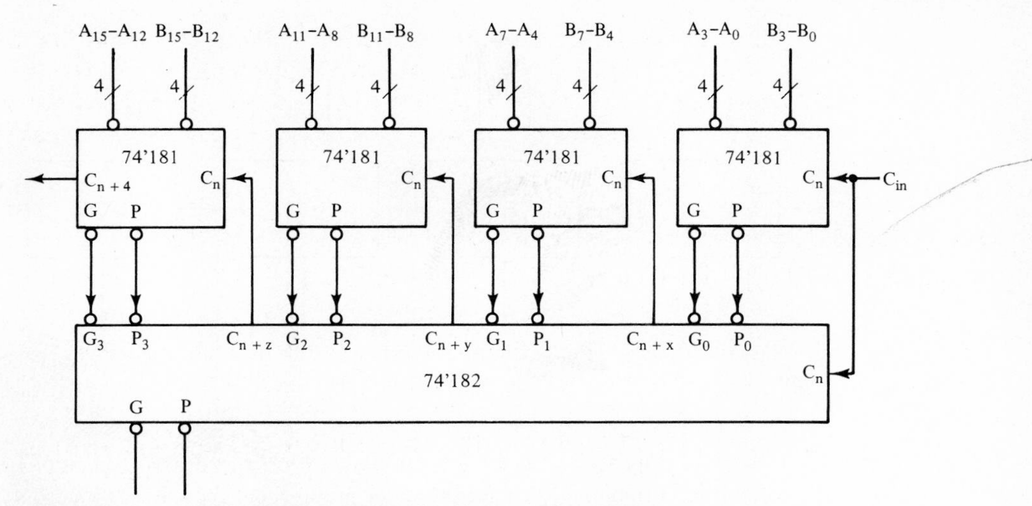

Figure 6.6 16-bit ALU with multilevel look-ahead carry structures.

As shown in Fig. 6.7, additional levels of look-ahead carry generators can be used to realize larger multibit ALUs. What is the propagation delay for this 64-bit ALU for the addition operation? (See Problem 6.8.) In general, with this multilevel look-ahead carry scheme, the addition operation delay is determined by the propagation delays through the levels of look-ahead circuitry. Certainly this delay is less than for the combinational approach of Fig. 6.5, where the carry is rippled down a chain of ALUs. Also, although the delay is somewhat greater, the hardware required is far less than that of an N-bit adder that employs full look-ahead carry across all N bits.

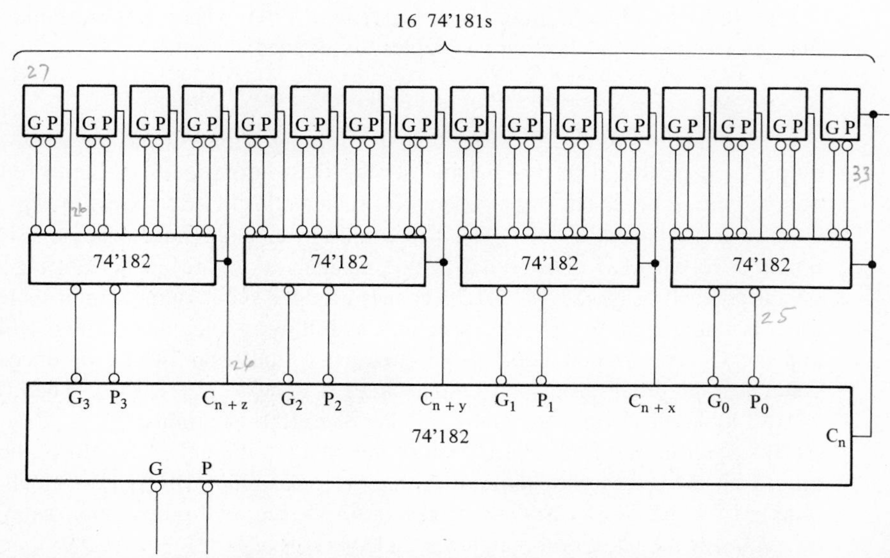

Figure 6.7 A 64-bit ALU with multilevel look-ahead carry structures.

This discussion of the various approaches to look-ahead carry provides a good illustration of the typical design trade-off between performance and hardware complexity. Which approach to be taken is, of course, a function of the performance requirements of the particular digital circuit along with other constraints such as cost, and also the quantities of the parts that are to be manufactured.

6.4 PROGRAMMABLE LOGIC ARRAY (PLA) AND PROGRAMMABLE ARRAY LOGIC (PAL)

Consider a digital circuit with ten inputs and eight outputs, as shown in Fig. 6.8(a). Suppose that each of the outputs is some combinational function of the ten inputs:

$$Z_i = f(X_0, X_1, X_2, \ldots, X_9)$$

A typical output, Z_i, can be realized with a two-level, sum-of-products, AND-OR gate structure, as shown in Fig. 6.8(b). Consequently, the entire circuit of Fig. 6.8(a) can be

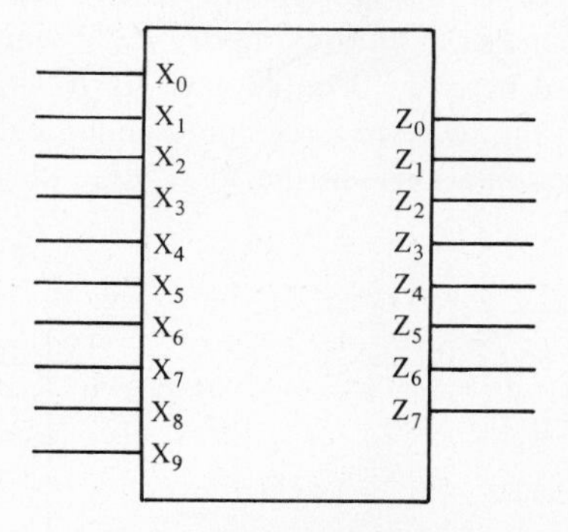

(a) Functional block diagram

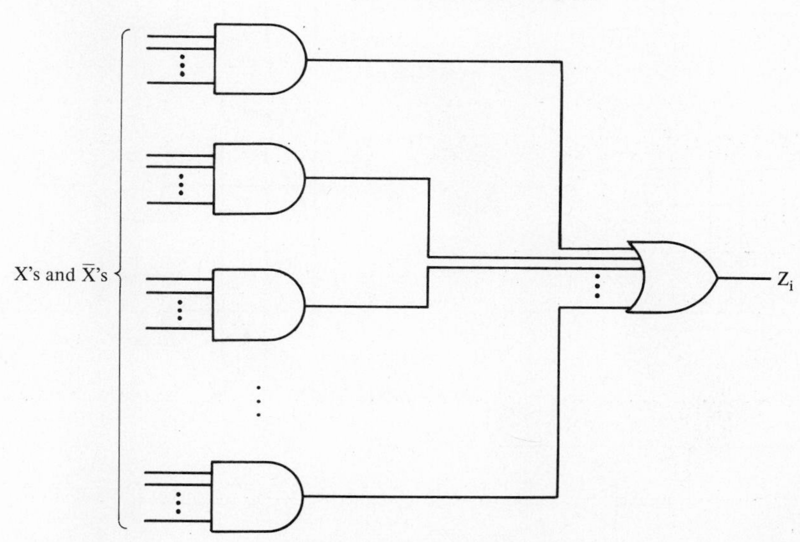

(b) Two-level AND-OR structure

Figure 6.8 Conventional realization of a multiinput and multioutput circuit.

realized with eight of these structures. If this circuit is realized with discrete SSI circuit elements, such as the AND and OR gates of Chapter 3, then the package count for the circuit would be substantial. Fortunately, there are attractive alternatives.

In this section we will study two circuit elements, programmable logic arrays (PLAs) and programmable array logic devices (PALs), that can be used as alternatives to discrete logic circuit elements for the realizations of multi-input, multioutput combinational circuits. Conceptually, PLAs and PALs are straightforward circuit elements that simply realize sum-of-products gate structures in a systematic manner. The power of the PLAs and PALs is that a large number of these sum-of-products structures can be integrated on a *single* IC.

6.4.1 Programmable Logic Array

As is illustrated by the one shown in Fig. 6.8(b), logic diagrams are convenient for representing small logic functions. They can, however, become cumbersome for the large logic functions that are typically used with PLAs. Consequently, it is desirable to devise a shorthand notation to simplify logic diagrams for use with PLAs. The notation that will be used here has been adopted by IC manufacturers.

Shown in Fig. 6.9 are three common logic diagram representations and their equivalent PLA representations. In the PLA logic diagram of Fig. 6.9(a), the AND gate for

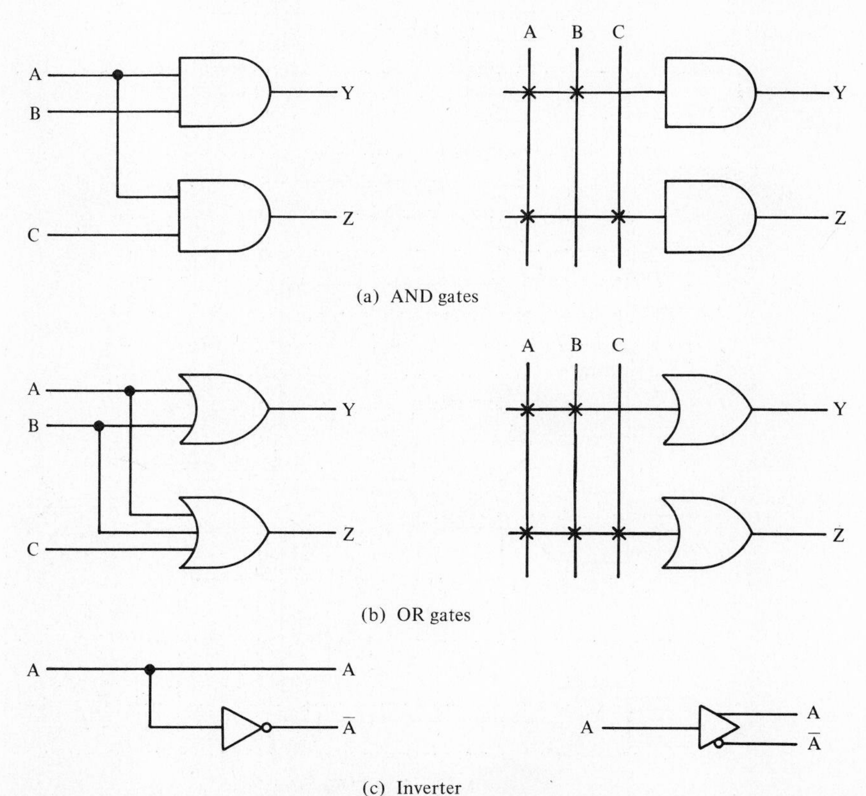

Figure 6.9 Equivalent PLA logic diagrams.

output Y has three inputs. But only two of the inputs are connected to A and B, as indicated by the two "x's," and one input is left unused (absence of an "x"). Similarly, the OR gates in the PLA logic diagram in Fig. 6.9(b) have three inputs, and an x indicates a connection to an input. The PLA diagram of Fig. 6.9(c) for an inverter is self-explanatory.

Example 6.1 PLA Equivalent of a Two-Level AND-OR Circuit

Using the notation just described for the circuit diagram of Fig. 6.10(a), we can obtain the equivalent PLA circuit diagram of Fig. 6.10(b). Note that the full capacity of the

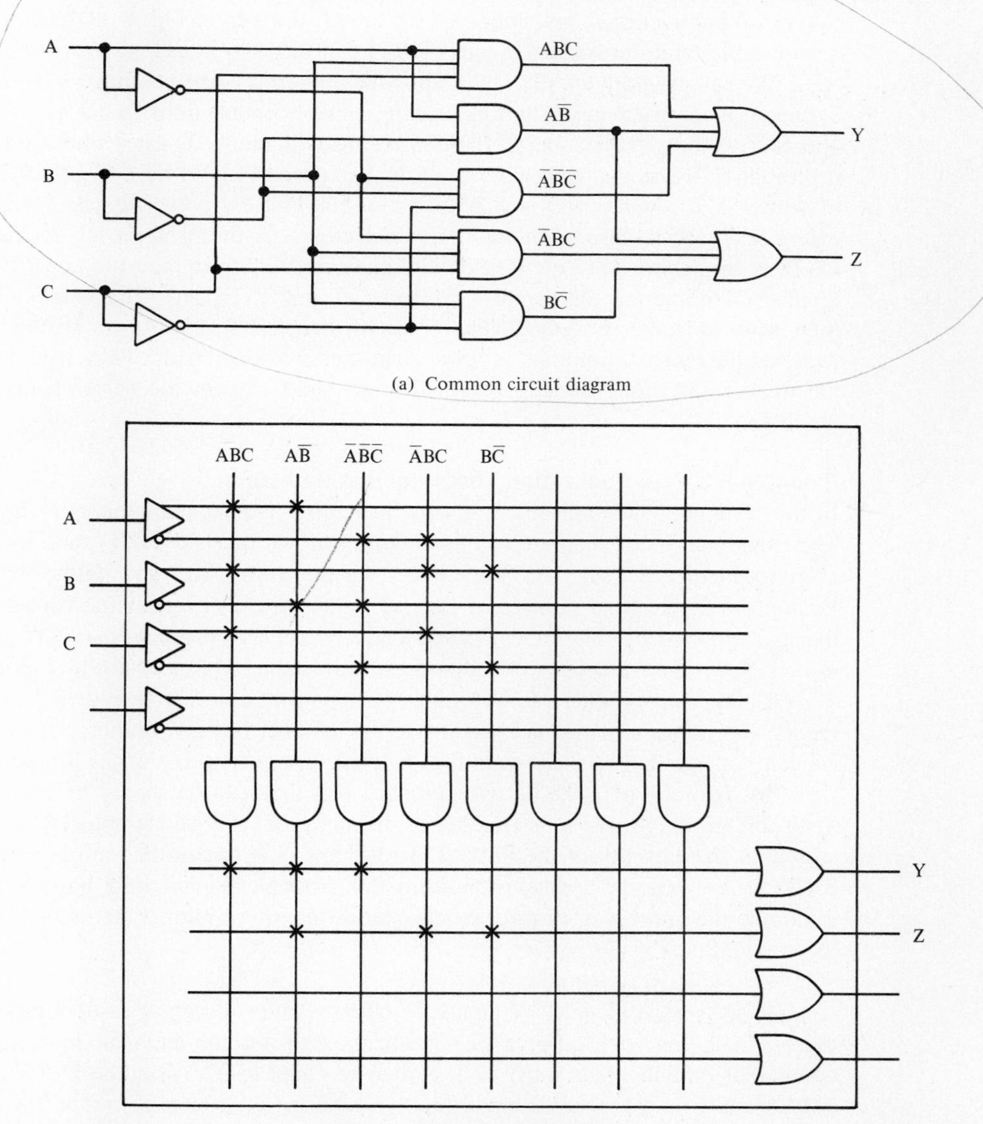

Figure 6.10 PLA circuit diagram for Example 6.1.

PLA structure is not utilized. This PLA has four inputs, four outputs, and eight product terms (AND terms). Thus it can realize up to four logic functions (outputs of the OR gates), but only two of them, Y and Z, are used. Also, each of the OR gates can have up to eight inputs, and so OR up to eight product terms each. And, each of the AND gates can have up to eight inputs (actually four inputs or their complements). ■ ■

Example 6.2 PLA Realization of a Truth Table Specification

In this example, the PLA of Fig. 6.10(b) is used to realize a combinational logic circuit of four inputs and two outputs, the functional block diagram of which is shown in Fig. 6.11(a). The functional description of the circuit is given in Fig. 6.11(b) in the form of a truth table, and the corresponding PLA circuit diagram is in Fig. 6.11(c).

We can program the PLA by finding the minterms corresponding to 1s in the output columns of the truth table, and by making suitable connections (the x's), as should be apparent. And, we can associate these minterms with the AND gates which are numbered 0 through 7. Note that minterm 0 is $\overline{X}_0\overline{X}_1\overline{X}_2X_3$, and is used by OR gate 0, the output of which is Z_0. Minterm 1 is $\overline{X}_0\overline{X}_1X_2\overline{X}_3$, and is used by both OR gate 0 (Z_0) and OR gate 1 (Z_1), and so forth with the other connections. With this approach we can program the PLA connections in a straightforward manner. Note from the truth table that the total number of minterms for the two functions Z_0 and Z_1 is ten. The number of *distinct* minterms, however, is only seven. Three of them (1, 2, and 4) are shared. Since the required number of minterms is fewer than that provided by the PLA, we can program the PLA connections directly from the truth table without the need of any reduction process. ■ ■

Example 6.3 PLA Realization That Requires Reduction

In this example, the same PLA is used to realize another combinational logic circuit. The functional block diagram of this circuit is shown in Fig. 6.12(a), and its truth table is given in Fig. 6.12(b). As is evident from the truth table, and unlike the circuit of Example 6.2, the total number of distinct minterms (12) exceeds the capacity of eight that is supported by this PLA. Consequently, we cannot program the PLA connections directly from the truth table, but first must reduce the number of product terms.

Using conventional minimization techniques, as shown in Fig. 6.12(c), we can reduce the number of distinct product terms to nine, but this number still exceeds the capacity of the PLA. As shown in Fig. 6.12(d), however, some of the K-map implicants [(a), (b), (c), and (d)] can be grouped such that they can be shared by both functions. With this grouping, the total number of distinct product terms is reduced to seven, and so within the capacity of the PLA. The resultant PLA circuit diagram is shown in Fig. 6.12(e). An important conclusion from this example is that it is more important to minimize the number of distinct product terms than to minimize the number of gates in the conventional sense. ■ ■

In general, a PLA has N inputs, M outputs, and supports K distinct product terms, as shown in Fig. 6.13. For a PLA realization of a combinational digital circuit, the number of circuit inputs must be less than or equal to N. Also, the number of circuit outputs must be less than or equal to M. Furthermore, if after reduction the number of distinct product terms still exceeds K, then a PLA with a larger capacity is required.

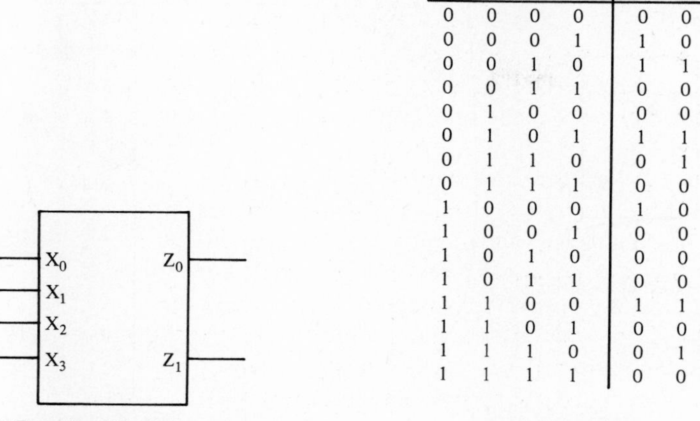

X_0	X_1	X_2	X_3	Z_0	Z_1
0	0	0	0	0	0
0	0	0	1	1	0
0	0	1	0	1	1
0	0	1	1	0	0
0	1	0	0	0	0
0	1	0	1	1	1
0	1	1	0	0	1
0	1	1	1	0	0
1	0	0	0	1	0
1	0	0	1	0	0
1	0	1	0	0	0
1	0	1	1	0	0
1	1	0	0	1	1
1	1	0	1	0	0
1	1	1	0	0	1
1	1	1	1	0	0

(a) Functional block diagram

(b) Truth table

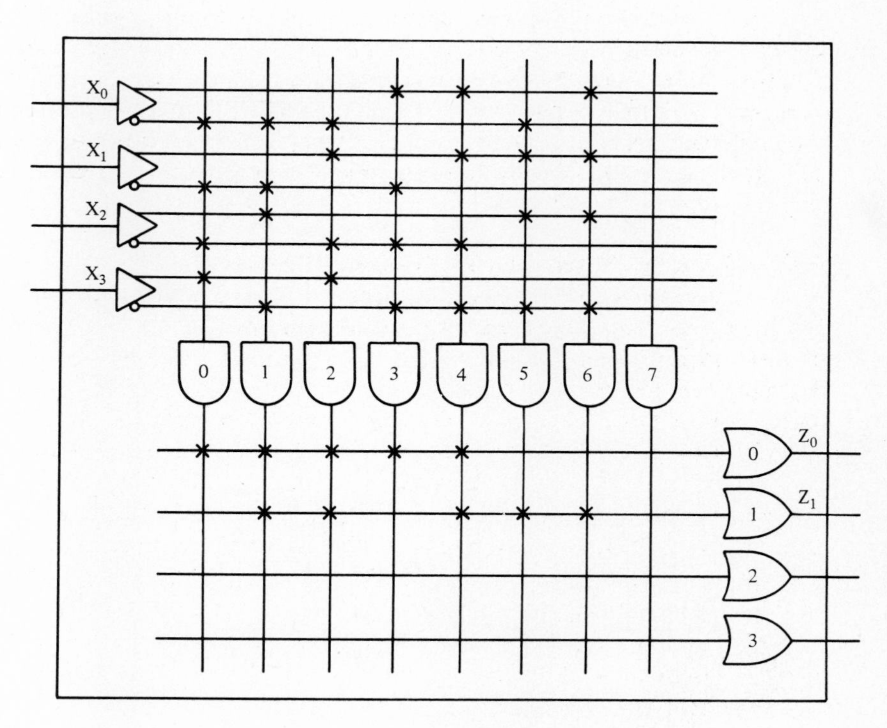

(c) PLA realization

Figure 6.11 Illustration for Example 6.2.

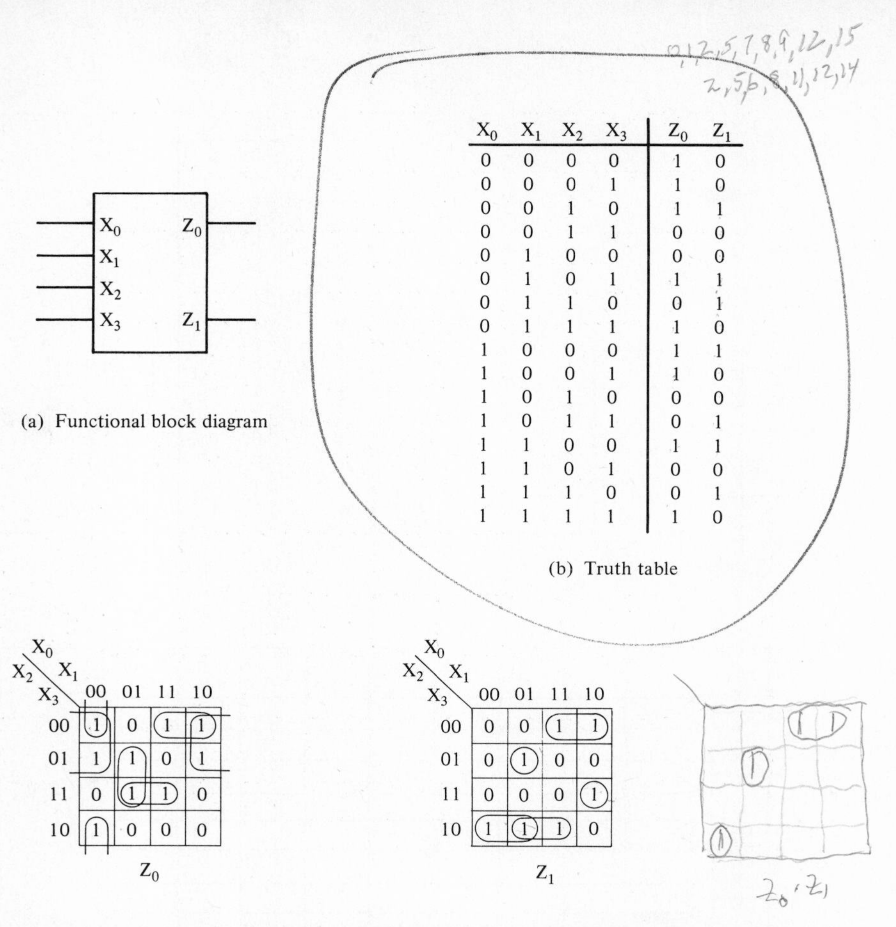

(a) Functional block diagram

X_0	X_1	X_2	X_3	Z_0	Z_1
0	0	0	0	1	0
0	0	0	1	1	0
0	0	1	0	1	1
0	0	1	1	0	0
0	1	0	0	0	0
0	1	0	1	1	1
0	1	1	0	0	1
0	1	1	1	1	0
1	0	0	0	1	1
1	0	0	1	1	0
1	0	1	0	0	0
1	0	1	1	0	1
1	1	0	0	1	1
1	1	0	1	0	0
1	1	1	0	0	1
1	1	1	1	1	0

(b) Truth table

(c) Conventional minimization results in a total of nine distinct AND terms

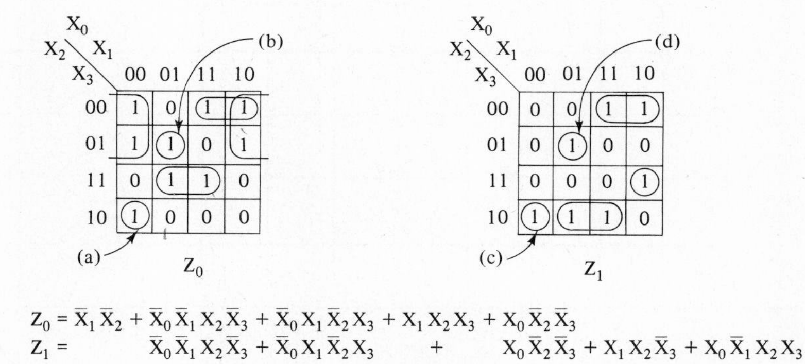

$$Z_0 = \overline{X}_1\overline{X}_2 + \overline{X}_0\overline{X}_1X_2\overline{X}_3 + \overline{X}_0X_1\overline{X}_2X_3 + X_1X_2X_3 + X_0\overline{X}_2\overline{X}_3$$

$$Z_1 = \overline{X}_0\overline{X}_1X_2\overline{X}_3 + \overline{X}_0X_1\overline{X}_2X_3 + X_0\overline{X}_2\overline{X}_3 + X_1X_2\overline{X}_3 + X_0\overline{X}_1X_2X_3$$

(d) Minimization for PLA realization – a total of seven distinct AND terms

Figure 6.12 Illustration for Example 6.3.

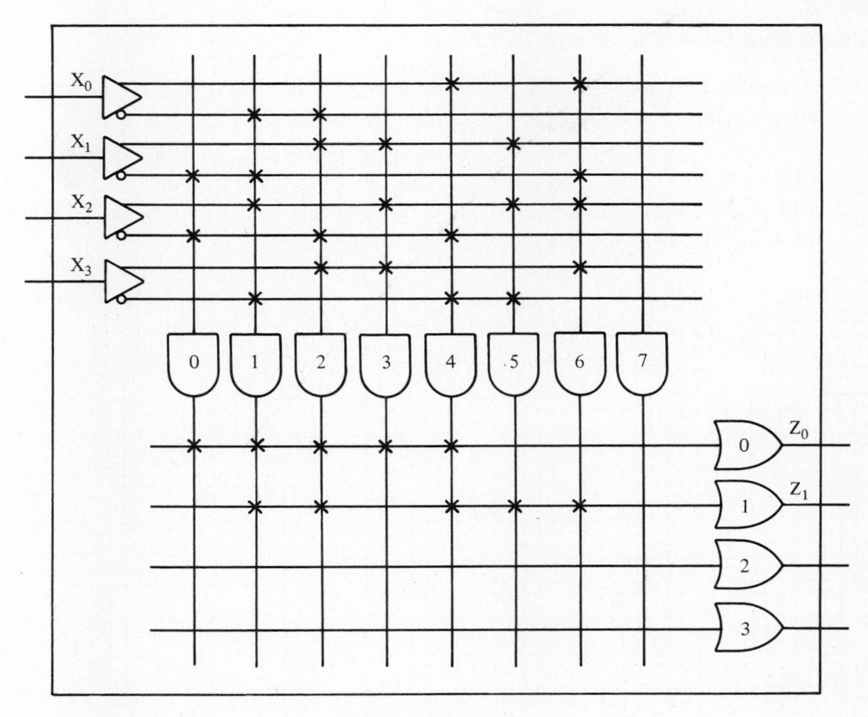

(e) PLA realization

Figure 6.12 *(cont.)*

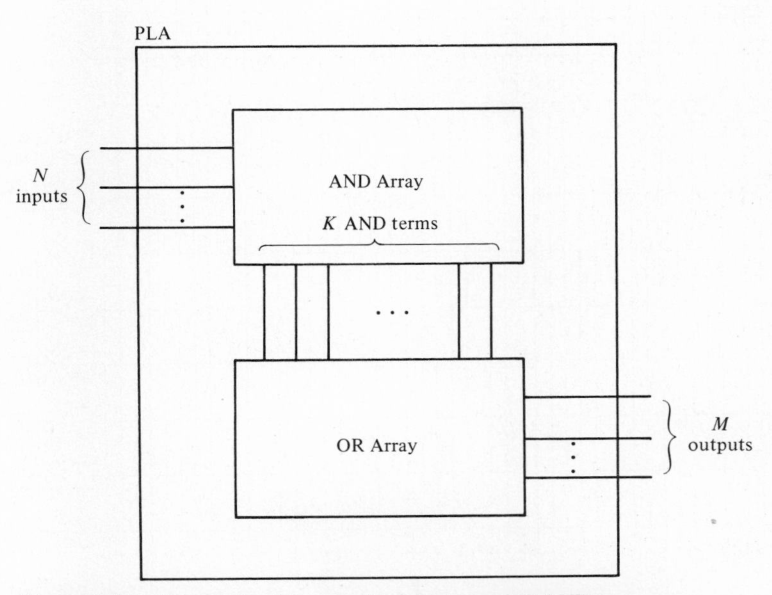

Figure 6.13 PLA with N inputs, M outputs, and K AND terms.

Commercially Available PLAs and FPLAs

Commercially available PLAs come in two forms: PLAs that are programmed by the IC manufacturer, and *field-programmable PLAs* (*FPLAs*) that can be programmed by users with FPLA programmers. An FPLA comes from the manufacturer with all the connections intact as integrated fuses. Using an FPLA programmer, a user can program an FPLA by leaving intact the desired connections (the x's in a PLA logic diagram) and blowing the fuses of the other unused connections.

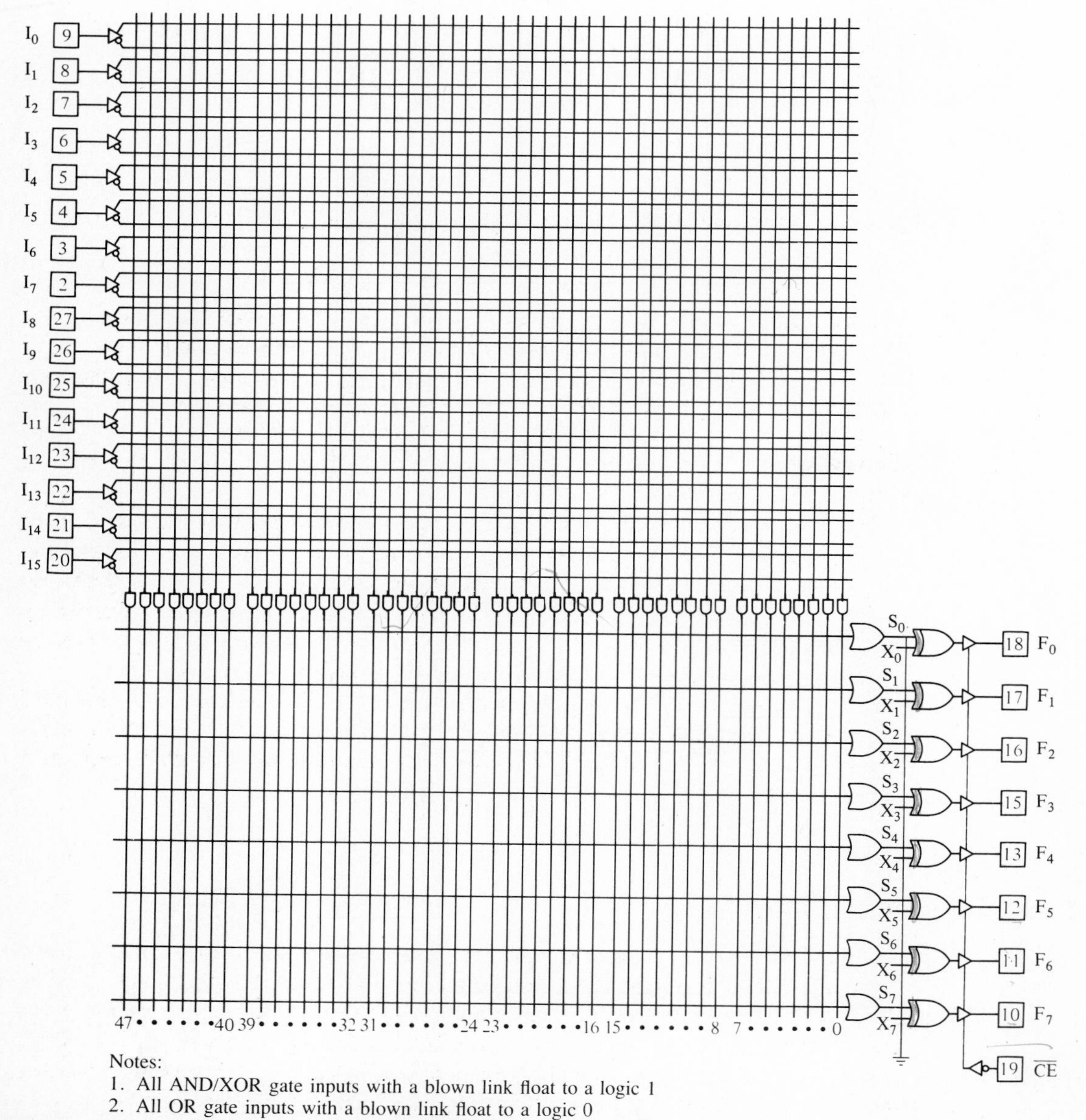

Figure 6.14 82S100 FPLA. (Courtesy of Signetics Corporation.)

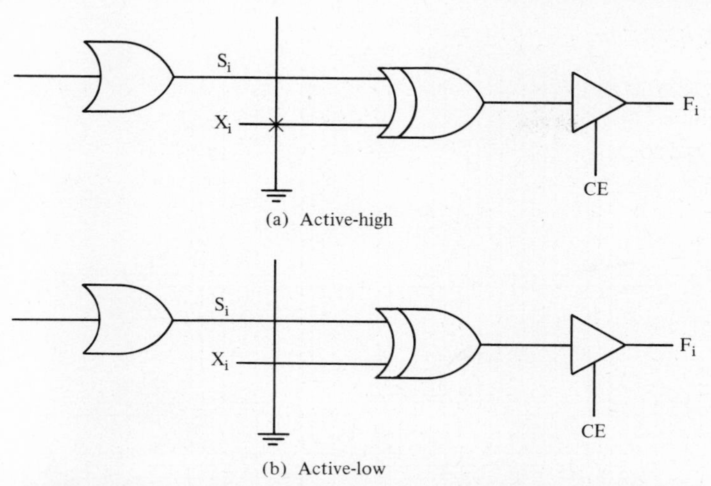

Figure 6.15 Programming the polarity of an FPLA output terminal.

A typical example of a commercially available FPLA is the 82S100 shown in Fig. 6.14. With its 16 inputs, 8 outputs, and support for up to 48 product terms, it is capable of replacing quite a few SSI circuit packages. In addition to the normal functions of a PLA, commercially available PLAs typically provide other functions as well. The 82S100, for example, provides a chip enable $\overline{CE}$ input (pin 19) that allows the outputs to be three-stated. Additionally, each output can be programmed to be active-high or active-low.

As shown in Fig. 6.15 for an 82S100, the programming of an output F_i to be active-high is obtained by leaving the fuse for X_i connected to ground as is graphically shown by the x. This ground provides a logic 0 to an input of the XOR gate, which also has an S_i input, thereby giving $S_i \oplus 0 = S_i$. On the other hand, for the programming of an output F_i to be active-low, the fuse for X_i is blown. This causes the input X_i to be left floating high, as shown in Fig. 6.15(b), thereby giving $S_i \oplus 1 = \overline{S_i}$.

Other features are also available in other commercially available PLAs. Included is the ability to program a terminal to be an input or an output terminal. Also, some PLAs have on-chip flip-flops for the realization of a single-chip ''state machine,'' such as those to be presented in Chapter 7.

6.4.2 Programmable Array Logic

The programmable array logic (PAL) also realizes sum-of-products gate structures in a systematic manner. It is a special case of the PLA, having a fixed-OR array instead of the PLA programmable OR array. Consequently, it is sometimes called a fixed-OR array. PALs are commercially available in various sizes for providing various functions. Shown in Fig. 6.16 is a relatively small PAL, the PAL14H4, which has 14 inputs, with internal inverters to provide the respective complements. It also has four active-high outputs, each from an OR gate that can accommodate only four product terms.

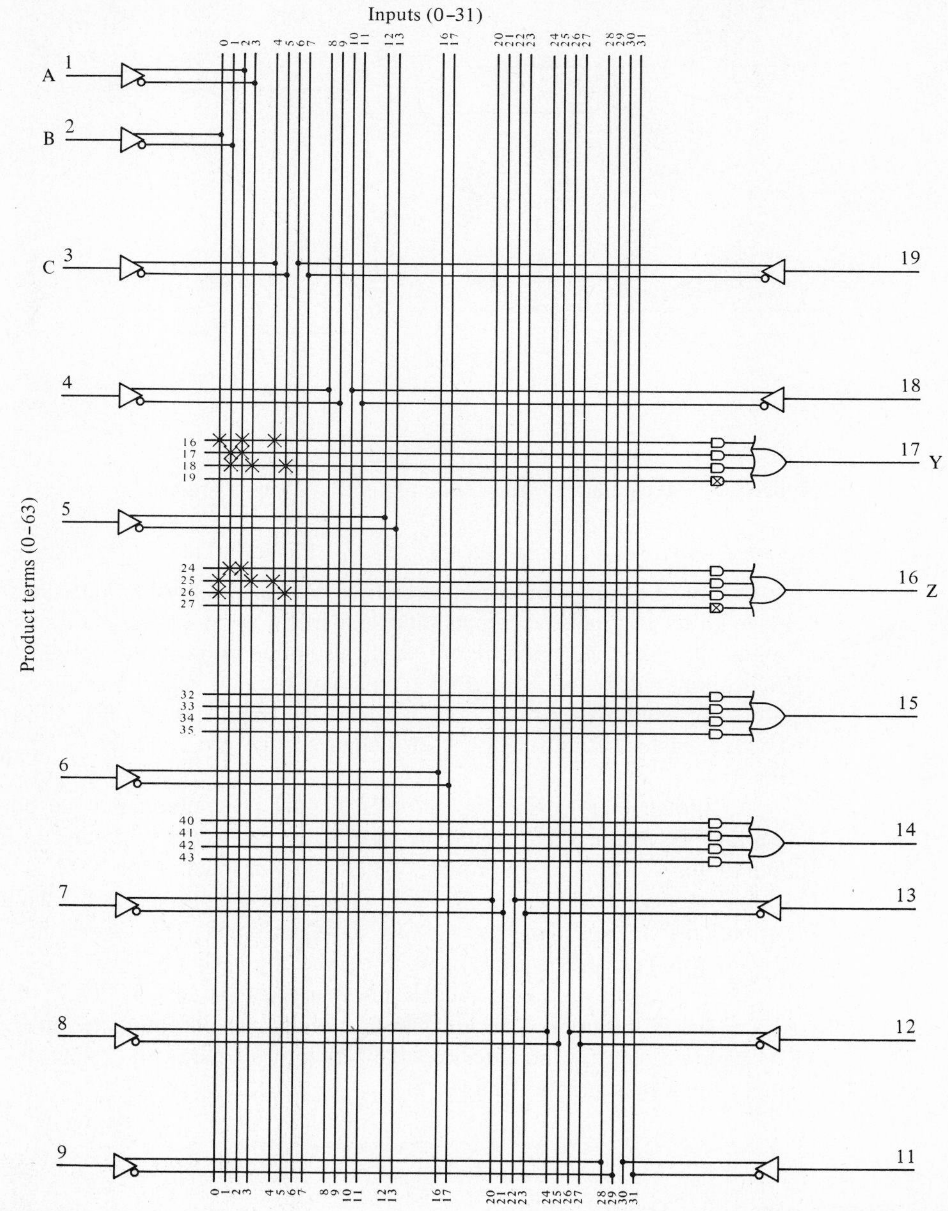

Figure 6.16 PAL14H4 realization of $Y = ABC + A\overline{B} + \overline{A}\,\overline{B}\,\overline{C}$ and $Z = A\overline{B} + \overline{A}BC + B\overline{C}$ from Fig. 6.10(a) of Example 6.1.

Figure 6.16 also illustrates the use of a PAL. The shown x's provide the realization of the combinational circuit in Fig. 6.10(a) of Example 6.1. Note the x's in two of the AND gate symbols. When an input of an OR gate is *not* used, an x is graphically placed in the corresponding AND gate symbol. As shown in Fig. 6.17(a), the x is simply a shorthand notation to designate that all the inputs (including the complement values) are

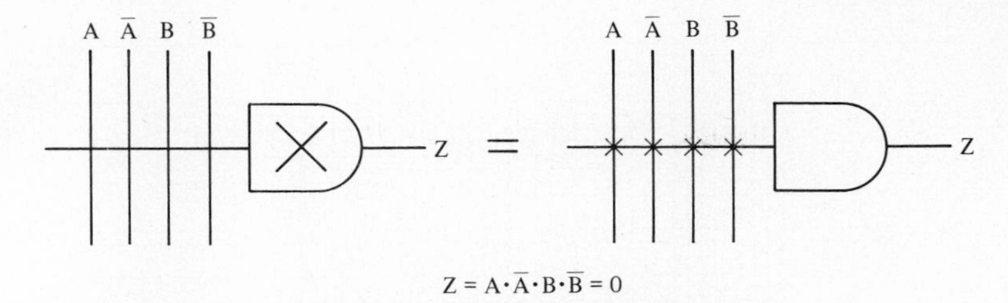

(a) All inputs are connected

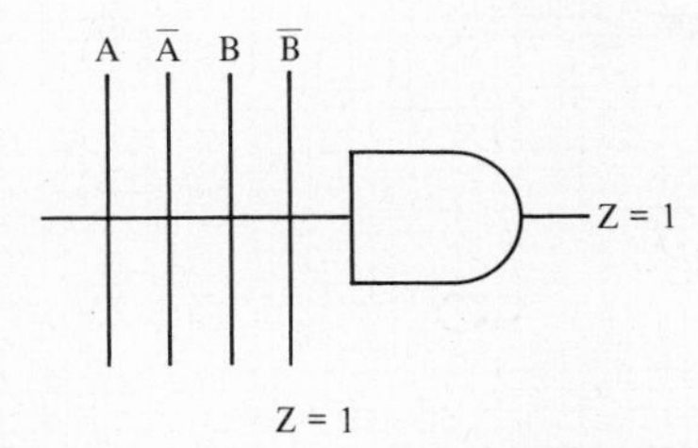

(b) All inputs are disconnected and left floating high

Figure 6.17 PAL shorthand notation.

left connected. As a result, the output of the AND gate is false (L) and so will not affect the function of the following OR gate. Note also that since the OR array is fixed (four AND gates are permanently assigned), the $A\overline{B}$ product term cannot be shared by the outputs, as it was in the PLA realization of Fig. 6.10(b). Instead, the product term $A\overline{B}$ has to be generated for both Y and Z.

In addition to providing the normal functions of a PAL, commercially available PALs typically provide other functions as well. Shown in Fig. 6.18 is the PAL16L8, which illustrates some of these additional functions. It has ten dedicated inputs (pins 1, 2, 3, 4, 5, 6, 7, 8, 9, and 11), and two dedicated active-low outputs (pins 12 and 19). Additionally, there are six I/O pins (pins 13–18), each of which can be programmed to be either an input or an output pin, as controlled by a three-state inverting buffer. Each output can accommodate up to seven product terms. With the programmable pins, the PAL16L8 can have up to 16 inputs and 2 outputs, or 10 inputs and 8 outputs, or any combination in between.

For an I/O pin to be programmed as an input, the output of the AND gate that controls the three-state buffer must be false (L). As is shown in Fig. 6.19(a), this is obtained by leaving connected all the inputs to that AND gate. For the programming of an I/O pin to be an output, the output of the controlling AND gate must be true (H). As is shown in Figs. 6.17(b) and 6.19(b), this is obtained by disconnecting all the inputs to that AND gate.

For an illustration of the use of a PAL, connections are shown in Fig. 6.18 for realizing the functions specified in the truth table in Fig. 6.12(b) of Example 6.3. The inputs X_0, X_1, and X_2 are assigned to dedicated input pins, but input X_3 is assigned to an I/O pin that is programmed as an input pin. The output Z_0 is assigned to a dedicated output pin, and output Z_1 is assigned to an I/O pin that is programmed to be an output pin.

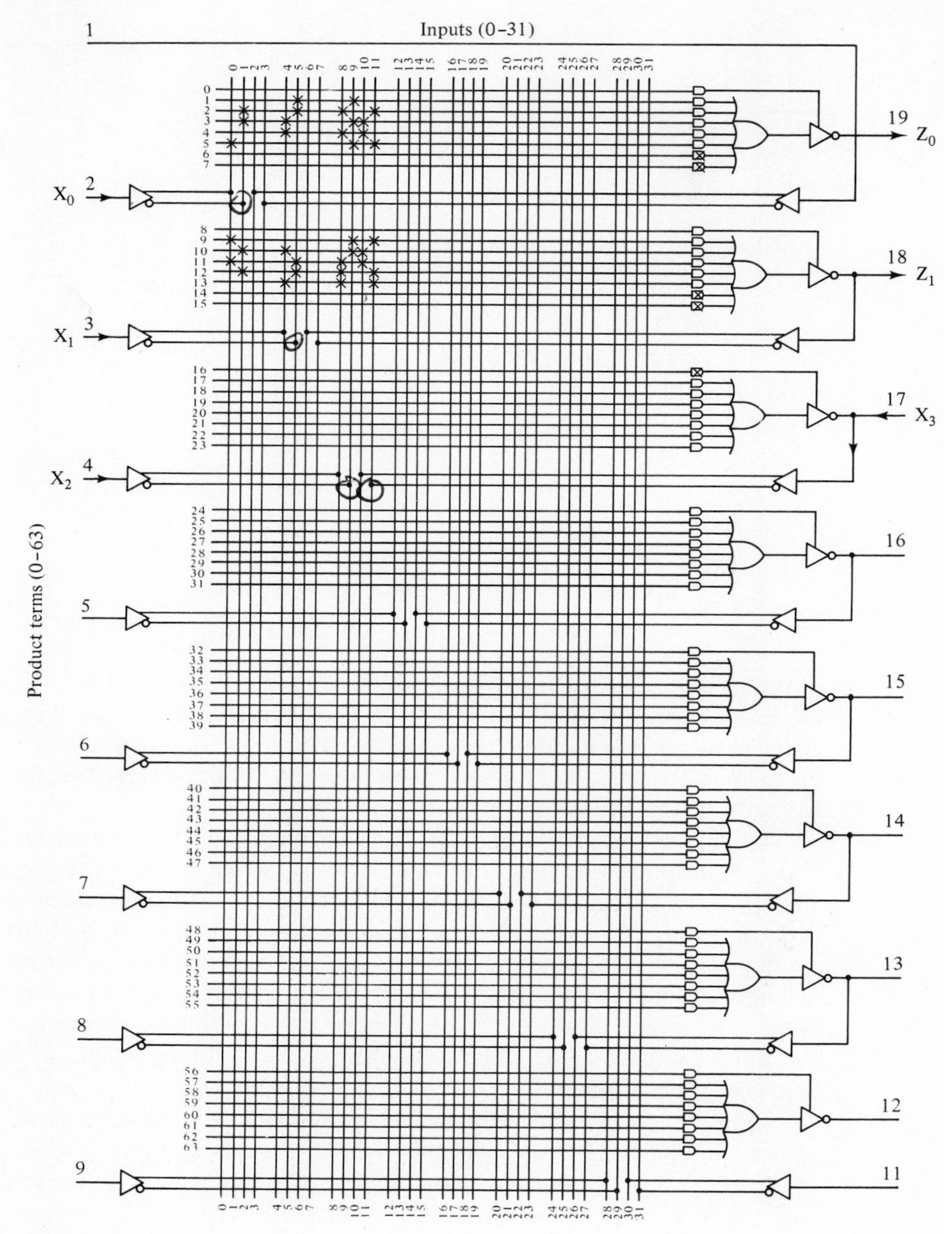

From Fig. 6.12(c):

$$Z_0 = \overline{X}_1 \overline{X}_2 + \overline{X}_0 \overline{X}_1 X_2 \overline{X}_3 + \overline{X}_0 X_1 \overline{X}_2 X_3 + X_1 X_2 X_3 + X_0 \overline{X}_2 \overline{X}_3$$

$$Z_1 = X_0 \overline{X}_2 \overline{X}_3 + \overline{X}_0 X_1 \overline{X}_2 X_3 + X_0 \overline{X}_1 X_2 X_3 + \overline{X}_0 \overline{X}_1 X_2 \overline{X}_3 + X_1 X_2 \overline{X}_3$$

Figure 6.18 PAL16L8 realization of Example 6.3.

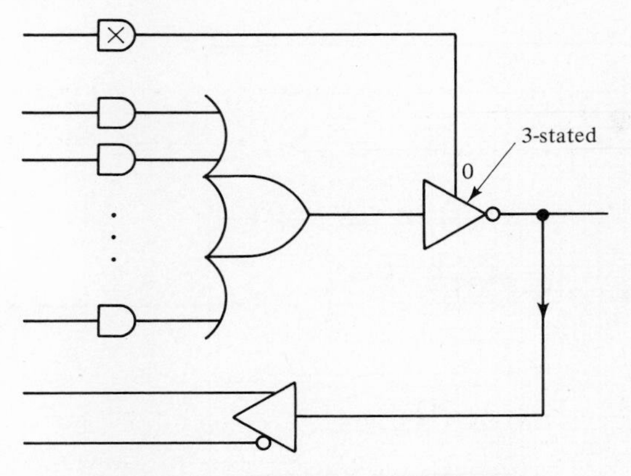

(a) PAL I/O terminal programmed as an input

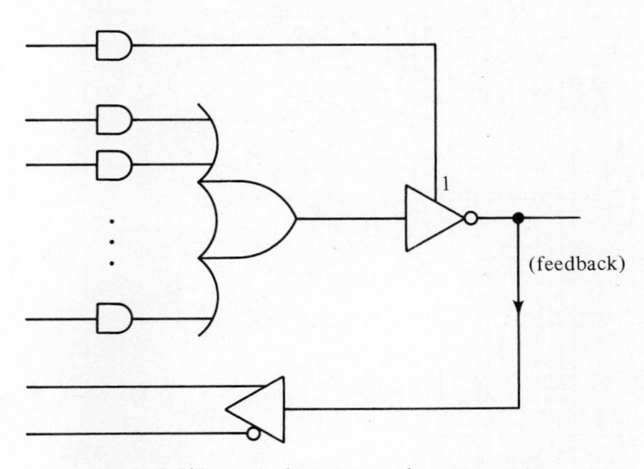

(b) PAL I/O terminal programmed as an output

Figure 6.19 Programming of an I/O terminal of a PAL.

Observe that Z_0 and Z_1 are active-low outputs. If active-high outputs are required, there are three ways of obtaining them. External inverters can be used to change the polarity of the outputs. Alternatively, DeMorgan's laws can be used to convert the logic functions to obtain $\overline{Z_0}$ and $\overline{Z_1}$, and then these can be realized with the PAL. [Recall that $\overline{Z}$.L is equal to Z.H. (See Problem 6.14.)] Finally, of course, another comparable PAL with active-high outputs can be used.

In summary, a PAL can be used to realize sum-of-products expressions in a systematic manner, and it is a special case of the PLA. Since the number of product terms for a PAL is fixed to a limited number for each output, a PAL is more restrictive in use than a PLA. However, when applicable, a PAL is less expensive and is generally easier to program than a PLA. These attributes make the PAL an attractive alternative to discrete SSI gates as the basic components of a digital system. According to PAL manufacturers, a single PAL package can realize the equivalent logic of 4 to 12 SSI and MSI packages.

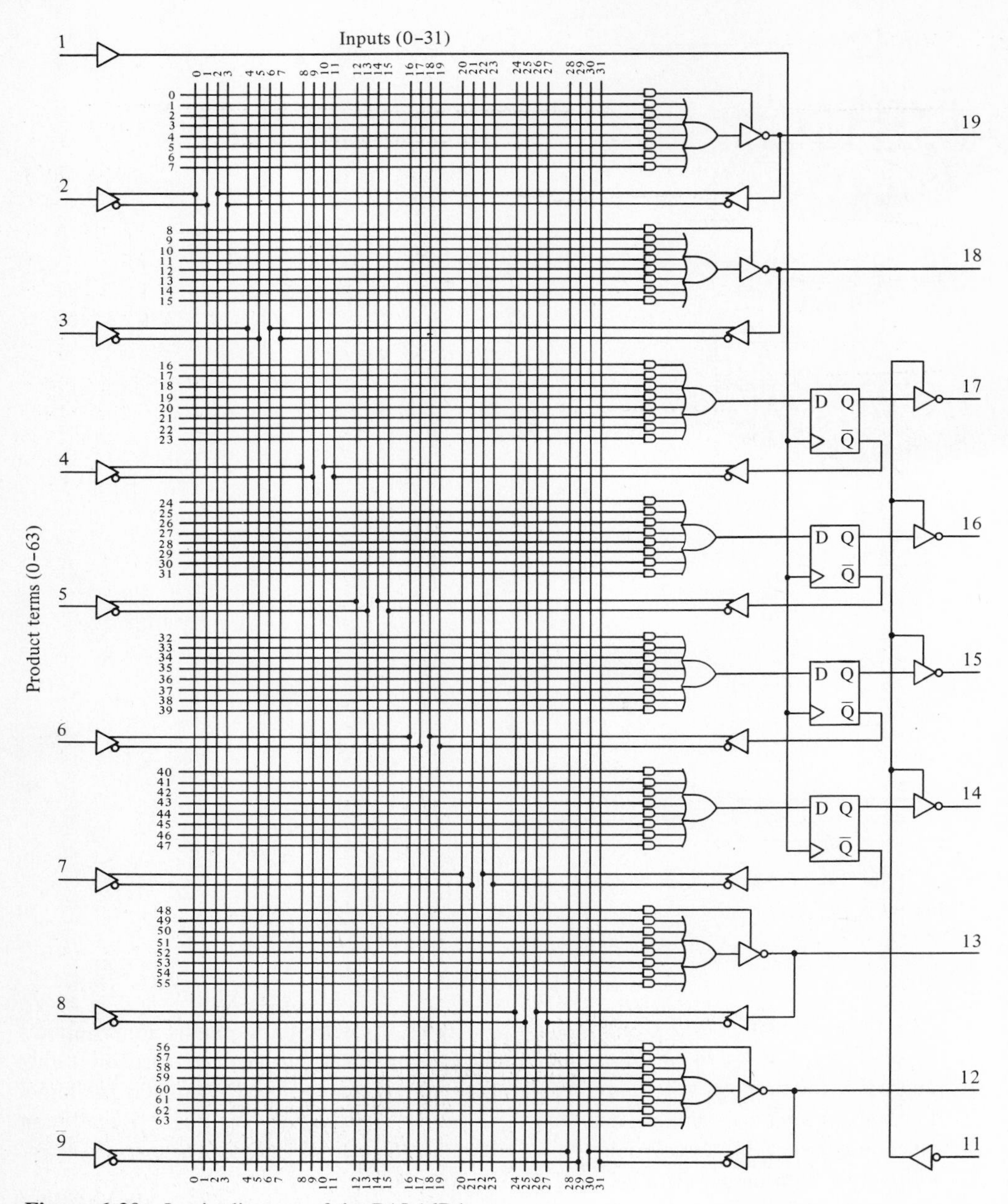

Figure 6.20 Logic diagram of the PAL16R4.

Additionally, other PALs, such as the PAL16R8, the PAL16R6, and the PAL16R4, have on-chip flip-flops along with the PAL arrays, as is shown in Fig. 6.20 for the PAL16R4. With one of these it is possible with a single IC to realize a state machine, such as one of those that are discussed in Chapter 7.

6.5 MEMORIES

Memories are circuit elements that are used in digital circuits for storing large amounts of information. A general model of a memory circuit element is shown in Fig. 6.21(a). Conceptually, it is a collection of 2^n addressable storage registers, each of which contains m bits. Associated with each storage register, which is called a *memory location,* is a unique *memory address*. As shown in Fig. 6.21(a), the address of the first memory location is 0, that of the second memory location is 1, and so forth up to the last memory location which has an address of $2^n - 1$. With the specification of an n-bit address at the ADDRESS inputs, the contents of any of the 2^n memory locations can be accessed directly (i.e., randomly), without the need to sequentially traverse the preceding locations to get to the specified location. For this reason, this type of memory is commonly called a *random access memory*. The m-bit data is transferred to and from a memory location through m bidirectional (input/output) DATA lines of the memory unit.

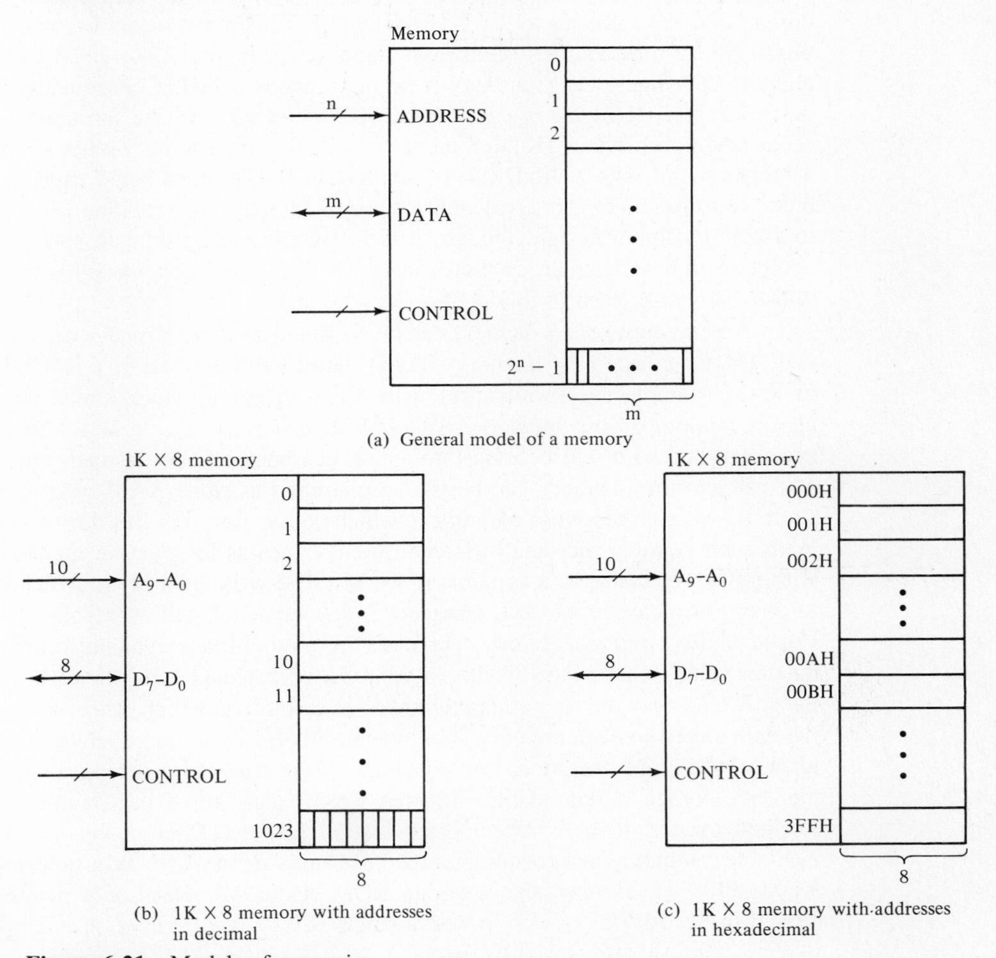

(a) General model of a memory

(b) 1K X 8 memory with addresses in decimal

(c) 1K X 8 memory with addresses in hexadecimal

Figure 6.21 Models of memories.

In general, two types of memory operations can be performed: memory *read* and memory *write*. For a read operation, data is retrieved from one of the memory locations by specifying an n-bit address at the ADDRESS inputs, and applying appropriate control signals at the CONTROL inputs to cause the contents to be read from the specified location. After a time equal to the access time of the memory, the m-bit contents of that memory location are available on the DATA outputs. For a write operation, data is stored into one of the memory locations by specifying an n-bit address at the ADDRESS inputs and applying the m-bit data to be stored at the DATA inputs. At the same time, appropriate control signals are applied at the CONTROL inputs to cause the m-bit data to be stored at the specified memory location.

The capacity of a memory unit is characterized by the number of memory locations that it contains and the number of bits per memory location. The capacity of a memory unit can be determined from the number of its ADDRESS input lines and the number of its DATA lines. For an illustration, consider the memory unit of Fig. 6.21(b) which has ten ADDRESS lines and eight DATA lines. With a 10-bit address, we can generate unique addresses for up to $2^{10} = 1024$ different memory locations, which means that this memory unit has this number of memory locations. Also, since the memory has eight DATA lines, each memory location contains 8 bits. Consequently, the memory unit of Fig. 6.21(b) has a capacity of 1024×8 bits. In the terminology of random access memory, it is a $1\text{K} \times 8$ memory unit, in which K represents 1024. In general, a memory unit with n ADDRESS lines and m DATA lines has a capacity of $2^n \times m$ bits. Note that in Fig. 6.21(b), the memory addresses are specified in decimal, from 0 to 1023. In digital design, though, it is frequently more useful to specify the memory addresses in hexadecimal, as shown in Fig. 6.21(c). Hex notation for memory addresses will be generally used in this book.

Any random access memory can be classified as either a *read-write* memory (RWM or RAM) or a *read-only* memory (ROM). Read-write memory is commonly referred to as RAM (random access memory), which is a misnomer since a read-only memory is also a random access memory. Although a misnomer, the term RAM is universally accepted, and so it will be used throughout this book to refer to read-write memory.

Read-write memory can be further classified as *static* RAM or *dynamic* RAM. A static RAM is a read-write memory in which data is stored in flip-flop storage elements. With such storage, the data bits retain their values as long as the memory is supplied with power. In contrast, a dynamic RAM is a read-write memory in which data is stored as charges on capacitors. Left unattended, any capacitor will eventually lose its charge. Consequently, periodic refresh operations are required in a dynamic RAM to retain the data bit values. Static RAM is discussed in Sec. 6.5.1, and dynamic RAM in Sec. 6.5.3.

A read-only memory is a random access memory in which, under normal operation, the data stored in each memory location can be read by a read operation, but cannot be altered by a write operation. An advantage of the read-only memory over RAM is that the data storage is nonvolatile. In other words, data stored in a read-only memory is retained even if there is a temporary loss of power. Different versions of ROM are available, including masked programmed read-only memory (ROM), field-programmable ROM (PROM), and erasable program ROM (EPROM). Read-only memories are discussed in Sec. 6.5.2.

6.5.1 Static RAM

A general model of the static RAM is shown in Fig. 6.22(a). It is essentially the same block diagram as that of Fig. 6.21(a) except for having a detailed specification of the control inputs, which are the (normally) active-low inputs WE and CE. The functions of these inputs are given in the table of Fig. 6.22(b).

For a read operation, the *n*-bit address is specified at the ADDRESS inputs, and the chip-enable (CE) input is made true (L) and the write-enable (WE) input is made false (H). After a time equal to the access time of the memory, the *m*-bit contents of the specified location become available at the DATA outputs. For a write operation, the *n*-bit address is specified at the ADDRESS inputs. Additionally, the *m*-bit data to be stored is applied at the DATA inputs. Also, the CE and WE inputs are made true (L). Then, after a time equal to the access time of the memory, the data is stored in the specified memory location.

Static RAMs are commercially available in various sizes, such as, for example, 256 × 4, 1K × 1, 1K × 4, 4K × 1, 2K × 4, 2K × 8, and up. Shown in Fig. 6.23 is a typical example of a commercially available 1K × 4-bit static RAM with three-state outputs. As shown in the block diagram of Fig. 6.23(a), it has ten address lines (A_9–A_0) and four data lines (D_3–D_0). There are also two active-low control inputs: WE (write enable) and CS (chip select). The controls for the operations of the RAM are summarized in the table of Fig. 6.23(b). The WE input specifies the operation, and is false (H) for a memory read and true (L) for a memory write. The chip is functional only if the chip-select input (CS) is true (L). Otherwise, the data lines, D_3–D_0, are put into a high-impedance (high-Z) state that electrically disconnects them from the data outputs.

Often in the design of a digital system, the desired memory module requires a capacity greater than that provided by any commercially available memory chip. Then it is necessary to construct this memory module from several memory modules of smaller sizes. Shown in Fig. 6.24 is a technique for realizing a memory module with an additional number of *bits* per memory location. For the realization of the 1K × 8 memory of Fig.

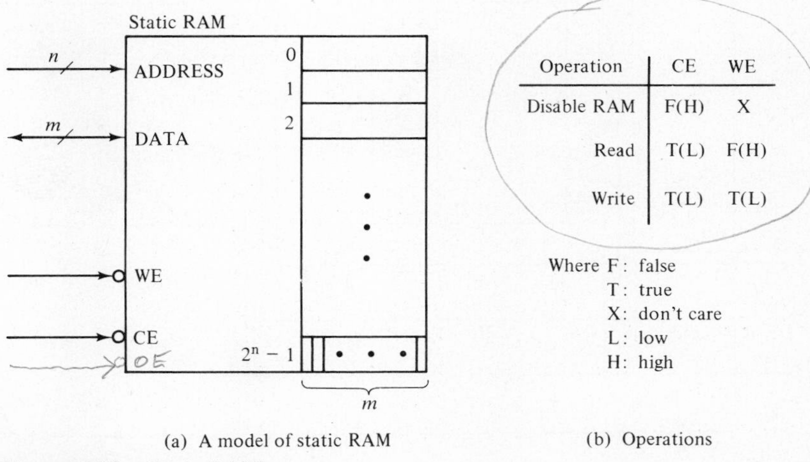

Operation	CE	WE
Disable RAM	F(H)	X
Read	T(L)	F(H)
Write	T(L)	T(L)

Where F: false
T: true
X: don't care
L: low
H: high

(a) A model of static RAM

(b) Operations

Figure 6.22 Static RAM.

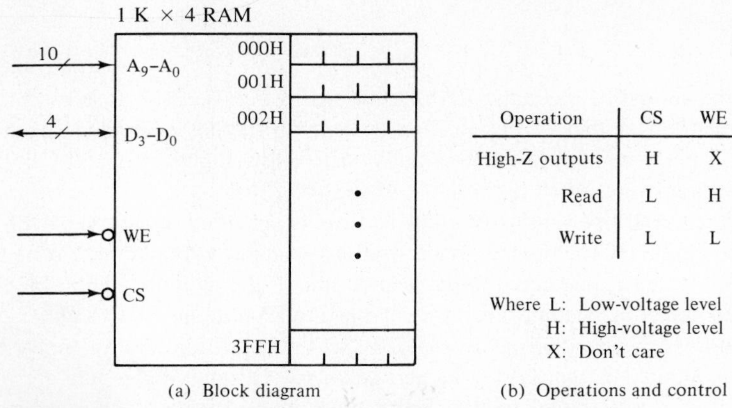

Operation	CS	WE
High-Z outputs	H	X
Read	L	H
Write	L	L

Where L: Low-voltage level
H: High-voltage level
X: Don't care

(a) Block diagram (b) Operations and control

Figure 6.23 Typical commercially available 1K × 4 RAM.

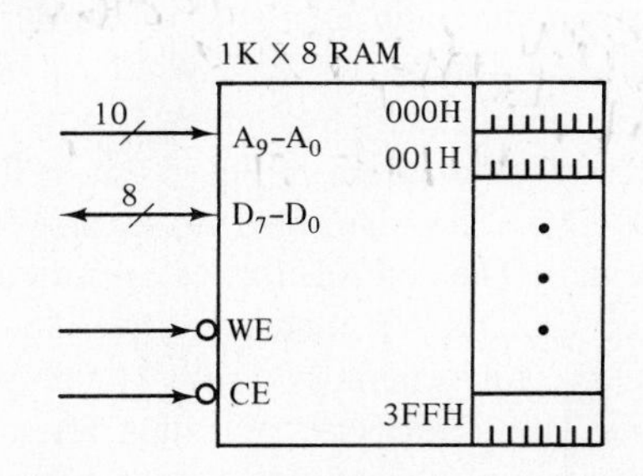

(a) Block diagram of a 1K × 8 RAM

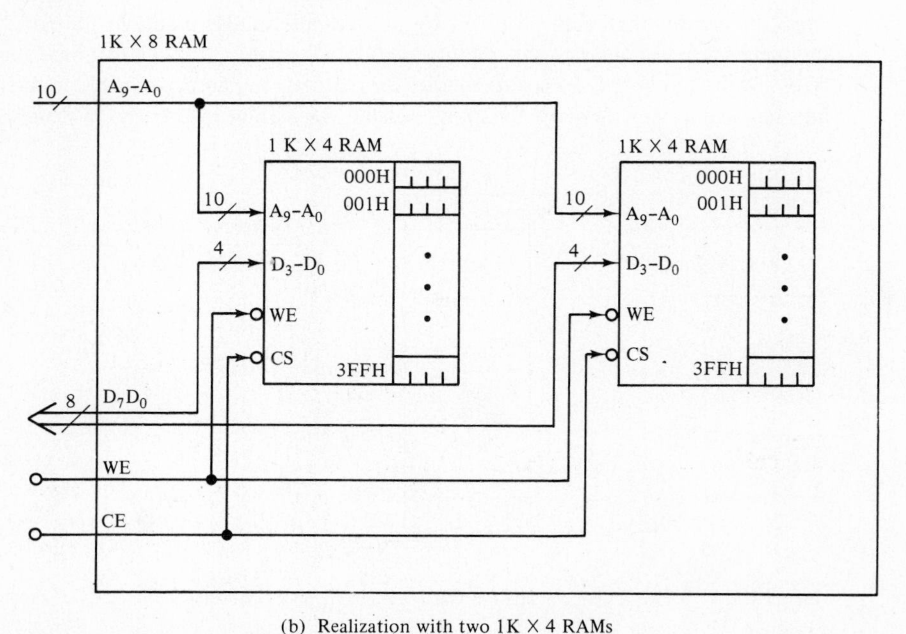

(b) Realization with two 1K × 4 RAMs

Figure 6.24 Realization of a 1K × 8 RAM.

6.24(a) with 1K × 4 RAMs, two of these RAMs are required, as shown in Fig. 6.24(b). Note that the inputs to the two sets of address lines are identical. Consequently, when a 10-bit address (A_9–A_0) is applied, the same relative memory locations for both memory chips are accessed, with the high nibble being found in one memory chip and the low nibble in the other. Together they form the 8-bit data for that address.

In Fig. 6.25 is shown a technique for realizing a memory module having an additional number of *memory locations*. For the realization of the 2K × 4 RAM of Fig. 6.25(a) with 1K × 4 RAMs, two of these RAMs are again required, as shown in Fig. 6.25(b). In this figure note the convention used for labeling the addresses of the memory

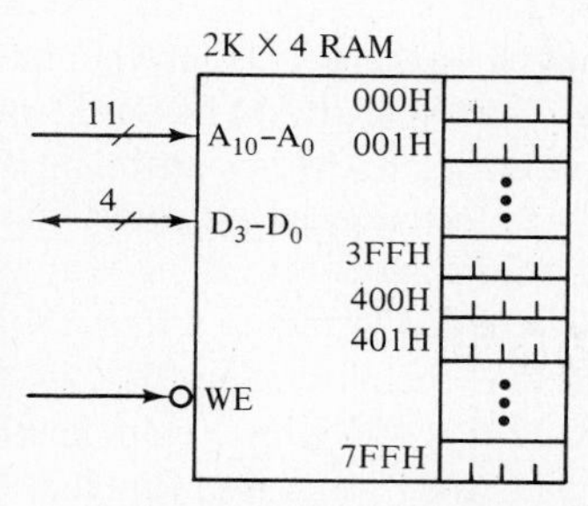

(a) Block diagram of a 2K X 4 RAM

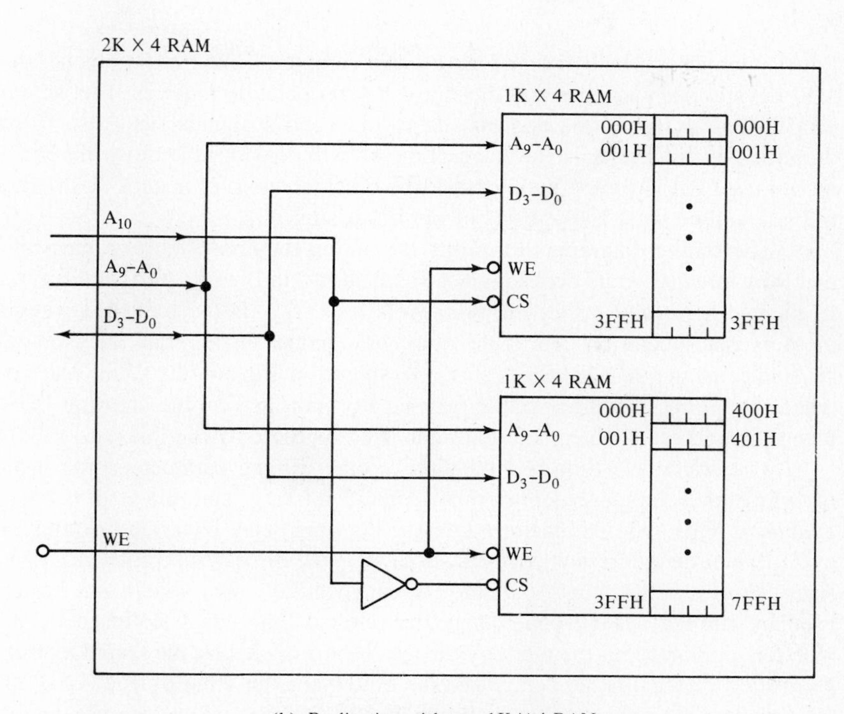

(b) Realization with two 1K X 4 RAMs

Figure 6.25 Realization of a 2K × 4 RAM.

locations. The 000H to 3FFH addresses labeled on the inside of each 1K × 4 RAM are the addresses of the locations with respect to that particular 1K × 4 RAM. But the 000H to 7FFH addresses labeled on the outside of the 1K × 4 RAMs are the addresses of the locations with respect to the entire 2K × 4 RAM.

Observe that the chip-select (CS) inputs of the two 1K × 4 RAMs are controlled by the high-order address bit A_{10}. A value of $A_{10} = 0$ enables the top 1K RAM, representing the first 1K block of memory ($\underline{0}$0000000000B to $\underline{0}$1111111111B). At the same time, this value disables the bottom 1K RAM, representing the second 1K block of memory ($\underline{1}$0000000000B to $\underline{1}$1111111111B), and three-states it from the external data lines. Conversely, a value of $A_{10} = 1$ enables the bottom 1K RAM and disables the top 1K RAM. As a result, each location of the 2K module has a unique 11-bit address even though the corresponding locations of the two 1K RAMs have the same 10-bit address. For example, the first location of the top 1K RAM has an 11-bit address of $\underline{0}$0000000000B and the first location of the bottom 1K RAM has an 11-bit address of $\underline{1}$0000000000B.

With a combination of these techniques, a memory module of any reasonable size can be realized with smaller memory modules, along with some external circuitry. For example, a realization of a 2K × 8 RAM requires four 1K × 4 RAMs and an inverter (see Problem 6.19). A 4K × 4 RAM requires four 1K × 4 RAMs and a 4-to-2 decoder (see Problem 6.20). And a 2K × 4 RAM with a chip-enable input requires two 1K × 4 RAMs along with inverters and AND gates (see Problem 6.21).

RAM Timings

RAMs, which are LSI devices, have more complex timing requirements than SSI and MSI devices such as gates and flip-flops. RAM operation requires strict adherence to the proper sequencing of the address, data, and control signals, and also to the required durations of these signals. In this section we will consider the most important and commonly used RAM timing parameters. The block diagram of a static RAM shown in Fig. 6.26(a) will be used as the basis of our discussion.

The timing diagrams illustrating the timing requirements for a *memory read cycle* and for a *memory write cycle* of a RAM are shown in Figs. 6.26(b) and (c), respectively. In the memory read cycle, the *read-cycle time,* t_{RC}, is the total time required for the memory read operation, and is the minimum amount of time that the n-bit address must be stable on the ADDRESS inputs. As shown in Fig. 6.26(b), the read cycle begins when the address becomes stable (graphically indicated by the crossing lines) and ends when the address is changed (again indicated by the crossing lines).

The read-cycle time is a function of other timing parameters, the most important of which are the *access-time-from-address,* t_A(AD), and the *access-time-from-chip-enable,* t_A(CE). The parameter t_A(AD) is the time delay from the beginning of the read cycle, when the address is changed, until the time the data becomes valid on the DATA lines. Consequently, for a certainty of data validity, data should not be accessed and used by another system component until after a time equal to this parameter t_A(AD), which is published on the memory device sheet provided by the manufacturer. The other parameter, t_A(CE), is the delay from the time that a CE input of true (L) is applied, until the time that the data becomes valid on the DATA lines. Consequently, enabling the CE input earlier will result in the data being valid earlier, thereby reducing the read-cycle

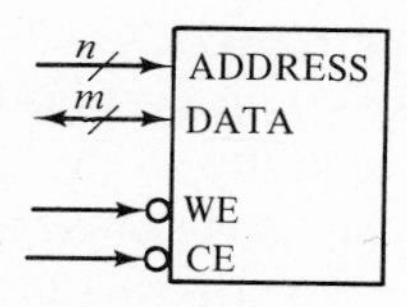

(a) Block diagram of a RAM

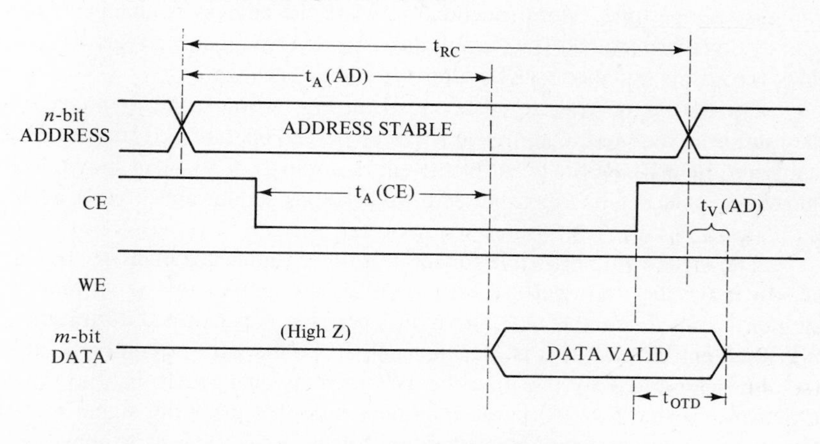

(b) Timing diagram for a memory read cycle

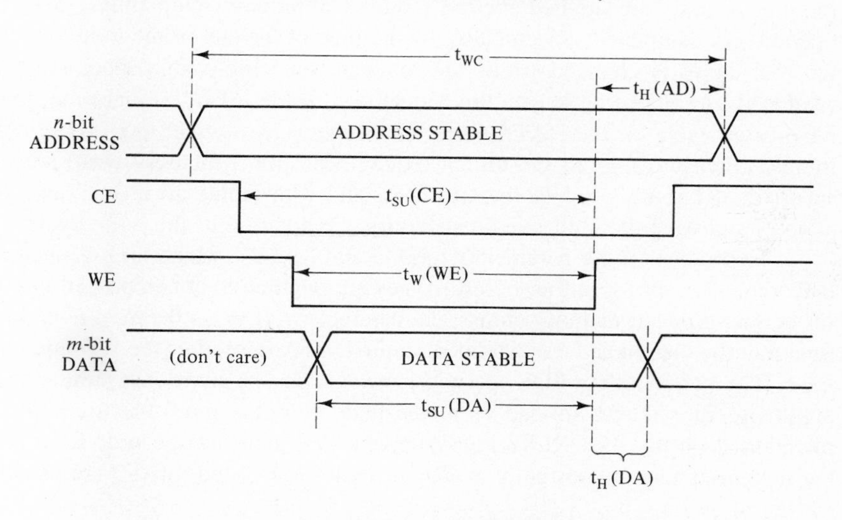

(c) Timing diagram for a memory write cycle

Figure 6.26 Memory read and memory write cycles.

time t_{RC}. As a rule of thumb, for the minimization of t_{RC}, it is best to apply the CE input at the same time as the address and to keep it stable for the entire duration of the read cycle. Of course, the write enable WE should be false (H) for the duration of the read cycle. Minimizing the read-cycle time improves the performance of the digital system since this time t_{RC} determines the maximum rate at which the memory can be read.

The two timing parameters t_{OTD} and $t_V(AD)$, which are of less importance, are also shown in Fig. 6.26(b). The parameter t_{OTD} is the time required for the data *outputs to three-state from deselection*. More specifically, if the CE input is changed to false before the address is removed, then the chip is no longer enabled. The data, however, still remains available on the DATA lines for a period of time equal to t_{OTD} until the data outputs become three-stated. The other parameter $t_V(AD)$ is the *output-valid-after-address-change* time. More specifically, when the address is changed, then the current read cycle is terminated. The current data, however, will still remain valid on the DATA lines for a time equal to $t_V(AD)$ after the address change.

For the memory write cycle shown in Fig. 6.26(c), the *write-cycle time,* t_{WC}, is the total time required to complete a memory write operation. The n-bit address must be stable on the ADDRESS lines for the entire duration of t_{WC}. As shown in Fig. 6.26(c), the write cycle begins when the address becomes stable, and it ends when the address is changed.

The write-cycle time is a function of other timing parameters, the most important of which are the *write-pulse width,* $t_W(WE)$, the *chip-enable-setup time,* $t_{SU}(CE)$, and the *data-setup time,* $t_{SU}(DA)$. The write operation is performed during the time that the WE input is true (L)—that is, during $t_W(WE)$. So the write operation must be completed and the data stored by the time the WE input is changed from true (L) to false (H), which means that $t_W(WE)$ is the minimum time that the write signal must be true (L). The most important timing consideration for the write cycle is to make certain that the required signals are applied for the required durations (setup times) before the write operation is completed. Specifically, by the time of the end of the write operation, when the WE input is changed from true to false, the chip-enable input (CE) must have remained true (L) for at least a time equal to $t_{SU}(CE)$. Also, by that time, the data must have been stable on the DATA lines for a time at least equal to $t_{SU}(DA)$. As a rule of thumb, to ensure that all the timing requirements for a memory write are going to be satisfied, it is best to apply the CE, WE, and data inputs all at the same time as the address and have them all remain stable for the duration of the write cycle.

Two other timing parameters $t_H(DA)$ and $t_H(AD)$, which are of much less importance, are also shown in Fig. 6.26(c). They are required to obtain proper write operations for certain types of memory chips. The parameter $t_H(DA)$ is the *data hold time*. It is the time that the data must be maintained on the DATA lines after the WE input has become false. The parameter $t_H(AD)$, the *address hold time,* is somewhat similar except that it applies to the address instead of to the data. It is the time that the address must be maintained on the ADDRESS lines after the WE input has become false. For most of the memories that are currently available, both $t_H(DA)$ and $t_H(AD)$ are usually zero.

6.5.2 Read-Only Memory

Functionally, a read-only memory, such as is shown in Fig. 6.27, is a special case of a read-write memory. For the normal operation of a read-only memory, the data that is stored in each memory location can be read with a read operation but cannot be altered by a write operation. As already stated, the principal advantage of the read-only memory over RAM is that the data storage is nonvolatile. More specifically, unlike for a RAM, the data stored in a read-only memory will be retained even after a loss of power.

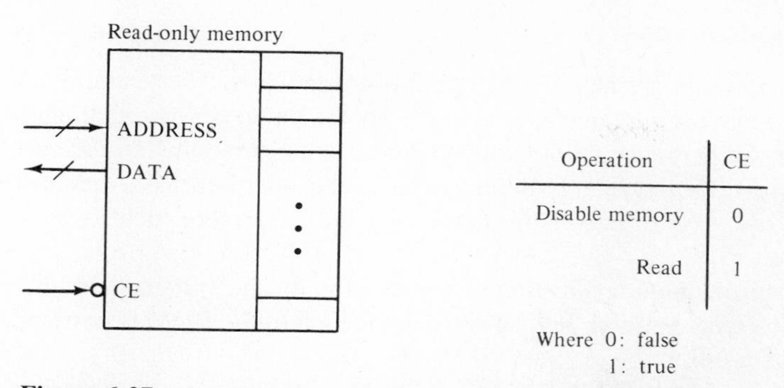

Figure 6.27 A model of a read-only memory under normal operation.

Therefore, a temporary loss of power will not cause a loss of data—a feature that is important in many applications.

The different types of read-only memories can be distinguished by the manner in which the data is originally stored. For a *masked programmed read-only memory* (ROM), the data is permanently stored by the manufacturer during the fabrication of the chip. Once stored, the contents of a ROM are fixed and cannot be altered. Since a custom mask has to be produced for each design of a ROM, mask programmed ROMs are economical only if manufactured in large quantities.

A *field-programmable read-only memory* (PROM) is the functional equivalent of a ROM under normal operating conditions. However, the one-time programming of a PROM is performed by the *user,* using a PROM programmer, rather than by the manufacturer. Physically, the PROM construction is based on the same integrated fuse technology used for field-programmable PLAs, as discussed in Sec. 6.4.1. A PROM comes from the manufacturer with all the connections intact as fuses. The user, using a PROM programmer, programs the PROM by leaving intact the fuses representing the 1-value bits and blowing the fuses representing the 0-value bits. The flexibility and relatively low cost of PROMs make them attractive for the small quantity production of parts that require read-only memories.

The *erasable programmable read-only memory* (EPROM) is another type of read-only memory. Under normal operation, an EPROM is the functional equivalent of a ROM or a PROM. Like a PROM, an EPROM can be programmed by the user, by means of an EPROM programmer. Unlike the ROM and PROM, however, the programming of an EPROM is not irreversible. The stored data, although nonvolatile, can be erased by placing the EPROM in an EPROM eraser. During the erasing process, ultraviolet radiation slowly releases the charge that has been stored as data and thereby restores the EPROM to its original state. The flexibility of the EPROM makes it ideal for developmental prototype implementation, and also for low-volume production in applications where the contents of a read-only memory need to be changed.

As mentioned, the functions of the various types of read-only memories are equivalent under normal operating conditions. For convenience, therefore, we will subsequently use the generic term read-only memory (ROM) to refer to all types of read-only memories (ROM, PROM, and EPROM), unless otherwise specified.

Read-Only Memory Applications

Read-only memories are required for applications in which a large amount of information needs to be stored in a nonvolatile manner. By far the most important application of the ROM is for the permanent storage of microprocessor programs and fixed tables of data for a microprocessor system. Microprocessors and microprocessor-based design are the subjects of the second half of this book. We will, therefore, defer this application till then.

Another common application of the ROM is for the systematic realization of complex combinational circuits. Perhaps this application of the ROM is best understood from a simple example.

Example 6.4 BCD-to-7-Segment Conversion

Consider the design and realization of a combinational circuit for converting a 4-bit BCD number to a 7-bit number that corresponds to the seven segments of a 7-segment display. The problem statement is summarized in Fig. 6.28(a), and the resultant truth table is shown in Fig. 6.28(b). Since this is strictly a combinational circuit, it can be realized in a straightforward manner with the techniques presented in Chapters 2 and 3, and so it will not be shown. Obviously, though, the realization would have seven sum-of-product AND-OR structures, with a substantial package count.

Alternatively, this combinational circuit can be realized with a single ROM that has a capacity greater than or equal to 16×7 bits. The result is shown in Fig. 6.28(c). Note that the right-hand side of the truth table of Fig. 6.28(b) is simply stored as the contents of the 16×7 ROM. Then when a particular set of inputs is applied, this set of incoming 4 bits is used as the address for looking up the contents of the corresponding ROM location, whose contents correspond to the valid outputs for that set of inputs.

From an input-output point of view, the operation of this ROM is indistinguishable from that of a traditional combinational circuit realization. For example, for an input of $B_3B_2B_1B_0 = 0101$, a traditional realization with discrete SSI components would output

$$Z_6Z_5Z_4Z_3Z_2Z_1Z_0 = 1011011$$

For the ROM realization of Fig. 6.28(c) the resultant output would also be

$$D_6D_5D_4D_3D_2D_1D_0 = Z_6Z_5Z_4Z_3Z_2Z_1Z_0 = 1011011$$

And for any other of the 16 possible sets of inputs, the same outputs would be obtained for either realization. Consequently, if the circuits resulting from the two different realization techniques were put into two separate ''black boxes,'' they would be *functionally* indistinguishable. ■■

It should be obvious that a ROM can be used for the realization of any combinational logic function. One important application is the realization of a microprogrammed controller, which is one of the subjects of the next chapter. This realization consists of a relatively complex combinational circuit, along with a number of flip-flops. For multi-input and output combinational realizations, the ROM method is generally superior to the traditional method, as is apparent from the power of, for example, the commercially available $8K \times 8$ EPROM, which can be used for the realization of a combinational

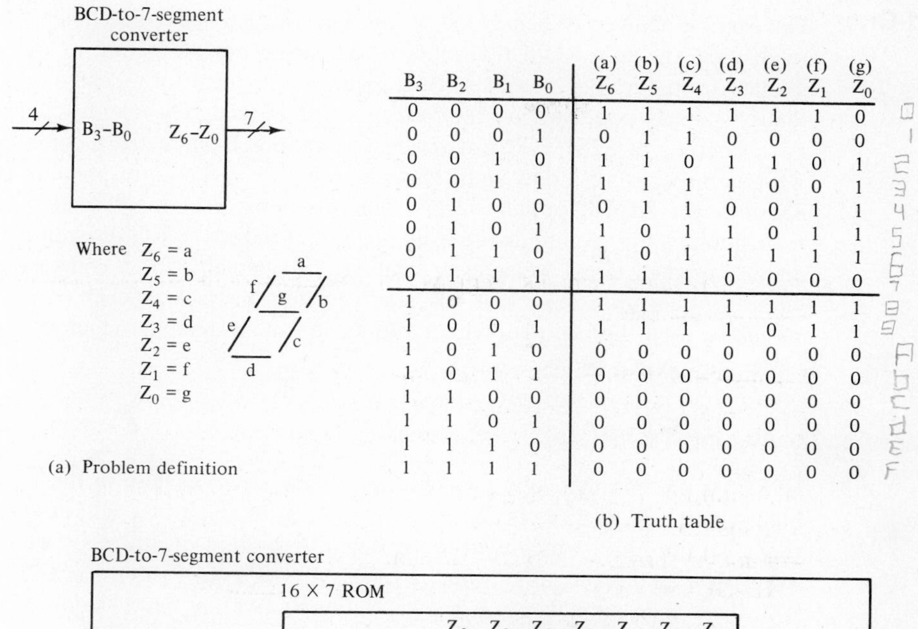

(a) Problem definition

B_3	B_2	B_1	B_0	(a) Z_6	(b) Z_5	(c) Z_4	(d) Z_3	(e) Z_2	(f) Z_1	(g) Z_0
0	0	0	0	1	1	1	1	1	1	0
0	0	0	1	0	1	1	0	0	0	0
0	0	1	0	1	1	0	1	1	0	1
0	0	1	1	1	1	1	1	0	0	1
0	1	0	0	0	1	1	0	0	1	1
0	1	0	1	1	0	1	1	0	1	1
0	1	1	0	1	0	1	1	1	1	1
0	1	1	1	1	1	1	0	0	0	0
1	0	0	0	1	1	1	1	1	1	1
1	0	0	1	1	1	1	1	0	1	1
1	0	1	0	0	0	0	0	0	0	0
1	0	1	1	0	0	0	0	0	0	0
1	1	0	0	0	0	0	0	0	0	0
1	1	0	1	0	0	0	0	0	0	0
1	1	1	0	0	0	0	0	0	0	0
1	1	1	1	0	0	0	0	0	0	0

(b) Truth table

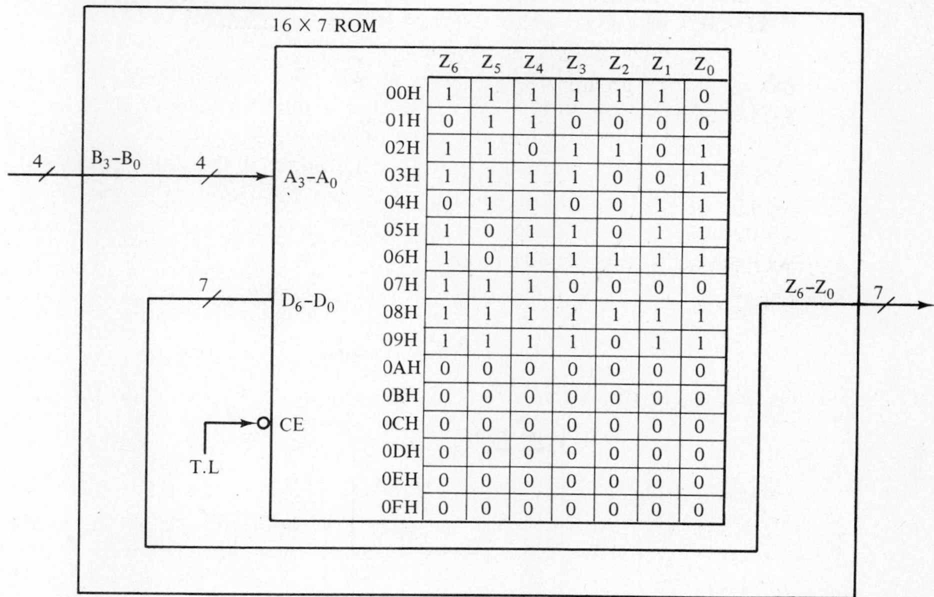

	Z_6	Z_5	Z_4	Z_3	Z_2	Z_1	Z_0
00H	1	1	1	1	1	1	0
01H	0	1	1	0	0	0	0
02H	1	1	0	1	1	0	1
03H	1	1	1	1	0	0	1
04H	0	1	1	0	0	1	1
05H	1	0	1	1	0	1	1
06H	1	0	1	1	1	1	1
07H	1	1	1	0	0	0	0
08H	1	1	1	1	1	1	1
09H	1	1	1	1	0	1	1
0AH	0	0	0	0	0	0	0
0BH	0	0	0	0	0	0	0
0CH	0	0	0	0	0	0	0
0DH	0	0	0	0	0	0	0
0EH	0	0	0	0	0	0	0
0FH	0	0	0	0	0	0	0

(c) ROM realization

Figure 6.28 ROM realization of a BCD-to-7-segment converter.

circuit having as many as 13 inputs and 8 outputs. The disadvantage of the ROM method is speed. The access time for a ROM, as for any memory, is slow compared to the speed of operation of a discrete logic realization.

The ROM method has the same advantage of package count reduction as does the PLA/PAL method of combinational circuits presented in Sec. 6.4. A ROM realization, however, is slower. When speed is not a constraint, then the choice of a PLA/PAL realization versus a ROM realization is dependent on the nature of the design. As is evident from Sec. 6.4, the PLA/PAL realization is ideal for a design with a truth table that is sparsely populated by 1s, corresponding to relatively few distinct product terms. For example, the 82S100 PLA described in Sec. 6.4.1 can realize a combinational circuit with as many as 16 inputs and 8 outputs, but only if the corresponding truth table has 48 or fewer distinct product terms. For this number of inputs and outputs, a ROM realization would require an absurdly huge (2^{16}) 64K × 8 ROM, even if the number of distinct product terms were fewer than 48. In summary, for a complex combinational circuit with a truth table that has relatively few distinct product terms, a PLA/PAL should be used. For a circuit with a truth table that has many product terms, then a ROM is the appropriate choice.

Commercially Available EPROMs

EPROMs are commercially available in various sizes, usually in the form of nK × 8 bits, where n = 2, 4, 8, 16, 32, 64, ... As an example, we will describe the original EPROM, the 2716, which is from the Intel Corporation. It illustrates the most important features of a commercially available EPROM. The 2716 EPROM is a 2K × 8 EPROM with three-state outputs. As is shown in Fig. 6.29(a), it has 11 address lines (A_{10}–A_0) and 8 data lines (D_7–D_0). Also, it has an output-enable input (OE), a chip-enable/EPROM program input (CE/PRG), and inputs V_{cc} and V_{pp} for two voltage supplies.

As is summarized in the table of Fig. 6.29(b), the 2716 EPROM has five main operations. Under normal operating conditions, if the chip-enable (CE/PRG) and the output-enable (OE) inputs are true (L), then the operation is a memory read. But if the chip-enable input is false, then the EPROM outputs (D_7–D_0) are three-stated, and the EPROM is in the standby mode.

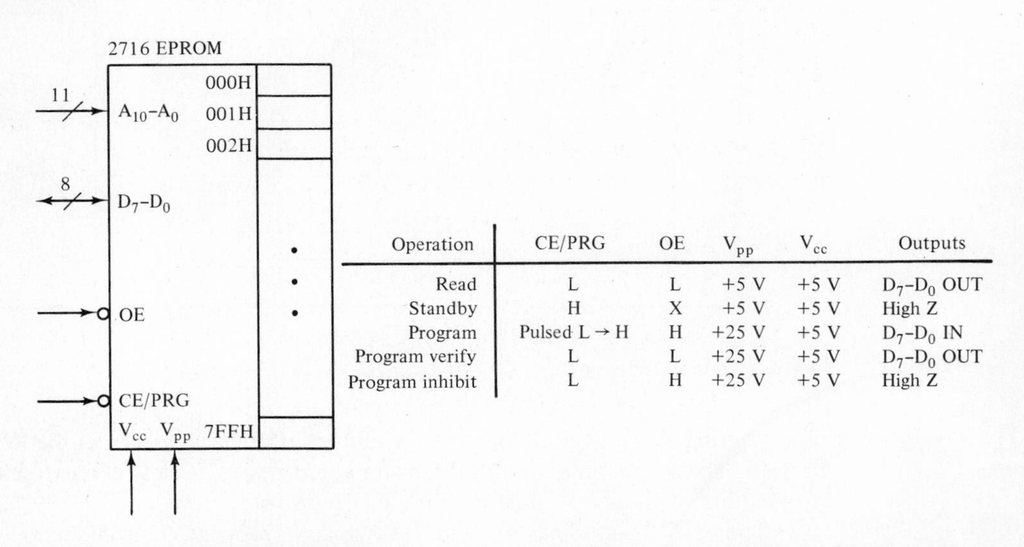

Operation	CE/PRG	OE	V_{pp}	V_{cc}	Outputs
Read	L	L	+5 V	+5 V	D_7–D_0 OUT
Standby	H	X	+5 V	+5 V	High Z
Program	Pulsed L → H	H	+25 V	+5 V	D_7–D_0 IN
Program verify	L	L	+25 V	+5 V	D_7–D_0 OUT
Program inhibit	L	H	+25 V	+5 V	High Z

(a) Block diagram of the 2716 EPROM (b) Table of operations

Figure 6.29 The 2716 EPROM.

The other three operations are for initially programming or for reprogramming the contents of the EPROM. An EPROM is usually programmed through the use of an EPROM programmer, a design tool that is widely available. The sequencing and timing of the various programming operations are the responsibility of the EPROM programmer. The steps for programming a 2716 EPROM are as follows: The 2716 EPROM is first put into the program-inhibit mode by setting CE/PRG to L and OE to H, and by raising the V_{pp} voltage supply to 25 volts. Also, the address of and the contents for the specified memory location are applied at the address and data lines, respectively. Next, the 2716 EPROM is put into the program mode by applying at the CE/PRG input an L-to-H pulse, which causes the data to be stored into the specified location. At the end of the L-to-H pulse, the return of the CE/PRG and OE inputs to L puts the 2716 into the program-verify mode, during which the contents of the just-programmed location are available on the data lines where they can be verified by the EPROM programmer, if desired. Finally, the OE input is changed to false (H) and so the program-inhibit mode is entered again. This programming cycle must be repeated for each location of the EPROM.

Again, the sequencing and timing of the various programming operations are the responsibility of the EPROM programmer. Consequently, EPROM programming is relatively easy for a digital designer. Because of its flexibility and ease of use, the EPROM is a very important tool for developmental and prototype implementations.

6.5.3 Dynamic RAM

As has been stated, read-write memory can be classified as either *static* RAM or *dynamic* RAM. A static RAM is a read-write memory in which data bits are stored in flip-flop storage elements. With this type of storage, the data bits retain their values as long as the memory is supplied with power. On the other hand, a dynamic RAM is a read-write memory in which data bits are stored as charges on capacitors. Unfortunately, left unattended, any capacitor will eventually lose its charge. Consequently, a dynamic RAM requires periodic refresh operations to retain the values of the stored data bits.

Dynamic RAMs offer several advantages over static RAMs. First, dynamic RAMs are approximately four times as dense as static RAMs. The result is a fourfold reduction of board space required for a given amount of memory. Second, and also as a result of the density, the cost per bit of dynamic RAMs is approximately one-fourth that of static RAMs. Finally, the power consumption of dynamic RAMs is significantly less than that of static RAMs.

On the other hand, one disadvantage of dynamic RAMs is that they are generally slower in operation than static RAMs. Another disadvantage of dynamic RAMs is the complexity of the functions required to support their operations. The additional support circuitry is justified only if a large memory module is required. Static RAMs are more cost-effective for smaller memory modules.

A general model of a dynamic RAM (DRAM) is shown in Fig. 6.30(a). It has n address lines and separate m-bit data-in (D_{IN}) and data-out (D_{OUT}) lines. For a reduction in the number of address pins, the n address lines are used to specify a $2n$-bit address. This is accomplished by time-multiplexing the required $2n$-bit address into two halves over the same n-bit address lines. The first half is commonly called the *row* address, and the second half is commonly called the *column* address. The row address is applied first at the address lines, and is latched internally by the dynamic RAM at the trailing edge of the RAS (row address strobe) input. Then the column address is applied at the address lines, and latched internally at the trailing edge of the CAS (column address strobe) input. Together, they form a $2n$-bit address that is capable of addressing up to 2^{2n} memory locations.

Shown in Fig. 6.30(b) is a typical example of a commercially available dynamic RAM. It is a relatively small 16K × 1 dynamic RAM; for it, $n = 7$ and $m = 1$.

In general, a dynamic RAM has three basic operations: *memory read, memory write,* and *memory refresh*. For a read or a write operation, the row and column addresses must be applied sequentially, and then latched, respectively, by the negative going edges of the RAS and CAS inputs. For a read operation, the WE (write enable) input must be made false (H), usually before the CAS input is made true (L), as shown in Fig. 6.31(a). After a time equal to the memory access time (measured from the trailing edge of CAS), the m-bit contents of the specified location are available on the D_{OUT} outputs. For a write operation, the row and column addresses are also latched. Additionally, the m-bit data

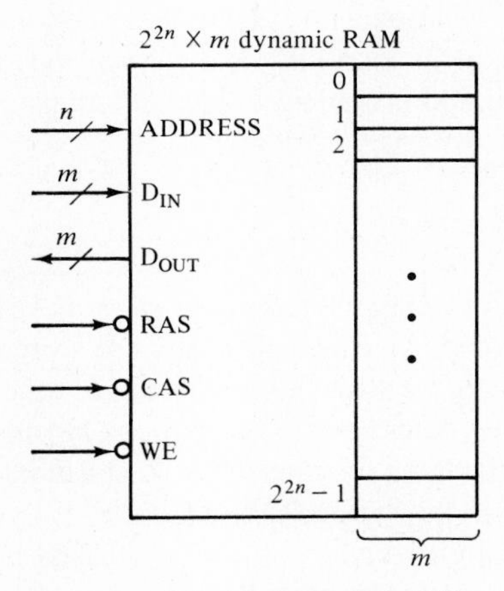

(a) General model of a dynamic RAM

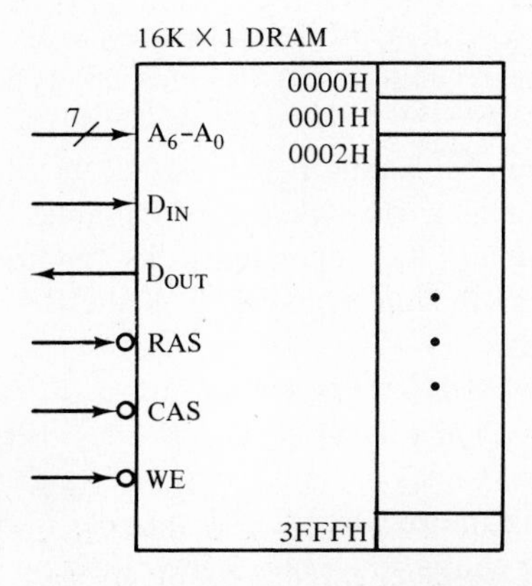

(b) 16K × 1 dynamic RAM

Figure 6.30 Dynamic RAM.

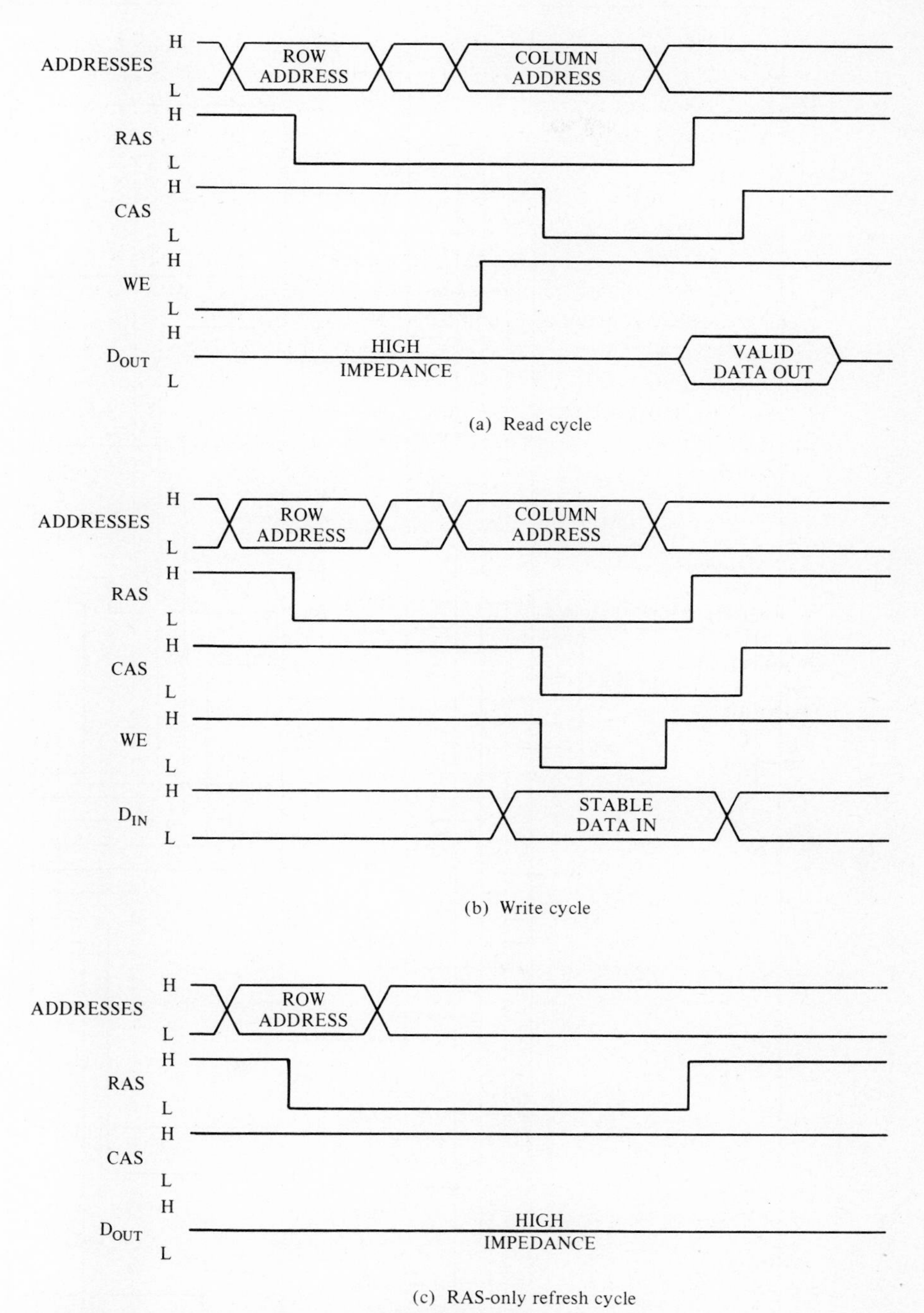

Figure 6.31 Basic operations for a dynamic RAM.

to be stored is applied at the D_{IN} inputs, and the WE input is made true (L). Then the data is latched by the dynamic RAM at the trailing edge of the CAS or WE signal, whichever occurs last, and is stored in the specified location.

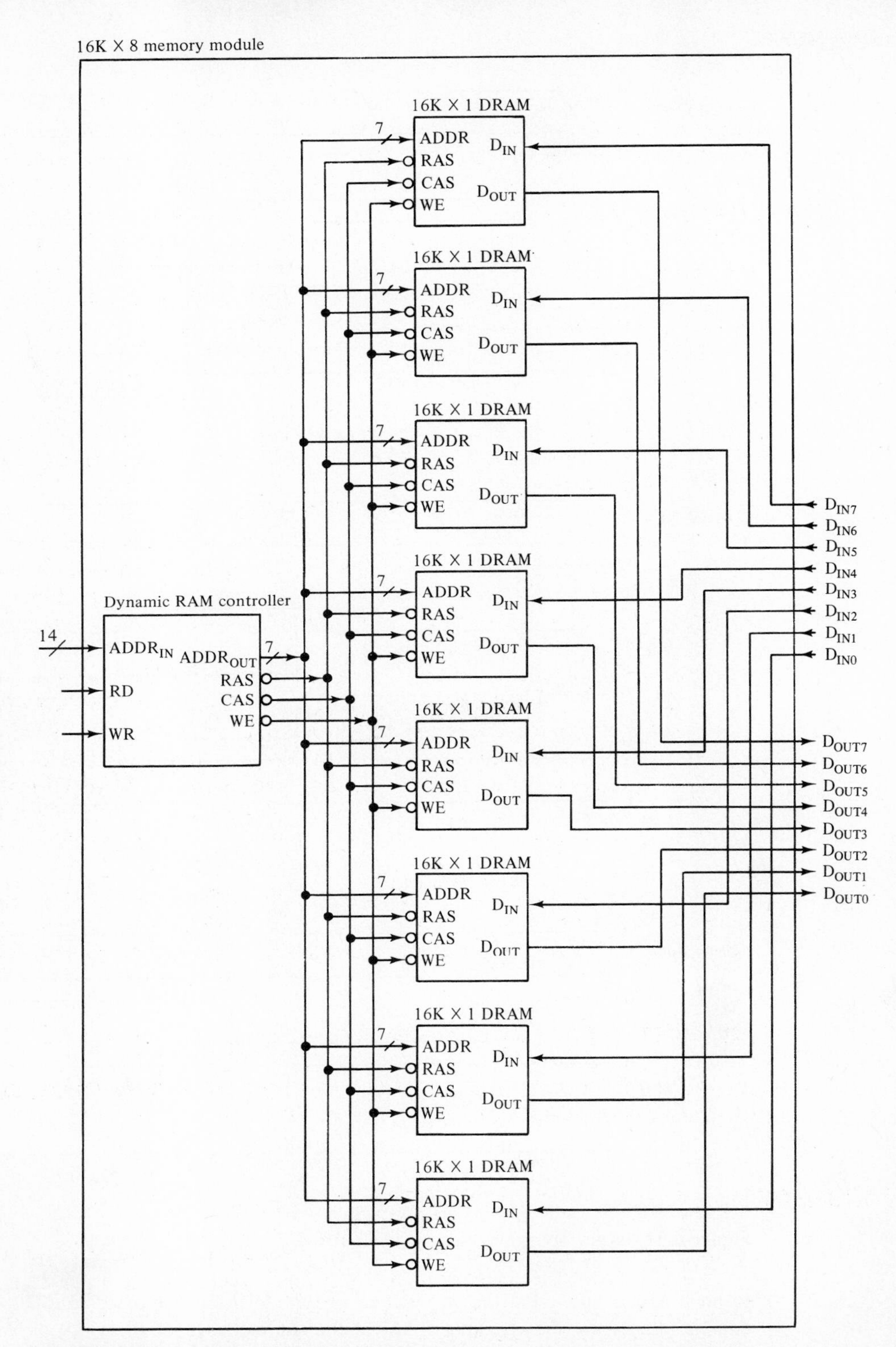

Figure 6.32 A 16K × 8 memory module constructed from dynamic RAMs.

As mentioned earlier, the data bits in a dynamic RAM are stored as charges on capacitors, and these charges must be periodically refreshed. For current commercially available dynamic RAMs, each bit-cell must be refreshed approximately every two milliseconds or less. Fortunately, dynamic RAMs are designed such that an entire row of bit-cells can be refreshed at once during the latching of the row address by the RAS input. Consequently, either the memory read operation or the memory write operation can be used to refresh a row of bit-cells. If we can be certain that every row of the memory will be accessed every 2 milliseconds or less by either a read or a write operation, then we can be certain that the memory will be properly refreshed. This is, of course, not generally the case. Therefore, a special memory refresh operation is required. The refresh cycle for a dynamic RAM is shown in the timing diagram of Fig. 6.31(c). It consists simply of latching the row addresses by the RAS input. Each refresh cycle refreshes one row of the dynamic RAM. Consequently, in order to refresh the entire dynamic RAM, a dynamic RAM controller needs to step through every row address within 2 milliseconds.

Shown in Fig. 6.32 is an example of a 16K × 8-bit memory module constructed from eight 16K × 1 dynamic RAMs and a dynamic RAM controller. Note that the inputs to the address lines of all eight 16K × 1 DRAMs are identical. Consequently, when an address is specified, the same relative memory location for all eight memory chips are accessed, with bit i being found in memory chip i. Together, the 8 bits from the eight chips form the 8-bit data for that address.

The inputs to the 16K × 8 memory and the dynamic RAM controller comprise a 14-bit address along with the RD and WR signals. The controller has the responsibility of time-multiplexing the 14-bit address into a 7-bit row address and a 7-bit column address. The controller must also generate and properly sequence the RAS, CAS, and WE signals as is required for the read and write operations. Furthermore, the controller also has the responsibility for generating and sequencing the signals that are required for refreshing the dynamic RAM bit-cells. Because of their complexity and common use, dynamic RAM controllers are commercially available in IC form.

A dynamic RAM controller is a *state machine* similar to those that will be studied in the next chapter. Therefore, the design methods of the next chapter will provide the reader with some insight into the design and realization of such a digital circuit.

SUPPLEMENTARY READING (see Bibliography)

[Blakeslee 79], [Intel-A], [Kline 83], [Mano 84], [Monolithic], [Motorola], [Short 81], [Signetics], [Texas Instruments]

PROBLEMS

6.1. What are the advantages of using LSI circuit elements in a digital circuit as compared to using MSI and SSI circuit elements?

6.2. A 16-bit ALU is to be realized by interconnecting four 74'181 ALUs.

(a) Draw the circuit diagram.

(b) Given that the propagation delay for a 74'181 to perform an add operation is

t_p(181ADD), how long does it take your ALU to perform a 16-bit add operation? Explain.

(c) Given that the propagation delay for a 74'181 to perform a logic operation is t_p(181LOG), how long does it take your ALU to perform a 16-bit logic operation? Explain.

6.3. Using a 74'181 and any additional logic that is required, design and realize the simplified ALU shown in block diagram form in Fig. 6.33. This ALU produces an output F that is the result of some operation on the inputs A and B. The particular operation depends on the control word SEL, as follows:

SEL	Operation	Definition
0	Add	F ← A plus B
1	Subtract	F ← A minus B
2	Increment	F ← A plus 1
3	Decrement	F ← A minus 1
4	Complement	F ← NOT A
5	OR	F ← A + B
6	XOR	F ← A ⊕ B

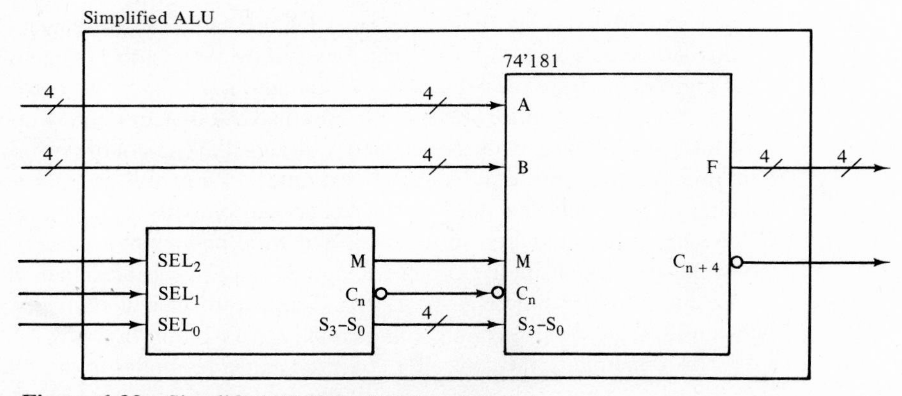

Figure 6.33 Simplified ALU for Problem 6.3.

6.4. Repeat Problem 6.3 using the block diagram of the simplified ALU shown in Fig. 6.34 and also the active-low view of the 74'181.

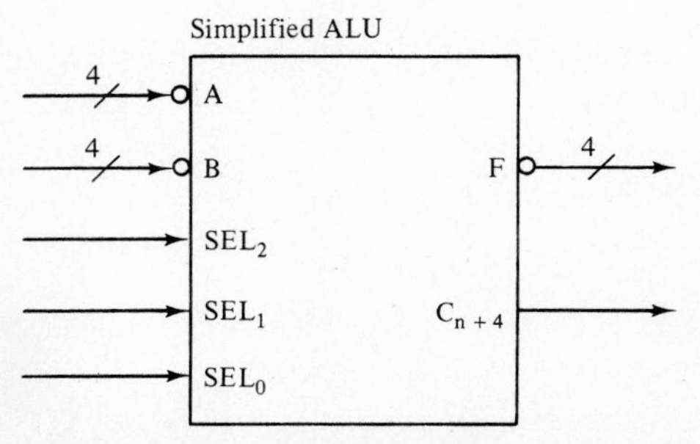

Figure 6.34 Simplified ALU for Problem 6.4.

6.5. For the chain of full adders shown in Fig. 6.2, what is the logic equation for the look-ahead carry circuit for C_2?

6.6. The following 2-bit numbers A and B are to be added:

Stage	N	$N-1$	$N-2$	$N-3$	$N-4$	$N-5$	$N-6$	. . .
A	0	1	0	0	1	0	1	. . .
B	1	0	0	1	1	1	0	. . .

Find the values of the carry-*outs* produced by the following stages: (a) $N-4$, (b) $N-2$, (c) $N-3$, (d) $N-1$, and (e) $N-5$. Explain your answers.

6.7. Convert the ripple adder circuit shown in Fig. 6.2 into a 4-bit adder with look-ahead carry circuitry, using a 74'182 and any additional logic that is required. (*Hint:* The carry-in of each adder stage will be generated by the 74'182.)

6.8. Determine the propagation delay required by an add operation for the 64-bit ALU with multilevel look-ahead carry structure shown in Fig. 6.7. Assume the following delay values:

$$t_p(181\text{PG}) = 33 \text{ ns to produce G and P}$$
$$t_p(181\text{ADD}) = 27 \text{ ns to perform an add operation}$$
$$t_p(182\text{PG}) = 25 \text{ ns to produce the } P_i \text{ and } G_i$$
$$t_p(182C_{xyz}) = 26 \text{ ns to produce } C_{n+x},\ C_{n+y}, \text{ and } C_{n+z}$$

6.9. Transform the logic diagram of Fig. 6.35 into a PLA circuit diagram similar to the one shown in Fig. 6.10(b).

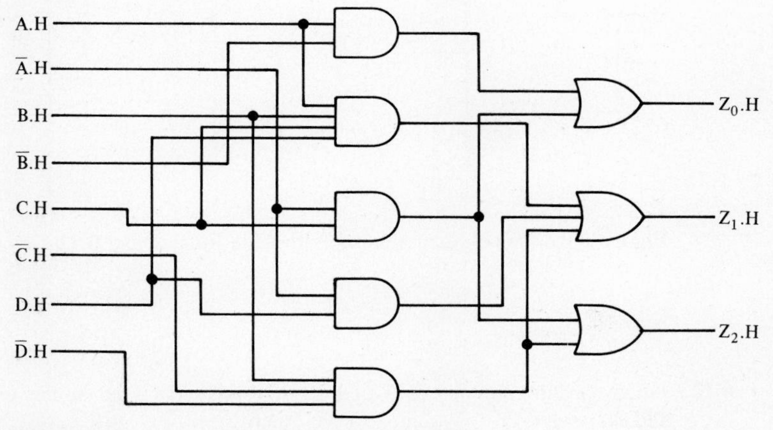

Figure 6.35 Logic diagram for Problem 6.9.

6.10. Given the truth table of Fig. 6.36 for the combinational circuit shown in block diagram form, realize the combinational circuit with a PLA that is similar to the one shown in Fig. 6.10(b) (i.e., one with four inputs, four outputs, and supporting eight product terms).

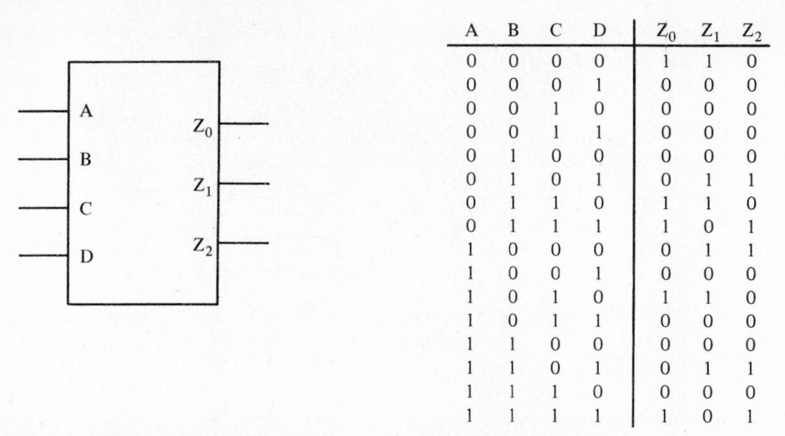

A	B	C	D	Z_0	Z_1	Z_2
0	0	0	0	1	1	0
0	0	0	1	0	0	0
0	0	1	0	0	0	0
0	0	1	1	0	0	0
0	1	0	0	0	0	0
0	1	0	1	0	1	1
0	1	1	0	1	1	0
0	1	1	1	1	0	1
1	0	0	0	0	1	1
1	0	0	1	0	0	0
1	0	1	0	1	1	0
1	0	1	1	0	0	0
1	1	0	0	0	0	0
1	1	0	1	0	1	1
1	1	1	0	0	0	0
1	1	1	1	1	0	1

Figure 6.36 Block diagram and truth table for Problem 6.10.

6.11. Repeat Problem 6.10 for the block diagram and truth table shown in Fig. 6.37.

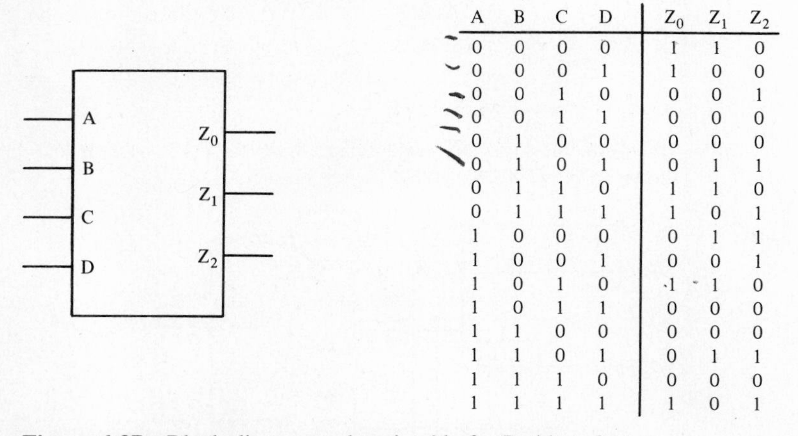

A	B	C	D	Z_0	Z_1	Z_2
0	0	0	0	1	1	0
0	0	0	1	1	0	0
0	0	1	0	0	0	1
0	0	1	1	0	0	0
0	1	0	0	0	0	0
0	1	0	1	0	1	1
0	1	1	0	1	1	0
0	1	1	1	1	0	1
1	0	0	0	0	1	1
1	0	0	1	0	0	1
1	0	1	0	1	1	0
1	0	1	1	0	0	0
1	1	0	0	0	0	0
1	1	0	1	0	1	1
1	1	1	0	0	0	0
1	1	1	1	1	0	1

Figure 6.37 Block diagram and truth table for Problem 6.11.

6.12. Given the programmed PLA of Fig. 6.38 with functions similar to those of the 82S100 FPLA,

(a) What are the logic/voltage assignments (i.e., active-high or active-low) for the inputs A, B, C, D, and CE and for the outputs Z_0, Z_1, and Z_2?

(b) Draw the mixed-logic block diagram for the corresponding combinational circuit.

(c) Determine the logic equations for Z_0, Z_1, and Z_2.

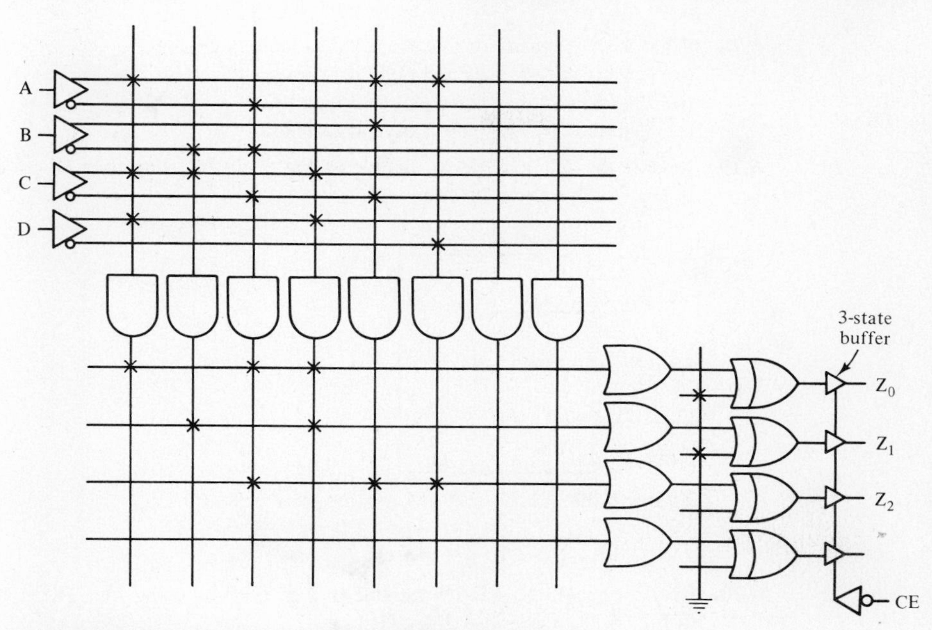

Figure 6.38 Programmed PLA for Problem 6.12.

6.13. What is the main difference between a PLA and a PAL?

6.14. Using the same PAL16L8, realize the following logic equations and have the realizations based on the following assignments:

Inputs: X_1.H is assigned to pin 2, X_2.H is assigned to pin 3, S_1.H is assigned to pin 17, and S_2.H is assigned to pin 16.

Outputs: Z_1.L is assigned to pin 19, Z_2.L is assigned to pin 18, Z_3.L is assigned to pin 12, and Z_4.H is assigned to pin 13.

All the other pins are not to be used unless specified otherwise.

(a) $Z_1 = S_2 \cdot X_1$ (*Hint:* Pin 16 needs to be programmed as an input.)

(b) $Z_2 = S_1 + S_1 \cdot \overline{X_2}$

(c) $Z_3 = X_2 \cdot (S_2 + S_1 \cdot \overline{X_1})$ (*Hint:* You can use pin 11 also if necessary.)

(d) $Z_4 = X_1 + X_2$ (*Hint:* Since Z_4 is active-high, you may need to use DeMorgan's laws.)

6.15. Using a PAL16L8, realize a BCD-to-7-segment decoder similar to the one shown in Fig. 6.28. However, the outputs a, b, c, d, e, f, and g are to be active-low.

6.16. Using a PAL16R4, realize a 4-bit decade counter with a synchronous CLEAR input. Compare your realization with the one obtained in Problem 5.26.

6.17. Draw block diagrams corresponding to the following static RAM module specifications. Specify the number of address lines and data lines.
(a) 64 × 4 bits **(b)** 4096 × 8 bits **(c)** 64K × 8 bits

6.18. What is the capacity of a static RAM module that has

(a) Seven address lines and eight data lines?

(b) Fourteen address lines and four data lines?

(c) Ten address lines and sixteen data lines?

6.19. Realize the 2K × 8 RAM module of Fig. 6.39 by using four 1K × 4 RAMs, as shown in Fig. 6.23, and an inverter.

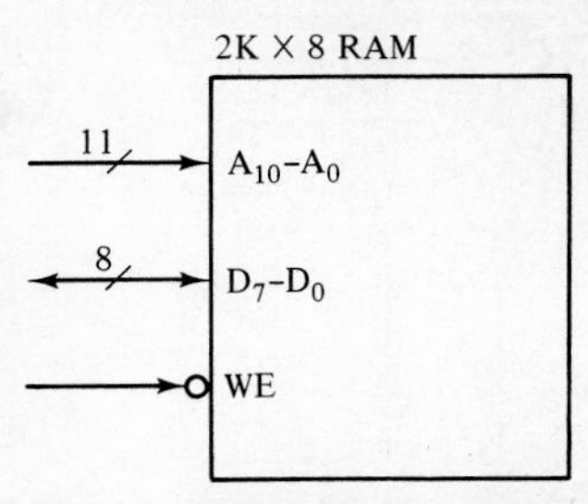

Figure 6.39 2K × 8 RAM module for Problem 6.19.

6.20. Realize the 4K × 4 RAM module of Fig. 6.40 by using four 1K × 4 RAMs, as shown in Fig. 6.23, and a 4-to-2 decoder.

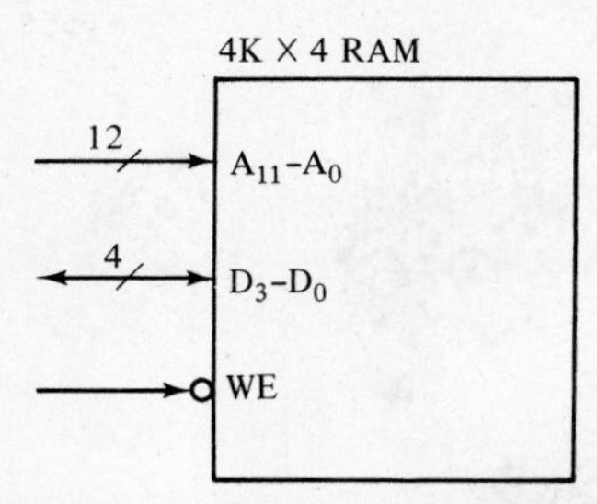

Figure 6.40 4K × 4 RAM module for Problem 6.20.

6.21. Realize the 2K × 4 RAM module with chip-select input of Fig. 6.41 by using 1K × 4 RAMs, as shown in Fig. 6.23, and any additional logic that is necessary.

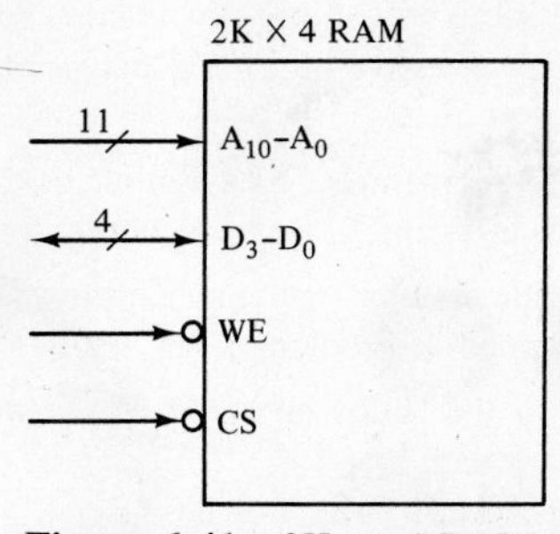

Figure 6.41 2K × 4 RAM module for Problem 6.21.

6.22. Realize the memory module of Fig. 6.42 by using a 1K × 4 RAM, as shown in Fig. 6.23, and any additional logic that is necessary. Note that the bidirectional data lines of the 1K × 4 RAM become two sets of data lines, DIN and DOUT. (*Hint:* Use three-state buffers.)

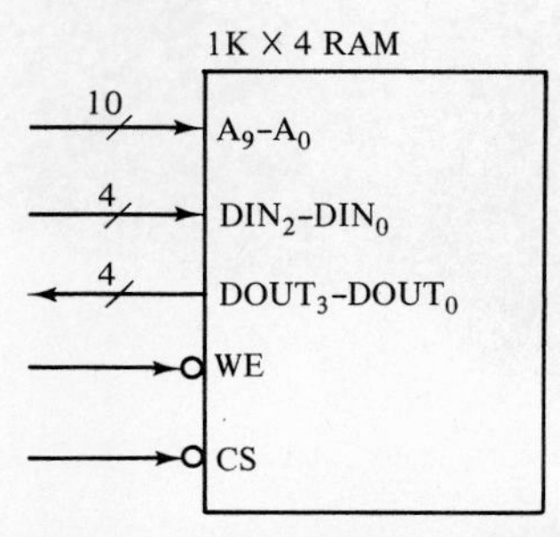

Figure 6.42 Memory module for Problem 6.22.

6.23. The static RAM chip shown in Fig. 6.26(a) has the following timing parameter values:

$$t_{RC} = 100 \text{ ns minimum}$$
$$t_A(AD) = 100 \text{ ns minimum}$$
$$t_A(CE) = 75 \text{ ns minimum}$$

At $t = 0$ s, a valid address is applied and the WE signal is set to false (H).

(a) If the chip-enable signal (CE) is applied at $t = 10$ ns, then when is the time t at which the data first becomes valid?

(b) If the chip-enable signal (CE) is applied at $t = 50$ ns, then when is the time t at which the data first becomes valid?

6.24. The static RAM chip shown in Fig. 6.26(a) has the following timing parameters:

$$t_{WC} = 100 \text{ ns minimum}$$
$$t_{SU}(CE) = 70 \text{ ns minimum}$$
$$t_W(WE) = 100 \text{ ns minimum}$$
$$t_{SU}(DA) = 70 \text{ ns minimum}$$

At $t = 0$ s, a valid address is applied and the WE signal is set to true (L).

(a) If CE and the data are applied at $t = 0$ s, then the WE signal must remain true (L) until a time t_x to ensure a valid write operation. What is this time t_x?

(b) If CE is applied at $t = 50$ ns and the data is applied at $t = 0$ s, then what is this time t_x?

(c) If CE is applied at $t = 0$ s and the data is applied at $t = 50$ ns, then what is this time t_x?

6.25. Discuss the similarities and differences among ROMs, PROMs, and EPROMs.

6.26. The hardware multiplier of Fig. 6.43 can multiply two 4-bit numbers (MCAND and MPLIER) and produce an 8-bit product (PRODUCT).

(a) Derive the truth table for this circuit. Use don't cares when convenient.

(b) If a ROM is used to realize this circuit, what must be the ROM capacity?

(c) Draw a block diagram of the ROM realization, specifying all connections to the address and data lines.

(d) What are the contents of the ROM? Explain in words.

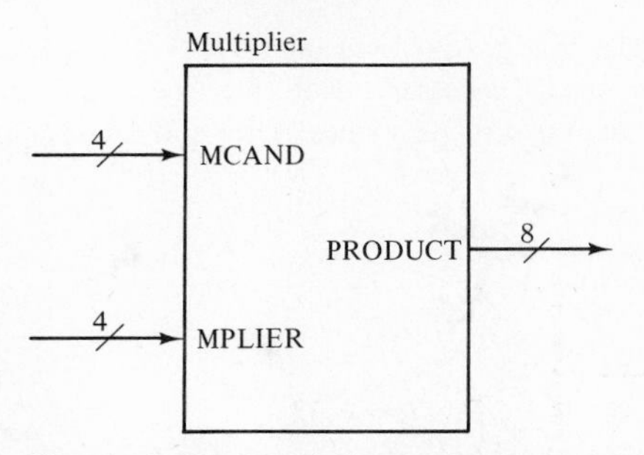

Figure 6.43 Multiplier for Problem 6.26.

6.27. Use a ROM to realize the four logic functions Z_1.L, Z_2.L, Z_3.L, and Z_4.H specified in Problem 6.14 as follows:

(a) Draw a block diagram design of the ROM realization, specifying all connections to the address and data lines.
(b) Specify in hexadecimal the contents of the ROM.
(c) Explain what an active-low output does to the corresponding contents of the ROM.

6.28. Consider a PLA with 12 inputs (actually 12 inputs and 12 complements), 8 outputs, and 64 AND gates. Can it be used to realize the following combinational circuits?

(a) A circuit with eight inputs and six outputs.
(b) A circuit with six inputs and eight outputs.
In each case answer yes, no, or maybe, and explain your answer.

6.29. Can you implement the logic equations of the following combinational circuits with a 128 × 8 ROM?

(a) A circuit with eight inputs and six outputs.
(b) A circuit with six inputs and eight outputs.
In each case answer yes, no, or maybe, and explain your answer.

6.30. Construct the memory module of Fig. 6.44 that provides 6K × 8 bits of EPROM and 2K × 8 bits of RAM. [*Hint:* Use three 2716 EPROMs and two 1K × 8 RAM modules (see Fig. 6.24), a 2-to-4 decoder, and any additional logic that is necessary.]

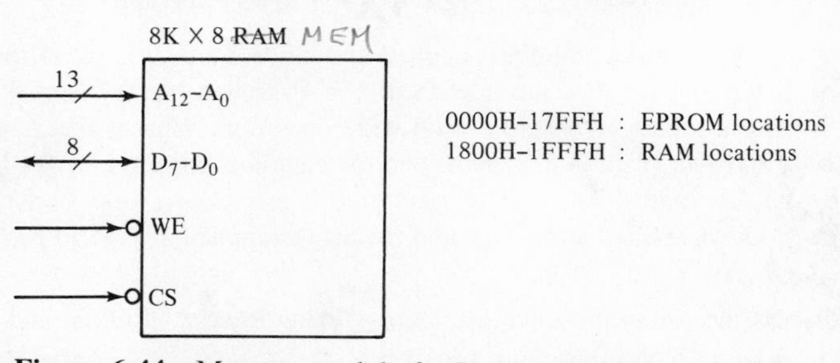

Figure 6.44 Memory module for Problem 6.30.

6.31. Discuss the advantages and disadvantages of using static RAMs versus dynamic RAMs in a digital circuit.

6.32. Consider the dynamic RAM of Fig. 6.45 that functions similarly to the one shown in Fig. 6.30(b).

(a) What is the capacity of this DRAM?

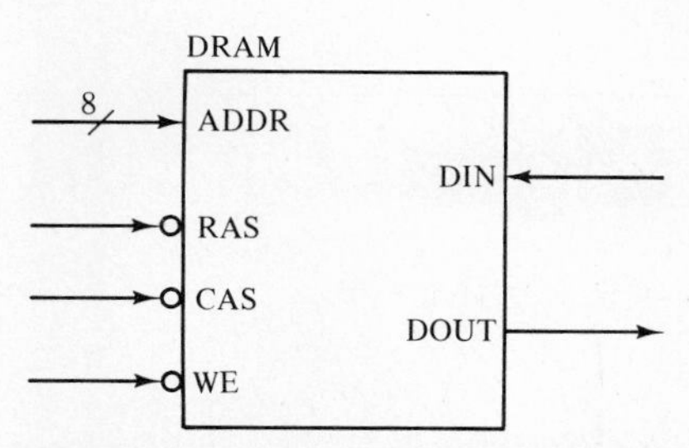

Figure 6.45 Dynamic RAM for Problem 6.32.

(b) Explain in words the sequence of steps (in terms of signals and order of events) that are required to perform a memory read operation.

(c) Explain in words the sequence of steps that are required to perform a memory write operation.

(d) Explain in words the sequence of steps that are required to perform a memory refresh operation.

6.33. Construct the 16K × 4 memory module of Fig. 6.46 by using four 16K × 1 dynamic RAMs [as shown in Fig. 6.30(b)], a DRAM controller (similar to the one shown in Fig. 6.32), and any additional logic that is needed. Note that the data lines of the memory module are bidirectional, whereas the data lines of the 16K × 1 dynamic RAMs are divided into DIN and DOUT. (*Hint:* Use three-state buffers.)

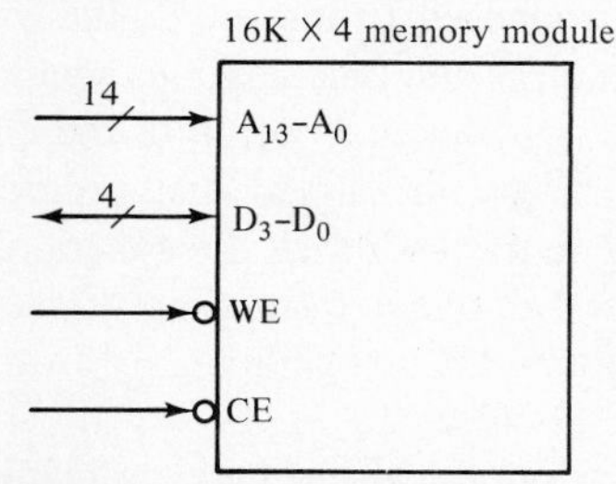

Figure 6.46 Memory module for Problem 6.33.

Chapter 7

Digital Circuit Design

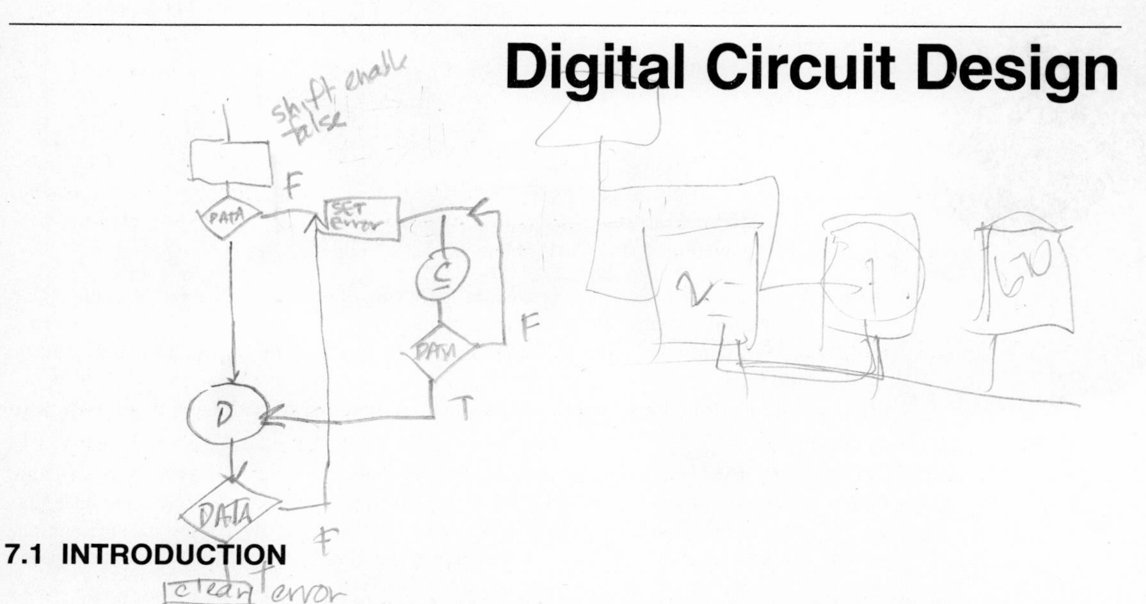

7.1 INTRODUCTION

The digital *circuit* design process, as opposed to that of digital *system* design (to be discussed in Chapter 8), begins with a clear and unambiguous requirement specification of the digital circuit. The final product is a detailed design of the circuit. In between is the design process. Like any design process, the digital design process requires a combination of creativity, experience, and understanding of the general design principles. In particular, one does not become a good designer from simply reading a textbook. On the other hand, mindlessly designing digital circuits without any awareness of the general design principles often produces mindless results.

Our purpose in this chapter is to present the general principles of digital circuit design and to discuss various techniques that are useful in the design process, thereby providing a solid foundation on which to build a knowledge of design. But providing the necessary creativity or experience is beyond the scope of this book.

In this chapter we will study the design of *sequential* circuits using circuit elements that were introduced in the preceding chapters. Sequential circuits are classified into two main types: *synchronous* and *asynchronous*. In a synchronous sequential circuit, the circuit elements respond to input signals only at discrete instants of time—at the active transitions of the clock signal. A sequential circuit having this feature is called a *clocked sequential circuit,* as was stated in Chapter 5. All the digital circuits considered in this chapter are clocked sequential circuits.

In an asynchronous sequential circuit, each circuit element operates at its own rate, and there are no clock signals to synchronize operation. Consequently, asynchronous sequential circuits can operate at faster rates than can synchronous sequential circuits. However, there can be serious operational problems because the outputs of the circuit elements depend on the *order* of the change in the input signals. As a result, the element outputs can be transiently unstable and unpredictable. For these and other problems relating to timing, the design of asynchronous sequential circuits is much more difficult than that of synchronous sequential circuits and is not considered in this introductory text.

We begin with a discussion of the digital circuit design fundamentals: the concepts of a controller and the controlled circuit elements, and the various phases of the design process. The next topic is Algorithmic State Machine (ASM) fundamentals, along with the various techniques for directly translating ASM charts into hardware controller circuits. (ASM charts are useful tools in the design and implementation of the controller of a digital circuit.) Finally, a series of detailed design examples are given to illustrate the digital circuit design concepts presented.

7.2 A MODEL FOR DIGITAL CIRCUIT DESIGN

As mentioned, the digital *circuit* design process begins with a clear and unambiguous requirement specification of the digital circuit. The final product is a detailed design of the circuit. In between is the design process. In its most unrefined sense, the digital design process can be viewed as being merely the selecting of the appropriate circuit elements and the interconnecting of them such that they function as specified.

We can formalize this idea into the concept of a *controller* and the *controlled circuit elements*. In other words, a digital circuit design is conceptually divided into two parts, as shown in the general circuit design of Fig. 7.1. The controlled circuit elements comprise a set of circuit elements, like those presented in the preceding chapters, that are selected to implement the functions that are specified for the digital circuit. The controller provides these circuit elements with the appropriate input control signals at every moment in time so that the circuit elements properly implement the specified functions and produce the required external output signals. In this manner, the controller functions as the "brain" of the digital circuit.

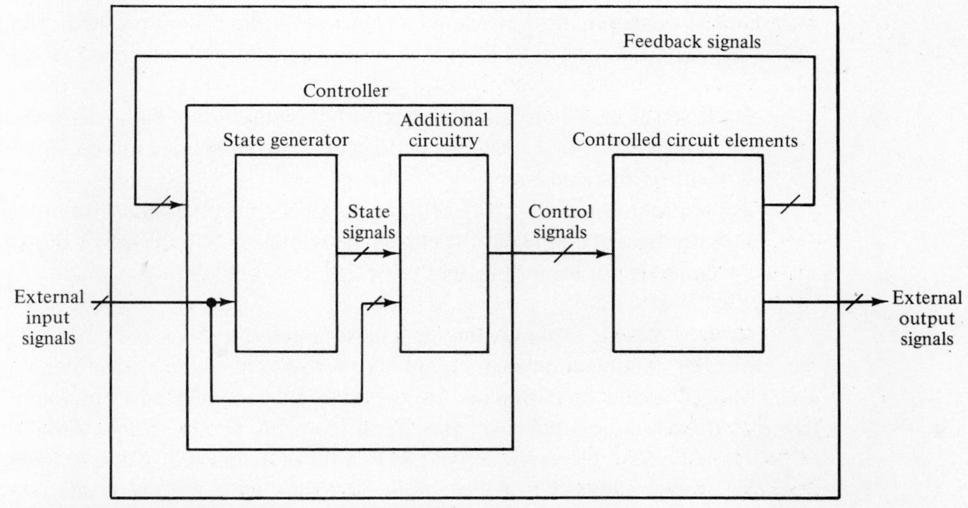

Figure 7.1 General model for digital circuit design.

The controller, itself a digital circuit, consists of a *state generator* and additional circuitry for producing the signals required for controlling the circuit elements. The inputs to the state generator are the external input signals and also feedback signals from the controlled circuit elements. The outputs of the state generator are the *state signals,* each representing a state of the controller. The function of the state generator is to place the controller in the appropriate state at the appropriate time so that it generates the appropriate control signals.

The concepts introduced above are among the most important and fundamental in digital design. Unfortunately, they are also among the most difficult ones to explain and comprehend. The remainder of this chapter will be devoted to making these concepts clearer. More details will be provided. Also, design techniques will be introduced, and design examples given.

7.3 DIGITAL CIRCUIT DESIGN PROCESS

The digital circuit design process can be divided into three major phases:

1. Preliminary design phase
2. Refinement phase
3. Realization phase

These phases occur in the indicated order. Before starting the preliminary design phase of a digital circuit, a designer must be given a well-defined requirement specification of the digital circuit that is to be designed.

In the *preliminary* design phase, the designer makes certain of obtaining a good understanding of the given requirement specification. The designer should develop a block diagram of the overall digital circuit, with the input and output signals well defined, and may use timing diagrams of relevant signals for further clarification. Given that the designer has conceptually formulated a solution for the design problem, the final product of the preliminary design phase is the preliminary design, consisting of the following.

1. A set of major circuit elements with the major data paths defined. The designer should, of course, have some idea of how the major circuit elements eventually will be realized.
2. A preliminary plan (algorithm) for the control of the circuit elements. This control algorithm can be stated in words, or, if the algorithm is sufficiently concrete, in a more formal representation such as a flowchart.

In the *refinement* phase, the designer iteratively refines the circuit design for both the controller and the controlled circuit element parts. For the circuit element part, circuit elements are added or eliminated as the solution becomes more in focus. During this iterative process, the signals of the circuit elements become more defined, and the set of controlled circuit elements converges to a set of actual ICs. Correspondingly, for each iteration of the refinement of the circuit elements, the control algorithm itself becomes more refined. The number of states becomes more stable and the control signals become more defined. Also, the timing among the signals increases in importance. During all this, the flowchart of the earlier design steps is converging to an ASM chart. Of course, the number of iterations required in the refinement phase depends on the complexity of the design. The end product of this phase is the following:

1. A set of detailed circuit elements with completely defined functions and completely defined signals. At this point of the design the circuit elements are sufficiently defined that they can be realized with available ICs in a straightforward manner.
2. The control algorithm in the form of an ASM chart in which the timing is unambiguously represented.

The final phase of the design process is the *realization* phase. At this point of the design, the hard work is over, and the realization of the detailed circuit elements with available ICs is straightforward. As we will see shortly, the hardware realization of an ASM chart is also straightforward, employing the techniques to be discussed in Sec. 7.5.

Examples are given at the end of this chapter to illustrate these design concepts and the entire design process. Before considering those examples, however, we will study some design tools and techniques that are necessary for digital design.

7.4 ALGORITHMIC STATE MACHINE (ASM)

The design of the controller, and the state generator section in particular, is the most difficult part of the digital circuit design process. If the complexity of the controller is nontrivial, then trying to realize the controller through trial and error is not desirable, even if possible. A systematic design procedure is necessary. Central to such a procedure is an unambiguous notation for representing the control algorithm. This notation enables the designer to bridge the gap between the conceptual control algorithm and the actual hardware realization of that algorithm.

Two characteristics are essential for this notation:

1. For the designer to use it effectively, the notation must provide a clear description of the algorithm, and in terms to which the designer can relate.
2. The notation must support a direct translation into a hardware realization of the control algorithm.

Traditional state diagram methods, such as the Mealy and Moore state machines discussed in Sec. 7.7, satisfy the second condition. Translation from a traditional state diagram to a hardware realization is straightforward. Unfortunately, though, with such diagrams it is difficult to represent complex control algorithms clearly. Moreover, representing a control algorithm with more than a limited number of input and output signals can be unwieldy.

A notation that has both essential characteristics is the Algorithmic State Machine (ASM) chart. Translation from an ASM chart to a hardware realization is practically identical to that for the traditional state diagram. And since the syntax of an ASM chart is very similar to that of a software flowchart, the control algorithm can be expressed in terms that are familiar to the designer.

The most important concept in the algorithm of a controller is the concept of a *state*. The term ''state'' refers to a stable condition of the controller over a fixed period of time. In terms of a sequential digital circuit, a state is represented by the binary information stored in the memory elements during that period of time. The notation of state will become clearer with the use of the term.

On an ASM chart, a state is represented by a *state box,* which is a rectangle with the name of the state encircled and placed at the upper left corner or at the side of the rectangle. In the ASM chart example of Fig. 7.2(a), there are four states: A, B, C, and D. This chart is for a controller that has two input signals: IN.BIT and BUF.FUL, and three output control signals: COUNT.EN, REG.LD, and OUT.FLAG.

Over time the controller, guided by the control algorithm, moves through a sequence of states. The state transition from the *present* state of the algorithm to the *next* state of the algorithm occurs at the active edge (leading edge in this case) of the system clock signal. In between state transitions the controller is stable.

There are two types of state transitions: *unconditional* and *conditional*. For an unconditional state transition the next state depends only on the present state of the controller and not on any input signals. As an illustration, in Fig. 7.2(a) the transition from state B to state A is unconditional. In other words, if the present state of the controller is state B, then the next state is state A regardless of the input signals IN.BIT and BUF.FULL. Similarly, the transition from state D to state A is unconditional.

In a conditional state transition the next state depends not only on the present state but also on the present values of the input signals. In an ASM chart a conditional state transition is represented by a *decision diamond*. Note from Fig. 7.2(a) that if the controller present state is state A, then the next state is either state B or state C, depending on the value of the input signal IN.BIT. As you can see from Fig. 7.2(b), the decision is made at the *end* of the present clock cycle, at the next active edge (leading edge) of the system clock signal. For example, for state time T1 in Fig. 7.2(b) the value of IN.BIT is false at the end of T1, and so the next state is state B. On the other hand, at the end of T3 the value of IN.BIT is true, and therefore the next state is state C.

ASM charts also specify controller output values. There are two types of controller outputs: *unconditional* and *conditional*. They differ in that the value of an unconditional output depends only on the present state, but the value of a conditional output depends not only on the present state but also on the present values of the input signals. Unconditional outputs are specified within the state boxes as shown for state boxes A, C, and D in Fig. 7.2(a). As will be described in Sec. 7.7, unconditional outputs are essentially Moore state machine outputs. Conditional outputs are specified in an oval associated with the state and its decision diamond. For the example of Fig. 7.2(a), the only conditional output is the COUNT.EN associated with state C. Conditional outputs are essentially Mealy state machine outputs.

Note from Fig. 7.2(a) and (b) that the unconditional outputs COUNT.EN and REG.LD are true every time the controller is in state A (T1, T3, and T7), and that unconditional output REG.LD is also true every time the controller is in state C (T4 and T5). Further, unconditional output OUT.FLAG is true when the controller is in state D (T6). Finally, the conditional output COUNT.EN is true when the controller is in state C only if the input signal BUF.FULL is true (T5 and not T4).

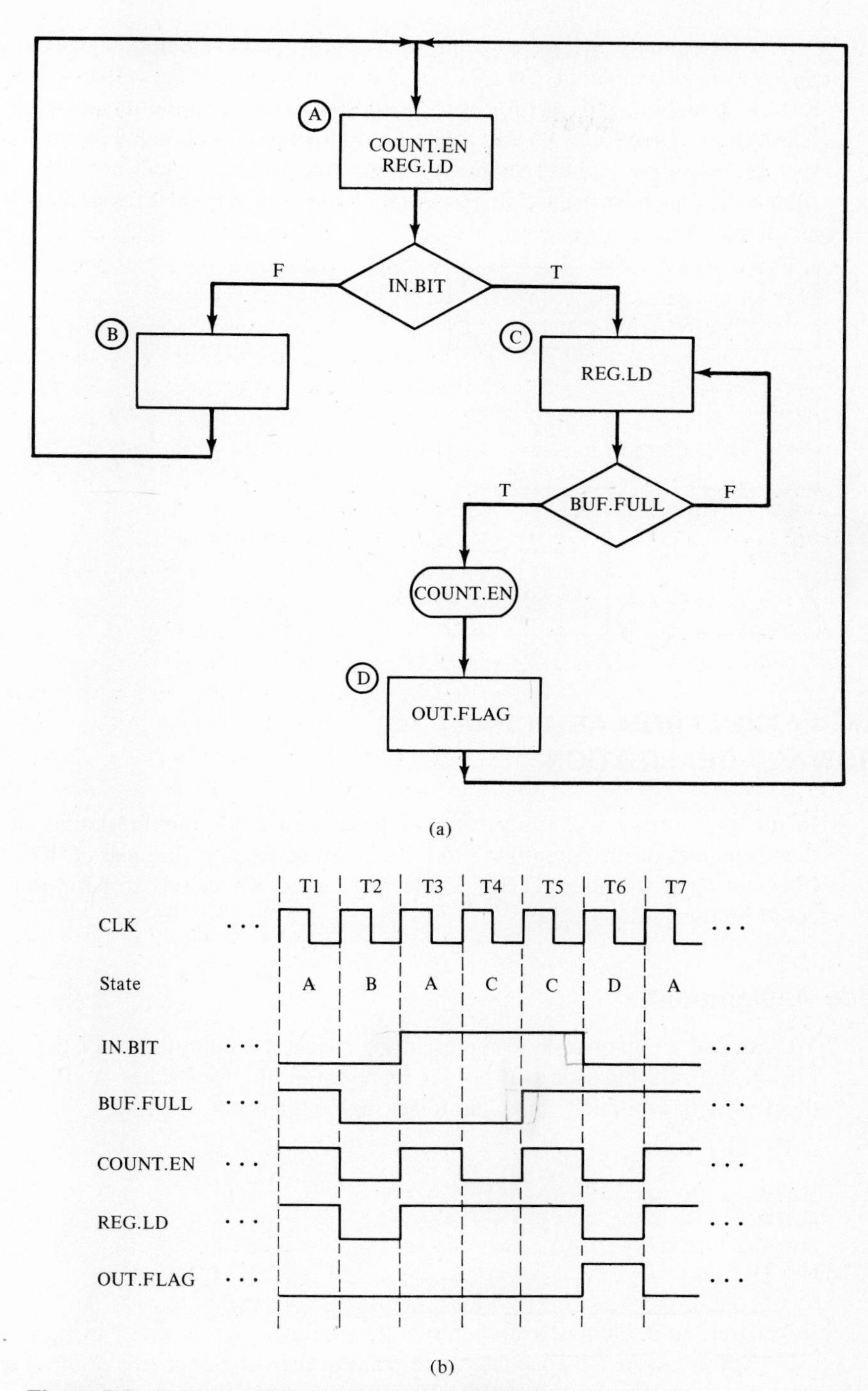

Figure 7.2 Example ASM chart and timing diagram.

Conceptually, we must specify the values for every output signal for every state. So, for the example of Fig. 7.2(a), we should specify the values of COUNT.EN, REG.LD, and OUT.FLAG for *every* state. But to do so would unnecessarily clutter the ASM chart. Therefore, we will adopt the following convention. For each state we will specify only those control signals having a true value; we will not show those output signals having false values in that state. Following are some examples of equivalent notations:

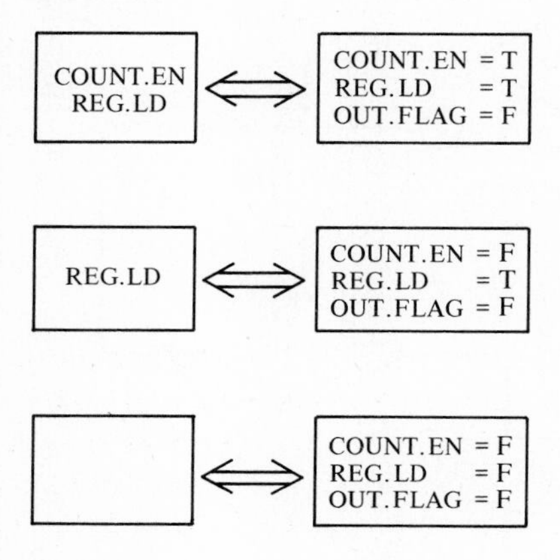

7.5 TRANSLATION FROM ASM CHART TO HARDWARE REALIZATION

In this section we will study some systematic methods for translating an ASM chart representation of the controller into a hardware realization. In most of the examples that illustrate these methods, D flip-flops will be used. Other types of flip-flops, however, could be used just as well.

7.5.1 Code Assignment

We can use a binary code to represent the states of a controller. For the ASM chart of Fig. 7.2(a), a 2-bit code suffices for representing the four states A, B, C, and D. For them we will arbitrarily make the following assignments:

	C1	C0
State A:	0	0
State B:	0	1
State C:	1	0
State D:	1	1

In general, an N-bit code can represent 2^N states.

The 2 bits of the code can correspond to the outputs of two D flip-flops, as in the state generator circuit of Fig. 7.3. The output signal C1 of one flip-flop corresponds to the second bit of the code, and the output signal C0 of the other flip-flop corresponds to

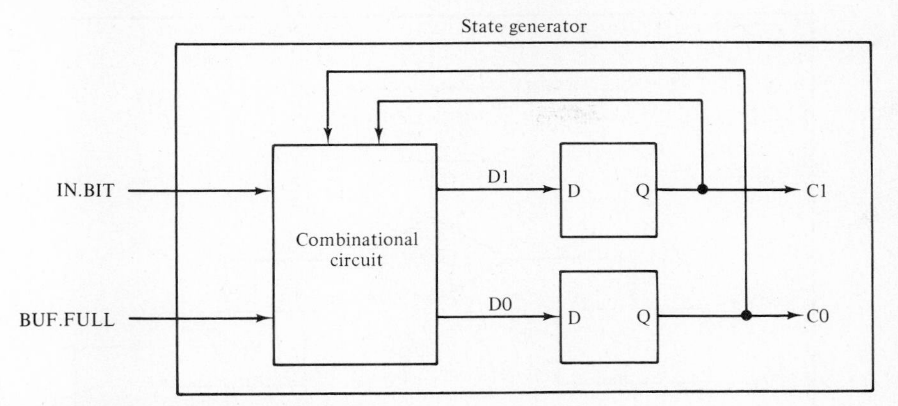

Figure 7.3 State generator circuit.

the first bit. From the ASM chart of Fig. 7.2(a) we can determine how this circuit must function. If the present state of the controller is state A and the input IN.BIT is true at the end of the present state time, then from Fig. 7.2(a) the next state of the controller is state C. Thus, for the state generator circuit of Fig. 7.3, if the present state of the controller is state A (C1 = 0, C0 = 0), and the input IN.BIT is 1 (true) at the end of the present clock cycle, then the combinational circuit should be designed such that D1 = 1 and D0 = 0, so that at the next active edge of the clock signal, C1 = 1 and C0 = 0. Referring back to Fig. 7.2(a), if the present controller state is C and the input BUF.FULL is true at the end of the present state time, then the next controller state is state D. Consequently, for the state generator circuit of Fig. 7.3, if the present state of the controller is state C (C1 = 1, C0 = 0), and the input signal BUF.FULL is 1 (true), then the combinational circuit should produce D1 = 1 and D0 = 1, so that at the next active edge of the clock cycle, C1 = 1 and C0 = 1, and so forth. In the following sections, we will consider three systematic methods for designing and realizing such a combinational circuit.

7.5.2 Traditional Method with D Flip-Flops

In the traditional method of ASM realization we follow the same procedure we used in Chapter 5 for the design of counters and shift registers. Basically, we just derive a next-state table for the state generator circuit, and then determine the input equations for the flip-flops. We will use the ASM chart of Fig. 7.2(a) to illustrate this method. This chart is shown in Fig. 7.4 along with the arbitrary code assignments for each state specified at the upper right-hand corner of each state box: state A is 00, state B is 01, state C is 10, and state D is 11.

Using information obtained from this ASM chart, we can form the next-state table of Table 7.1 for the state generator circuit. Each row of Table 7.1 contains the present values of the flip-flop output signals C1 and C0, the present values of the input signals IN.BIT and BUF.FULL, and the desired values for the flip-flop outputs ($C1^+$ and $C0^+$) immediately after the next active clock edge. For example, the first row shows that when

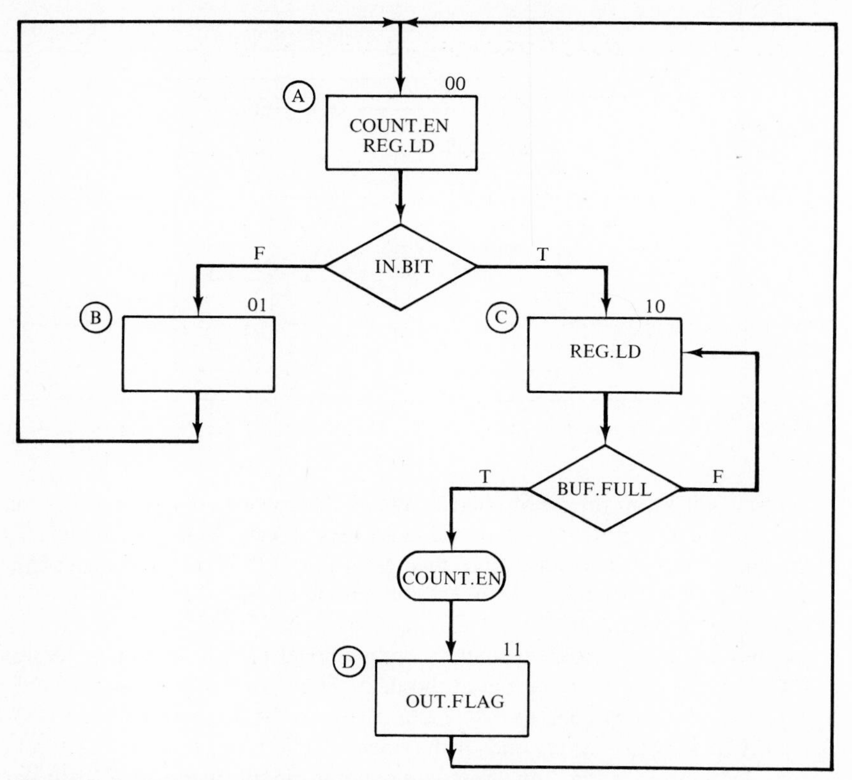

Figure 7.4 Example ASM chart with state assignments.

the present state of the controller is A (C1 = 0 and C0 = 0), and the inputs are IN.BIT = 0 and BUF.FULL = 0, then the desired next state of the controller is B (C1 = 0 and C0 = 1). Similarly, the third row shows that if the present state of the controller is A (C1 = 0 and C0 = 0) and the inputs are IN.BIT = 1 and BUF.FULL = 0, then the desired next state is C (C1 = 1 and C0 = 0). Finally, the last four rows of the table show that if the present state is D (C1 = 1 and C0 = 1), then the next state is A (C1 = 0 and C0 = 0), regardless of the values of IN.BIT and BUF.FULL.

To design the combinational circuit of Fig. 7.3, we need the logic equations for the inputs D1 and D0 of the D flip-flops. As in the designs of the counters in Chapter 5, we can get these equations by making a new table from Table 7.1, using the excitation table for the D flip-flop:

D	Q	Q^+
0	0	0
0	1	0
1	0	1
1	1	1

characteristic table

rearrange ⟶

Q	Q^+	D
0	0	0
0	1	1
1	0	0
1	1	1

excitation table

The result is shown in Table 7.2.

TABLE 7.1 NEXT-STATE TABLE FOR THE STATE GENERATOR CIRCUIT

Present-state code		Inputs		Next-state code	
C1	C0	IN.BIT	BUF.FULL	C1$^+$	C0$^+$
0	0	0	0	0	1
0	0	0	1	0	1
0	0	1	0	1	0
0	0	1	1	1	0
0	1	0	0	0	0
0	1	0	1	0	0
0	1	1	0	0	0
0	1	1	1	0	0
1	0	0	0	1	0
1	0	0	1	1	1
1	0	1	0	1	0
1	0	1	1	1	1
1	1	0	0	0	0
1	1	0	1	0	0
1	1	1	0	0	0
1	1	1	1	0	0

TABLE 7.2 NEXT-STATE TABLE WITH CORRESPONDING D FLIP-FLOP INPUTS

C1	C0	IN.BIT	BUF.FULL	C1$^+$	C0$^+$	D1	D0
0	0	0	0	0	1	0	1
0	0	0	1	0	1	0	1
0	0	1	0	1	0	1	0
0	0	1	1	1	0	1	0
0	1	0	0	0	0	0	0
0	1	0	1	0	0	0	0
0	1	1	0	0	0	0	0
0	1	1	1	0	0	0	0
1	0	0	0	1	0	1	0
1	0	0	1	1	1	1	1
1	0	1	0	1	0	1	0
1	0	1	1	1	1	1	1
1	1	0	0	0	0	0	0
1	1	0	1	0	0	0	0
1	1	1	0	0	0	0	0
1	1	1	1	0	0	0	0

For Table 7.2, we can readily determine the D1 and D0 entries by recalling from Chapter 5 that each row of the D flip-flop excitation table indicates the value required for the D input for the desired D flip-flop output transition. As indicated at the top of Table 7.2, the values of D1 are derived from those of C1 and $C1^+$ in agreement with the D flip-flop excitation table. For example, in row 3, for the transition from $C1 = 0$ to $C1^+ = 1$, the value of D1 must be 1. In row 13, for the transition from $C1 = 1$ to $C1^+ = 0$, the value of D1 must be 0, and so forth. Similarly, the values of D0 are derived from the values of C0 and $C0^+$.

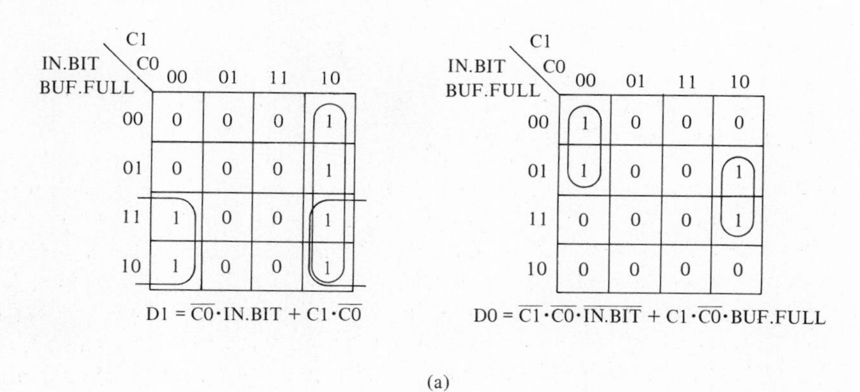

(a)

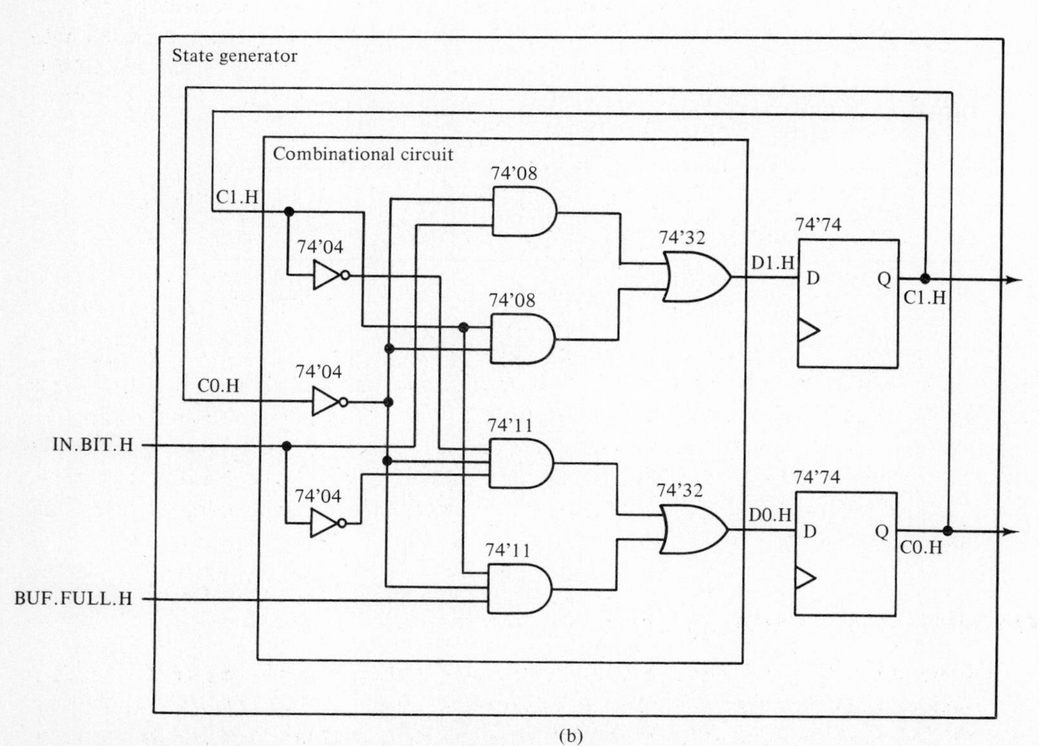

(b)

Figure 7.5 D flip-flop state generator realization.

Using Table 7.2 we can derive the logic equations for D1 and D0 as functions of C1, C0, IN.BIT, and BUF.FULL. The K-maps for D1 and D0 are given in Fig. 7.5(a) and the resulting circuit diagram in Fig. 7.5(b).

To complete the controller design, we must have a circuit for decoding the state code and producing the control signals REG.LD, OUT.FLAG, and COUNT.EN, which are the ASM outputs. From the ASM chart of Fig. 7.4 we see that REG.LD = 1 when the controller is in state A or state C. Consequently, REG.LD = A + C. Similarly, OUT.FLAG = D. The third output, COUNT.EN, is 1 if the controller is in state A or if it is in state C and if BUF.FULL = 1. Therefore, COUNT.EN = A + C·BUF.FULL. Using a 74'139 decoder, we can decode the state code. The complete controller circuit is shown in Fig. 7.6.

7.5.3 PLA/PAL Method of ASM Realization

In the traditional method of ASM realization that we have just considered, we first derive the next-state table of the state generator circuit (e.g., Table 7.2), then determine the input equations for the flip-flops [e.g., Fig. 7.5(a)], and finally realize the input equations with AND, OR, and NOT gates [e.g., Fig. 7.5(b)]. Note that the combinational circuit of Fig. 7.5(b) is a two-level AND-OR realization. Recall from Sec. 6.4 that a programmable logic array (PLA) or a programmable array logic (PAL) is essentially a two-level AND-OR circuit element. Consequently, we can replace the combinational circuit in Fig. 7.5(b) with a *single* PLA or PAL circuit element. We will now do this with a PLA.

The design procedure for the PLA/PAL method of ASM realization is identical to that of the traditional method up to the realization step. In other words, the PLA/PAL method also involves the derivation of the next-state table, which is Table 7.2 here. From this table we see that the minterm expansions for D1 and D0 are

$$\begin{aligned}
\mathrm{D1} = {} & \overline{\mathrm{C1}}\cdot\overline{\mathrm{C0}}\cdot\mathrm{IN.BIT}\cdot\overline{\mathrm{BUF.FULL}} + \overline{\mathrm{C1}}\cdot\overline{\mathrm{C0}}\cdot\mathrm{IN.BIT}\cdot\mathrm{BUF.FULL} \\
& + \mathrm{C1}\cdot\overline{\mathrm{C0}}\cdot\overline{\mathrm{IN.BIT}}\cdot\overline{\mathrm{BUF.FULL}} + \mathrm{C1}\cdot\overline{\mathrm{C0}}\cdot\overline{\mathrm{IN.BIT}}\cdot\mathrm{BUF.FULL} \\
& + \mathrm{C1}\cdot\overline{\mathrm{C0}}\cdot\mathrm{IN.BIT}\cdot\overline{\mathrm{BUF.FULL}} + \mathrm{C1}\cdot\overline{\mathrm{C0}}\cdot\mathrm{IN.BIT}\cdot\mathrm{BUF.FULL} \\
\mathrm{D0} = {} & \overline{\mathrm{C1}}\cdot\overline{\mathrm{C0}}\cdot\overline{\mathrm{IN.BIT}}\cdot\overline{\mathrm{BUF.FULL}} + \overline{\mathrm{C1}}\cdot\overline{\mathrm{C0}}\cdot\overline{\mathrm{IN.BIT}}\cdot\mathrm{BUF.FULL} \\
& + \mathrm{C1}\cdot\overline{\mathrm{C0}}\cdot\overline{\mathrm{IN.BIT}}\cdot\mathrm{BUF.FULL} + \mathrm{C1}\cdot\overline{\mathrm{C0}}\cdot\mathrm{IN.BIT}\cdot\mathrm{BUF.FULL}
\end{aligned}$$

We can directly realize these logic equations with a single PLA, as shown in Fig. 7.7.

The obvious advantage of this method is the replacement of several IC packages with a single IC. This advantage is more dramatic when the controller circuit is complex and many IC packages are required for the traditional method. This PLA/PAL method is even more attractive with PLAs and PALs that have built-in flip-flops on-board the chips. With one of these PLAs or PALs, we can realize an entire controller with a single chip.

7.5.4 ROM Method of ASM Realization

Recall from Sec. 6.5.2 that a read-only memory (ROM) can also be used to realize random logic. Obviously, we can replace the combinational circuit in Fig. 7.5(b) with a ROM just as we did with a PLA in the last section. In addition, we can use the same ROM to generate the controller outputs. We will now show how to do this.

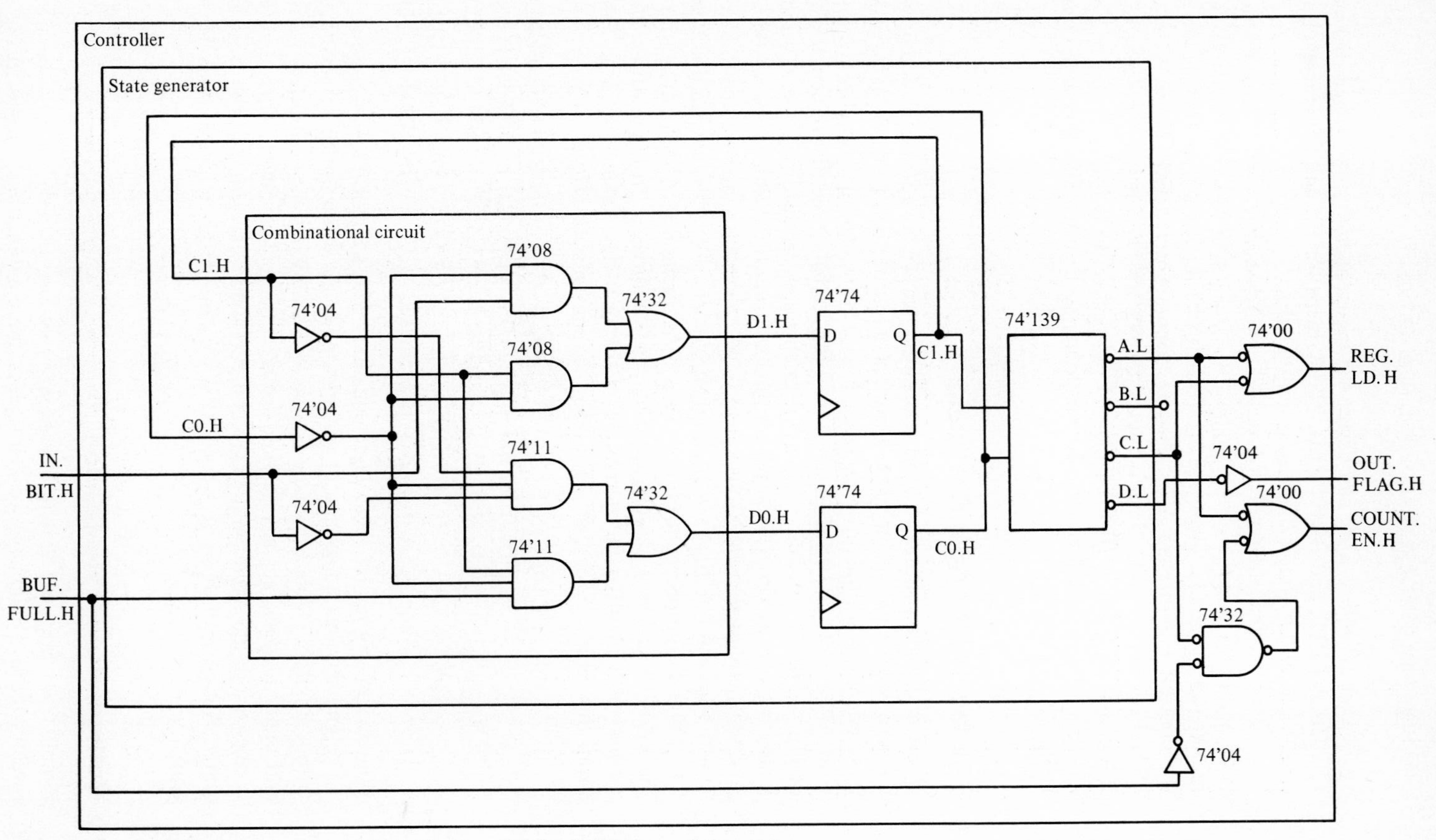

Figure 7.6 Controller realization with D flip-flops for the ASM chart of Fig. 7.4.

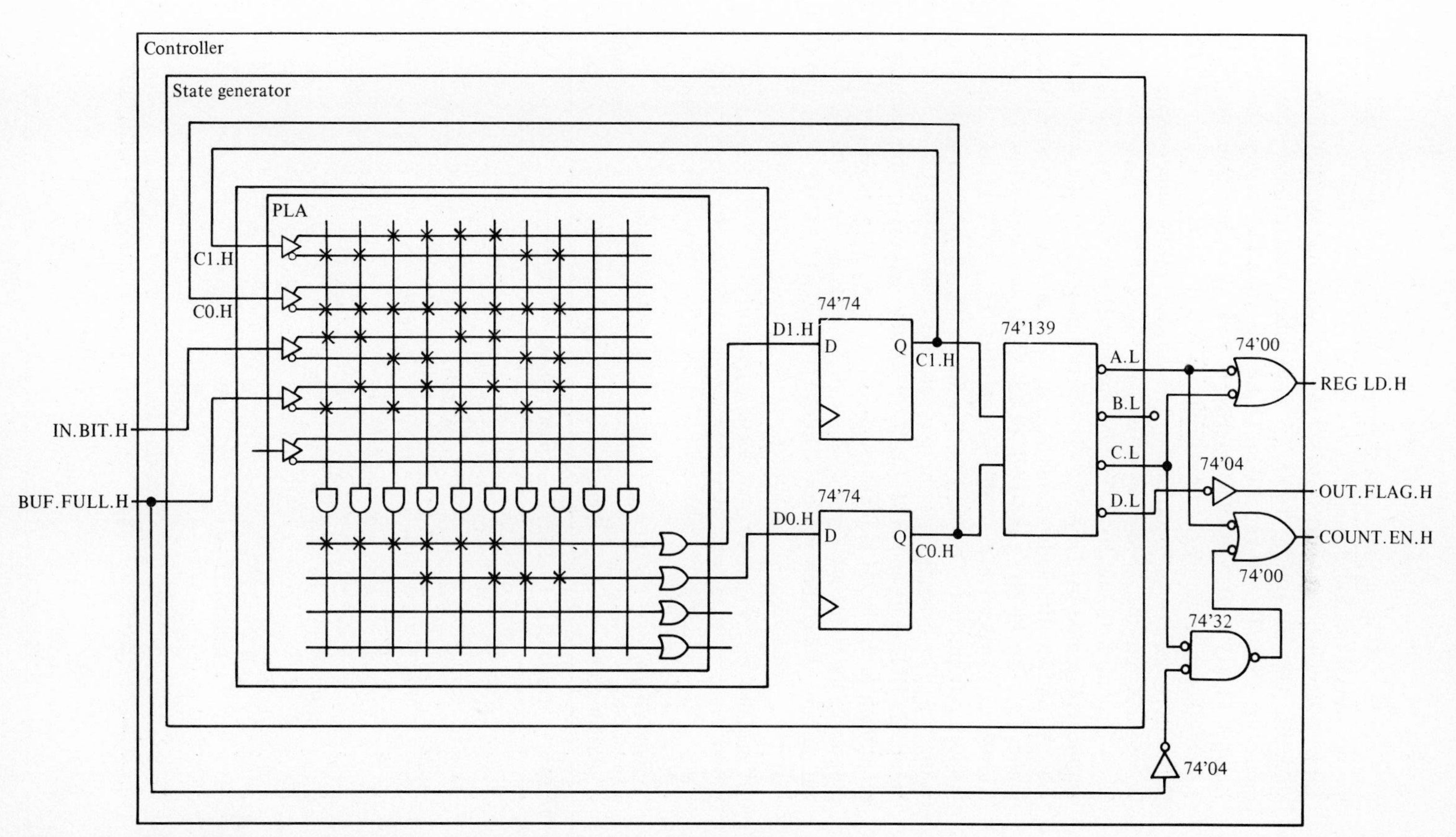

Figure 7.7 Controller with PLA state generator realization for the ASM chart of Fig. 7.4.

In the ROM method of ASM realization, we follow the same initial steps as for the traditional method. Doing this for the ASM chart of Fig. 7.4, we derive, as before, Table 7.2, which is reproduced in Table 7.3 with the addition of the controller output columns. The first row of Table 7.3 shows that if the present-state code is 00 and the inputs are IN.BIT = 0 and BUF.FULL = 0, then the outputs should be REG.LD = 1, OUT.FLAG = 0, and COUNT.EN = 1, as is evident from the ASM chart of Fig. 7.4. Furthermore, the desired next-state code is 01, which implies that the inputs to the D flip-flops are D1 = 0 and D0 = 1. Likewise, all the other entries of Table 7.3 can be verified from the ASM chart of Fig. 7.4.

Using a fictitious 16 × 5 ROM and two D flip-flops, we can realize the controller circuit represented by Table 7.3. The circuit diagram for the controller is given in Fig. 7.8. As shown, the address bits A_3, A_2, A_1, and A_0 correspond, respectively, to the inputs C1, C0, IN.BIT, and BUF.FULL. Also, the stored bits Z_4, Z_3, Z_2, Z_1, and Z_0 correspond, respectively, to the outputs REG.LD, OUT.FLAG, and COUNT.EN and the D flip-flop inputs D1 and D0. The shown stored values are easy to determine from Table 7.3. For example, for the present-state code of 00 (C1 = 0, C0 = 0) and the inputs IN.BIT = 0 and BUF.FULL = 0, the *address* of the memory location to be referenced in the ROM is 0000 ($A_3 = 0$, $A_2 = 0$, $A_1 = 0$, $A_0 = 0$). In this case, the outputs of the ROM are $Z_4 = 1$, $Z_3 = 0$, $Z_2 = 1$, $Z_1 = 0$, and $Z_0 = 1$. Consequently, the outputs of the controller are REG.LD = 1, OUT.FLAG = 0, and COUNT.EN =

TABLE 7.3 INPUT AND OUTPUT VALUES FOR ROM REALIZATION

Present-state code		Input values for the present state		Output values for the present state			Next-state code		D flip-flop input values	
C1	C0	IN.BIT	BUF.FULL	REG.LD	OUT.FLAG	COUNT.EN	$C1^+$	$C0^+$	D1	D0
0	0	0	0	1	0	1	0	1	0	1
0	0	0	1	1	0	1	0	1	0	1
0	0	1	0	1	0	1	1	0	1	0
0	0	1	1	1	0	1	1	0	1	0
0	1	0	0	0	0	0	0	0	0	0
0	1	0	1	0	0	0	0	0	0	0
0	1	1	0	0	0	0	0	0	0	0
0	1	1	1	0	0	0	0	0	0	0
1	0	0	0	1	0	0	1	0	1	0
1	0	0	1	1	0	1	1	1	1	1
1	0	1	0	1	0	0	1	0	1	0
1	0	1	1	1	0	1	1	1	1	1
1	1	0	0	0	1	0	0	0	0	0
1	1	0	1	0	1	0	0	0	0	0
1	1	1	0	0	1	0	0	0	0	0
1	1	1	1	0	1	0	0	0	0	0

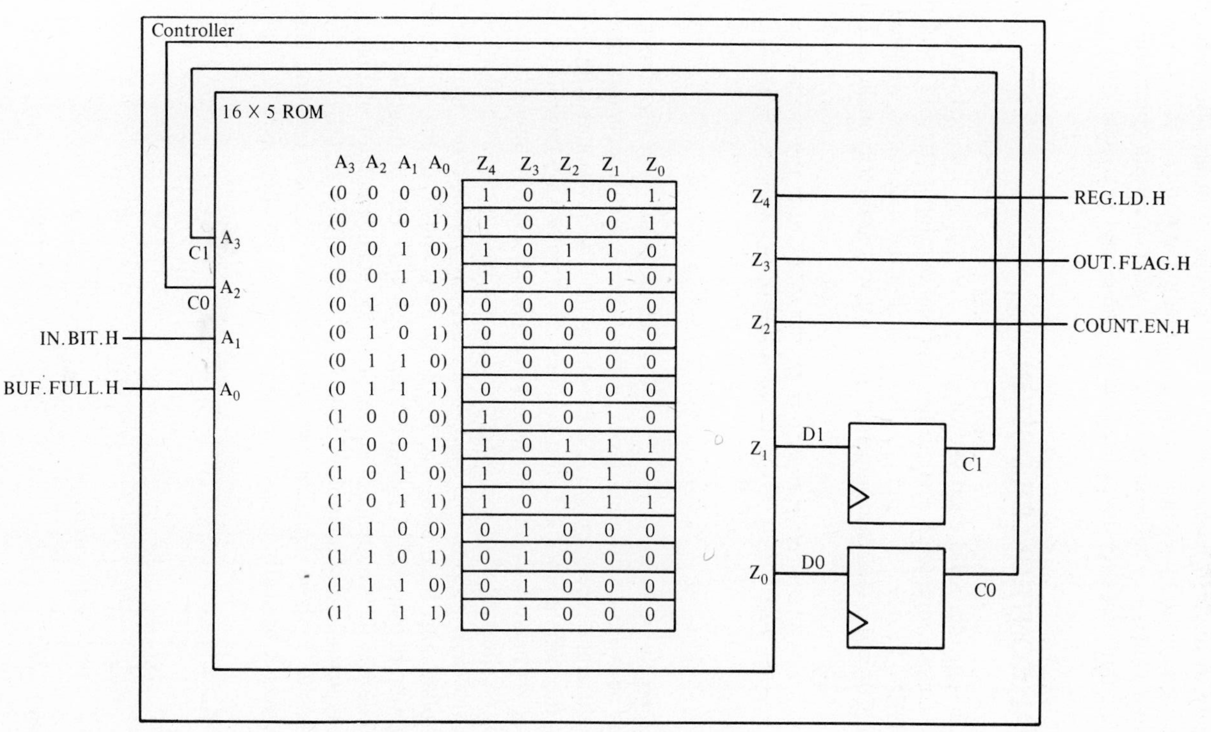

Figure 7.8 ROM realization of the controller for the ASM chart of Fig. 7.4.

1, the same as for the circuit of Fig. 7.6. Other combinations of inputs and present-state codes produce equivalent results. In fact, the circuit of Fig. 7.8 is functionally equivalent to that of Fig. 7.6. In other words, if both circuits were in black boxes, we could not functionally distinguish them—we could not tell one from the other. In effect, instead of designing a circuit by using the traditional method to *realize* Table 7.3, we have *stored* the required values in a table in a ROM. Then when a certain combination of inputs is given, we simply *look up* the desired output values and next-state code.

7.6 AN ADDITIONAL CONTROLLER DESIGN

For this design, the ASM chart shown in Fig. 7.9 will be realized using J-K flip-flops. This ASM chart represents a controller that has one input (IN.BIT) and three outputs (COUNT.EN, FLAG.SET, and COUNT.LD). For the five states, we need a 3-bit code (C2, C1, C0) and a code assignment, which we will arbitrarily make as follows:

State	C2	C1	C0
S0	0	0	0
S1	0	0	1
S2	0	1	0
S3	0	1	1
S4	1	0	0

This assignment is shown in Fig. 7.9. Now we are ready to realize this chart.

Just as we used different types of flip-flops to realize counters in Chapter 5, we can use different types for an ASM realization. The procedure with J-K flip-flops is basically the same as for the D flip-flop procedure, which we have just considered. But, of course, we must determine the logic equations for J and K inputs instead of D inputs.

For an illustration of the J-K procedure we will use J-K flip-flops in a realization of the ASM chart of Fig. 7.9. Since this chart has a 3-bit code, the state generator for the controller must have three J-K flip-flops, as is shown in Fig. 7.10. Of course, the three flip-flop outputs C2, C1, and C0 represent the 3 bits of the code.

To design the combinational circuit of the state generator of Fig. 7.10, we need the logic equations for the flip-flop inputs C2(J), C2(K), C1(J), C1(K), C0(J), and C0(K). As in Chapter 5, these equations are obtained by using the excitation table for the J-K flip-flop, which basically is just a rearrangement of the characteristic table.

J	K	Q	Q^+
0	0	0	0
0	0	1	1
0	1	0	0
0	1	1	0
1	0	0	1
1	0	1	1
1	1	0	1
1	1	1	0

characteristic table

rearrange →

Q	Q^+	J	K
0	0	0	X
0	1	1	X
1	0	X	1
1	1	X	0

excitation table

As is explained in Chapter 5, each row of the excitation table specifies the J and K inputs for the desired flip-flop output transitions.

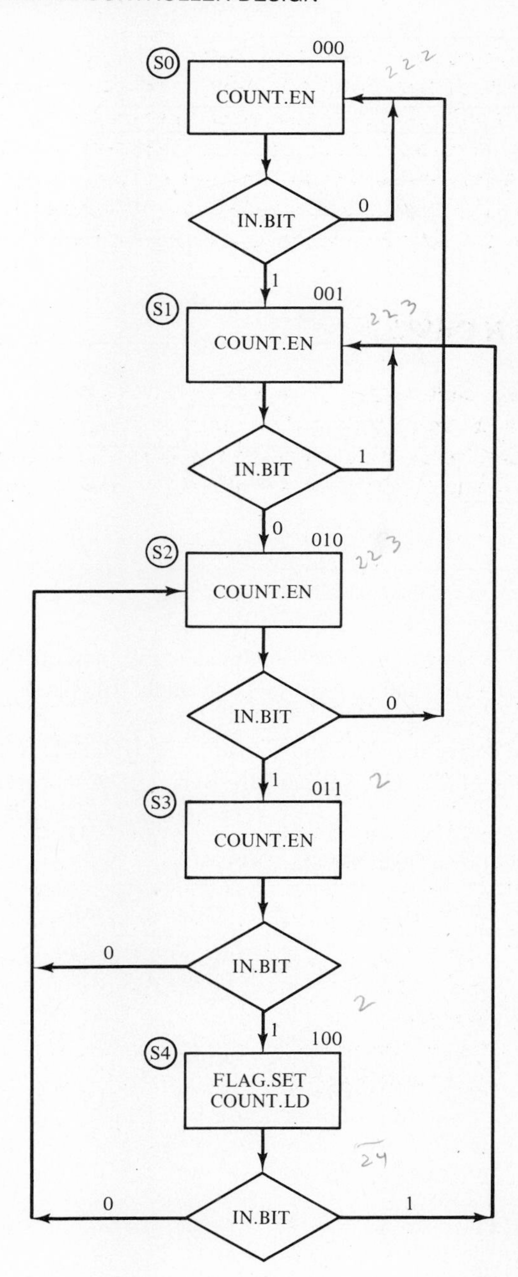

Figure 7.9 ASM chart for a controller that has one input and three outputs.

Using information from the ASM chart of Fig. 7.9, we can readily determine the values of $C2^+$, $C1^+$, and $C0^+$ as a function of IN.BIT and the values of C2, C1, and C0, as shown in Table 7.4. Then, we can use the J-K excitation table to derive the values of C2(J) and C2(K) from the values of C2 and $C2^+$. Similarly, we can derive the values of C1(J) and C1(K) from the values of C1 and $C1^+$, and the values of C0(J) and C0(K)

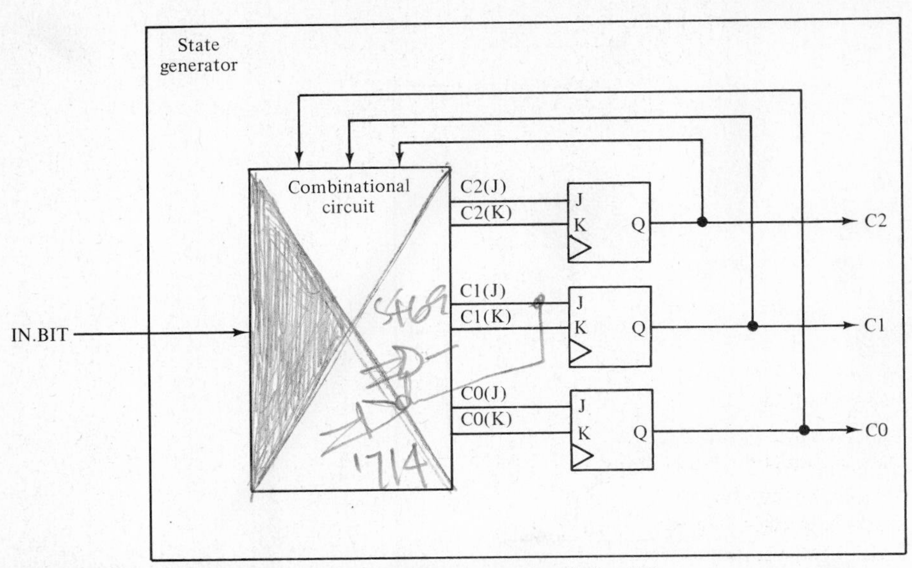

Figure 7.10 State generator circuit for the controller of Fig. 7.9.

from C0 and C0$^+$. The result is shown in Table 7.4. Incidentally, note in Table 7.4 that for the unused state codes 101, 110, and 111 (rows 11 through 16), the next-state code is arbitrarily specified as 000.

Using Table 7.4 we can derive the logic equations for C2(J), C2(K), C1(J), C1(K), C0(J), and C0(K) as functions of C2, C1, C0, and IN.BIT. The K-maps are shown in

TABLE 7.4 J AND K INPUTS FOR PRODUCING THE STATE TRANSITIONS

C2	C1	C0	IN.BIT	C2$^+$	C1$^+$	C0$^+$	C2(J)	C2(K)	C1(J)	C1(K)	C0(J)	C0(K)
0	0	0	0	0	0	0	0	X	0	X	0	X
0	0	0	1	0	0	1	0	X	0	X	1	X
0	0	1	0	0	1	0	0	X	1	X	X	1
0	0	1	1	0	0	1	0	X	0	X	X	0
0	1	0	0	0	0	0	0	X	X	1	0	X
0	1	0	1	0	1	1	0	X	X	0	1	X
0	1	1	0	0	1	0	0	X	X	0	X	1
0	1	1	1	1	0	0	1	X	X	1	X	1
1	0	0	0	0	1	0	X	1	1	X	0	X
1	0	0	1	0	0	1	X	1	0	X	1	X
1	0	1	0	0	0	0	X	1	0	X	X	1
1	0	1	1	0	0	0	X	1	0	X	X	1
1	1	0	0	0	0	0	X	1	X	1	0	X
1	1	0	1	0	0	0	X	1	X	1	0	X
1	1	1	0	0	0	0	X	1	X	1	X	1
1	1	1	1	0	0	0	X	1	X	1	X	1

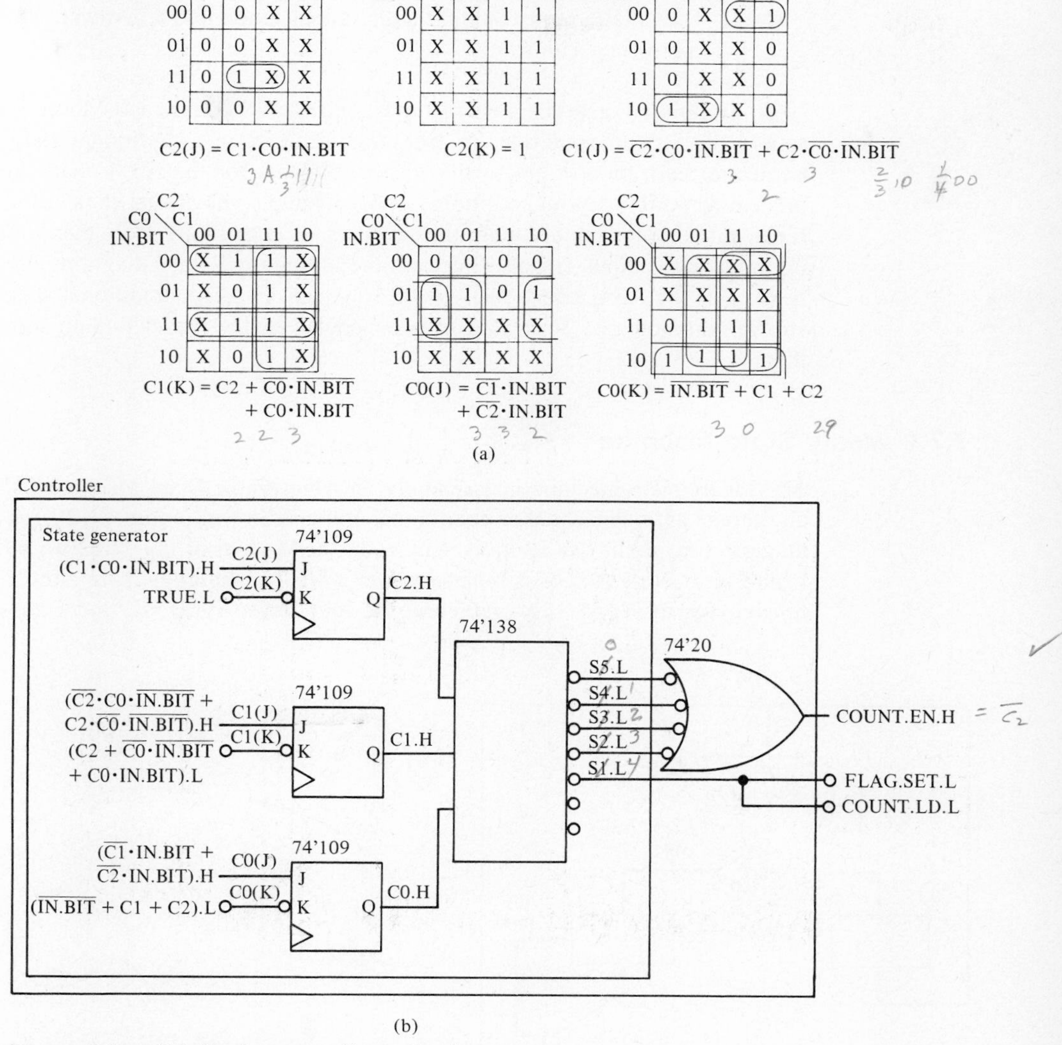

(b)

Figure 7.11 J-K flip-flop controller realization.

Fig. 7.11(a), and the resulting circuit diagram, along with the decoded state signals and ASM outputs, is given in Fig. 7.11(b).

7.7 TRADITIONAL STATE MACHINES

As stated in Sec. 7.4, two characteristics are essential for the notation used to represent the control algorithm of a digital circuit:

1. For the designer to effectively use it, the notation must provide a clear description of the algorithm, and in terms to which the designer can relate.
2. The notation must support a direct translation into a hardware realization of the control algorithm.

Traditional state-diagram methods, such as the Mealy and Moore state machines, satisfy the second condition. In fact, translation from a traditional state diagram to a hardware realization is practically identical to that for the ASM chart presented in the preceding section. With traditional state diagrams, however, it is difficult to clearly represent complex control algorithms, especially if there are more than a limited number of input and output signals. Furthermore, traditional state diagrams are often not as flexible as the ASM charts, as we will soon see. Even so, traditional state diagrams are still in common use. Therefore, it is beneficial for the reader to gain some exposure to them.

7.7.1 Mealy State Machine

The Mealy state machine is essentially an Algorithmic State Machine (ASM) in which all outputs are represented as conditional outputs. Shown in Fig. 7.12(b) is a Mealy state diagram (or graph) that is equivalent to the ASM chart of Fig. 7.12(a), which is copied from Fig. 7.2(a). A state is represented in a Mealy state graph by a circle. Therefore the four circles in Fig. 7.12(b) represent the four states: A, B, C, and D. As in the ASM

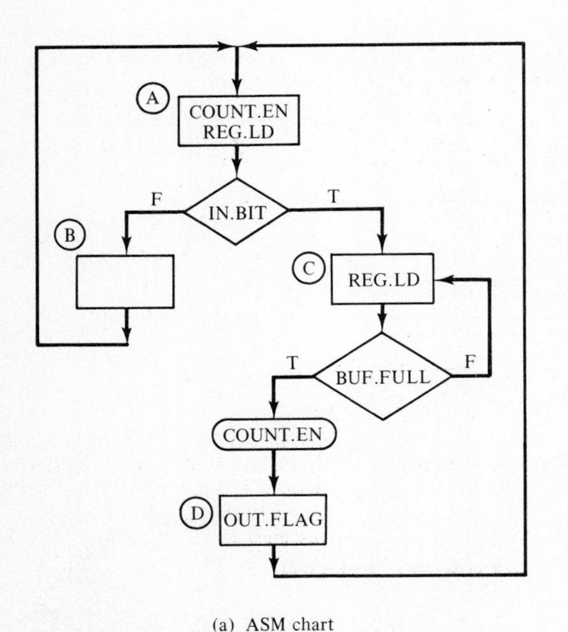

(a) ASM chart

(b) Corresponding Mealy state graph

Figure 7.12 Mealy state graph illustration.

chart, a state transition is represented by an arrow from one state to a next state. Further, specified for each transition arrow are the present values of every input signal and every conditional output signal, with the input values specified on the top and the output values on the bottom. The correspondences of these values to the actual signals are implicit through the ordering. Here, the ordering is IN.BIT, BUF.FULL on each top, and COUNT.EN, REG.LD, OUT.FLAG on each bottom. Again, note that all outputs for the Mealy state diagram are represented as conditional outputs. Therefore, to represent unconditional outputs such as COUNT.EN and REG.LD for state A, we have to specify those output values at both transition arrows from state A, as shown.

7.7.2 Moore State Machine

The Moore state machine is essentially an ASM in which all outputs are represented as unconditional outputs. Shown in Fig. 7.13(b) is a Moore state diagram that is equivalent to the ASM chart of Fig. 7.13(a). Just as for a Mealy state diagram, a state in a Moore state diagram is represented by a circle, and a state transition is represented by an arrow. However, the specification for each transition arrow has only the present values of the input signals used to determine the next state; there are no output signal values. Instead, the outputs are associated unconditionally with states, being specified inside each state circle under the state name. As a result, a Moore state machine cannot be used to represent the ASM chart of Fig. 7.2 or 7.12(a), where conditional and unconditional outputs are present.

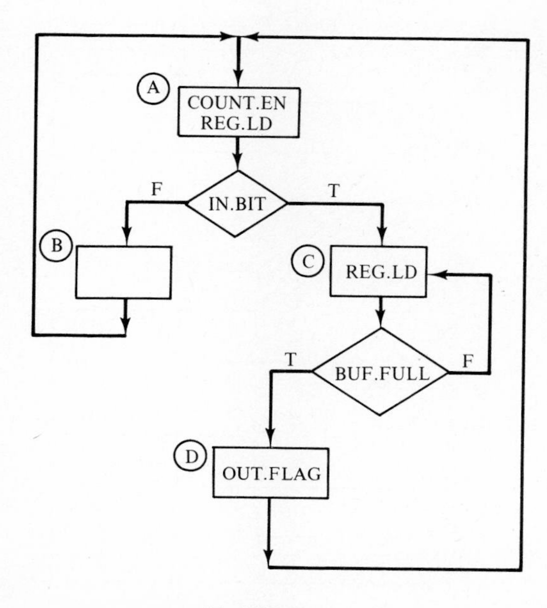

(a) ASM chart

Inputs: IN.BIT, BUF.FULL
Outputs: COUNT.EN, REG.LD, OUT.FLAG

A 110
0X
1X
B 000
XX
C 010
X0
X1
D 001
XX

(b) Corresponding Moore state graph

Figure 7.13 Moore state graph illustration.

7.8 DESIGN EXAMPLES

In this section we will consider design examples to illustrate the digital circuit design fundamentals that have been presented in this chapter. In studying these examples the reader is particularly urged to keep in mind the two main concepts in the design of a digital circuit:

1. A digital circuit is conceptually divided into the following:
   (a) A set of circuit elements.
   (b) A controller that controls the inputs to these circuit elements.
2. A digital circuit design is stepwise refined through the following phases:
   (a) The preliminary design phase—which results in block diagrams of the set of circuit elements and a flowchart of the controller.
   (b) The refinement phase—which results in a set of detailed circuit elements with completely defined functions and signals, and an ASM chart of the controller.
   (c) The realization phase—which results in a hardware realization with commercially available ICs of the circuit elements and the ASM chart.

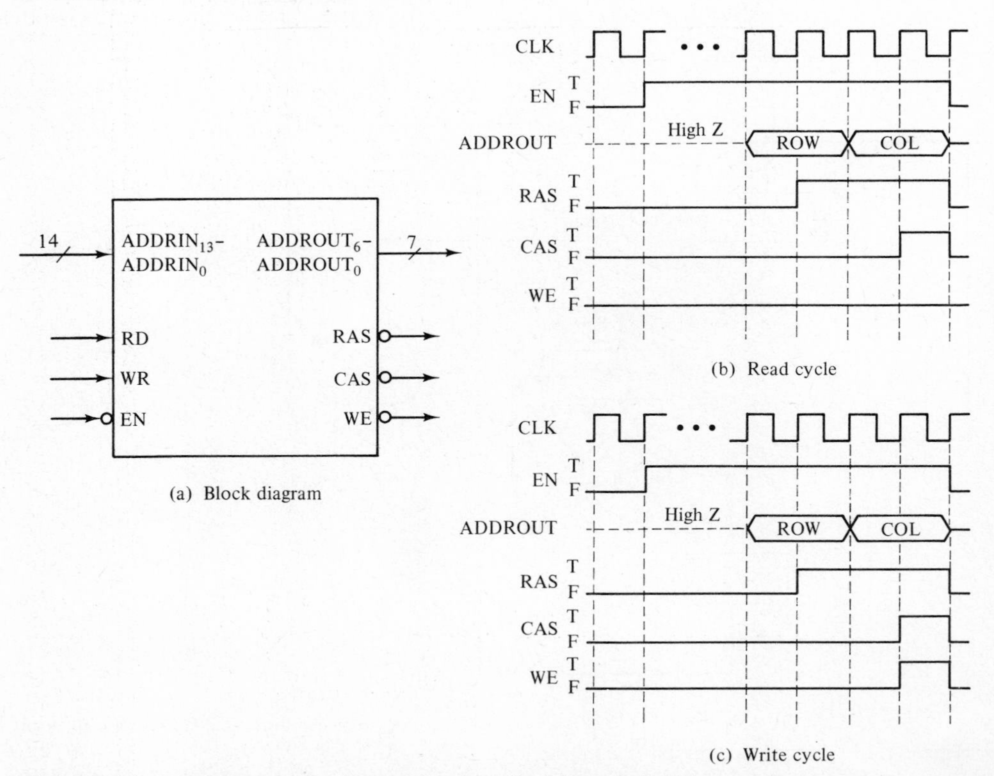

Figure 7.14 Problem specification for the simplified dynamic RAM controller.

7.8.1 Simplified Dynamic RAM Controller

For this example we are to design a simplified dynamic RAM controller circuit, the block diagram of which is shown in Fig. 7.14(a). This circuit has the following specifications: The inputs to the circuit are a 14-bit address ($ADDRIN_{13}$–$ADDRIN_0$), a read signal (RD), a write signal (WR), and an enable signal (EN). This circuit does not function until EN is true. When EN becomes true, the 14-bit ADDRIN is loaded in as a row address ($ADDRIN_{13}$–$ADDRIN_7$) and a column address ($ADDRIN_6$–$ADDRIN_0$). Also when EN becomes true, the values of RD and WR are latched. Subsequently, the row address is outputted (at ADDROUT) along with the row address strobe (RAS) signal (at RAS). Then, the column address is outputted (at ADDROUT) along with the column address strobe (CAS) signal (at CAS). Finally, if the operation is a write operation (RD = false, WR = true), then the WE output is true. Otherwise, for a read operation (RD = true, WR = false), the WE output remains false.

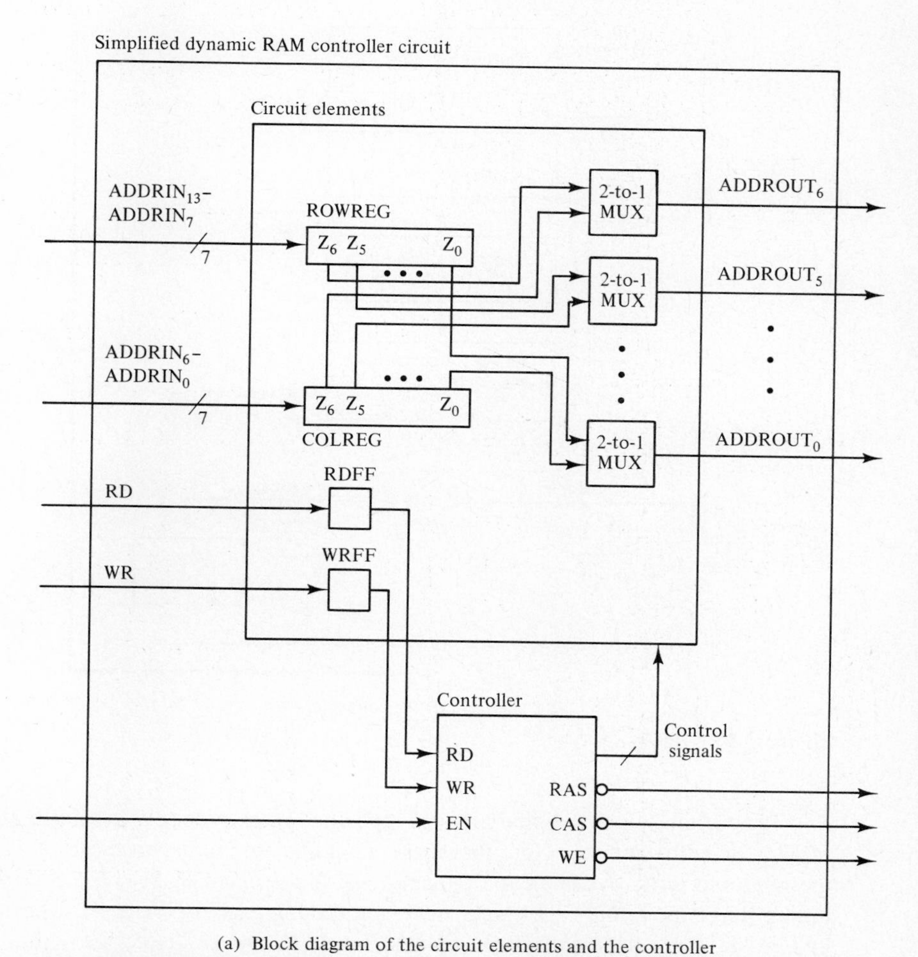

(a) Block diagram of the circuit elements and the controller

Figure 7.15 Preliminary design for the dynamic RAM controller circuit.

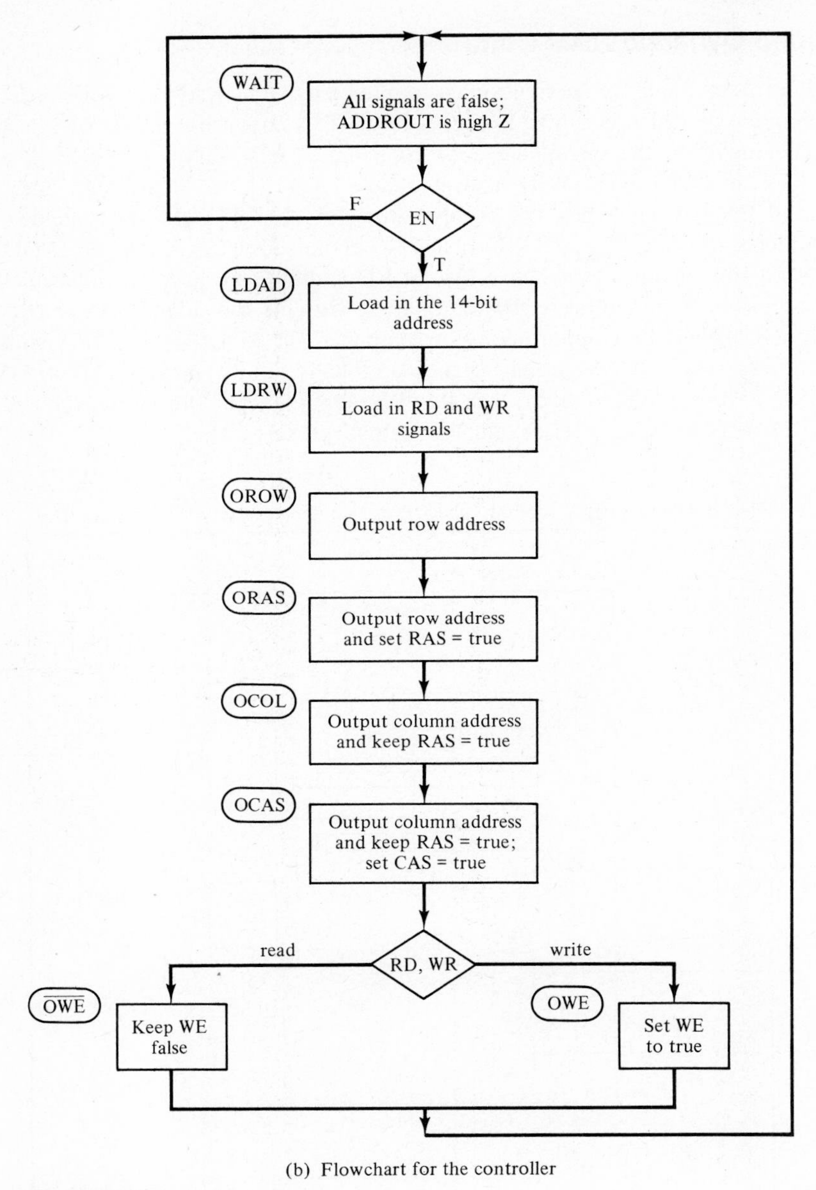

(b) Flowchart for the controller

Figure 7.15 *(cont.)*

The timing of this digital circuit is specified more precisely in the timing diagrams of Figs. 7.14(b) and (c). Note that these diagrams are similar to those of Fig. 6.31 for the read and write cycles of the dynamic RAM presented in Sec. 6.5.3. However, the timing diagrams in Fig. 7.14 are specified at the logic level (T and F), whereas the ones in Fig. 6.31 are specified at the voltage level (H and L). Note also from the timing diagrams in Figs. 7.14(b) and (c) that the row address is outputted at ADDROUT some time after EN becomes true. The circuit should be designed so that this time delay is as small as possible.

A preliminary design of the simplified dynamic RAM controller circuit is shown in Fig. 7.15. It should be obvious that we need two 7-bit storage registers to store the row address (ROWREG) and the column address (COLREG), respectively. Also needed are two 1-bit storage registers (flip-flops) to latch the values of RD and WR. All these registers should have a synchronous load input so that when EN becomes true, the controller can generate the load signals to these registers to load in the values. After the row and column addresses are loaded, they are time-multiplexed at the ADDROUT outputs. More specifically, the row address is outputted first, and then the column address, with both appearing at the same outputs (ADDROUT). As a result, seven 2-to-1 multiplexers are needed to select either the row address or the column address to be outputted. It is the function of the controller to perform this selection (via the select inputs of the MUXs) along with outputting the appropriate RAS and CAS outputs. Additionally, the controller should output WE = true for a write operation and WE = false for a read operation. The control algorithm for the preliminary design is specified in the form of a flowchart in Fig. 7.15(b).

The next phase of the design process is the refinement phase. The first step in it is the refining of the definitions of the circuit elements. At this point of the design we have a good idea of the functions that are required for each of the circuit elements. Consequently, a detailed definition of them is possible. The result is shown in Fig. 7.16(a). Next, we see that the control signals outputted from the controller are now

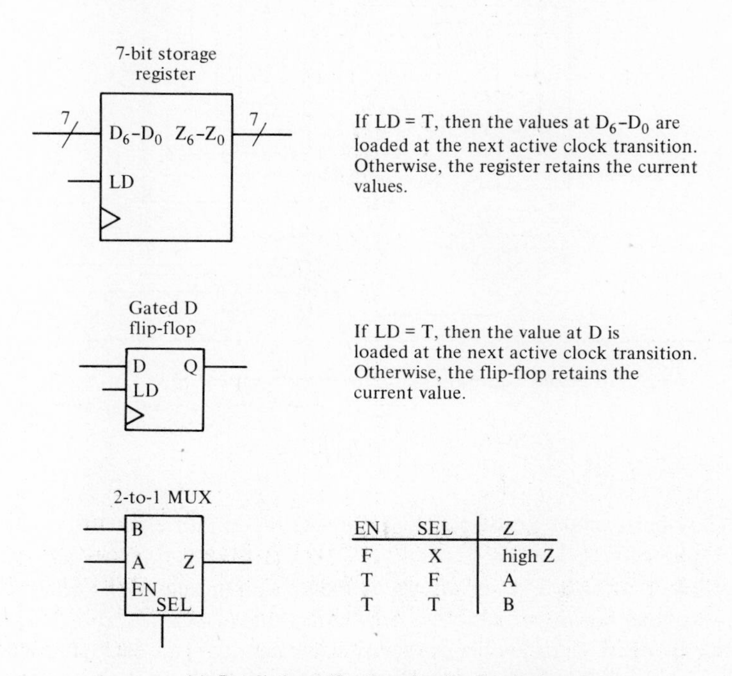

EN	SEL	Z
F	X	high Z
T	F	A
T	T	B

(a) Detailed specifications for the circuit elements

Figure 7.16 Refined design of the dynamic RAM controller circuit.

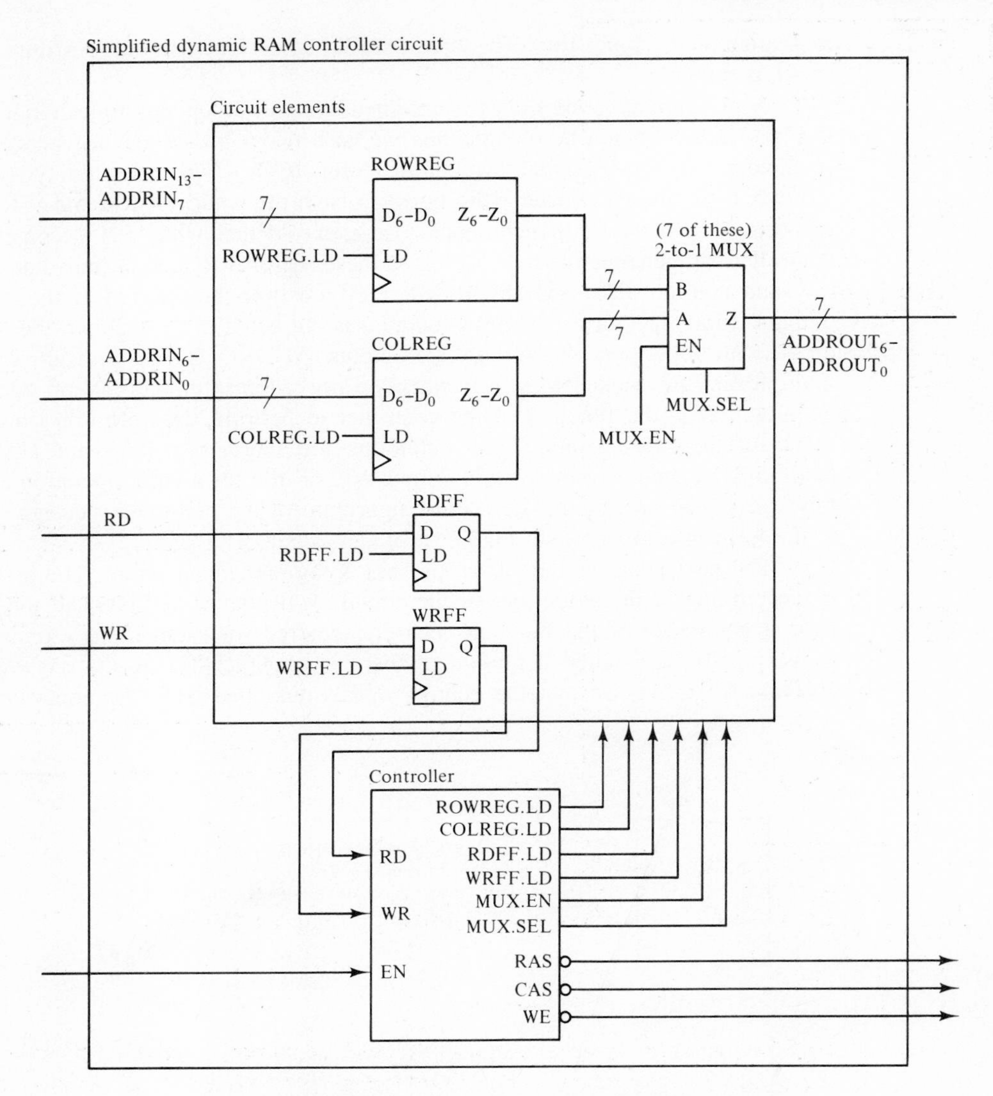

(b) Block diagram of the circuit elements and controller

Figure 7.16 *(cont.)*

defined. They correspond to the control inputs of the circuit elements as shown in Fig. 7.16(b).

Based on the refined set of circuit elements shown in Fig. 7.16(b), the flowchart of Fig. 7.15(b) can be converted to the ASM chart shown in Fig. 7.17(a). Different from a flowchart, an ASM chart precisely specifies the timing, with each state corresponding to a clock period. Additionally, unconditional and conditional control outputs are speci-

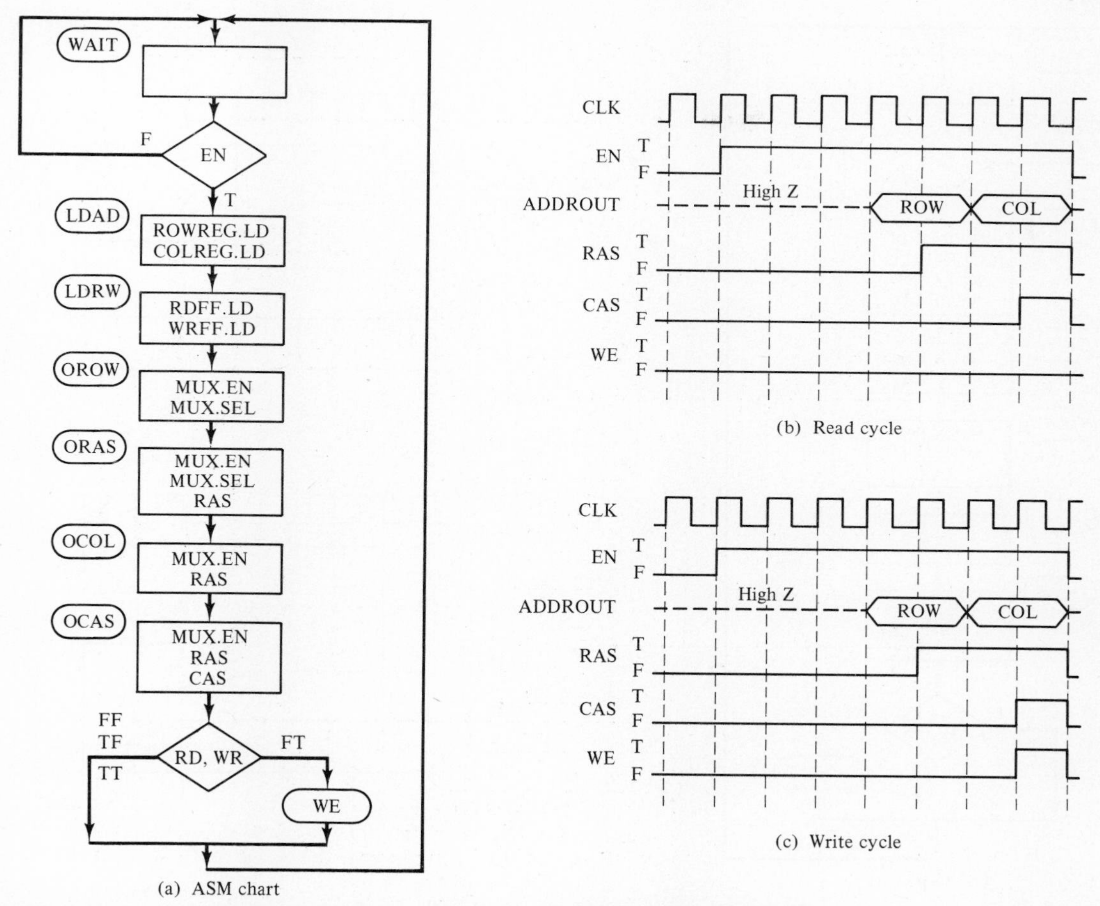

(a) ASM chart

(b) Read cycle

(c) Write cycle

Figure 7.17 An ASM chart for the controller of the dynamic RAM controller circuit.

fied for each state to accomplish the required functions for that state. For example, in the state LDAD, the row address and column address are synchronously loaded into ROWREG and COLREG, respectively, by applying true load signals ROWREG.LD and COLREG.LD during that clock cycle. Note that the state OWE in the original flowchart is changed in the ASM chart to a conditional output that is associated with the state OCAS. This is necessary because we want to output WE = T in the same clock cycle as the CAS output. Also note that in the state OCAS, if RD, WR = FF or TT, there is a default to a read operation (WE = F). This is a designer's choice. We could have just as well decided to make the default a write operation. The precise timing of the ASM chart in Fig. 7.17(a) is shown in the corresponding timing diagrams of Figs. 7.17(b) and (c). Note that for a controller realized from this ASM chart, the row address is not outputted until three clock cycles after EN becomes true.

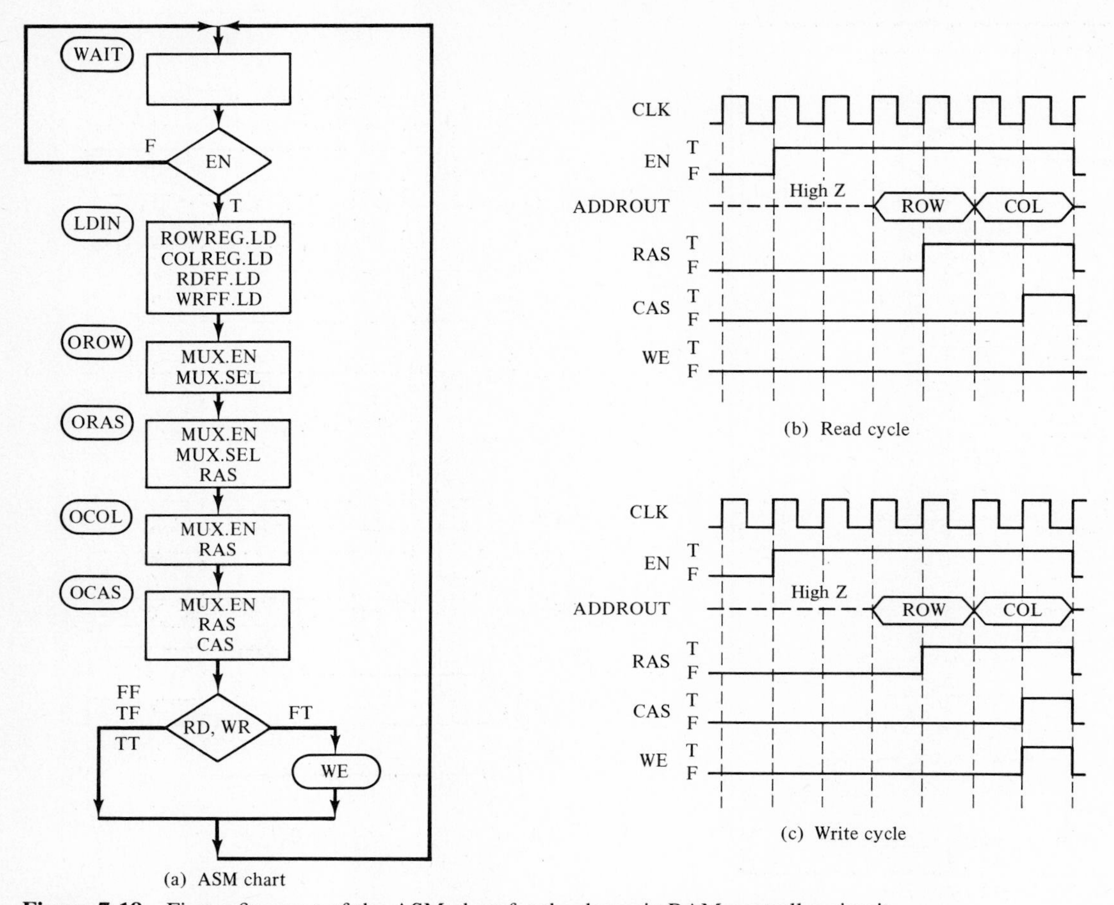

Figure 7.18 First refinement of the ASM chart for the dynamic RAM controller circuit.

Figures 7.18 and 7.19 illustrate stepwise refinements of the controller ASM chart. Note in the ASM chart of Fig. 7.18(a) that the states LDAD and LDRW, from Fig. 7.17(a), are combined into a single state LDIN. The result is that both the row and column addresses are loaded during the same state as the RD and WR signals. From the circuit elements shown in Fig. 7.16(b) we see that this is physically possible and does not introduce any timing problem. Clearly, this joint loading is desirable since it reduces the delay before the row address can be outputted, as shown in Figs. 7.18(b) and (c).

The ASM chart can be further optimized by eliminating the state LDIN by associating conditional outputs with the state WTLD as shown in Fig. 7.19(a). Again, this elimination does not introduce any timing problem. In addition, it reduces the row address output delay by one more clock pulse, as shown in Figs. 7.19(b) and (c).

One more refinement for the digital circuit can be made. We observe that the four load signals ROWREG.LD, COLREG.LD, RDFF.LD, and WRFF.LD are applied in the

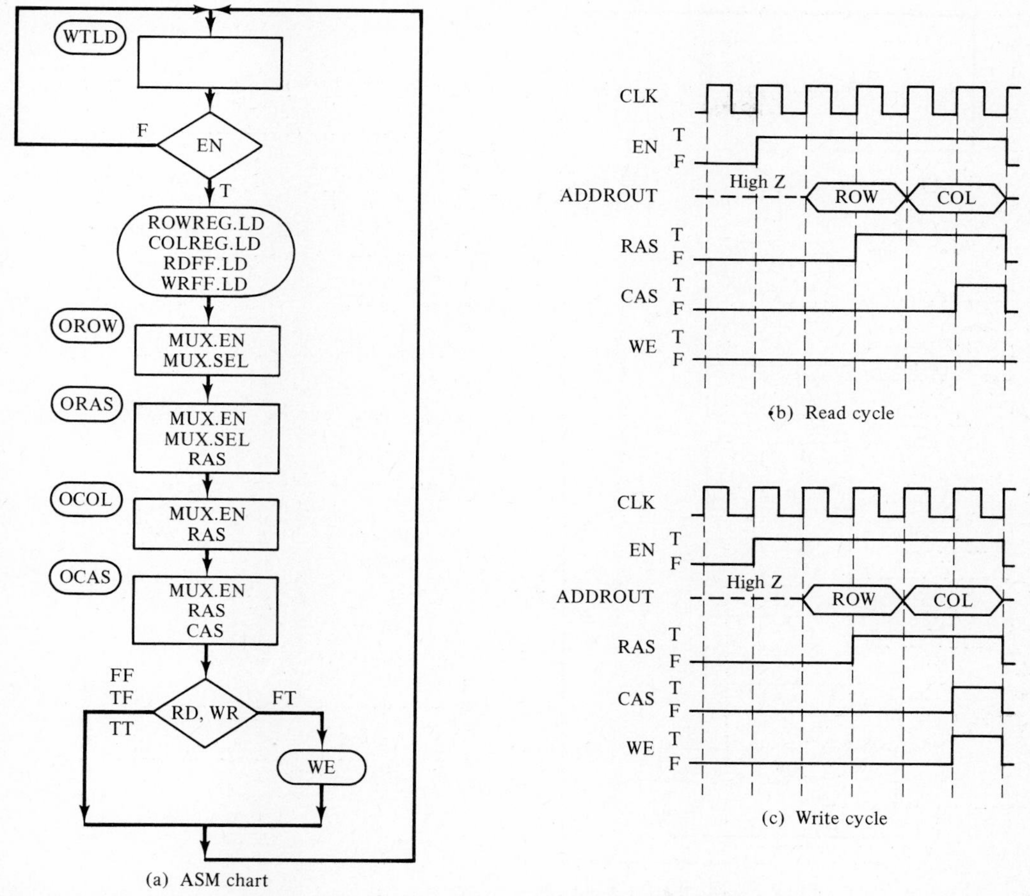

Figure 7.19 Final refinement of the ASM chart for the dynamic controller circuit.

same state and under the same condition. Therefore, they can be replaced by a single load signal, LOAD. So, ROWREG.LD = COLREG.LD = RDFF.LD = WRFF.LD = LOAD. The final ASM chart for the controller is shown in Fig. 7.20.

All the ASM charts in Figs. 7.17(a), 7.18(a), 7.19(a), and 7.20 will function as specified in the original problem statement. However, the ASM charts of Figs. 7.19(a) and 7.20 provide the best performances.

Having gone from the flowchart of Fig. 7.15(b) to the final ASM chart of Fig. 7.20, we can now appreciate the general method for designing the controller of a digital circuit. First, the algorithm for the controller is specified in the form of a flowchart. Then, the flowchart is converted into an ASM chart in which the timing problems are resolved. Finally, the ASM chart is refined and optimized by combining and removing states and by using conditional outputs wherever possible.

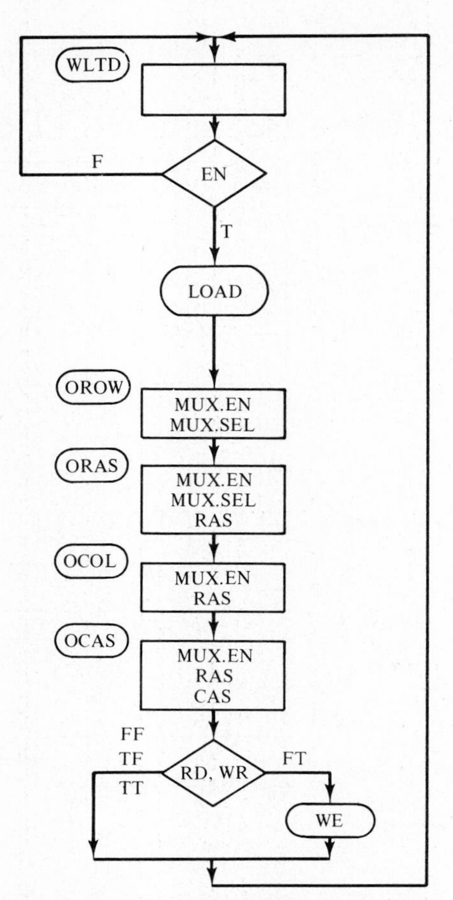

Figure 7.20 Final ASM chart for the dynamic RAM controller.

The final phase of the design process is the realization phase. For this circuit, this phase consists of using commercially available ICs to realize the circuit elements in Fig. 7.16(b) and the controller represented by the ASM chart in Fig. 7.20. As has been mentioned, the realization phase is straightforward. At this point of the design process, the hard work is over. The realization of the simplified dynamic RAM controller circuit is summarized in Figs. 7.21 and 7.22.

As shown in Fig. 7.21, the circuit element ROWREG is realized with two 74'163 counters utilized as storage registers, with enable inputs of ET = EP = false to prevent counting. The circuit element COLREG is realized in the same manner. The 2-to-1 multiplexers with three-state outputs are directly available commercially as ICs (74'257). The RDFF and WRFF are D flip-flops with a synchronous load input. This type of flip-flop is not directly available as a commercial IC; nevertheless, a gated D flip-flop can be easily designed and, in fact, was designed as an example in Sec. 5.3.4 and shown in

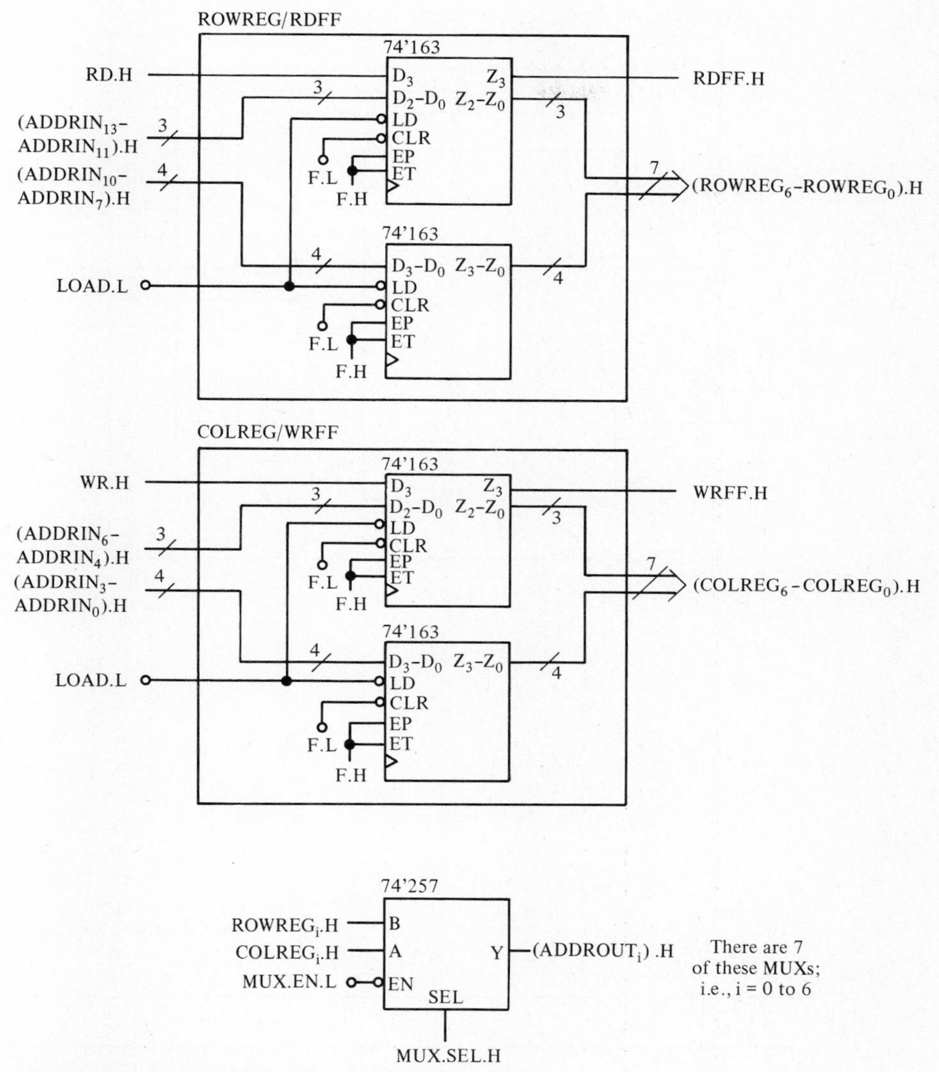

Figure 7.21 Realization of the circuit elements.

Fig. 5.10. This is, however, not necessary since in the 74'163 storage registers used for ROWREG and COLREG there are two unused cells that are perfectly suitable for realizing the RDFF and WRFF.

Any of the ASM realization methods presented in this chapter can be used to realize the ASM chart of Fig. 7.20. As an illustration, the ROM method will be used. The result is summarized in Fig. 7.22. The block diagram of the controller and the state assignments are shown in Fig. 7.22(a). The next-state and output table, derived from the ASM chart of Fig. 7.20, is shown in Fig. 7.22(b). Finally, the corresponding contents of the ROM are given in Fig. 7.22(c). Of course, the ROM addresses correspond to C_2, C_1, C_0, RD,

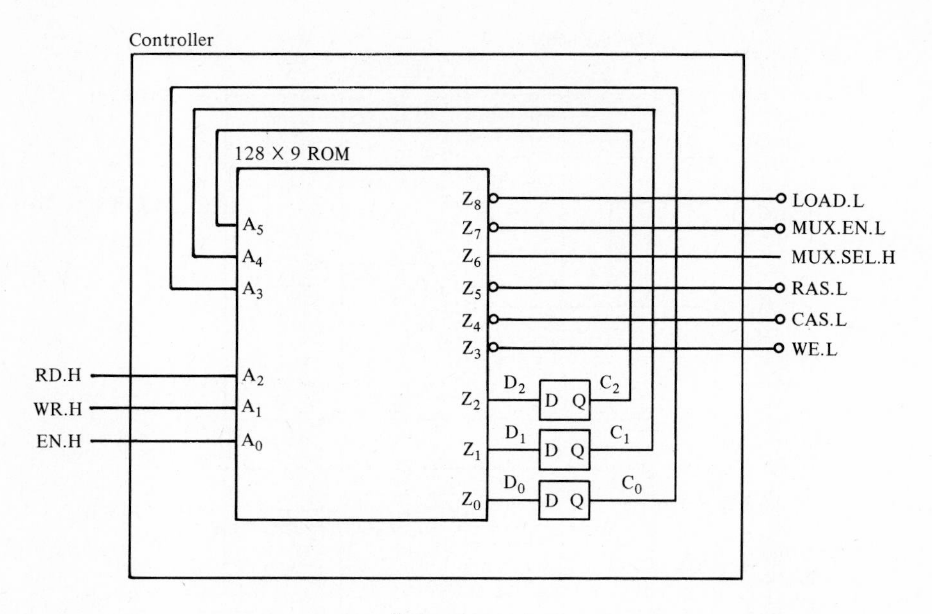

State assignments

	C_2	C_1	C_0
WTLD	0	0	0
OROW	0	0	1
ORAS	0	1	0
OCOL	0	1	1
OCAS	1	0	0

(a) Block diagram and state assignments

Present-state code			Input values for the present state			Output values for the present state						Next state			D flip-flop input values		
C_2	C_1	C_0	RD	WR	EN	LOAD	MUX. EN	MUX. SEL	RAS	CAS	WE	C_2^+	C_1^+	C_0^+	D_2	D_1	D_0
0	0	0	X	X	0	0	0	0	0	0	0	0	0	0	0	0	0
0	0	0	X	X	1	1	0	0	0	0	0	0	0	1	0	0	1
0	0	1	X	X	X	0	1	1	0	0	0	0	1	0	0	1	0
0	1	0	X	X	X	0	1	1	1	0	0	0	1	1	0	1	1
0	1	1	X	X	X	0	1	0	1	0	0	1	0	0	1	0	0
1	0	0	0	0	X	0	1	0	1	1	0	0	0	0	0	0	0
1	0	0	0	1	X	0	1	0	1	1	1	0	0	0	0	0	0
1	0	0	1	X	X	0	1	0	1	1	0	0	0	0	0	0	0
1	0	1	X	X	X	0	0	0	0	0	0	0	0	0	0	0	0
1	1	X	X	X	X	0	0	0	0	0	0	0	0	0	0	0	0

LOAD, MUX. EN: Active low; MUX. SEL: Active high; RAS, CAS, WE: Active low

Where 0 = false, 1 = true, and X = don't care

(b) Next-state and output table

Figure 7.22 Realization of the controller.

ROM address in decimal	ROM address in hexadecimal	Contents in binary (where 0 = low, 1 = high) $Z_8 \cdots Z_0$								
0	0 0	1	1	0	1	1	1	0	0	0
1	0 1	0	1	0	1	1	1	0	0	1
2	0 2	1	1	0	1	1	1	0	0	0
3	0 3	0	1	0	1	1	1	0	0	1
4	0 4	1	1	0	1	1	1	0	0	0
5	0 5	0	1	0	1	1	1	0	0	1
6	0 6	1	1	0	1	1	1	0	0	0
7	0 7	0	1	0	1	1	1	0	0	1
8–15	08–0F	1	0	1	1	1	1	0	1	0
16–23	10–17	1	0	1	0	1	1	0	1	1
24–31	18–1F	1	0	0	0	1	1	1	0	0
32–33	20–21	1	0	0	0	0	1	0	0	0
34–35	22–23	1	0	0	0	0	0	0	0	0
36–39	24–27	1	0	0	0	0	1	0	0	0
40–63	28–3F	1	1	0	1	1	1	0	0	0

(c) ROM contents

Figure 7.22 *(cont.)*

WR, EN and the ROM contents to LOAD, MUX.EN, MUX.SEL, RAS, CAS, WE, D_2, D_1, D_0. The correspondences between the rows of the next-state and output table and the rows of the ROM table are apparent from an inspection. For example, the first two rows of the next-state and output table correspond in an alternate fashion to the ROM address rows 0 through 7. Row 3 of the next-state and output table corresponds to ROM address rows 8 through 15, and so on. Note that for the active-low outputs (LOAD, MUX.EN, RAS, CAS, and WE), the ROM content values are inverted. In other words, the value true is 0 (low) and the value false is 1 (high).

7.8.2 Modified Counter

For this example, we are to design the digital circuit shown in block diagram form in Fig. 7.23(a). Overall, the circuit is a modified 4-bit counter with outputs Z_3–Z_0. There is also a FLAG output. Under normal operation, the circuit counts with the frequency of the input clock signal, and produces a binary output Z_3–Z_0 corresponding to . . . , 0, 1, 2, . . . , 14, 15, 0, 1, ... An input IN.BIT can, however, interrupt the normal counting sequence and cause the count to jump to a number corresponding to another input: D_3–D_0. Specifically, when IN.BIT, which is a serial string of 0s and 1s, contains the pattern 1011, the circuit does the following at the *next* active transition (leading edge in this case) of the clock signal:

1. Parallel loads the input D_3–D_0 into the counter component of the circuit so that the subsequent count sequence begins from there
2. Sets the output signal FLAG to true for the next clock period, after which the signal returns to false

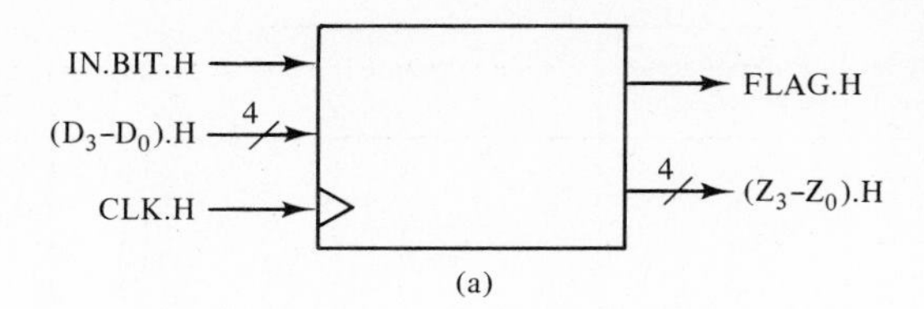

(a)

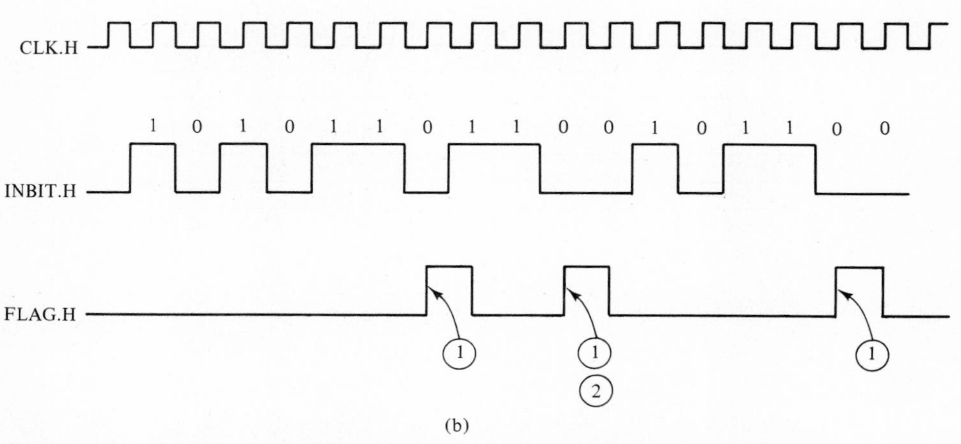

(b)

Figure 7.23 Block diagram and timing diagram specifications for the counter example.

For a better understanding of the desired operation, consider Fig. 7.23(b), which is a timing diagram of a sample input string for IN.BIT and of the resulting FLAG output of the circuit. Note that at the times designated by ① a 1011 pattern is detected. Thus the circuit loads D_3–D_0 into the counter component, which means that D_3–$D_0 \rightarrow Z_3$–Z_0. Also, the circuit causes the output FLAG to go from false to true. Further note that at time ② this circuit action occurs even though there is no separate preceding 1011 IN.BIT sequence, which means that activating 1011 sequences can overlap.

With a little thought, we see that we need a 4-bit binary counter with a parallel load capability for producing the Z_3–Z_0 output and for loading the D_3–D_0 input. Also, we can use a flip-flop for generating the FLAG output. Finally, we need a sequence detector for examining the serial input IN.BIT and for generating a feedback signal FOUND.SEQ when it detects an input 1011 sequence. As is shown in Fig. 7.24(a), our identification of these three main components completes the selection of the circuit elements for the preliminary design of the circuit.

As is indicated by the flowchart of Fig. 7.24(b), the controller action is quite simple. Under normal operation, when the sequence detector has not detected the sequence 1011, the detector output is false (FOUND.SEQ = F). In this case the counter counts normally and the FLAG flip-flop has a false output. But when the sequence detector detects the sequence 1011, its output becomes true (FOUND.SEQ = T). Then, at the next active clock transition, the counter loads in D_3–D_0. Also at this transition, the FLAG flip-flop is set to true.

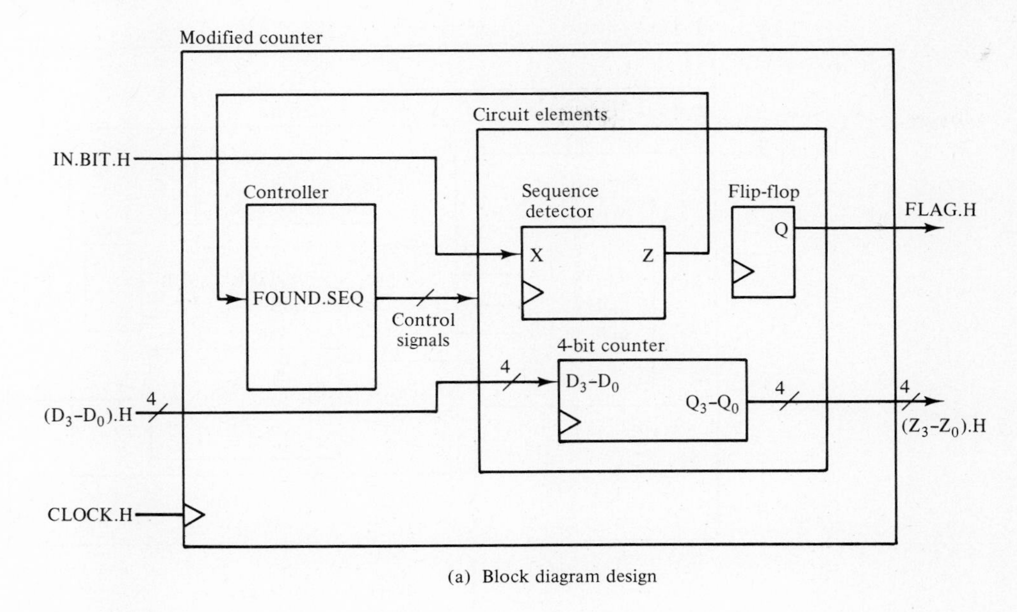

(a) Block diagram design

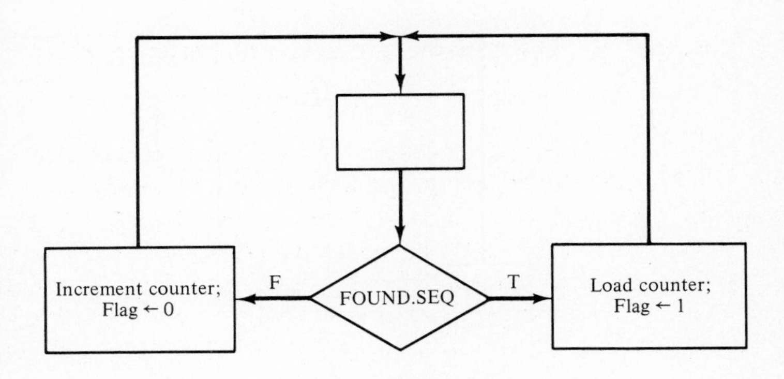

(b) Flowchart for the controller

Figure 7.24 Preliminary design for the modified counter.

In the refinement phase of the design we define the circuit elements in more detail. The result is shown in Fig. 7.25(a). As is typical, the counter has enable (EN) and load (LD) inputs, in addition to the clock input. For a normal count, the control inputs are EN = true and LD = false. And for the loading of data from D_3–D_0, these inputs are just the opposite: EN = false and LD = true. Also, the FLAG flip-flop has a SET input in addition to the clock input. To set this flip-flop we make SET = true before the next active clock transition. And to clear it we make SET = false before the next active clock transition. For the sequence detector we will specify that it makes FOUND.SEQ = true when it detects the sequence 1011. Note in Fig. 7.25(a) that we do not assign voltage

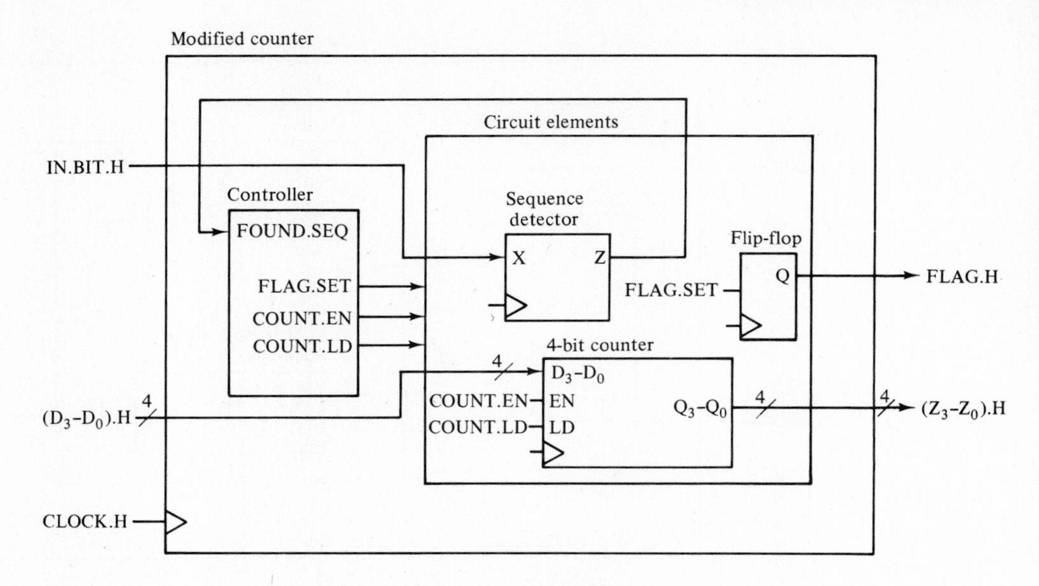

(a) Block diagram design

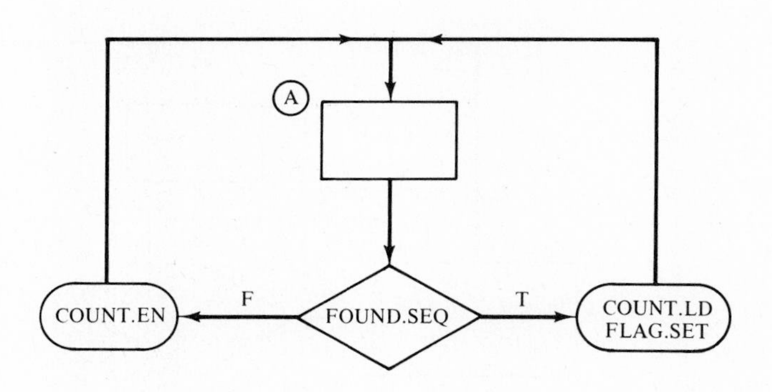

(b) ASM chart for the controller

Figure 7.25 Refined design for the modified counter.

representations for the signals FOUND.SEQ, FLAG.SET, COUNT.EN, and COUNT.LD because at this point in the design we do not know whether active-high or active-low signals are required.

Now having the refined circuit elements of Fig. 7.25(a), we can convert the preliminary controller flowchart of Fig. 7.24(b) into the ASM chart shown in Fig. 7.25(b). Note the use of conditional outputs. In this case the conditional outputs are not used simply for performance (as in the case of the dynamic RAM controller example), but also are necessary to achieve the required timing. The reader is urged to verify the timing of this ASM chart by comparing it with the timing diagram shown in Fig. 7.23(b).

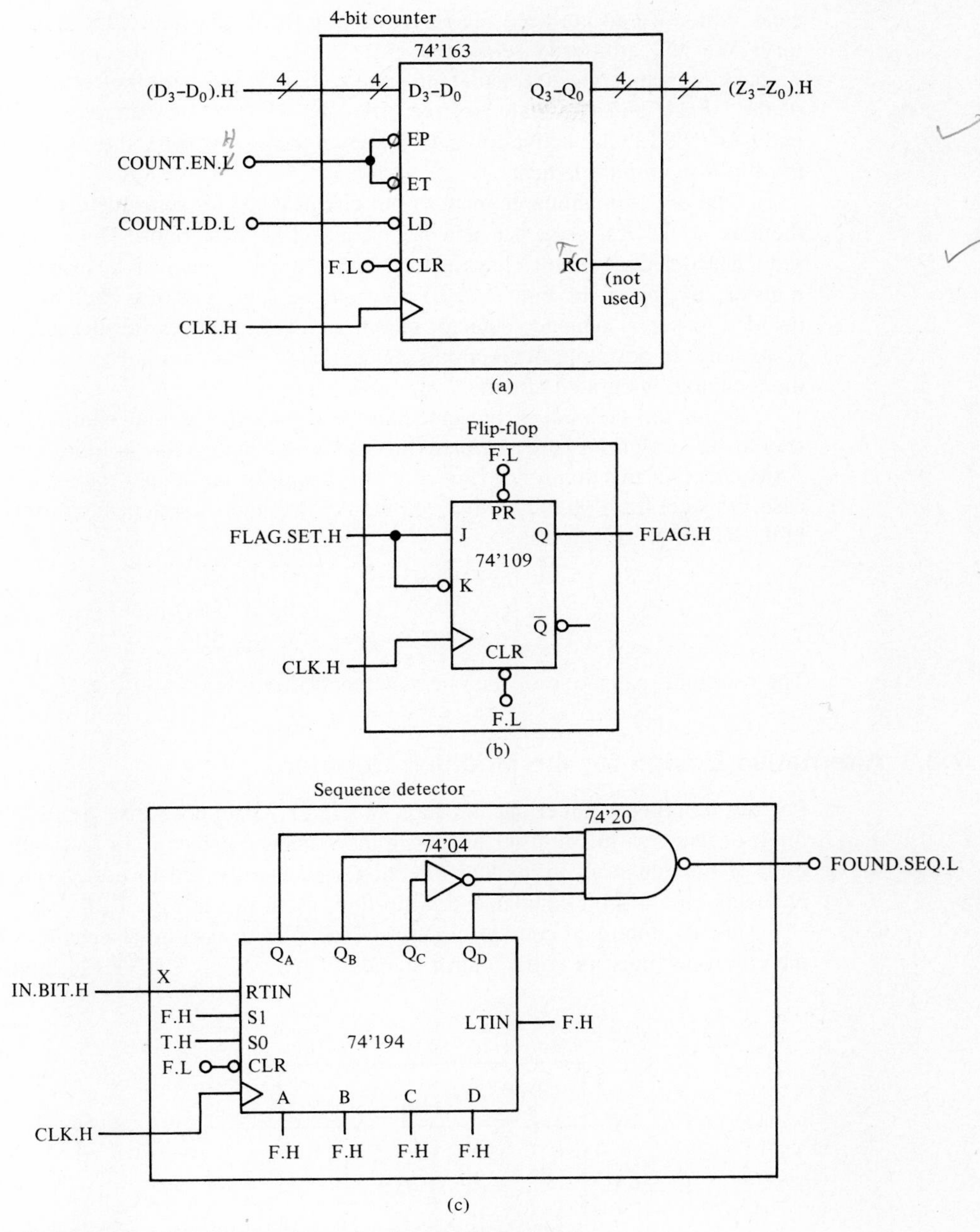

Figure 7.26 Circuit components for the modified counter.

The final phase of the design process is the realization phase. The realization of the circuit elements is summarized in Fig. 7.26. From a TTL data book we can find that the 74'161 and the 74'163 are both 4-bit counters with parallel load capability. They differ only in that the 74'161 has an asynchronous clear and the 74'163 a synchronous clear. Since we do not need the clear function for this design, either counter is satisfactory. We will arbitrarily select the 74'163, which requires the signals shown in Fig. 7.26(a). Note that once the actual integrated circuit is selected, the voltage representations of the signals and terminals are fixed. For the 74'163, for example, both COUNT.EN and COUNT.LD are active-low. As is shown in Fig. 7.26(b), the 74'109 is suitable for the flip-flop circuit element.

The only remaining element of our circuit is the sequence detector. Unfortunately, there is no 74'XX sequence detector in any TTL data book. But we can realize the sequence detector circuit element in a rather obvious manner by using a 74'194 shift register, as shown in Fig. 7.26(c). We should have had this fact in mind when we decided to use a sequence detector circuit element in our earlier design steps, for there is no sense in designing with circuit elements that we do not have or cannot realize with the available integrated circuits.

From the techniques presented in this chapter, the realization of the controller should be straightforward. In fact, this particular realization is quite simple since the ASM chart for this controller [Fig. 7.25(b)] has only one state—state A. For this special case, no state flip-flop is required and the ASM outputs are dependent only on the input FOUND.SEQ, as follows:

$$\begin{aligned} \text{COUNT.EN} &= \overline{\text{FOUND.SEQ}} \\ \text{COUNT.LD} &= \text{FOUND.SEQ} \\ \text{FLAG.SET} &= \text{FOUND.SEQ} \end{aligned}$$

The resultant realization of the one-state controller is shown in Fig. 7.27.

7.8.3 Alternative Design for the Modified Counter

For the modified counter specification of Fig. 7.23 suppose that we did not happen to think of using a shift register to realize the sequence detector. In fact, suppose that our train of thought went in another direction, and we decided to use as our major circuit elements only a 4-bit counter and a flip-flop, as shown in Fig. 7.28.

For the modified counter circuit of Fig. 7.28 to operate as specified in Fig. 7.23, the controller must provide control signals FLAG.SET, COUNT.EN, and COUNT.LD

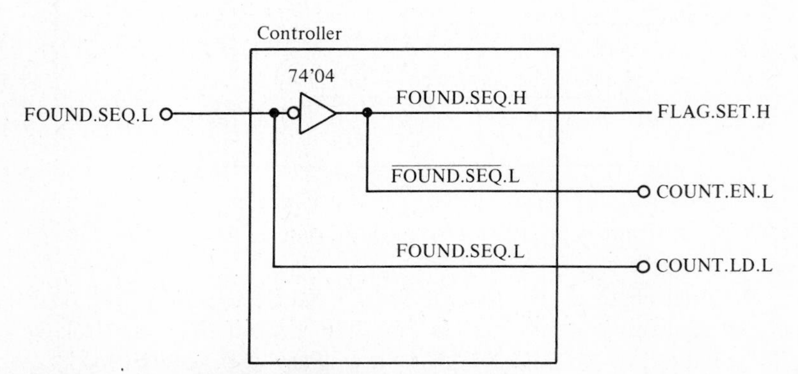

Figure 7.27 Controller for the modified counter.

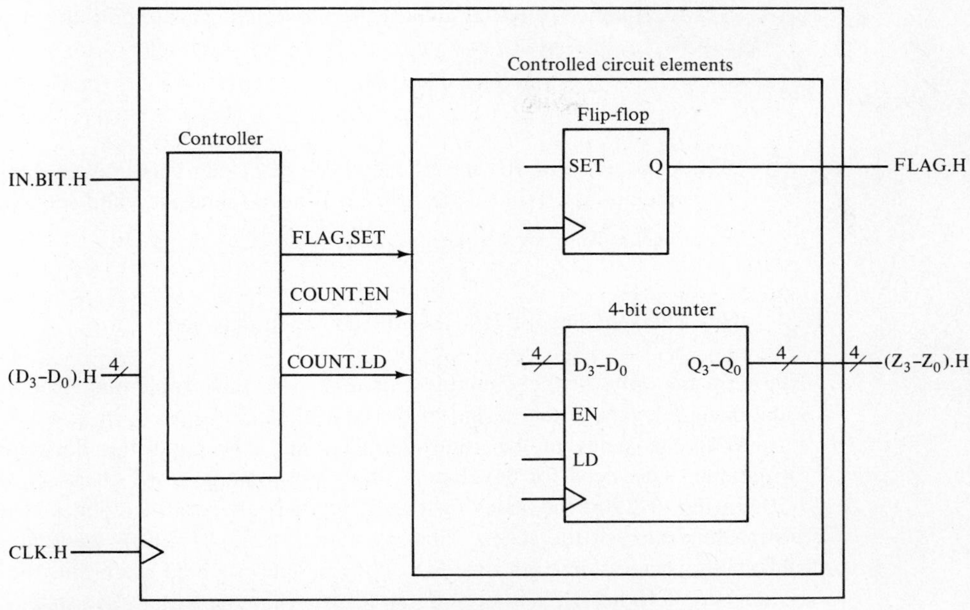

Figure 7.28 Preliminary design for the alternative modified counter.

that are correct for each moment in time. In other words, for every moment in time, the functions of the controller are as follows:

1. To determine the current situation of the input sequence, based on the present value of IN.BIT along with a memory of the past values of IN.BIT leading up to the present. (Note that we no longer can use the signal FOUND.SEQ since we no longer are using the sequence detector circuit element.)
2. For that situation, provide the appropriate input control signals COUNT.EN and COUNT.LD for the counter and FLAG.SET for the flip-flop.

Obviously, the controller for this circuit will be more complex than that of the original design. First, we need to determine all the *distinct* situations for the input sequence. Fortunately, there are only a finite number of them:

(S0) No part of the valid sequence has been detected. Example of such a sequence:

. . . XXXXX00 (where X is either a 0 or a 1)
↑
last value of IN.BIT received

(S1) A sequence of 1 has already been detected. This implies that this could be the beginning of a valid sequence. Example:

. . . XXXABC1 in which ABC $\neq$ 101 and BC $\neq$ 10
↑

(S2) A sequence of 10 has already been detected. This implies that a valid sequence is possible within two more clock cycles. Example:

. . . XXXXX10
↑

(S3) A sequence of 101 has already been detected. This implies that if the current value of IN.BIT is 1, then we will have found the valid sequence. Example:

. . . XXX101
↑

(S4) The sequence of 1011 has already been detected.

These are the only distinct situations, or *states,* for this problem environment. No other situation is relevant to the design of the controller.

With the states of the controller identified, the controller design becomes less formidable. The steps for developing the corresponding ASM chart are shown in Fig. 7.29. In Fig. 7.29(a) the states for the ASM chart are identified and a state box is used to represent each of the states. Then as shown in Fig. 7.29(b), the state transition for each state is determined and specified. For example, state S2 represents the situation that a sequence of 10 has already been detected. If, while the circuit is in state S2, the current input IN.BIT is a 1, then the resulting partial sequence becomes 101, which is represented by state S3. Consequently, the state transition from state S2 is to state S3. If, however, when the circuit is in state S2, the current input IN.BIT is a 0, then the resulting sequence becomes 100, which is no longer a part of the valid sequence 1011. Consequently, the state transition is to state S0, the reset state, where no part of the valid sequence has been detected. The state transitions for each of the states can be determined in this manner.

Next, the control outputs for each state of the ASM chart have to be determined and specified, which is quite straightforward for this controller. The result is shown in Fig. 7.29(c). In state S4, where the valid sequence has been detected, the counter needs to be loaded (COUNT.LD) and the flip-flop needs to be set (FLAG.SET) at the next active clock transition. For the remainder of the states, where the valid sequence has not been detected, all that is needed is to increment the counter (COUNT.EN).

Finally, the ASM chart is reviewed to determine whether the timing is correct and whether it can be optimized. This is normally accomplished through the use of conditional outputs. In this case, however, the ASM chart shown in Fig. 7.29(c) is the final one.

The final phase of the design process is the realization phase. The realization of the circuit elements of this circuit (counter and flip-flop) is the same as that shown in Figs. 7.26(a) and (b). The realization of the ASM chart has already been accomplished in Sec. 7.6 and Fig. 7.11.

7.8.4 Hardware Multiplier

In this example we are to design a hardware multiplier circuit that can multiply two 4-bit unsigned binary numbers to produce an 8-bit product. The block diagram is shown in Fig. 7.30. The inputs to this circuit are a 4-bit multiplicand

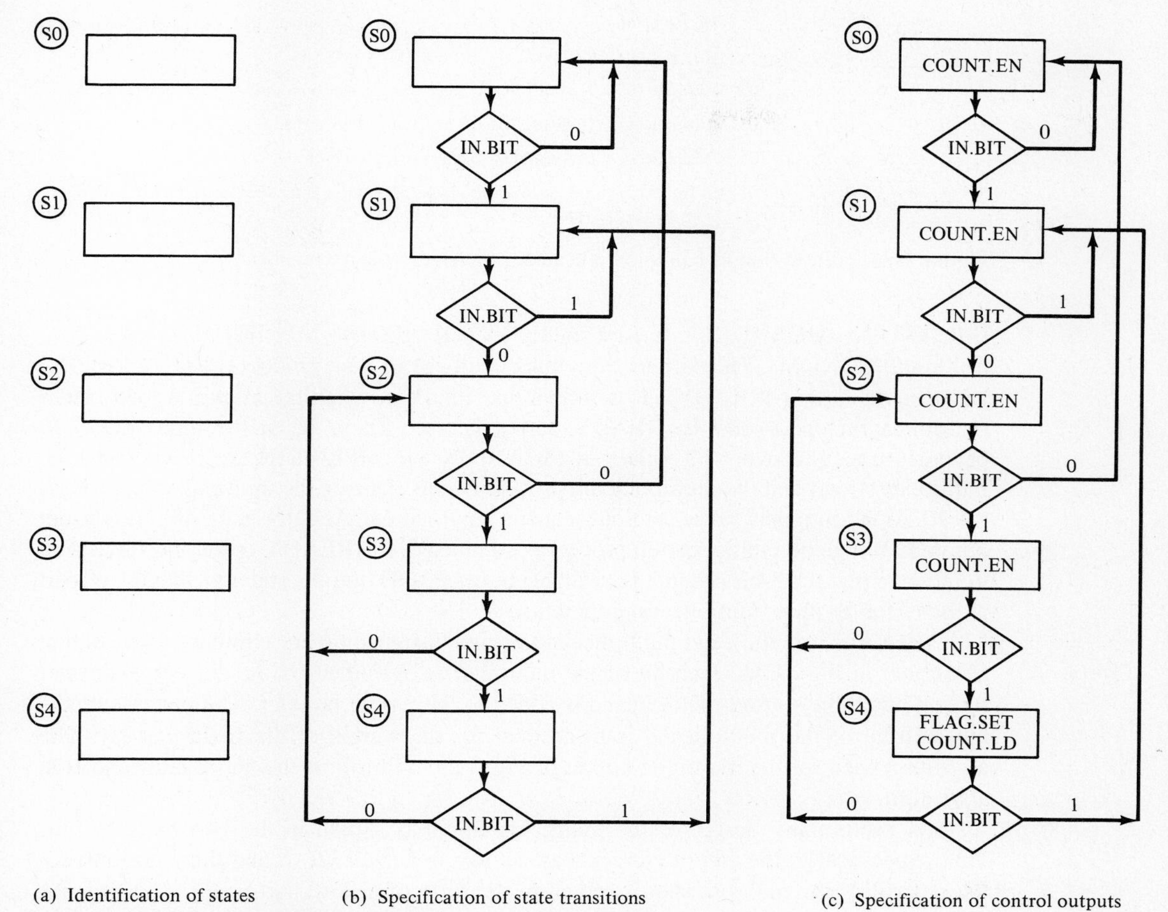

(a) Identification of states (b) Specification of state transitions (c) Specification of control outputs

Figure 7.29 Development of the ASM chart for the controller.

4-bit multiplier

4

MCANDIN$_3$–MCANDIN$_0$

PROD$_7$–PROD$_0$

8

4

MPLIERIN$_3$–MPLIERIN$_0$

READY

LDNUM

Figure 7.30 Block diagram of the multiplier.

				1	0	1	0	(Multiplicand)
			×	1	0	1	1	(Multiplier)
0	0	0	0	1	0	1	0	Multiplier(0) = 1; multiplicand is shifted 0 times and added to the partial product
0	0	0	1	0	1	0	0	Multiplier(1) = 1; multiplicand is shifted 1 time and added to the partial product
0	0	0	0	0	0	0	0	Multiplier(2) = 0; 0 is added to the partial product
0	1	0	1	0	0	0	0	Multiplier(3) = 1; multiplicand is shifted 3 times and added to the partial product
0	1	1	0	1	1	1	0	(Product)

Figure 7.31 Illustration of hand multiplication of unsigned binary numbers.

($MCANDIN_3$–$MCANDIN_0$), a 4-bit multiplier ($MPLIERIN_3$–$MPLIERIN_0$), and a control input (LDNUM). The outputs from this circuit are an 8-bit product ($PROD_7$–$PROD_0$) and a status output (READY). It is further specified that when the circuit is ready for a multiplication operation, the READY output is true. Then, when the input LDNUM becomes true, the two 4-bit values at MCANDIN and MPLIERIN are loaded into the multiplier circuit and the multiplication process begins. During the multiplication process the READY output is false, and subsequent inputs at MCANDIN and MPLIERIN are ignored. When the multiplication process is completed, the READY output becomes true to indicate that the 8-bit product is available at the PROD outputs and also that the circuit is ready for another multiplication operation.

Recall that in the hand multiplication of two unsigned binary numbers, such as that illustrated in Fig. 7.31, each bit of the multiplier is examined. If the current multiplier bit is 1, then the shifted multiplicand is added to the partial product. The purpose of the left-shifting of the multiplicand is to account for the weight of the multiplier bit. The algorithm used for the multiplier circuit of Fig. 7.30 will follow this basic multiplication algorithm.

A preliminary design of the multiplier circuit is shown in the two parts of Fig. 7.32. Specifically, the circuit elements are shown in Fig. 7.32(a), and the algorithm for the control of the multiplication is given in the form of a flowchart in Fig. 7.32(b). As can be seen from the flowchart, with references to the circuit diagram, in the WAIT state the circuit outputs READY = true to indicate that the circuit is ready for another multiplication operation. And, the circuit remains in the WAIT state as long as the LDNUM input is false. But when LDNUM becomes true, the values at the inputs MCANDIN and MPLIERIN are loaded into two 4-bit registers MCAND and MPLIER, respectively. Also, the circuit is initialized to begin the multiplication by clearing the 8-bit PROD register and a 2-bit counter. This PROD register holds the partial product during the multiplication and then the final product at the end. For reasons that will be seen, it is divided into two parts: PRODHI ($PROD_7$–$PROD_4$) and PRODLO ($PROD_3$–$PROD_0$). The 2-bit counter is used to ensure in each multiplication that the shift-add process is performed for exactly four times (COUNT = 00, 01, 10, 11). The algorithm for the multiplication process is based on the basic algorithm for hand multiplication as illustrated in the example shown in Fig. 7.31. However, the left-shifting of the multiplicand of the hand multiplication is replaced by the functionally equivalent right-shifting of the partial product. Then, only a 4-bit adder is required instead of an 8-bit adder. The reader is urged to work through a simple multiplication example based on the algorithm in Fig. 7.32(b) to verify the correctness of this algorithm.

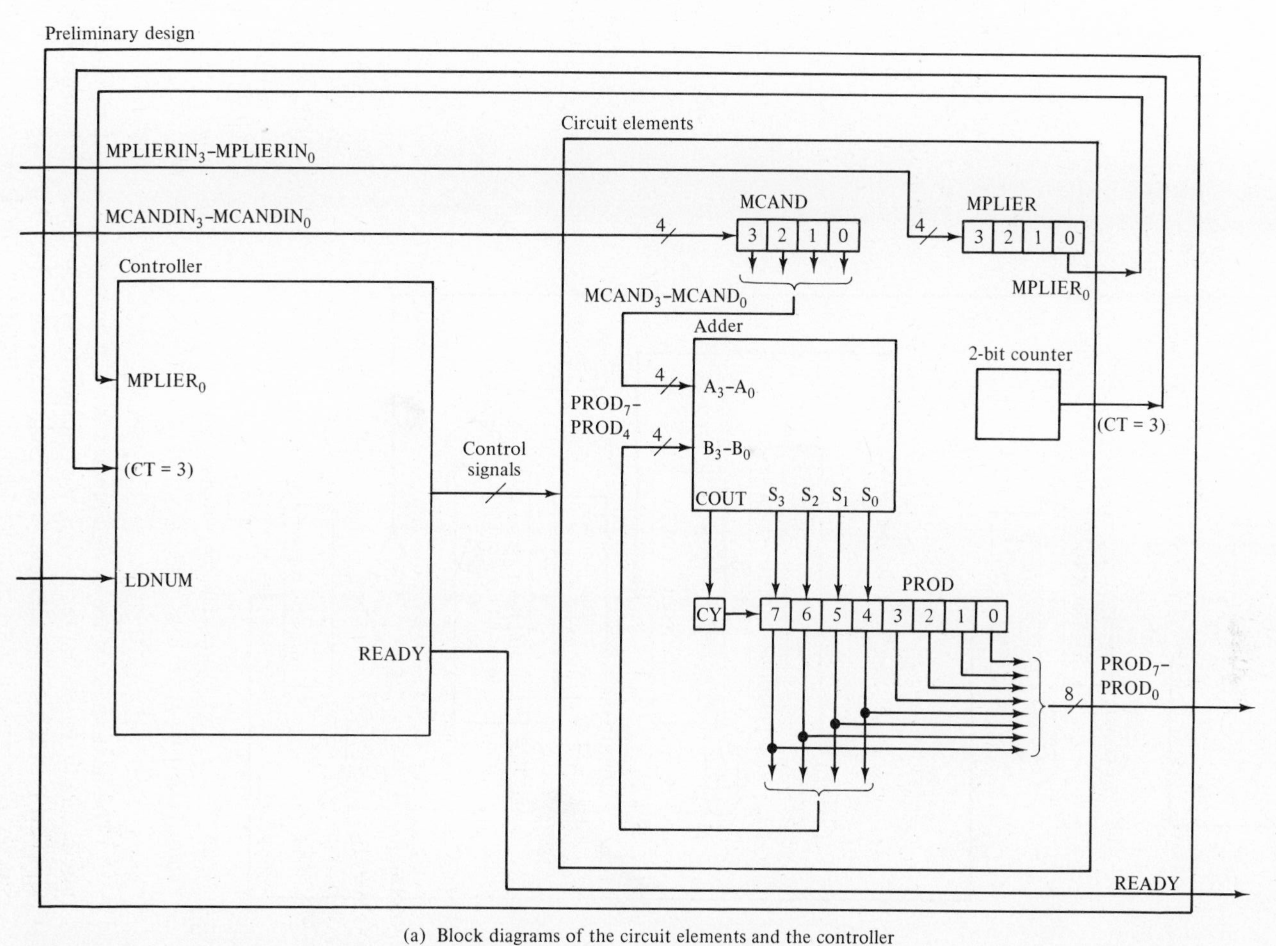

(a) Block diagrams of the circuit elements and the controller

Figure 7.32 Preliminary design of the multiplier circuit.

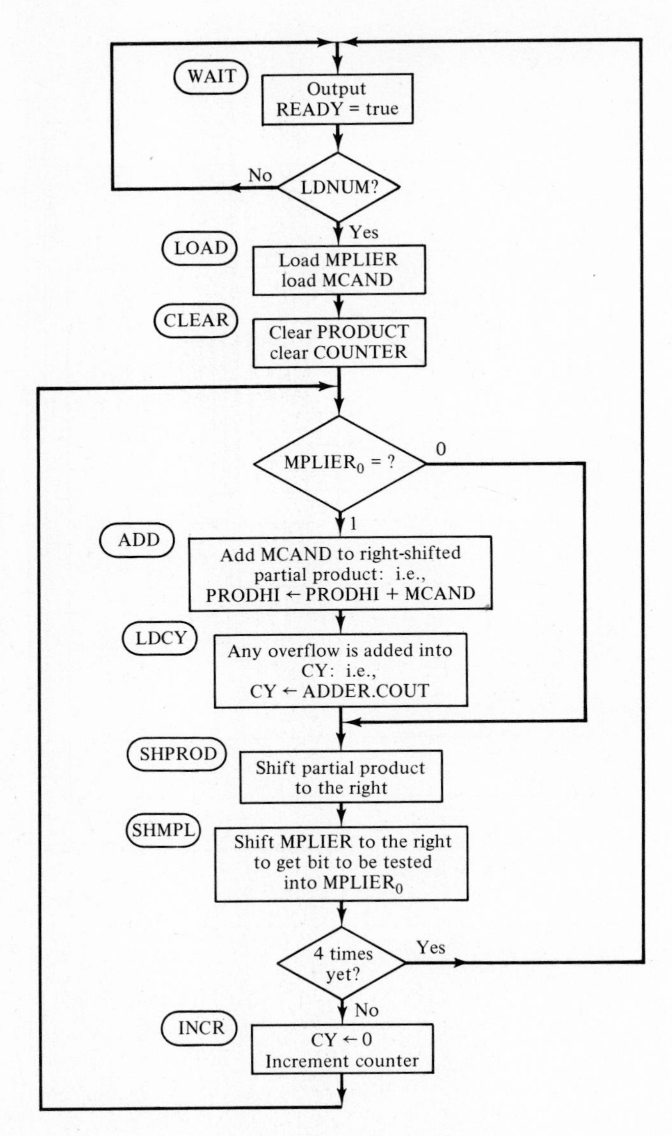

(b) Flowchart for the controller

Figure 7.32 *(cont.)*

The next phase of the design process is the refinement phase. After some thought and several iterations on the design, we can specify in more detail the functions that are required for each circuit element. The result is shown in Fig. 7.33. MCAND is a 4-bit storage register with a synchronous load input. PRODHI, PRODLO, and MPLIER all are shift-right registers, with the incoming bit value applied at the serial input (SI). Furthermore, each of these shift registers has a synchronous load and/or clear input, as dictated by the control algorithm. ADDER is a standard 4-bit adder. COUNT is a standard

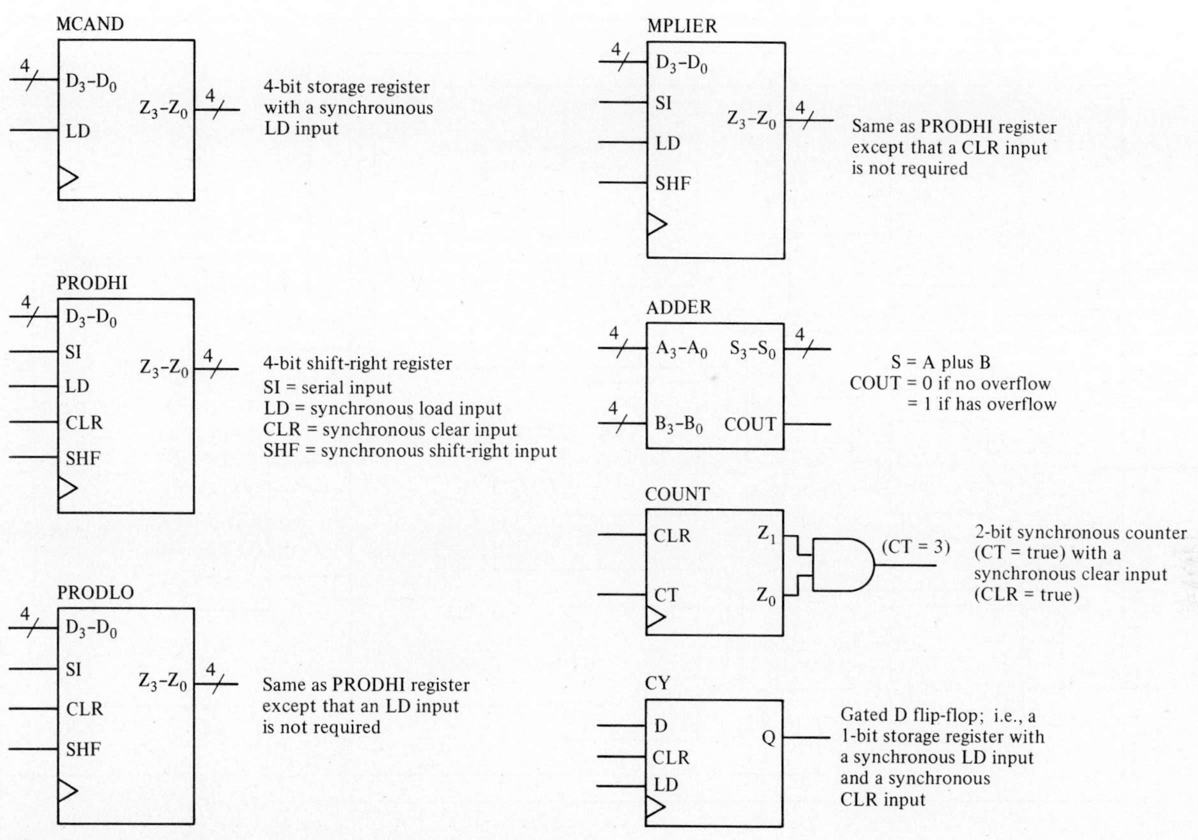

Figure 7.33 Detailed specifications for the circuit elements.

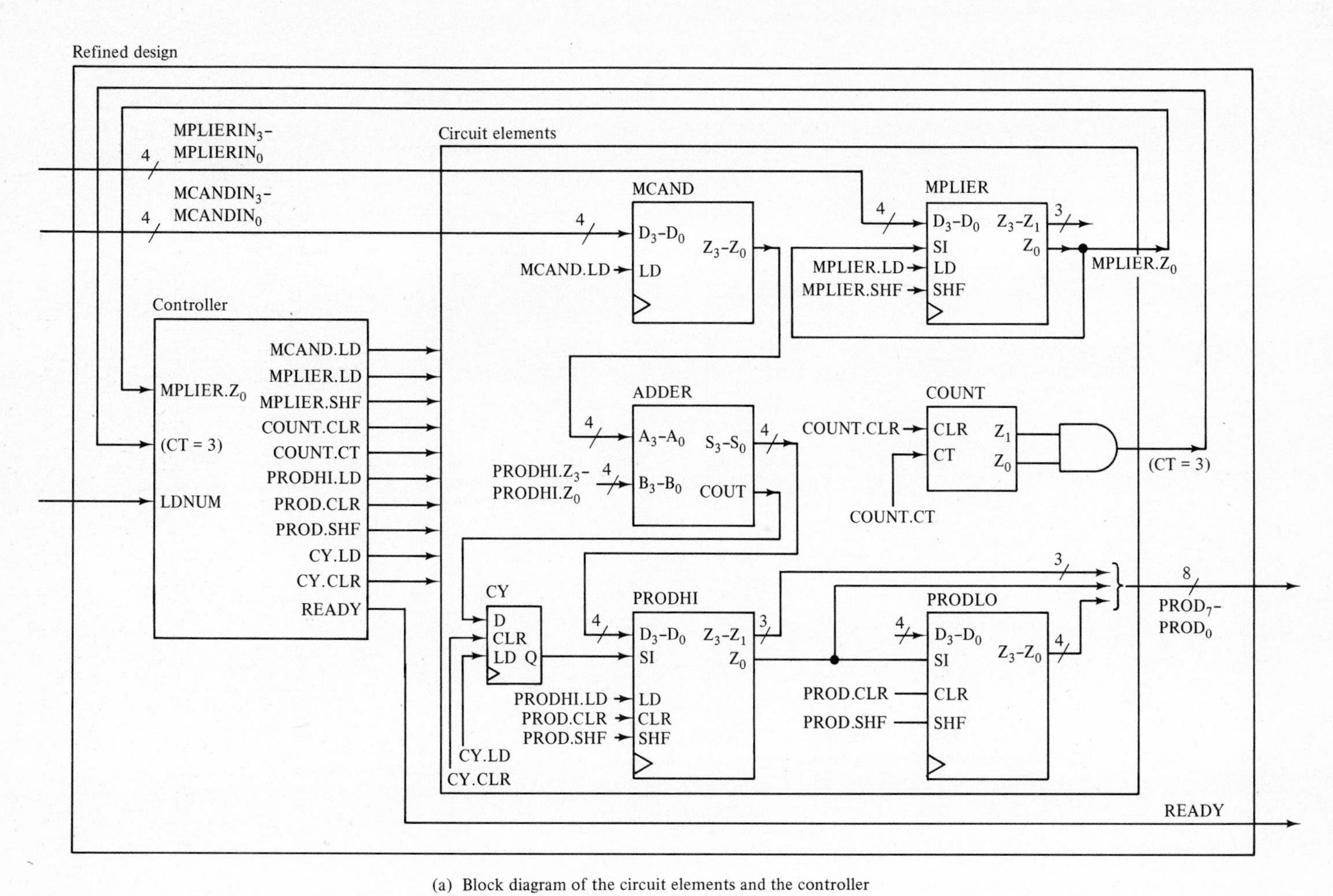

(a) Block diagram of the circuit elements and the controller

Figure 7.34 Refined design of the multiplier circuit.

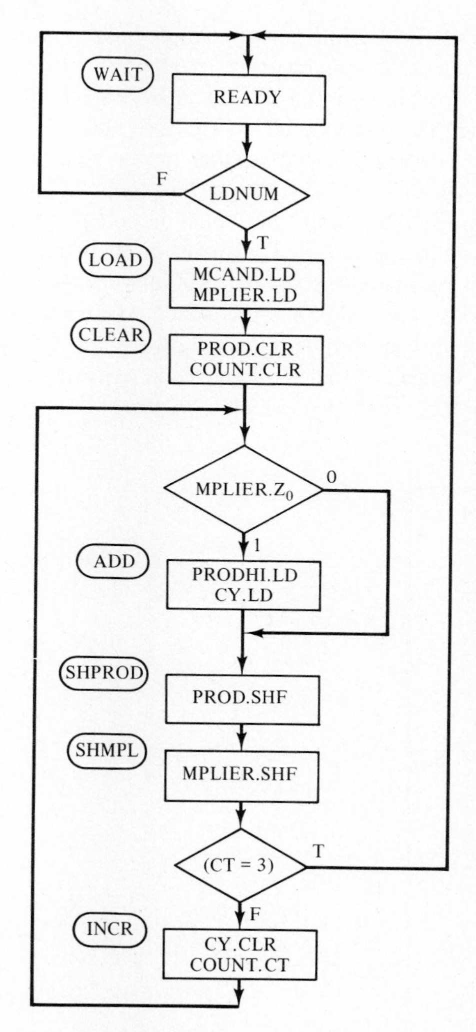

(b) ASM chart for the controller

Figure 7.34 *(cont.)*

2-bit binary counter plus an AND gate for generating the (CT = 3) signal. Finally, CY is a D flip-flop with synchronous load and clear inputs.

Using this detailed specification for the circuit elements, we can obtain the refined design of Fig. 7.34. As shown in Fig. 7.34(a), the control signals outputted from the controller are now defined. They correspond to the control inputs of the circuit elements. Based on the refined set of circuit elements shown in Fig. 7.34(a), the flowchart of Fig. 7.32(b) can be converted to the ASM chart shown in Fig. 7.34(b). Again, unlike the flowchart, the ASM chart precisely specifies the timing. Since the control outputs are now defined, their logic values can be specified for each state to accomplish the required functions for that state. While paying particular attention to the timing of the circuit, the reader is urged to work through a simple multiplication example based on the ASM chart of Fig. 7.34(b) to verify the correctness of this chart.

The ASM chart in Fig. 7.34(b) can be refined to the ASM chart in Fig. 7.35. We observe from Figs. 7.34(a) and (b) that the loading of MCAND and MPLIER and the clearing of PROD and COUNT can be performed together in one state. Consequently, the two states LOAD and CLEAR in Fig. 7.34(b) can be combined into one state. Furthermore, this state can be eliminated by using conditional outputs without introducing any timing problems, as is shown in Fig. 7.35. The result is an increase in the performance of the multiplier circuit by two clock cycles. Similarly, states SHPROD and SHMPL can be combined into one state, SHIFT. And states ADD and INCR can be eliminated through the use of conditional outputs. Then, however, a new state TSBIT has to be added. (Why?) In all, a net gain of four clock cycles in performance is achieved through the refinement of the original ASM chart of Fig. 7.34(b).

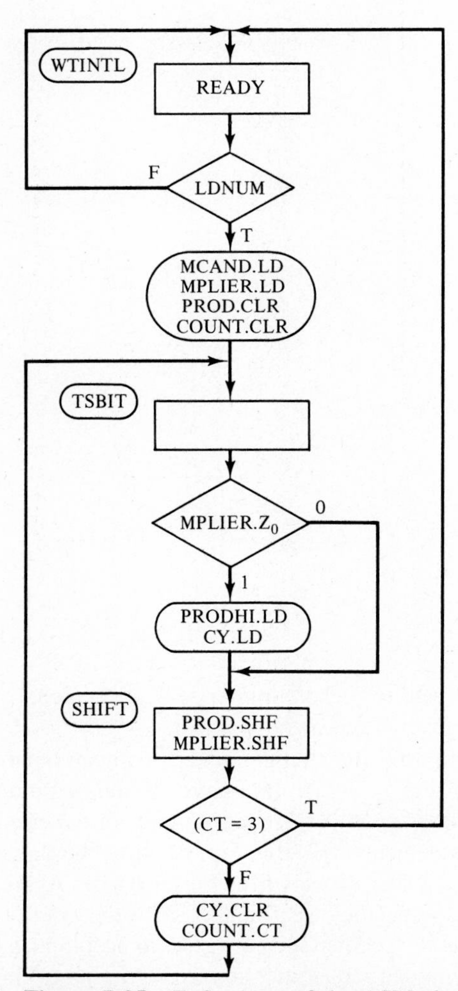

Figure 7.35 Refinement of the ASM chart for the controller.

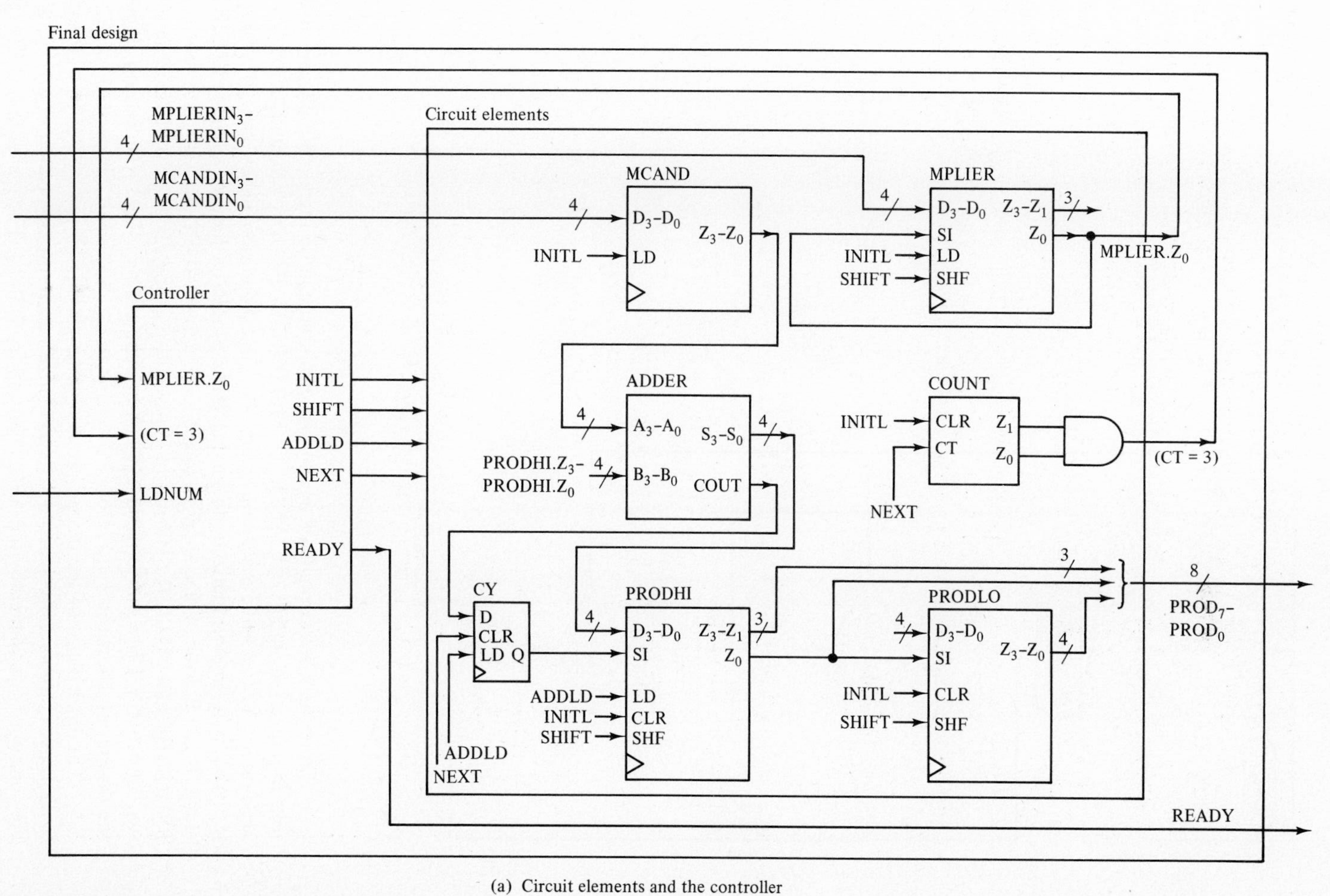

(a) Circuit elements and the controller

Figure 7.36 Final design for the multiplier circuit.

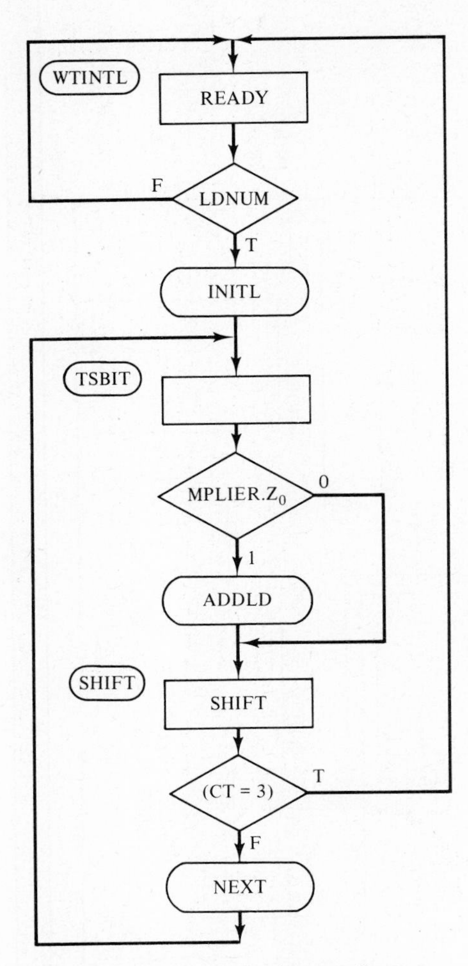

(b) ASM chart for the controller

Figure 7.36 *(cont.)*

Some final refinements of the multiplier circuit can be made. We observe that the four initialization signals (MCAND.LD, MPLIER.LD, PROD.CLR, and COUNT.CLR) are applied in the same state and under the same condition. They can, therefore, be replaced by a single initialization signal, INITL. Similarly, the two signals PRODHI.LD and CY.LD can be replaced by a single load signal, ADDLD. Also, the two signals PROD.SHF and MPLIER.SHF can be replaced by a single signal, SHIFT. Finally, the two signals CY.CLR and COUNT.CT can be replaced by NEXT. These refinements in the design are shown in Fig. 7.36.

The final phase of the design process is the realization phase. For this circuit, we need to use commercially available ICs to realize the circuit elements in Fig. 7.36(a) and the controller represented by the ASM chart in Fig. 7.36(b). The realization of the circuit elements is summarized in Fig. 7.37. As shown, MCAND is realized by using a 74'163 configured as a storage register (EP = ET = false). PRODHI is realized by using a

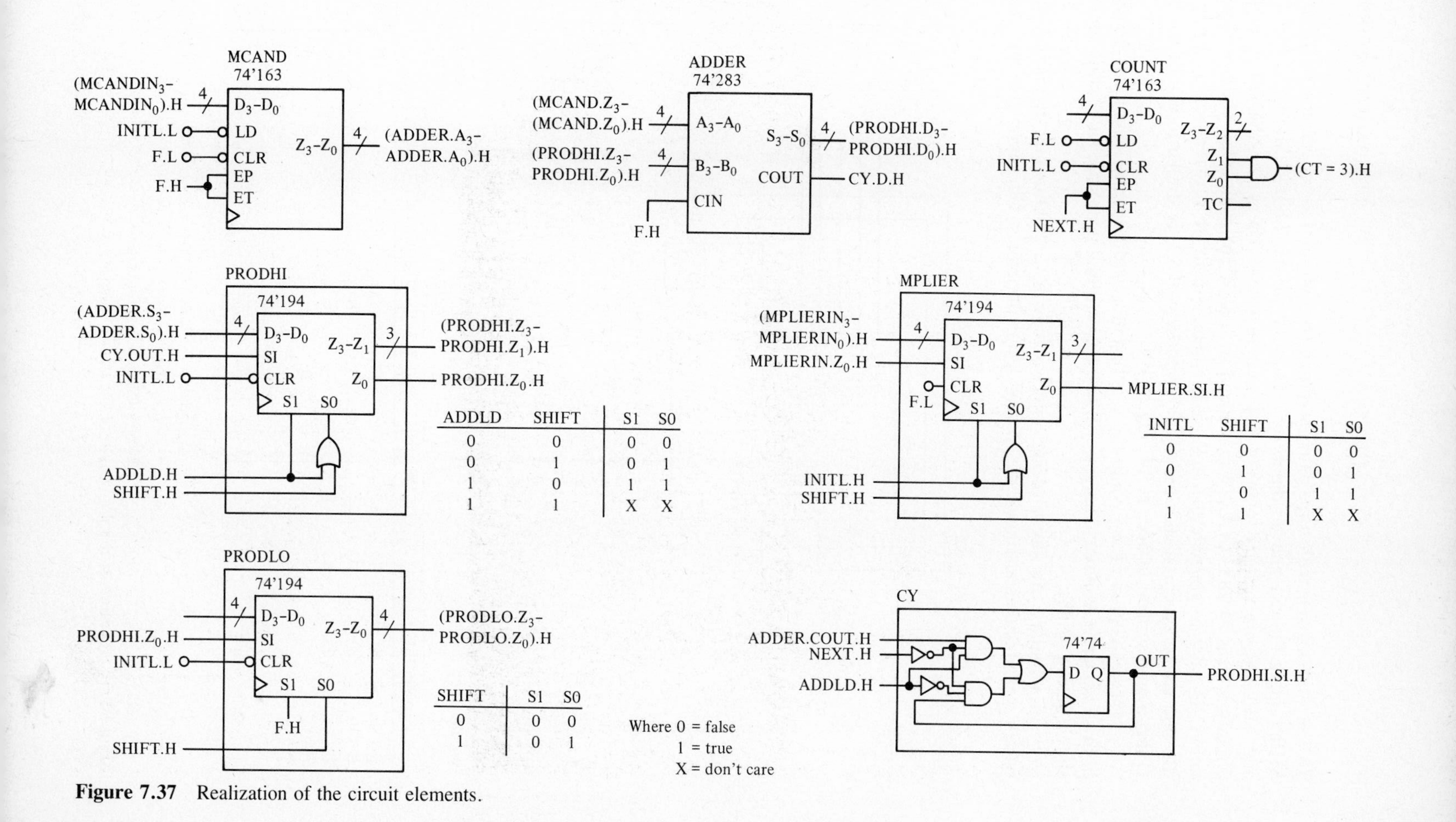

ADDLD	SHIFT	S1	S0
0	0	0	0
0	1	0	1
1	0	1	1
1	1	X	X

INITL	SHIFT	S1	S0
0	0	0	0
0	1	0	1
1	0	1	1
1	1	X	X

SHIFT	S1	S0
0	0	0
1	0	1

Where 0 = false
1 = true
X = don't care

Figure 7.37 Realization of the circuit elements.

74'194 shift register with some external gating (an OR gate) to convert the ADDLD and SHIFT input into S1 and S0 inputs as is required for the 74'194. Similarly, PRODLO and MPLIER are realized with 74'194 shift registers. The 4-bit ADDER is realized directly with a 74'283. COUNT is realized with a 74'163 with some additional gating (an AND gate) to generate the (CT = 3) signal. Finally, CY is realized as a gated D flip-flop with a synchronous clear input. The design of CY is similar to that of the gated D flip-flop described in Sec. 5.3.4 (Fig. 5.10).

The realization of the controller is straightforward with the techniques presented in this chapter. The block diagram of it and the resulting next-state and output table are shown in Fig. 7.38.

In summary, the digital design process requires a combination of creativity, experience, and understanding of the general design principles. Our purpose in this chapter

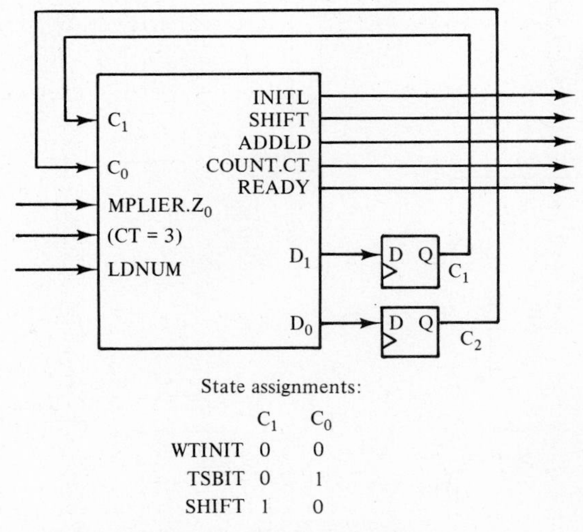

(a) Block diagram and state assignment

present state		inputs			outputs					next state		D flip-flop inputs	
C_1	C_0	MPLIER.Z_0	(CT = 3)	LDNUM	INITL	SHIFT	ADDLD	COUNT.CT	READY	C_1^+	C_0^+	D_1	D_0
0	0	X	X	0	0	0	0	0	1	0	0	0	0
0	0	X	X	1	1	0	0	0	1	0	1	0	1
0	1	0	X	X	0	0	0	0	0	1	0	1	0
0	1	1	X	X	0	0	1	0	0	1	0	1	0
1	0	X	0	X	0	1	0	1	0	0	1	0	1
1	0	X	1	X	0	1	0	0	0	0	0	0	0
1	1	X	X	X	0	0	0	0	0	0	0	0	0

Where 0 = false
1 = true
X = don't care

(b) Next-state and output table

Figure 7.38 Outline of the realization of the controller.

has been to present and illustrate the general principles of digital design and to consider various tools that are useful in the design process. These design principles and tools form a solid foundation on which to build experience and to exercise creativity.

SUPPLEMENTARY READING (see Bibliography)

[Clare 73], [Fletcher 80], [Kline 83], [Mano 79], [Mano 84], [Peatman 80], [Prosser 87], [Wiatrowski 80]

PROBLEMS

7.1. Explain the general model for a digital circuit design shown in Fig. 7.1.

7.2. The digital design process can be divided into three major phases: the preliminary design phase, the refinement phase, and the realization phase. At the conclusion of each phase, what are the expected end products in terms of the controller and the controlled circuit elements?

7.3. Given the ASM chart of Fig. 7.39(a), complete the corresponding timing diagram of Fig. 7.39(b).

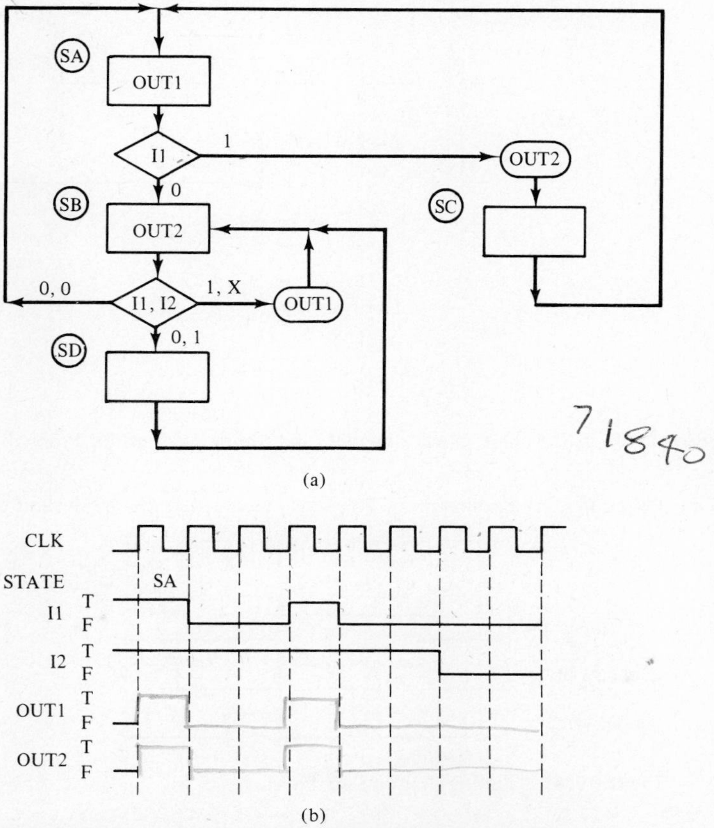

Figure 7.39 ASM chart and timing diagram for Problem 7.3.

7.4. Given the ASM chart and block diagram of Fig. 7.40(a), complete the corresponding timing diagram of Fig. 7.40(b).

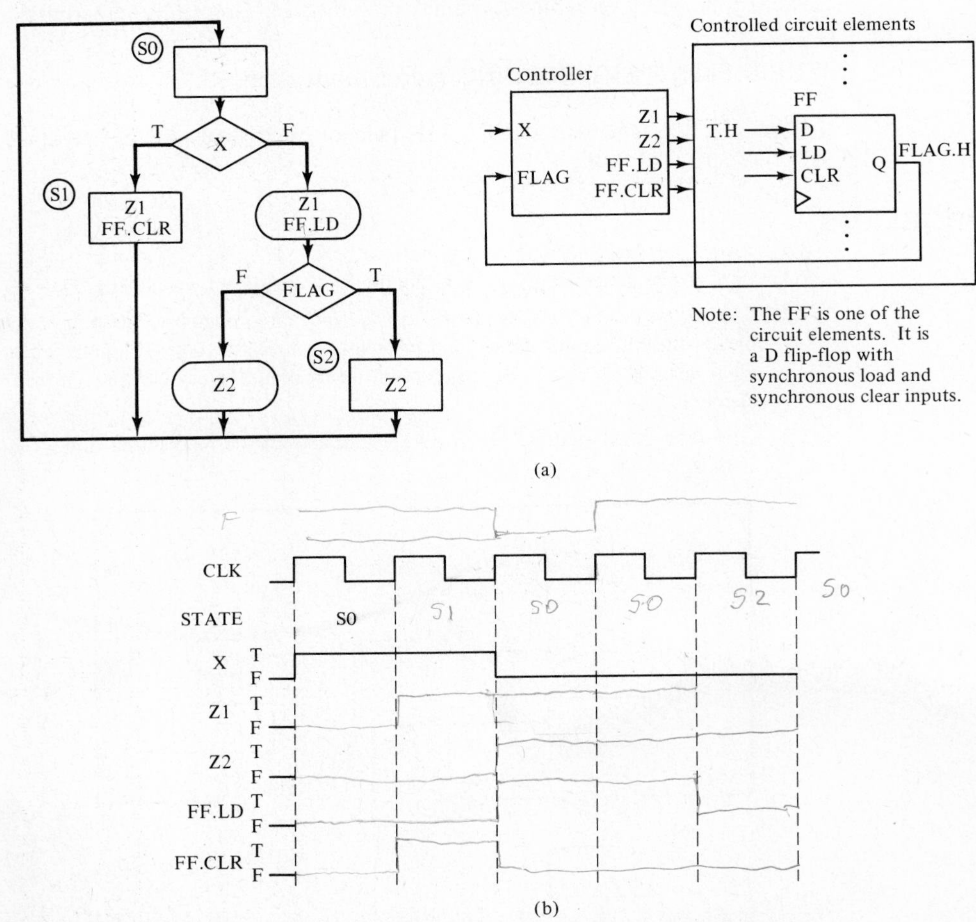

Figure 7.40 ASM chart, block diagram, and timing diagram for Problem 7.4.

7.5. Given the timing diagram of Fig. 7.41, reconstruct the ASM chart that corresponds to it.

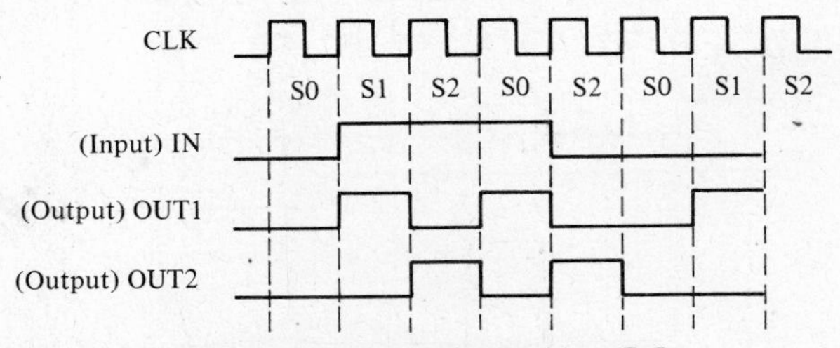

Figure 7.41 Timing diagram for Problem 7.5.

7.6. Given the timing diagram of Fig. 7.42, reconstruct the ASM chart that corresponds to it.

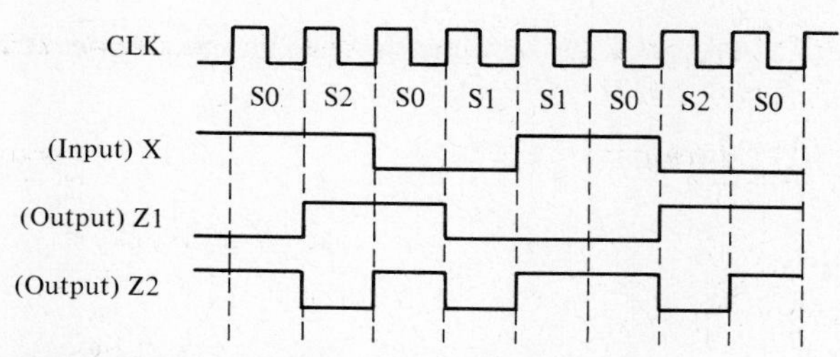

Figure 7.42 Timing diagram for Problem 7.6.

7.7. For the ASM chart specified in Fig. 7.40(a),

(a) Make a block diagram design (similar to the one shown in Fig. 7.3) of the corresponding state generator, specifying the inputs, outputs, and the state flip-flops (D type).

(b) Construct the next-state table for the state generator, given the following state code assignment:

	C1	C0
S0	1	0
S1	0	0
S2	0	1

(c) Realize the controller by the traditional method. Use D flip-flops, and draw a detailed circuit diagram of the controller.

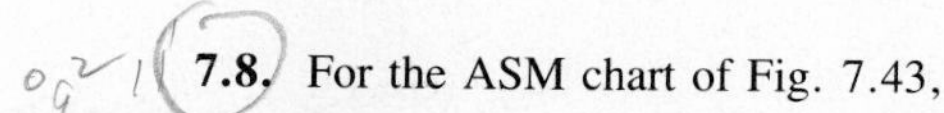

7.8. For the ASM chart of Fig. 7.43,

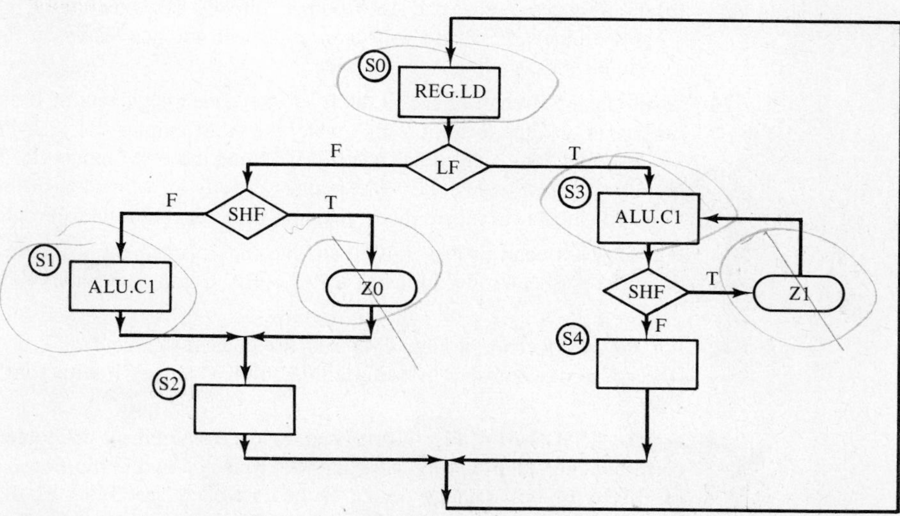

Figure 7.43 ASM chart for Problem 7.8.

(a) Make a block diagram of the corresponding controller, specifying the controller inputs and outputs.

(b) Make a block diagram design (similar to the one shown in Fig. 7.3) of the corresponding state generator, specifying the inputs, outputs, and the state flip-flops (D type).

(c) Determine the next-state table for the state generator, given the following state code assignment:

	C2	C1	C0
S0	0	0	0
S1	0	0	1
S2	0	1	0
S3	0	1	1
S4	1	0	0

(d) Realize the controller by the traditional method. Use D flip-flops, and draw a detailed circuit diagram of the controller.

7.9. Given the ASM chart of Fig. 7.4,

(a) Specify the ASM outputs as a function of the state code and the ASM inputs. In other words, complete the following truth table:

C1	C0	IN. BIT	BUF. FULL	COUNT. EN	REG. LD	OUT. FLAG
0	0	0	0			
0	0	0	1			
0	0	1	0			
		⋮				
1	1	1	1			

(b) Using gates, realize the ASM outputs directly as a function of the state code and the ASM inputs. Compare your realization with the one shown in Fig. 7.6, which has a circuit for decoding the state code.

7.10. Given the ASM chart of Fig. 7.40(a) and the code assignment of Problem 7.7(b),

(a) Specify, in the form of a truth table, the ASM outputs (Z1, Z2, FF.LD, and FF.CLR) as a function of the state code (C1, C0) and the ASM inputs (X, FLAG).

(b) Using gates, realize the ASM outputs directly as a function of the state code and the ASM inputs. Compare this realization with your solution to Problem 7.7.

7.11. For the ASM chart of Fig. 7.40(a) and the code assignment of Problem 7.7(b), realize the corresponding controller by using a PAL16R4. Compare it with your solution to Problem 7.7.

7.12. For the ASM chart of Fig. 7.43 and the code assignment of Problem 7.8, realize the corresponding controller by using a PAL16R4. Compare it with your solution to Problem 7.8.

7.13. For the ASM chart of Fig. 7.40(a) and the code assignment of Problem 7.7(b), realize the corresponding controller by using the ROM method and D flip-flops. That is,

(a) Make a block diagram design of the controller, specifying the state flip-flops and all the appropriate connections to the inputs and outputs of the ROM. In other words, make a block diagram design similar to the one shown in Fig. 7.8, but without the actual ROM contents.

(b) Specify the ROM contents in hexadecimal.

7.14. Repeat Problem 7.13 for the ASM chart of Fig. 7.43 and the code assignment of Problem 7.8(c).

7.15. Figure 7.44 shows a block diagram design of a controller based on the ROM method and with D flip-flops. The ROM contents are also given. Derive the corresponding ASM chart.

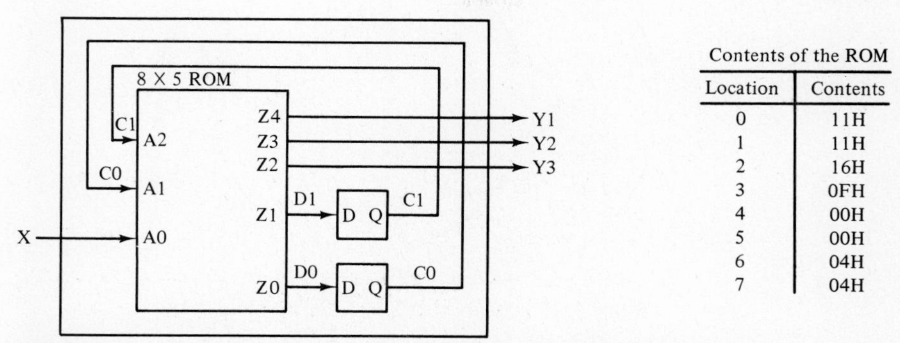

Contents of the ROM

Location	Contents
0	11H
1	11H
2	16H
3	0FH
4	00H
5	00H
6	04H
7	04H

Figure 7.44 Controller block diagram and ROM contents for Problem 7.15.

7.16. For the ASM chart of Fig. 7.39(a),

(a) Draw a block diagram of the corresponding controller, specifying the controller inputs and outputs.

(b) Make a block diagram design (similar to the one of Fig. 7.10) for the corresponding state generator, specifying the inputs, outputs, and the state flip-flops (J-K type).

(c) Determine the next-state table for the state generator, given the following state code assignment:

	C1	C0
SA	0	0
SB	0	1
SC	1	0
SD	1	1

(d) Realize the controller by the traditional method. Use J-K flip-flops, and draw a detailed circuit diagram of the controller.

7.17. For the ASM chart of Fig. 7.39(a) and the state code assignment of Problem 7.16(c), realize the corresponding controller by using the ROM method and J-K flip-flops; that is,

(a) Make a block diagram design of the controller, specifying the state flip-flops and all the appropriate connections to the ROM inputs and outputs.

(b) Specify the ROM contents in hexadecimal.

7.18. Convert the Mealy state graph of Fig. 7.45 into an equivalent ASM chart. The inputs are X1 and X2, and the outputs are Z1 and Z2.

7.19. Convert the Moore state graph of Fig. 7.46 into an equivalent ASM chart. The inputs are X1 and X2, and the outputs are Z1, Z2, and Z3.

7.20. Assume that the design of the dynamic RAM controller circuit described in Sec. 7.8.1 is to be modified. In addition to the already specified functions, if the RD and WR signals are both true, then the function of the circuit is to be described by the timing diagram of Fig. 7.47. Specifically,

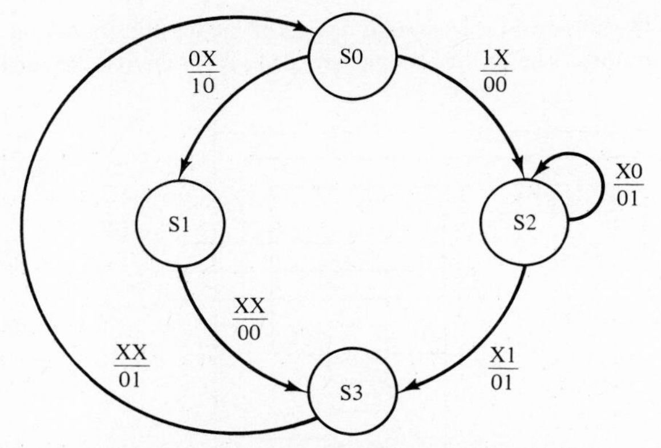

Figure 7.45 Mealy state graph for Problem 7.18.

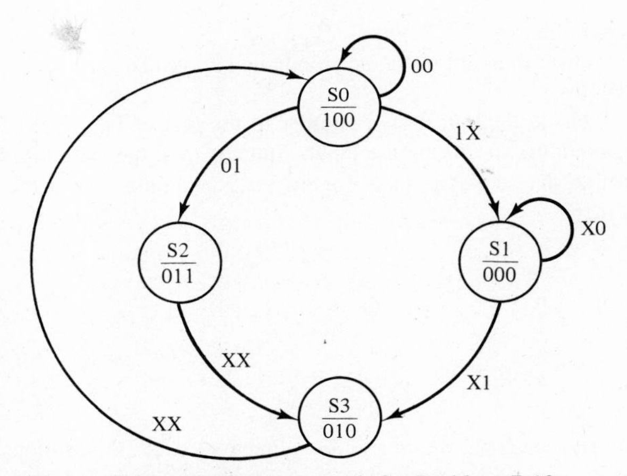

Figure 7.46 Moore state graph for Problem 7.19.

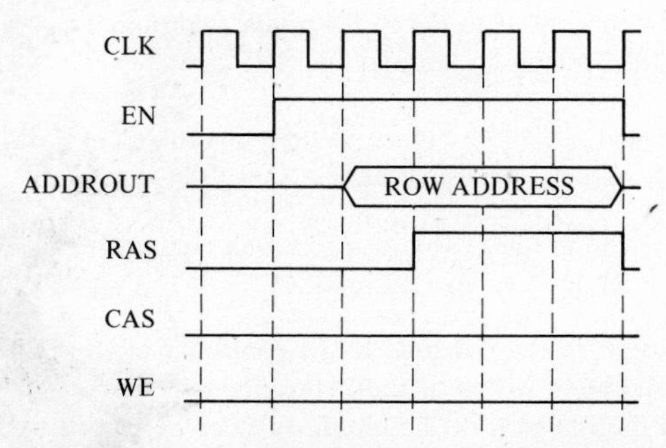

Figure 7.47 Timing diagram for Problem 7.20.

(a) Modify the ASM chart of Fig. 7.20 to include this function.
(b) Realize this modified ASM chart by using the ROM method and D flip-flops.

7.21. An 8-bit parallel-to-serial converter circuit, the block diagram of which is shown in Fig. 7.48, is to be designed and realized. This circuit remains in an idle state as long as the START input is false. But when this START input becomes true, the 8-bit data BYTE is loaded into the shift register and the right-shifting of the data begins. After the 8 bits are shifted out, the circuit returns to the idle state.
(a) The major circuit elements for the parallel-to-serial converter circuit are given in Fig. 7.48. Make any reasonable assumptions for the circuit elements that are required and derive the ASM chart for the controller. (*Hint:* A two-state ASM chart can be derived for this circuit.)
(b) Realize the circuit elements with commercially available chips.
(c) Realize the ASM chart by using any of the methods presented in this chapter.

7.22. A lock to a certain safe can be opened with a correct combination or with a key. If the combination feature is used, then the lock must be supplied with the correct combination within three attempts. Otherwise an alarm will sound. Specifications for the lock circuit are as follows:
1. After the first unsuccessful attempt, output message 1 (MSG1).
2. After the second unsuccessful attempt,
   (a) output message 2 (MSG2), and
   (b) wait for 10 seconds before being ready to be tried again (READY).
3. After the third successive unsuccessful attempt, sound an alarm.
4. When the safe is opened, reset to allow three more tries.
5. When the key is used, stop the alarm and reset to allow three more tries.

Your part in this problem is to design the module of Fig. 7.49. Make a detailed design of this module, using a counter for the major circuit element. Realize the controller with the ROM method and D flip-flops.

7.23. Shown in Fig. 7.50(a) are the circuit elements for a digital circuit that functions as follows:

If OP = 00, then REGB ← REGA; followed by REGA ← IN + 1.
If OP = 01, then REGB ← 0; REGA ← REGA + 1.
If OP = 10, REGA ← IN; REGB ← IN; followed by REGB ← REGB + 1.
If OP = 11, then the contents of REGA are doubled; REGB ← REGB + 1.

Notation: An arrow (←) designates to load the data at the next active clock transition. For example, REGA ← IN + 1 designates to add 1 to the value of IN and load the sum into REGA at the next active clock transition. In the function statements, more than one operation can be performed within a single state unless otherwise stated, as designated by "followed by."

The flowchart for the controller is given in Fig. 7.50(b). Derive the corresponding ASM chart for it. Optimize it by using the least number of states. In other words, if more than one operation can be performed within a single state, then do so. Also, make use of conditional outputs.

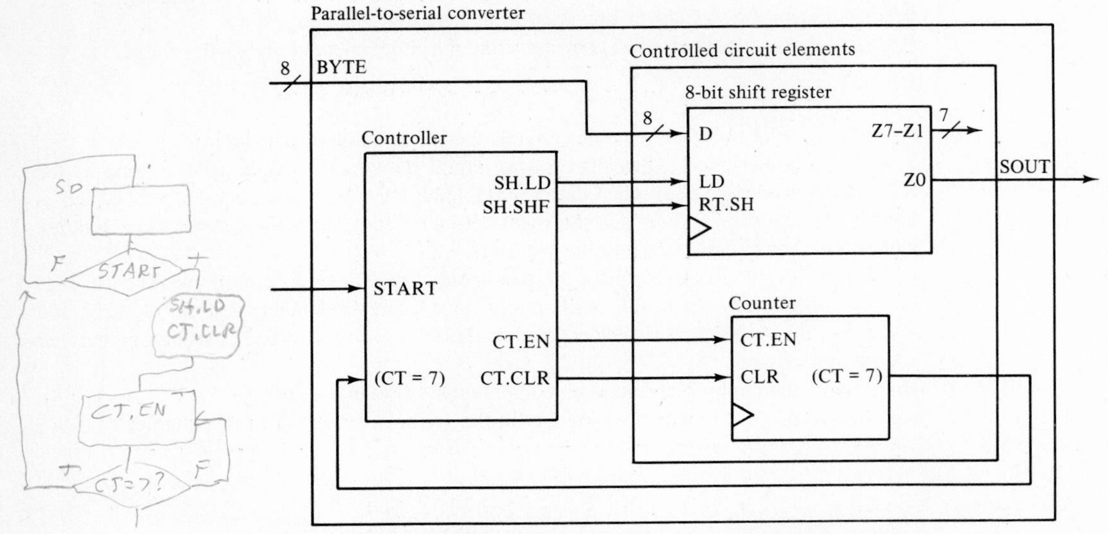

Figure 7.48 Parallel-to-serial converter circuit for Problem 7.21.

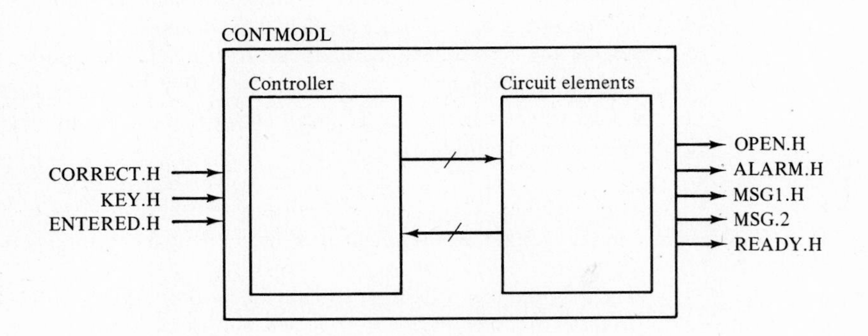

Signal specifications

CORRECT (from another comparator module): T = correct combination;
F = incorrect combination

KEY: T = a key is used to open the safe; F = no key is used

ENTERED: T = another combination is entered;
F = another attempt has not been made

OPEN: an output signal for another module to open the safe

ALARM: an output signal for another module to sound the alarm

MSG1: an output signal for another module to output message 1

MSG2: an output signal for another module to output message 2

READY: an output signal for another module to allow another attempt

Figure 7.49 Module for Problem 7.22.

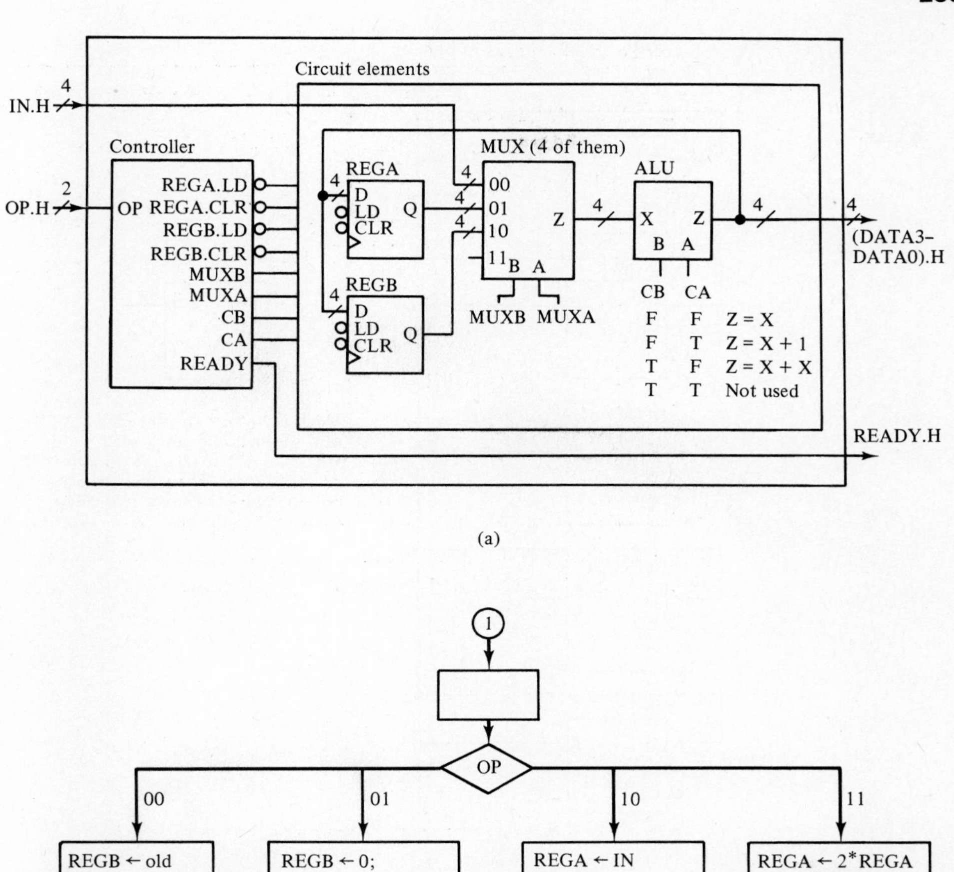

Figure 7.50 Circuit elements and flowchart for Problem 7.23.

7.24. Figure 7.51(a) shows circuit elements for a digital circuit that is to be designed. They are organized in a common bus structure.

(a) It is important not to have more than one set of Z outputs connected to the common bus at any one time. Why?

(b) With the shown connections of circuit elements, can we perform the following operations within a single state?

$$\text{REGA} \leftarrow \text{REGD} \quad \text{and} \quad \text{REGC} \leftarrow \text{REGA}$$

Explain your answer.

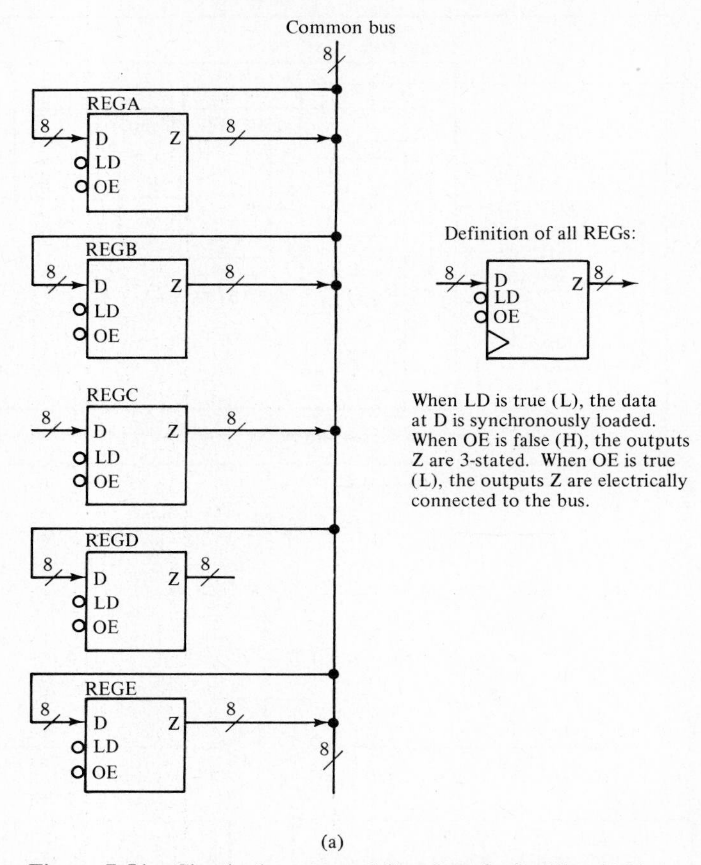

Figure 7.51 Circuit elements and flowchart for Problem 7.24.

(c) The flowchart for the controller is given in Fig. 7.51(b). Derive the corresponding ASM chart for it. Optimize it by using the least number of states. In other words, if more than one operation can be performed in a single state, then do so. Also, make use of conditional outputs.

7.25. **(a)** Given in Fig. 7.52(a) are the circuit elements for a digital circuit that is to be designed. They are organized in a common bus structure. Assume that the access time of MEM is 125 ns (nanoseconds), and that the clock period for the ASM is 100 ns. Then, how many states are required to perform a MEM read or MEM write operation?

(b) The flowchart for the controller is given in Fig. 7.52(b). Derive the corresponding ASM chart for it. Optimize the ASM chart by using the least number of states. In other words, if more than one operation can be performed in a single state, then do so. Also, make use of conditional outputs.

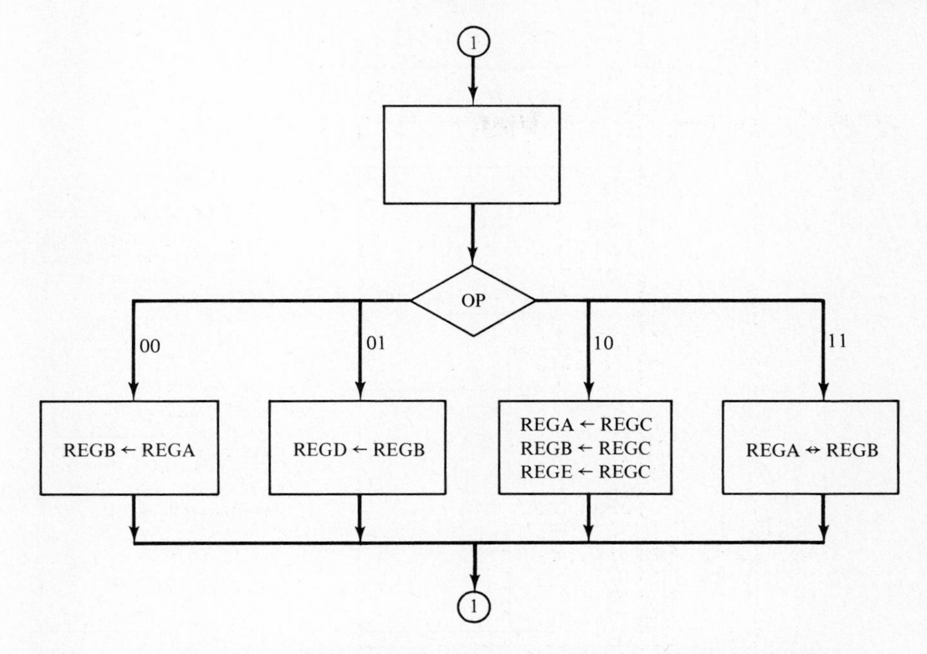

Note: REGA ↔ REGB means to interchange the contents of REGA and REGB

(b)

Figure 7.51 *(cont.)*

7.26. A hardware stack module, the block diagram of which is shown in Fig. 7.53, is to be designed and implemented. The function of this hardware stack module is defined as follows:

If STKENBL = 0, do nothing.
If STKENBL = 1, then there are four possible operations, depending on OP:

OP = 00	DEFINE a new top of stack; i.e., SP ← IN.
OP = 01	PUSH the stack; i.e., increment SP; MEM(SP) ← IN.
OP = 10	POP the stack; i.e., MEM(SP) is connected to OUT until STKENBL becomes 0; decrement SP.
OP = 11	READ the top of the stack; i.e., SP is connected to OUT until STKENBL becomes 0.

Notes:

1. SP contains the address (8 bits) of the top of the stack.
2. MEM(SP) is the memory content of that address.
3. Do not worry about the stack being empty or full.

(a) Using the circuit elements shown in Fig. 7.52(a), derive the ASM chart for the controller for the hardware stack module. Optimize it by using the minimum number of states.
(b) Using any commercially available chips, realize your design.

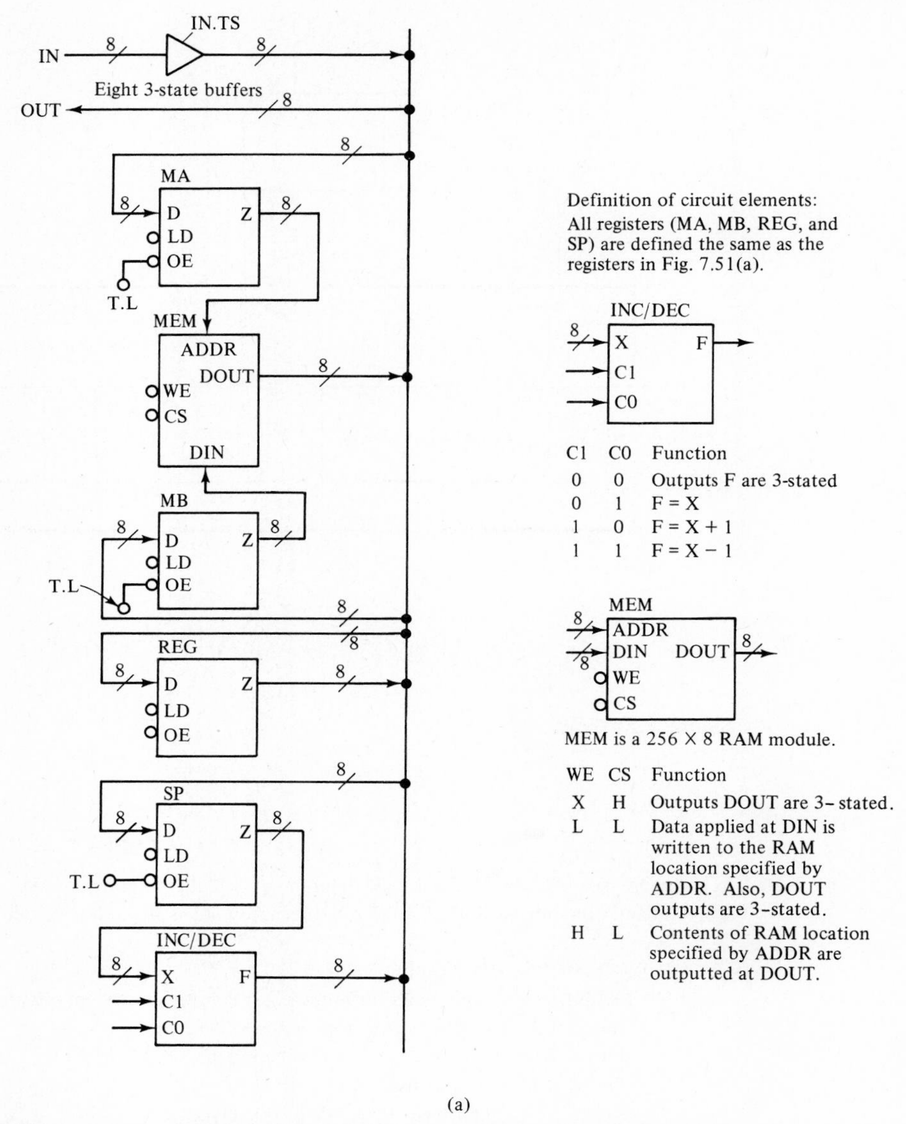

Figure 7.52 Circuit elements and controller flowchart for Problem 7.25.

7.27. Repeat Problem 7.26 and take care of the problem of whether the stack is empty (for the POP operation) or full (for the PUSH operation).

7.28. Design and realize the hardware multiplier of Sec. 7.8.4, modified as follows: Instead of using two separate 4-bit registers (MPLIER and PRODLO) to store the multiplier and the low-nibble product, as shown in Fig. 7.36(a), use PRODLO for storing both the multiplier and low-nibble product. Specifically, use PRODLO initially for storing the multiplier since

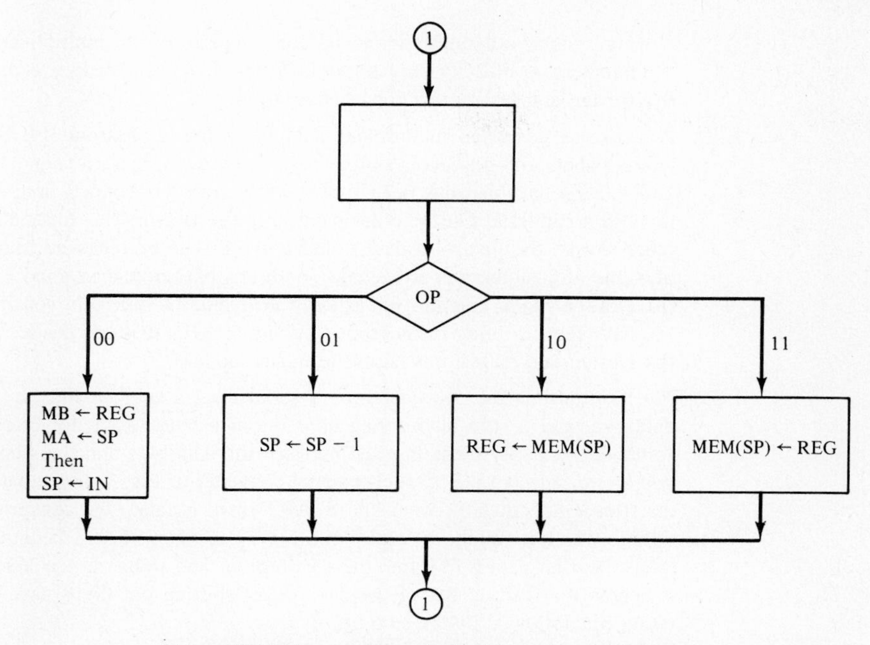

(b)

Figure 7.52 *(cont.)*

this register is not used initially to store the low-nibble product. As the multiplication process proceeds, have the multiplier shifted out of PRODLO one bit at a time. At the same time, have the low-nibble product shifted into PRODLO from PRODHI. This is a more elegant design that saves the use of a register.

7.29. **(a)** Draw a block diagram for an enhanced hardware multiplier and specify its functions. You have the flexibility of incorporating any features that you desire. For example, you

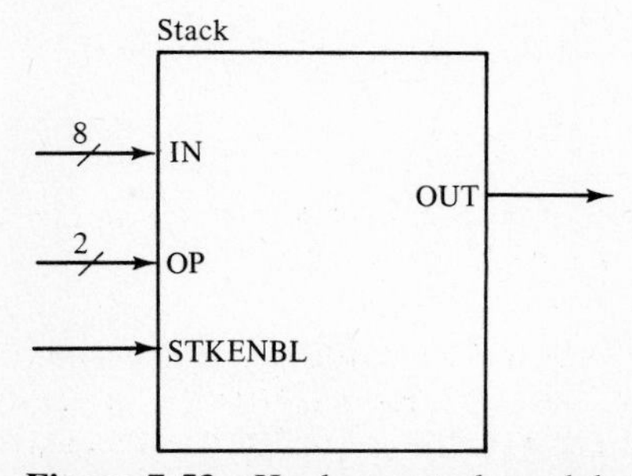

Figure 7.53 Hardware stack module for Problem 7.26.

may want your multiplier to be able to perform the multiplication of signed binary numbers, or of BCD numbers, or have additional handshaking capabilities, and so forth.

(b) Design and realize this enhanced multiplier.

7.30. A stack, as described in Problem 7.26, is a first-in, last-out (FILO) structure. In other words, when an 8-bit data is stored (pushed) onto the stack, it cannot be retrieved (popped) until all the data that has been subsequently stored is popped first, much like a stack of trays in a cafeteria. On the other hand, a queue is a first-in, first-out (FIFO) structure. In other words, the first 8-bit data stored is the first to be retrieved, much like the servicing of a line of customers in a cafeteria. With this background in mind,

(a) Draw a block diagram for a hardware queue module and specify its functions. You have the flexibility of incorporating any features that you desire.

(b) Design and realize this hardware queue module.

7.31. For the transmission of asynchronous serial data, a start bit must be inserted before each data byte and a stop bit inserted after the data byte. Also, for error-checking purposes, sometimes a parity bit is inserted between the data byte and the stop bit. In this problem you are to redesign the parallel-to-serial converter of Fig. 7.48 so that it will perform these insertions. Specifically, when the START input is false, the converter is to be in an idle state in which it outputs a ''high'' at SOUT. But when START becomes true, the converter loads the 8-bit data BYTE into the shift register and shifts out the start bit first, after which it begins the shifting out of the data. After shifting out the 8 data bits, it shifts out the parity bit, followed by the stop bit.

Notes:

1. The value of the start bit is L.

2. The value of the stop bit is H.

3. The value of the parity bit has to be determined as follows: the parity bit = L if the number of ones in the data byte is an even number; the parity bit = H if the number of ones in the data byte is an odd number.

Chapter 8

Life Cycle of a Digital System and Introduction to Microprocessor-Based Designs

8.1 LIFE CYCLE OF A DIGITAL SYSTEM

The preceding chapter presented systematic methods for the designs of digital *circuits* of limited complexity. In general, these digital circuits are components of a more complex digital *system*. The distinction between a digital circuit and system is, of course, only conceptual. Each digital circuit can be viewed as a system consisting of other components. Moreover, a digital system can be a subsystem of an even more complex system. In this book we will loosely define a digital *circuit* as a digital subsystem that is simple enough for its functional specification to be described easily and unambiguously. And we will loosely define a digital *system* to be a collection of related digital subsystems, which themselves can be collections of subsystems until at the lowest level of subdivision there are digital circuits. With this framework in mind, we will now proceed with a discussion of the digital system *life cycle,* which is the complete span from conception through utilization.

As is illustrated in Fig. 8.1, the life cycle of a digital system can be generally divided into the following phases:

Requirement analysis and specification phase

Design phase

Implementation and testing phase

Maintenance and field service phase

The first phase in any digital system development project is the requirement analysis and specification phase, in which we determine and specify in detail the function, per-

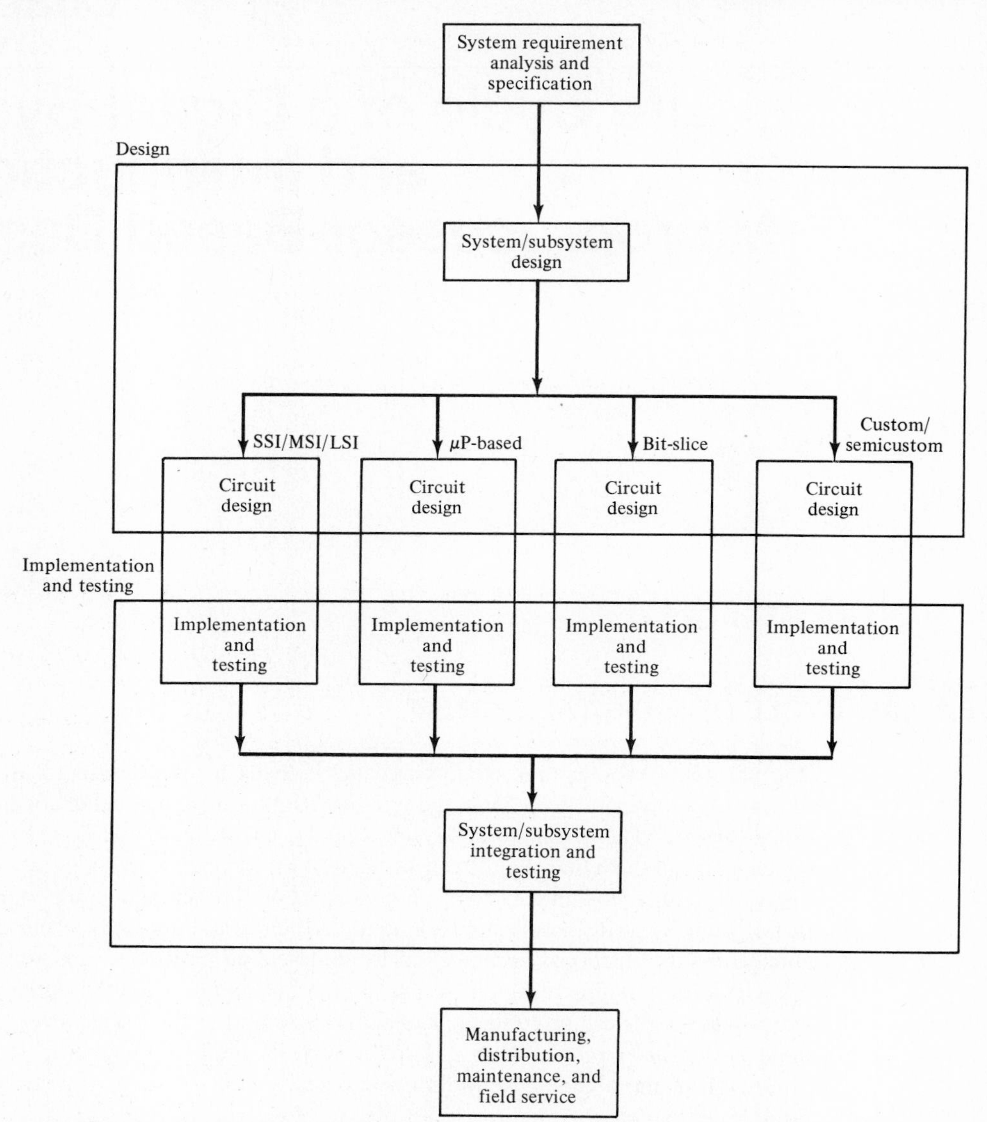

Figure 8.1 Life cycle of a digital system.

formance, and interface requirements of the system. At the digital system level, as opposed to the circuit level, the requirements and concepts are not well defined. Thus, we may be tempted to rush into the design phase with the hope of specifying the system while designing it. This is unfortunate because any improvement (or blunder) at this level has a great effect on the remainder of the development phases, greatly simplifying (or complicating) the design and implementation effort. Consequently, we should devote a great amount of time and effort on this phase of the project.

After the digital system is specified in detail, the next phase of the system development is the design phase. In the *top-down approach* to design, we partition the preliminary design of the digital system into more detailed designs of less complex subsystems, which in turn we partition until eventually we have the detailed designs of interrelated digital circuits in forms that permit ready physical implementation. The subsystem design at each level consists of the functional specification of the component subsystems at the next level and the interrelationships among these component subsystems.

At some point of the design process, we must decide on what implementation methods to use for the various digital subsystems. These methods can be classified as follows:

Combinational/sequential circuit design using catalogued SSI/MSI/LSI circuit elements

Microprocessor-based design

Bit-slice circuit-element-based design

Custom/semicustom LSI/VLSI circuit design

Each implementation method requires a different set of structured design techniques and design principles. The choice of implementation method depends, of course, on the application and performance constraints of the digital subsystem, along with factors such as physical size limitations and the quantities to be produced.

We have already considered the first implementation method, that of circuit design with catalogued SSI/MSI/LSI circuit elements. This method is appropriate, in general, for circuits that are of limited complexity and that have high-speed requirements.

In recent years, the second implementation method, that of microprocessor-based design, has replaced the first method for most digital circuit designs. There are several reasons for this. Microprocessors are inexpensive and powerful. Further, since software development replaces much of the more difficult hardware development, the development effort is far less than that of the traditional method. Modifications are also easier since software modifications are far simpler than hardware modifications. The principal disadvantage of a microprocessor-based digital circuit is its low speed. The sequential execution of the program statements is inherently slower than the parallel nature of operation of traditional combinational/sequential circuits.

For digital subsystems that are complex and that have high-speed requirements, one of the last two implementation methods should be used. Compared to a microprocessor circuit, a circuit based on the bit-slice method is faster and more flexible. This method can be used for building a digital subsystem of virtually any complexity, and any word size, and with a customized instruction set. The price paid for this flexibility is greater complexity of design and development. The fourth method of custom or semicustom LSI/VLSI circuit design also provides speed and flexibility. Its principal disadvantage is the extremely high cost of development. Custom LSI/VLSI circuits are justified only if they are manufactured in great quantities. Otherwise, the bit-slice method should be used.

As mentioned, the design process is top-down. The implementation and testing process is, however, bottom-up. For it, we first implement and test the individual digital circuits. Then, we integrate these circuits into digital subsystems and test them. We continue this process until we have synthesized the entire digital system and have carried out system tests to verify that the system requirement specifications have been satisfied. The final phase of the digital system life cycle involves the manufacture, distribution, maintenance, and field service of the digital system that we have developed.

In this text we are primarily concerned with the design and implementation phases of the digital system life cycle. Each of the four implementation methods requires a different set of structured design techniques and design principles. In the preceding chapters we provided the foundation for a systematic design method using catalogued SSI/MSI/LSI circuit elements. In the remainder of this book we will concentrate on the fundamentals of microprocessors and of microprocessor-based designs.

8.2 FUNCTIONAL COMPONENTS OF A MICROPROCESSOR SYSTEM

Shown in Fig. 8.2 is a block diagram of a microprocessor-based digital subsystem that we will, for convenience, refer to simply as a microprocessor system (or μP system). In Fig. 8.2 the notation (i) designates one typical component of possibly several similar components. A μP system is a digital subsystem in which the key processing element is a microprocessor that contains the arithmetic logic unit, the registers, and the control logic that are required for performing operations on data and for synchronizing and controlling transfer operations among the μP system components.

The sequence of operations performed by the microprocessor is directed by instructions of a *program* stored in the system memory, which usually comprises both a *read-only memory* (ROM) and a read-write *random access memory* (RAM). Typically, the ROM permanently stores the programs to be run, as well as fixed data tables. During operation of the μP system, instructions and data are read from the ROM. (These instructions and tables were stored in the ROM prior to the assembly of the μP system and cannot be changed.) The RAM, on the other hand, stores intermediate data during the running of the programs.

Like any digital subsystem, a μP system processes input data from the ''external world,'' performs any required transformations, and then transfers the resultant data to the external world. Unlike the digital circuits discussed earlier, a μP system does *not* function as fixed by the hardware, but rather as determined by the software program(s) stored in the ROM. More specifically, a μP system operates on input data provided by *input devices,* each of which can be an electronic or electromechanical device or another digital subsystem. Each input device is connected to the μP system through an *input interface circuit* that is unique to that device. Through its interface circuit, each input device provides input data for the appropriate *input port*(s) at appropriate times. Essentially, each input port is an external register from which the μP system can read input data. Similarly, each *output device,* which again is some electronic or electromechanical device or another digital subsystem, is connected to the μP system through a unique *output interface circuit*. Through its output interface circuit, each output device removes from the appropriate *output port*(s) data placed there by the μP system.

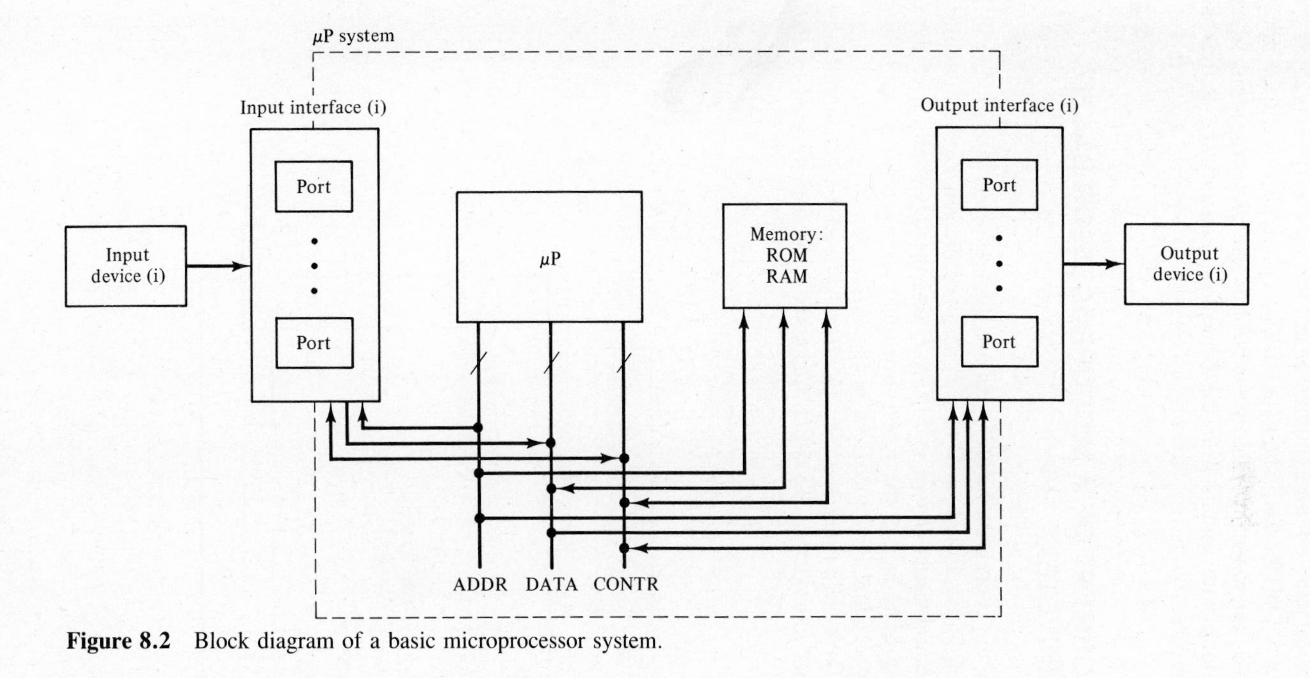

Figure 8.2 Block diagram of a basic microprocessor system.

The communication and data transfer among the component blocks of the μP system are handled through three buses: the *data bus,* the *address* (ADDR) *bus,* and the *control* (CONTR) *bus*. The data bus is a common path for the transfer of data between any two components. Under the control of the microprocessor, signals from the address bus, along with signals from the control bus, specify the source and destination of the data transfer.

8.3 THE DEVELOPMENT OF A MICROPROCESSOR-BASED DIGITAL SUBSYSTEM

Within the context of the life cycle of a digital system (refer to Fig. 8.1), the development of a μP-based digital subsystem consists of the selection and configuration of the *hardware* components shown in Fig. 8.2, and the development of the controlling *software* (programs) to satisfy the given functional specification of the subsystem. The development procedure is summarized in Fig. 8.3.

Given the *functional specification* for a μP-based digital subsystem, a μP system designer must determine which functions are best performed by the subsystem software

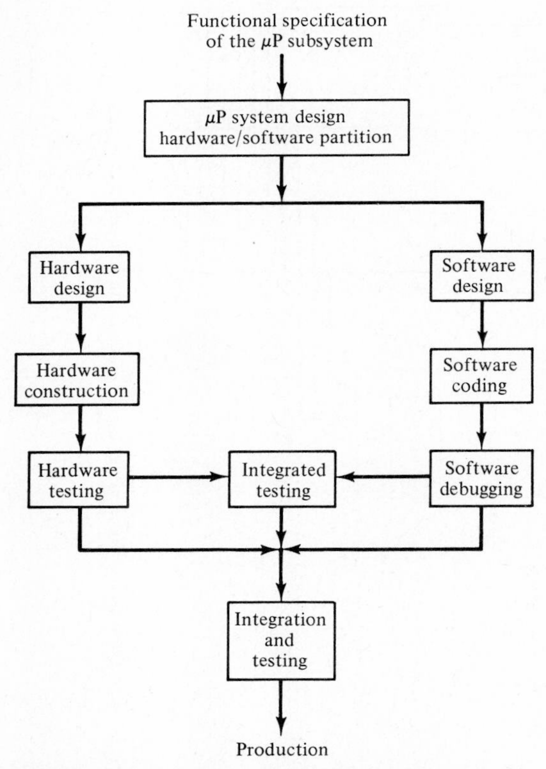

Figure 8.3 Procedure for microprocessor system development.

and which are best performed by the subsystem hardware. The result is a μP subsystem design consisting of the *design specifications* for the μP subsystem hardware and the μP subsystem software. In more detail, the functional specification specifies *what* the subsystem does, and the design specification specifies *how* the subsystem is to be implemented to satisfy the functional specification. In terms of the hardware, the design specification consists of the specification of the hardware modules that are required and the configuration of these modules. In terms of the software, the design specification consists of the specification of the program modules that are required to control the hardware modules, and the specification of the data structures that are necessary to support the program modules. After broadly determining the design specifications, the μP designer can, as shown in Fig. 8.3, proceed in parallel with the detailed design of the μP system software and hardware, with the design of the software constrained by the design specification of the hardware, and vice versa.

The implementation and testing process is again bottom-up. For the software, the program modules are individually coded and tested. Then they are integrated into larger program modules and tested. In a similar manner, the hardware modules are individually constructed and tested. Then they are integrated into larger hardware modules and tested. When possible, the integration and testing involve both hardware and software modules. The process continues until the whole μP subsystem is integrated and tested.

8.4 MICROPROCESSOR PRESENTATION

The organization of the remainder of this book parallels the development procedure illustrated in Fig. 8.3. Specifically, Chapters 9 and 10 explore the various μP software concepts:

- Assembly language programming
- Addressing modes
- Data transfer/data transformation
- Program control
- Program structures
- Program assembly
- Assembler and assembler directives
- Machine code
- Program execution
- Stacks and stack operations
- Subroutines

Chapters 11 and 12 present the hardware aspects of a microprocessor and provide the fundamentals for interfacing to a microprocessor. These include

- Model of a μP system
- External connections to a microprocessor

- Bus structure
- Machine cycles/timing
- Interfacing the memory module
- Interfacing an I/O interface module
- Standard I/O
- Memory-mapped I/O
- Polling/interrupt-driven synchronization
- Peripheral interface devices

Chapter 13 integrates the material in Chapters 9, 10, 11, and 12, and contains other topics that are related to microprocessors and the structured development of a microprocessor system, including

- Historical development of the microprocessor
- Single-chip microcomputers
- High-performance microprocessors
- Microprocessor selection
- Selection of the programming language
- Hardware/software trade-off
- Microprocessor system development
- Microprocessor development system

Throughout Chapters 9, 10, 11, and 12 we will use a specific microprocessor, the Intel 8085, and its supporting chips, to illustrate the various μP concepts. As stated in the preface of this book, the use of a commercially available microprocessor as a unifying thread throughout the discussion is a more effective approach than is the use of a nonexistent "generic" microprocessor.

The 8085 is not, of course, representative of the state of the art in μP technology. But, this is a problem with any microprocessor. Even the most current of the 16-bit or 32-bit microprocessors will be out of date in a relatively short time. Therefore, we are not concerned about selecting a current microprocessor, but rather one that is excellent for illustrating those concepts and fundamentals that will remain current. For this purpose, an 8-bit microprocessor, such as the 8085, is ideal for an introductory text. The 8085 is powerful enough to illustrate the important μP concepts, but simple enough to avoid obscuring these concepts with complex component details. Finally, it is our experience that once a specific microprocessor and its applications are mastered, it is then straightforward, with some training, to become proficient in the use of other microprocessors and their applications.

SUPPLEMENTARY READING (see Bibliography)

[Doty 79], [Fletcher 80], [Freedman 83], [Kline 83], [Mead 80], [Mick 80], [Peatman 77] [Rafiquzzaman 84], [Short 81], [Wiatrowski 80]

PROBLEMS

8.1. Explain the differences between a digital circuit and a digital system as defined in this book.

8.2. List the four phases of a life cycle of a digital system. Briefly explain each.

8.3. Name the four general implementation methods for a digital system. Briefly explain the advantages and disadvantages of each method.

8.4. Describe the major function of each of the following components of a μP system:
- **(a)** Microprocessor
- **(b)** Memory module
- **(c)** Input interface module
- **(d)** Output interface module
- **(e)** Address bus
- **(f)** Data bus
- **(g)** Control bus

8.5. Using Fig. 8.3, briefly describe the development of a μP system.

Chapter 9

Microprocessor Software Concepts I

9.1 INTRODUCTION

As for any digital subsystem, the function of a microprocessor-based digital subsystem (again, for convenience, we will just use the term μP system) is to process digital data from input devices, perform any data transformations that are required, and transfer the resultant data to output devices. However, unlike for the digital circuits considered in the preceding chapters, the tasks to be performed by a μP system are not frozen in the logic circuitry. Rather, the sequence of operations to be performed is directed by the instructions of a software program stored in the system memory module. The power and the flexibility of a μP system is a function of the sophistication of this program. Therefore, one of the most important tasks in the design of a μP system is the development of the software used to control the system. In this chapter and the next, we will explore the software concepts that are essential for fully exploiting the power of a μP system.

9.2 ASSEMBLY LANGUAGE PROGRAMMING

Programming a microprocessor in an assembly language is somewhat like using an electronic calculator. To use a calculator we must know what operations the calculator can perform. All common calculators can add, subtract, multiply, and divide. Some can perform more sophisticated mathematical operations such as exponentiation and evaluating logarithms and trigonometric functions. Still others can perform business-oriented operations. In effect, the set of operations that a calculator can perform is the ''instruction set'' of that calculator.

Further, to use an electronic calculator effectively we must know its register structure. Some calculators have only an accumulator in which the current result is stored (and displayed). Others have an additional memory register in which we can store and later retrieve a number. Still others have a memory stack of several registers. And, the most sophisticated calculators have tens, hundreds, or even thousands of memory registers. To make full use of an electronic calculator we must know its register structure as well as the set of operations that it can perform.

There are many similarities in the programming of a μP system using an *assembly language*. But, for a μP system the set of allowed operations, called an *instruction set,* is a great deal more extensive than that of a calculator, and includes operations such as MOVE, LOAD, SHIFT, and JUMP, in addition to ADD, SUBTRACT, and so on. Further, the register structure of a μP system is a great deal more extensive, having different types of internal registers and external registers, as well as memory registers. To make full use of a μP system, we must know its register structure and fully understand its instruction set.

9.2.1 Programming Model of a Microprocessor System

The register structure of a μP system as viewed by the programmer is called the *programming model* of that microprocessor. In other words, a programming model comprises all the registers in the μP system that are available for use by the programmer. Figure 9.1 shows the programming model for an Intel 8085 μP system.

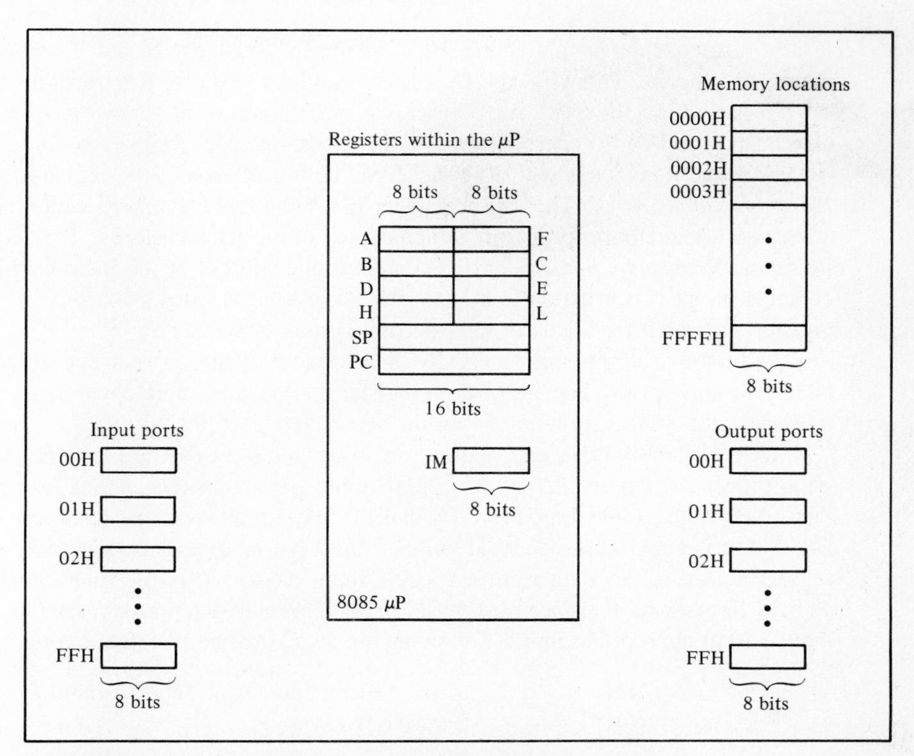

Figure 9.1 Programming model for the 8085 μP system.

Within the 8085 microprocessor itself, there are seven *general-purpose* registers. They are referred to as Registers A, B, C, D, E, H, and L. Each register is 8 bits wide, which means that each can store 8 bits, commonly called a *byte*. These registers are involved in the executions of many instructions. For example,

Instruction			Explanation
MOV	A,B	;	Move the contents of Register B to Register A.
MVI	C,00H	;	Move the hexadecimal number 00 into Register C.
ADD	D	;	Add the contents of Registers D and A, and place the sum in Register A.
RAL		;	Rotate the contents of Register A to the left by 1 bit.

Register A, often called the *accumulator*, is the most important general-purpose register in the 8085 microprocessor. In fact, Register A is involved in most of the instructions in the 8085 instruction set, as is demonstrated by the above example in which Register A is involved in three of the four instructions.

Register B and Register C are sometimes treated together as a single 16-bit *register pair* in some instructions. As an illustration,

PUSH	B	;	The contents of the Register Pair BC (i.e., both Registers B and C) are stored onto the top of the stack in the memory.

(Stacks are considered in Chapter 10.). Similarly, Registers D and E sometimes form the 16-bit Register Pair DE, and Registers H and L the 16-bit Register Pair HL.

Also within the 8085 microprocessor there are several *special-purpose* registers. One is an 8-bit Flag Register F in which the condition codes (flags) are stored and tested. Additionally, Register A with Register F can be treated as a 16-bit register called PSW, *Program Status Word*. The 16-bit register SP is the *Stack Pointer,* which contains the address of the memory location in which the top of the stack is stored. The 16-bit register PC is the *Program Counter,* and its content is the address of the memory location that contains the next instruction to be executed. The 8-bit register IM is the *Interrupt Mask* register, in which the status of the interrupt system is stored.

The above description is simply an overview of the register structure within the 8085 microprocessor. The use of each register will become more apparent as we proceed to discuss the 8085 instruction set in the subsequent sections.

Also available to a programmer are some registers that are external to the 8085 microprocessor. These include up to 256 input ports—namely, Input Port 00H, Input Port 01H, . . . , Input Port FFH. (Recall that FFH represents the hexadecimal number FF, which is equal to the decimal number 255.) An input port is essentially an external register that receives data from a specific input device. It is the function of the input device, through its interface circuit, to place the required input data into its designated input port at appropriate times. Following are some instructions that involve input ports.

IN	01H	;	Move the contents (8-bit data) of Input Port 01H to Register A.
IN	7EH	;	Move the contents of Input Port 7EH to Register A.

In addition to input ports, there are output ports, up to 256 of them. They are Output Port 00H, Output Port 01H, . . . , Output Port FFH. An output port is essentially an external register from which a specific output device can obtain data that was placed there by the μP. The output device, through its interface circuit, removes data from its designated output port at appropriate times. Some instructions that involve an output port are

OUT	02H	;	Move the contents (8-bit data) of Register A to Output Port 02H.
OUT	7EH	;	Move the contents of Register A to Output Port 7EH.

Also within the 8085 μP system, but external to the 8085 microprocessor, there can be up to 65,536 (2^{16}) memory registers. These are designated as Memory Location 0000H, Memory Location 0001H, . . . , Memory Location FFFFH. The numeric designation of a memory register is commonly known as the *address* of that memory location. As we will see in later sections, the memory registers are used for storing the controlling program as well as any data required for the program. Examples of instructions that involve memory registers are

LDA	0702H	;	Move the contents (8-bit data) of Memory Location 0702H to Register A.
STA	7EE2H	;	Move the contents of Register A to Memory Location 7EE2H.

Before immersing ourselves into a detailed discussion of the 8085 instruction set, we will consider a simple 8085 assembly language program to obtain a flavor or feel for what an assembly program is. Although we should not expect to fully understand the details of this example, we will obtain an overview, a perspective, that will give us a sufficient background for understanding the individual instructions of the 8085 instruction set that are presented in the following sections.

9.2.2 Assembly Language Program Example

Suppose we are to write an 8085 assembly language program to perform some monitoring function through an input device. This input device provides monitoring data to the μP system by continually storing an 8-bit data byte into Input Port 03H. The function of our monitoring program is to input that data byte from Input Port 03H, and compare it with an 8-bit test data byte stored in a memory location. If the incoming data byte is equal to or greater than the stored test byte, then the μP system outputs the bit pattern 00000000 to Output Port 10H. But if the incoming data byte is smaller than the stored test byte, then the μP system increments by one a counter stored in a memory location, and also outputs the bit pattern 11111111 to Output Port 10H. An 8085 assembly language program for performing these functions is given in Fig. 9.2.

Before considering this example in some detail, we should know that at this point of our discussion, we can view an assembly language program as comprising our commands to the 8085 microprocessor. In particular, a program consists of a series of

```
     Label     Operation  Operand(s)             Comment
 1   TESTV     EQU        0700H       ; Memory Location 0700H, which contains the 8-bit test value, is
                                      ; now also named TESTV.
 2   COUNT     EQU        0701H       ; Memory Location 0701H is now also named COUNT. (This
                                      ; location contains a counter for counting the number of times in
                                      ; which the incoming byte is less than the test byte.)
 3             ORG        0800H       ; This program is stored in successive memory locations, starting in
                                      ; Memory Location 0800H.
 4             LDA        TESTV       ; Move contents of Memory Location 0700H (also named TESTV)
                                      ; to Register A.
 5             MOV        C,A         ; Move contents of Register A to Register C.
 6             MVI        A,00H       ; Insert the value 0 (bit pattern 00000000B) into Register A.
 7             STA        COUNT       ; Move Register A contents (0 in this case) to Memory Location
                                      ; 0701H (also named COUNT).
 8   TEST3:    IN         03H         ; Move contents of Input Port 03H to Register A.
 9             CMP        C           ; Compare the value in Register C (test byte) with the value in
                                      ; Register A (incoming byte).
10             JC         ALERT       ; If the incoming byte in Register A is less than the test byte in
                                      ; Register C, then go to the instruction labeled ALERT. Otherwise,
                                      ; execute the next instruction.
11             MVI        A,00H       ; Insert the value 0 (bit pattern 00000000B) into Register A.
12             JMP        OUTPT       ; Go to the instruction labeled OUTPT.
13   ALERT:    LDA        COUNT       ; Move the value in Location 0701H into Register A.
14             INR        A           ; Increment the value in Register A by 1.
15             STA        COUNT       ; Move the incremented value back to Memory Location 0701H.
16             MVI        A,0FFH      ; Insert the bit pattern 11111111B into Register A.
17   OUTPT:    OUT        10H         ; Move the value in Register A to Output Port 10H.
18             JMP        TEST3       ; Go back to the instruction labeled TEST3; i.e., repeat the
                                      ; monitoring process.
```

Figure 9.2 8085 assembly language program example.

instructions that the microprocessor can "understand" and execute. However, for the microprocessor to understand our commands, each instruction must conform to a specific format. The general format for an executable 8085 instruction is

```
Label:    Operation    Operand(s)    ;    Comment
```

Each instruction must include, of course, the *operation* that the microprocessor is to perform. Any operation can be specified, provided it is a valid operation in the instruction set. Example of valid 8085 operations, taken from Fig. 9.2, are load Register A (LDA), store from Register A (STA), compare (CMP), and jump (JMP). Note that each operation code is an abbreviation of the corresponding command.

Each instruction may also include an operand (or operands) that is the object of the operation. An operand can be a source register, a destination register, a memory location, a port address, a label, a numeric value, or the like. Depending on the operation, an instruction can have one or two operands, or none at all.

Optionally, an instruction can have a *label* associated with it. A label is a symbolic name for an instruction, and is usually assigned when there is to be a jump from another instruction to that instruction. The name is usually selected so that it has some significance for the corresponding instruction and hence will serve as a memory aid. When associated with an executable instruction, a label is separated from the operation code by a colon. Also optionally, we can associate a *comment* with an instruction. A comment is separated from the rest of the instruction by a semicolon. Just as for a high-level programming language such as PASCAL or FORTRAN, a comment is *not* a command to the microprocessor. Rather, it serves simply as a reminder to the programmer or as an explanation to anyone reading the program of what the corresponding instruction does.

We will now consider in more detail the example assembly program of Fig. 9.2. Note that a statement number is assigned to each instruction. These numbers are not a part of the program; they serve only to facilitate our consideration of this example.

Statements 1, 2, and 3 contain *pseudo-instructions* that provide information about the program, but are *not* executed by the microprocessor. Those of Statements 1 and 2 name memory locations; that is, Memory Location 0700H is also to be named TESTV and Memory Location 0701H is also to be named COUNT. The pseudo-instruction ORG of Statement 3 causes the program to be stored in memory starting in Memory Location 0800H. Except for such pseudo-instructions, the microprocessor executes the instructions one at a time, in sequential fashion.

Statement 4 has the first instruction executed by the microprocessor. This instruction moves the contents of Memory Location 0700H (also named TESTV) to Register A. Here we assume that an 8-bit test value is already stored in Location 0700H. So, immediately after the execution of this instruction, Register A contains a copy of the value in Memory Location 0700H. Next, the microprocessor executes the instruction of Statement 5, which moves the contents of Register A into Register C. Note that the combined effect of the instructions of Statements 4 and 5 is to move the test byte from Memory Location 0700H into Register C. At this point, one might validly ask, why not instead simply move the contents of Memory Location 0700H directly into Register C? The answer is that there is no such instruction in the 8085 instruction set. Consequently, this moving has to be done in a roundabout manner, using Register A.

Continuing, the instructions of Statements 6 and 7 effectively initialize the contents of Memory Location 0701H (also named COUNT) to zero. Again, this is done using Register A, since it cannot be done directly. The instruction of Statement 8 brings into Register A the contents of Input Port 03H. Recall that Input Port 03H contains the 8-bit monitoring data value placed there by the input device through its interface circuit.

The instructions of Statements 9 and 10 direct the microprocessor to compare the incoming value from Input Port 03H with the test value stored in Register C (originally from Memory Location 0700H). If the incoming value is less than the test value, the next instruction executed is that of Statement 13, followed by those of 14, 15, 16, 17, 18, 8, ... But, if the incoming value is not less than the test value in Register C, then the next instruction executed is that of Statement 11, followed by those of 12, 17, 18, 8, ...

In the first case the instructions of Statements 13, 14, and 15, in effect, increment by one the value in Memory Location 0701H (also named COUNT). Then, the instruction of Statement 16 moves bit pattern 11111111 (which is equal to hexadecimal FFH) into Register A, and the instruction of Statement 17 outputs that bit pattern to Output Port 10H. Next, the instruction of Statement 18 directs the microprocessor to execute the instruction with the label TEST3, which is the instruction of Statement 8. Consequently, the monitoring cycle continues.

If the incoming value is not less than the test value in Register C, then the instruction of Statement 11 is executed after that of Statement 10. This instruction inserts bit pattern 00000000 into Register A. Next, the instruction of Statement 12 directs the microprocessor to execute the instruction of Statement 17, which outputs that bit pattern to Output Port 10H. Then, the instruction of Statement 18 instructs the microprocessor to next execute the instruction of Statement 8, and so the monitoring cycle continues.

We should begin to see that the power and flexibility of a μP system is a function of the sophistication of the stored program that controls the μP system. To develop effective programs a programmer must have a complete knowledge of the μP instruction set, and of the software techniques required to utilize the instructions fully. In the remainder of this chapter, and also in Chapter 10, we will use the instruction set of the 8085 to illustrate the software concepts that are important in the development of μP system software.

9.3 DATA TRANSFER BETWEEN REGISTERS

Basically, a μP system performs all its functions through a sequence of data transfers and data transformations. Instructions for data transfer within the 8085 μP system are considered in this section. Instructions for data transformation will be considered in subsequent sections.

The performance of a program can be greatly affected by the convenience with which the data can be transferred within the register structure of a μP system. Data transfer operations can be divided into the following categories:

1. Data transfer between two general-purpose registers within the microprocessor
2. Data transfer between an input or output (I/O) port register and a general-purpose register within the microprocessor
3. Data transfer between a memory register and a general-purpose register within the microprocessor

We will consider each separately.

9.3.1 Data Transfer Between Two General-Purpose Registers

We can use the MOV instruction to transfer data between any two general-purpose registers within the 8085 microprocessor. For such a transfer, this instruction has the general format specified in Fig. 9.3(a), in which r1 designates the *destination register* and r2 the *source register*. Execution of this instruction causes the contents of r2 to be moved to r1, as illustrated in Fig. 9.3(b). The parentheses in Fig. 9.3(b) mean "the contents of" and the arrow means "are transferred to." An example of the MOV instruction is given in Fig. 9.3(c) in binary form and in Fig. 9.3(d) in hexadecimal form. In our examples, we will, for convenience, use the hexadecimal form to illustrate the contents of a register, a port, or a memory location. A second example of the MOV instruction is given in Fig. 9.3(e).

Note from these examples that after the execution of the MOV instruction, the prior contents of the destination registers are destroyed. The source register contents,

(a) General format: MOV r1, r2 ;r1 (and r2) is Register A, B, C, D, E, H, or L

(b) Description: (r1) ← (r2)

(c) Example 1: MOV A, C

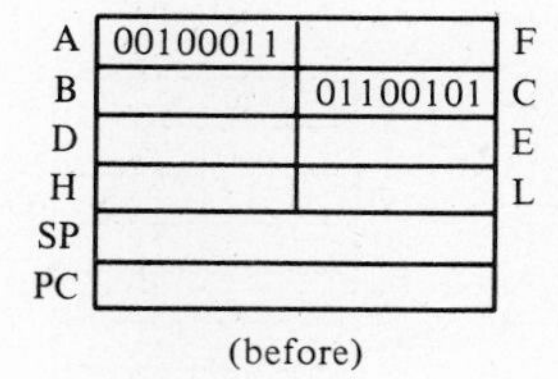

A	01100101		F
B		01100101	C
D			E
H			L
SP			
PC			

(after)

(d) Example 1 illustrated in hex notation for convenience:

A	23H		F
B		65H	C
D			E
H			L
SP			
PC			

(before)

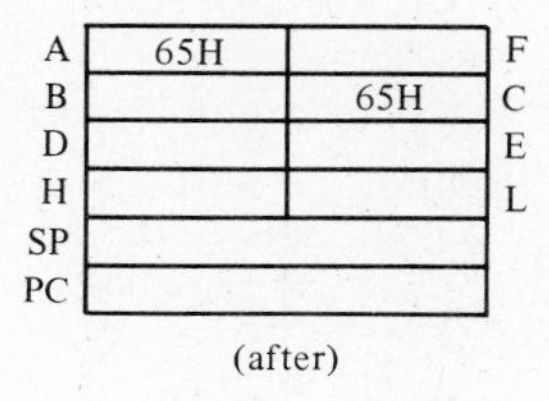

(e) Example 2: MOV H, B

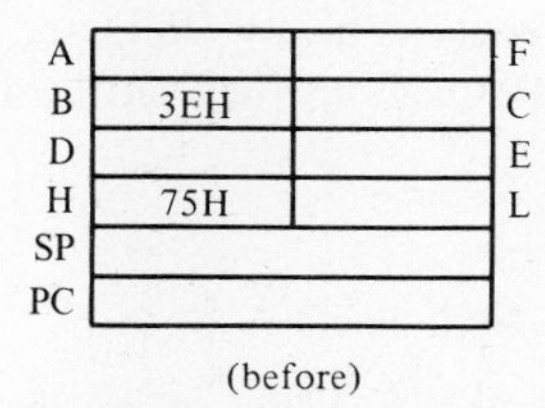

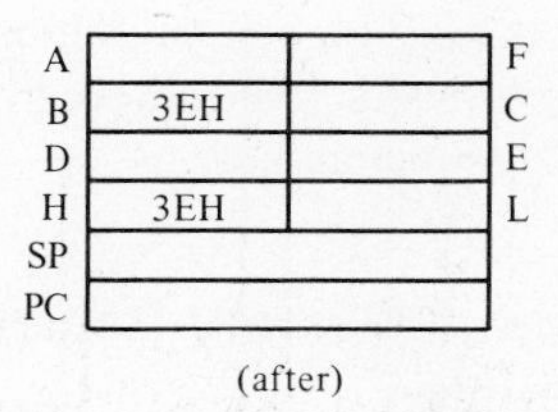

Figure 9.3 MOV instruction.

however, are not affected by the MOV instruction. Therefore, this instruction is effectively a "copy" instruction since execution of it causes the contents of the source register to be copied into the destination register. The same is true of the other data transfer instructions.

9.3.2 Data Transfer Between an I/O Port and a General-Purpose Register

In the 8085 μP system, the IN instruction is used to transfer data from an input port to Register A. Figure 9.4 shows the general format, description, and an example of the IN instruction. In Fig. 9.4(c), the XXH designates a don't-care situation. For the IN instruction, the destination register is always Register A, and the source register is the input port specified by the input port number in the IN instruction. Note that in the 8085 μP system, we cannot transfer data from an input port directly to any other register. Instead,

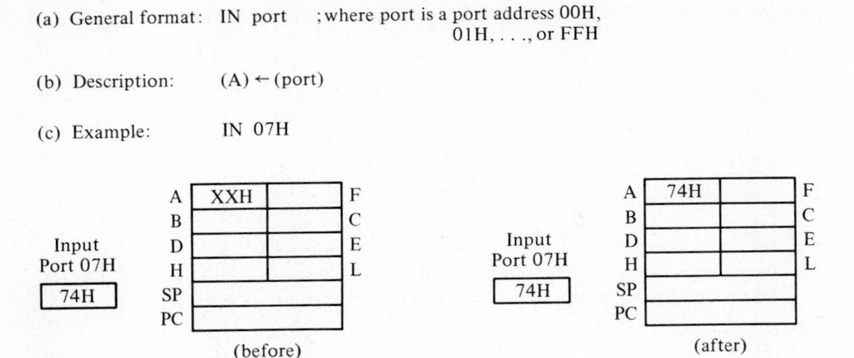

Figure 9.4 IN instruction.

we must use Register A. For example, to transfer data from Input Port 07H to Register C, we need two instructions:

```
IN      07H
MOV     C,A
```

We can transfer data from Register A to an output port with an OUT instruction. The general format, description, and an example of the OUT instruction are given in Fig. 9.5. Since for this instruction the source register is always Register A, we must use a MOV to Register A instruction followed by an OUT instruction to transfer data from any other register to an output port.

9.3.3 Data Transfer Between a Memory Register and a General-Purpose Register

The data required to drive a program is generally stored in the memory registers (i.e., locations in memory). During the execution of a program, data is frequently transferred, often in large quantities, between the memory locations and the general-purpose registers. Instructions for such transfers, commonly called *memory reference instructions,* have two operands, one of which may be implied. One of the two operands specifies a general-purpose register, and the other the *effective address* of a memory location. The manner of specifying the effective address is called the *addressing mode*. For the 8085 μP, the addressing mode is classified as *direct, indirect,* or *immediate*. We will now consider each of these modes.

Direct Addressing

For *direct addressing,* we specify, as a part of an instruction, the address of the memory location that is to be referenced. One instruction that employs direct addressing is the LDA instruction, which is used for transferring data from a memory location to Register A. The general format, description, and an example of the LDA instruction are given in Fig. 9.6. Note that there is no need to specify the destination register because for the

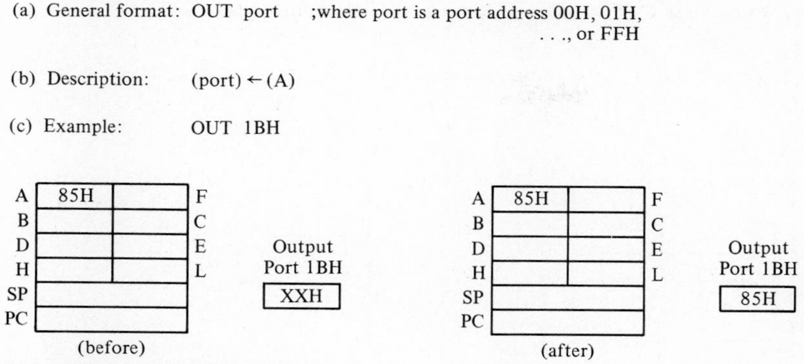

Figure 9.5 OUT instruction.

LDA instruction it is *implied* to be Register A. But there is a specification of the source register, and it is with a 16-bit memory address as the operand of the instruction. Execution of this instruction causes the movement into Register A of the contents of the memory location specified by the address, as illustrated in Fig. 9.6(f). Note that the 16-bit effective address is either a 16-bit binary number [Fig. 9.6(c)], a 4-digit hexa-

(a) General format: LDA addr ;where addr is a 16-bit memory address

(b) Description: (A) ← (addr)

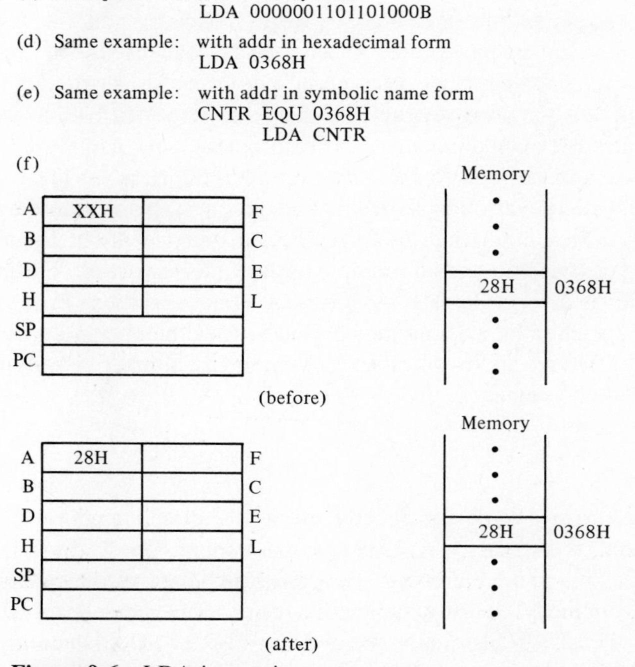

Figure 9.6 LDA instruction.

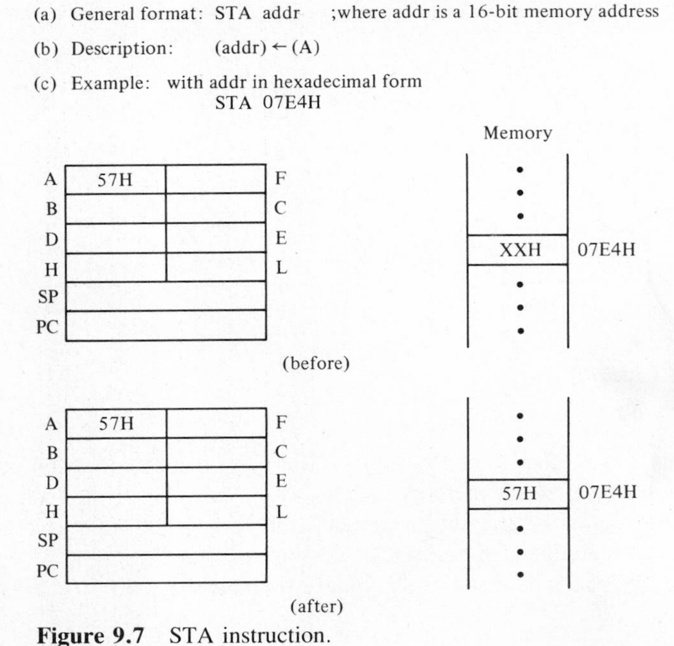

Figure 9.7 STA instruction.

decimal number [Fig. 9.6(d)], or a *symbolic address* [Fig. 9.6(e)]. The *pseudo-instruction* EQU makes it possible to use a symbolic name as a memory address. As will be seen later, this is often a convenient feature. At this point of our discussion, and as has been mentioned, we can view a pseudo-instruction simply as an instruction that provides additional information about the program but that is *not* executed by the microprocessor. Here, EQU equates the symbolic name CNTR to the memory address 0368H. In other words, it allows us to also refer to Memory Location 0368H as CNTR.

The STA instruction also employs direct addressing. This instruction transfers data from Register A, which is implied, to a specified memory location. Figure 9.7 has the general format, a description, and an example of the STA instruction. For this instruction the source register is always Register A, and so it is not specified. But, the destination register must be specified by a 16-bit memory address as the operand of the instruction. Again, the 16-bit effective address is either a 16-bit binary number, a 4-digit hexadecimal number, or a symbolic name.

Indirect Addressing

With *indirect addressing* we do *not* directly specify the effective address as a part of an instruction. Instead, we specify a register pair that contains the 16-bit effective address of the memory location to be referenced. The LDAX instruction uses indirect addressing, and is different from the LDA instruction in this respect. The general format, description, and examples of the LDAX instruction are given in Fig. 9.8. Note that only the Register Pairs BC and DE can be used with this instruction. Execution of this instruction causes movement into Register A of the contents of the memory location whose address is in

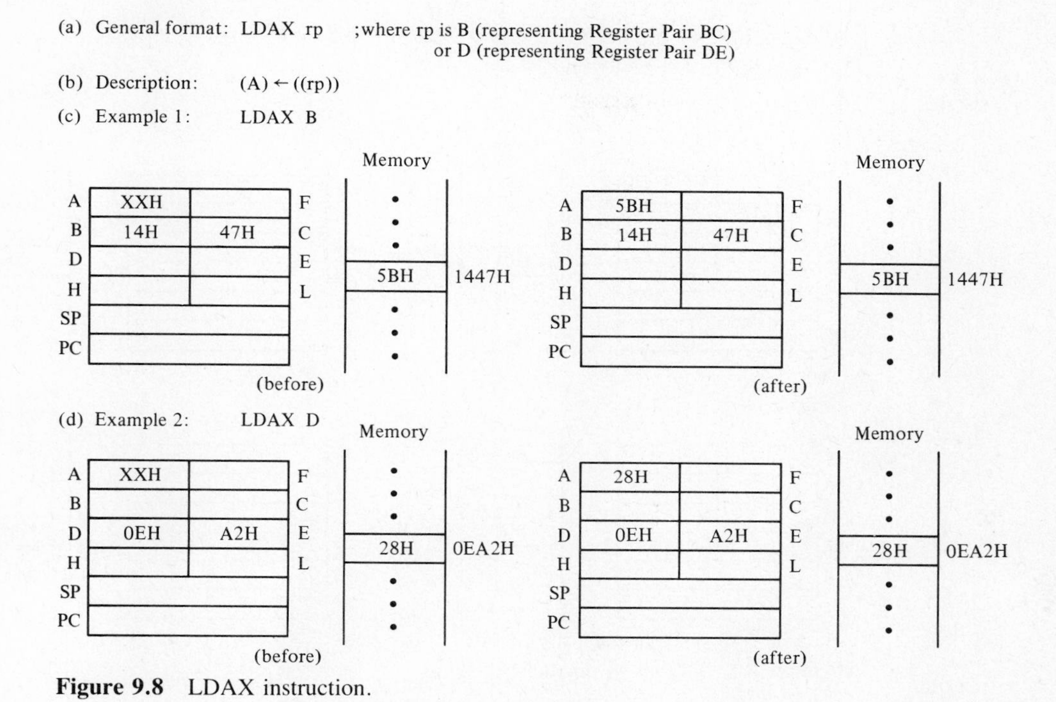

Figure 9.8 LDAX instruction.

the register pair rp. How to get this address to the register pair beforehand is discussed in the next section.

The instruction STAX is another instruction that uses indirect addressing. Execution of this instruction causes the contents of Register A to move to the memory location whose address is specified by the contents of a register pair rp. Again, only Register Pairs BC or DE can be used as a register pair for the address. The general format, description, and examples of the STAX instruction are given in Fig. 9.9.

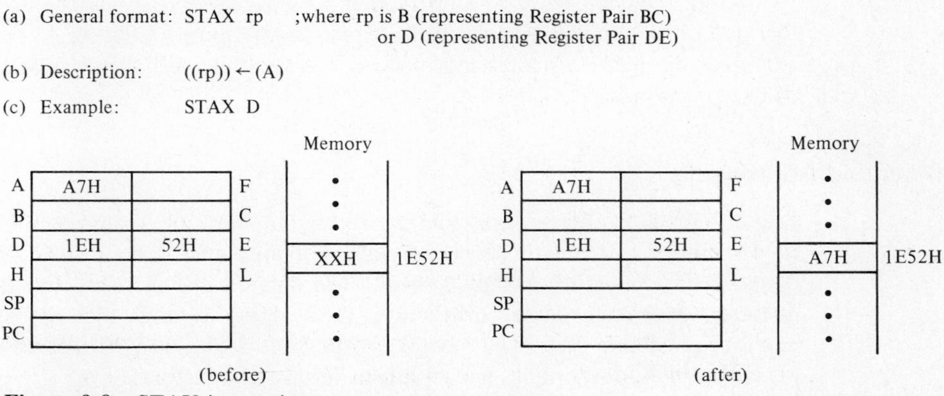

Figure 9.9 STAX instruction.

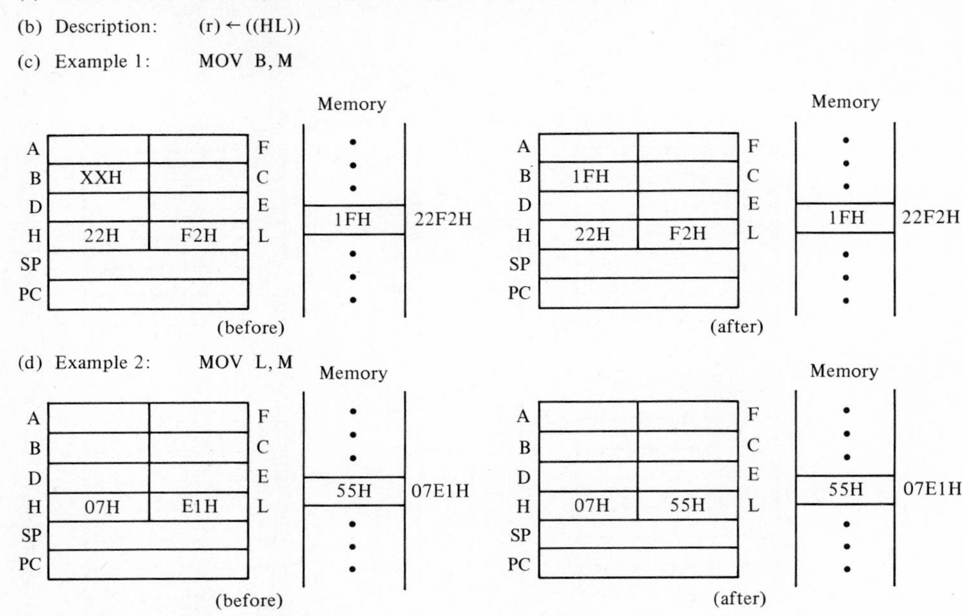

Figure 9.10 MOV r,M instruction.

Although Register Pair HL cannot be used with either the LDAX or STAX instruction, it can be used for indirect addressing with a form of the MOV instruction. Figures 9.10 and 9.11 show the general format, description, and examples of this use. Note that for indirect addressing with the MOV instruction, the letter M is either the source or destination register designator, and implies indirect addressing with the Register Pair HL. The MOV r,M instruction is similar to the LDAX instruction, except for the destination register. For the LDAX instruction the destination register is always Register A. But for the MOV r,M instruction, the destination register can be any of the general-purpose registers, including H or L. Note the effect of the MOV L,M instruction as illustrated in Fig. 9.10(d).

Similarly, the MOV M,r instruction is analogous to the STAX instruction, except that any of the general-purpose registers can be the source register. As shown in a later example, the indirect addressing mode is a very effective means for accessing a block of data from memory.

Immediate Addressing

In the immediate addressing mode, an instruction does not contain the effective address of a memory location to be referenced to obtain some data. Instead, the instruction contains the data itself, as an operand. The data can be an 8-bit quantity or a 16-bit quantity, depending on the instruction. In the 8085 the data transfer instructions that employ immediate addressing are MVI (for 8-bit data) and LXI (for 16-bit data). The general format, description, and examples of the MVI instruction are given in Fig. 9.12.

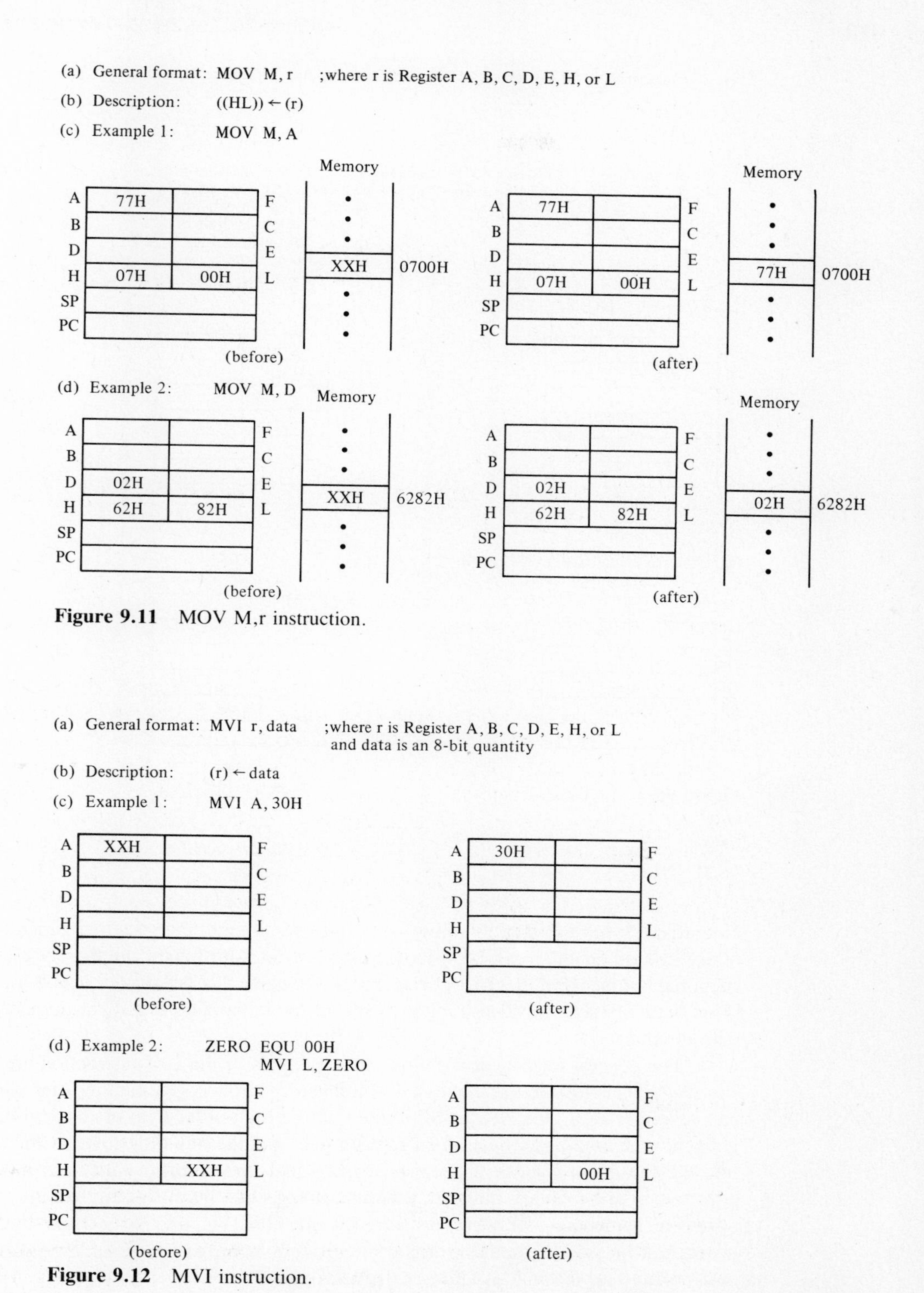

Figure 9.11 MOV M,r instruction.

Figure 9.12 MVI instruction.

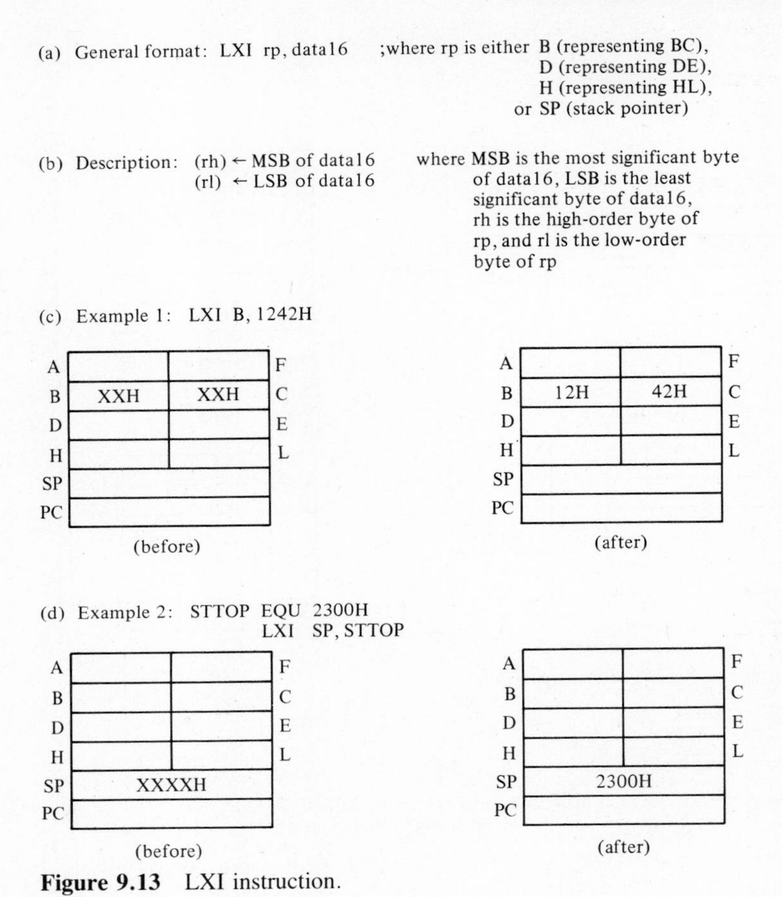

Figure 9.13 LXI instruction.

Note that the MVI instruction moves an 8-bit quantity into the specified general-purpose register. This quantity can be specified either as an 8-bit binary number, a 2-digit hexadecimal number [Fig. 9.12(c)], or as a symbolic name that is equated to an 8-bit quantity [Fig. 9.12(d)]. The MVI instruction is useful for initializing a general-purpose register with a value.

The general format, description, and examples of the LXI instruction are given in Fig. 9.13. This instruction moves a 16-bit quantity into the specified register pair, which can be either BC, DE, HL, or SP. Note that the high-order byte of the 16-bit quantity goes into the high-order byte of the register pair, and the low-order byte of the data goes into the low-order byte of the register pair. Again, the 16-bit quantity can be specified either as a 16-bit binary number, a 4-digit hexadecimal number [Fig. 9.13(c)], or as a symbolic name that is equated to a 16-bit quantity [Fig. 9.13(d)]. Uses for the LXI instruction include the initialization of a register pair for the indirect addressing mode of data transfer, or the initialization of Stack Pointer SP.

TABLE 9.1 DATA TRANSFER INSTRUCTIONS

Instruction	Description
Move registers	
MOV r1,r2	(r1) ← (r2)
Input/output	
IN port	(A) ← (port)
OUT port	(port) ← (A)
Memory reference	
Direct addressing	
LDA addr	(A) ← (addr)
STA addr	(addr) ← (A)
Indirect addressing	
LDAX rp′	(A) ← ((rp′))
STAX rp′	((rp′)) ← (A)
MOV r,M	(r) ← ((HL))
MOV M,r	((HL)) ← (r)
Immediate addressing	
MVI r,data	(r) ← data
LXI rp,data16	(rh) ← MSB of data16 (rl) ← LSB of data16
Indirect and immediate	
MVI M,data	((HL)) ← data

Notes:
- r1,r2,r: Register A, B, C, D, E, H, or L
- rp: B (representing BC), D (representing DE), H (representing HL), or SP (stack pointer)
- rp′: B (representing BC) or D (representing DE)
- rh: high-order byte of rp
- rl: low-order byte of rp
- addr: 16-bit memory address
- port: 8-bit port address
- data: 8-bit quantity
- data16: 16-bit quantity

9.3.4 Summary of the Data Transfer Instructions

Table 9.1 contains a summary of the basic data transfer instructions for the 8085. Note from the last entry that we can use the MVI instruction to move data directly into a memory location specified by the contents of Register Pair HL. This use is a combination of indirect and immediate addressing.

9.4 DATA TRANSFORMATIONS— BASIC ARITHMETIC OPERATIONS

As stated, a μP system performs its functions by a sequence of data transfers and data transformations. Having just considered data transfer instructions, we will now consider data transformation instructions. These instructions are generally divided into two categories: arithmetic and logic. This section describes and illustrates the use of the 8085

TABLE 9.2 ARITHMETIC INSTRUCTIONS

Instruction	Description	Flags affected
Add group		
(a) ADD r	(A) ← (A) + (r)	S, Z, AC, P, CY
(b) ADD M	(A) ← (A) + ((HL))	S, Z, AC, P, CY
(c) ADI data	(A) ← (A) + data	S, Z, AC, P, CY
Add-with-carry group		
(d) ADC r	(A) ← (A) + (r) + (CY)	S, Z, AC, P, CY
(e) ADC M	(A) ← (A) + ((HL)) + (CY)	S, Z, AC, P, CY
(f) ACI data	(A) ← (A) + data + (CY)	S, Z, AC, P, CY
Subtract group		
(g) SUB r	(A) ← (A) − (r)	S, Z, AC, P, CY
(h) SUB M	(A) ← (A) − ((HL))	S, Z, AC, P, CY
(i) SUI data	(A) ← (A) − data	S, Z, AC, P, CY
Subtract-with-borrow group		
(j) SBB r	(A) ← (A) − (r) − (CY)	S, Z, AC, P, CY
(k) SBB M	(A) ← (A) − ((HL)) − (CY)	S, Z, AC, P, CY
(l) SBI data	(A) ← (A) − data − (CY)	S, Z, AC, P, CY
Increment/Decrement group		
(m) INR r	(r) ← (r) + 1	S, Z, AC, P
(n) INR M	((HL)) ← ((HL)) + 1	S, Z, AC, P
(o) DCR r	(r) ← (r) − 1	S, Z, AC, P
(p) DCR M	((HL)) ← ((HL)) − 1	S, Z, AC, P
(q) INX rp	(rp) ← (rp) + 1	none
(r) DCX rp	(rp) ← (rp) − 1	none

Notes:

r: Register A, B, C, D, E, H, or L
rp: B (i.e., BC), D (i.e., DE), H (i.e., HL), or SP (stack pointer)
S: Sign Flag
Z: Zero Flag
AC: Auxiliary Carry Flag
P: Parity Flag
CY: Carry Flag
data: 8-bit quantity

basic arithmetic instructions, which are summarized in Table 9.2. As shown, they divide into five groups: Add, Add-with-carry, Subtract, Subtract-with-borrow, and Increment/Decrement.

For the Add group, an 8-bit quantity is added to the contents of Register A and the sum placed in Register A. The 8-bit quantity may be from another general-purpose register [Table 9.2(a)], indirectly from a memory location [Table 9.2(b)], or immediately as a part of the instruction [Table 9.2(c)]. Note from the rightmost column of Table 9.2 that execution of the Add instructions also affects *condition flags*. This is also true of the other arithmetic instructions.

There are five condition flags associated with the execution of the arithmetic instructions of the 8085. These flags are Sign (S), Zero (Z), Auxiliary Carry (AC), Parity (P), and Carry (CY). They are stored in Flag Register F in the format described in Fig. 9.14. Not all 8085 instructions affect the condition flags. For example, none of the data transfer instructions discussed earlier affect them. Furthermore, some instructions affect only some of the flags. Unless otherwise indicated, if an instruction does affect a flag, then it is in the following manner:

Sign: If the most significant bit (bit 7) of the result of an instruction operation is 1, then the Sign Flag is set (to 1); otherwise, it is reset (to 0).

Zero: If the result of an instruction operation is zero, then the Zero Flag is set; otherwise, it is reset.

Auxiliary Carry: If the instruction operation causes a carry out of bit 3 and into bit 4 of the resulting value, then the Auxiliary Carry Flag is set; otherwise, it is reset. This flag is for internal use only and is not available to the programmer.

Parity: If the result of an instruction operation has even parity, then the Parity Flag is set; otherwise (odd parity), it is reset.

Carry: If an instruction operation results in a carry (from addition) or a borrow (from subtraction or comparison) out of bit 7 of the resulting value, then the Carry Flag is set; otherwise, it is reset.

Examples of the instructions in the Add group and their effects on the condition flags are given in Fig. 9.15. Note in Example 2 that the leading 0 has no effect on the actual value of the number. In other words, the specified 0DDH is equal to DDH. As will be explained in Sec. 9.5, a leading zero is commonly added to a hex number whose leading digit is a letter (A, B, C, D, E, or F).

For the Add-with-carry group, an 8-bit quantity, along with the contents of the Carry Flag are added to the contents of Register A, and the sum placed in Register A. As before, the 8-bit quantity may be from another general-purpose register [Table 9.2(d)], indirectly from a memory location [Table 9.2(e)], or immediately as a part of the instruction [Table 9.2(f)].

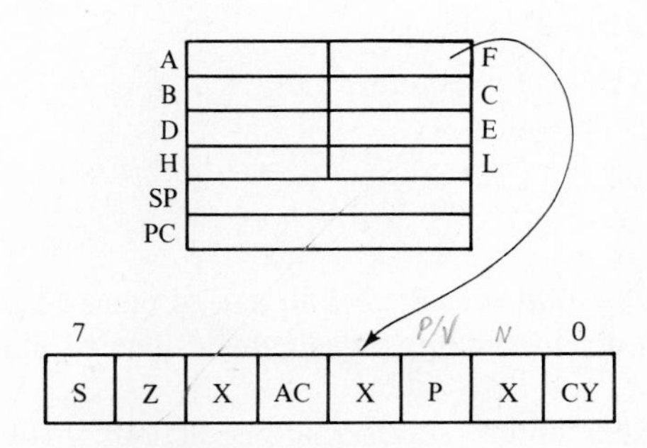

Flag Register F, where

S: Sign Flag
Z: Zero Flag
AC: Auxiliary Carry Flag
P: Parity Flag
CY: Carry Flag
X: Undefined (not used)

Figure 9.14 Format for Flag Register F.

(a) Example 1: ADD C

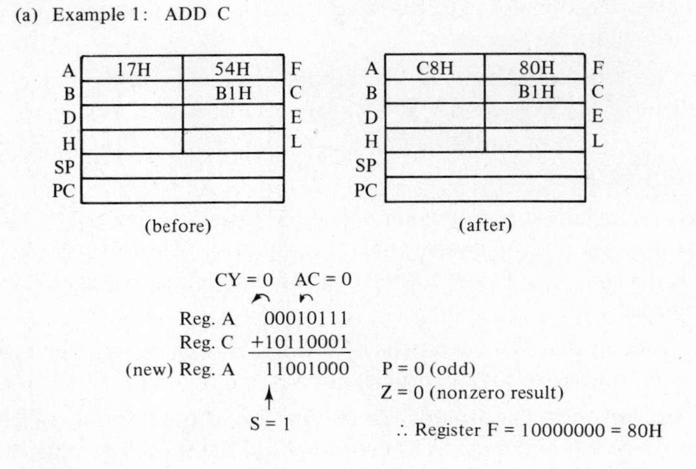

(b) Example 2: ADI 0DDH

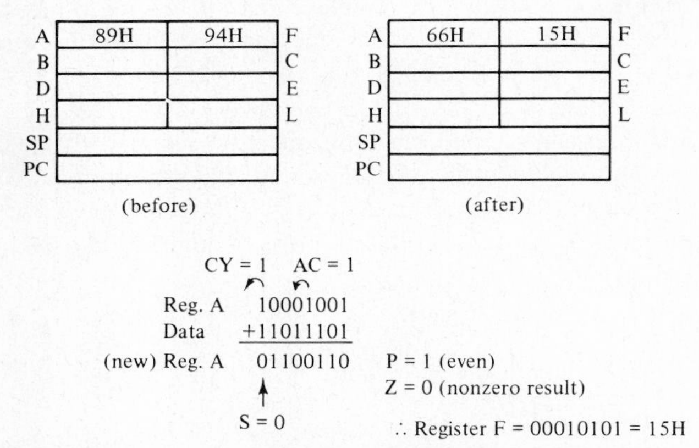

Figure 9.15 Examples of addition instructions.

The description of the Subtract group is analogous to that of the Add group. And, the description of the Subtract-with-borrow group is analogous to that of the Add-with-carry group.

As Table 9.2 shows, all the addition and subtraction instructions involve Register A, and all affect the five condition flags. The Increment/Decrement instructions, on the other hand, can involve other general-purpose registers [Table 9.2(m),(o)], memory locations [Table 9.2(n),(p)], or register pairs [Table 9.2(q),(r)]. Furthermore, none of the Increment/Decrement instructions affects the CY Flag. In fact, the 16-bit Increment/Decrement instructions do not affect any flag. Because of these different effects on the flags, care must be taken in testing the condition flags after the execution of arithmetic instructions.

9.5 ARITHMETIC PROGRAMMING EXAMPLES

Example 9.1 Single-Byte Add For the first arithmetic programming example, consider the design of a simple subroutine program for adding two single-byte (8-bit) numbers stored in memory. Assume that the calling program has stored these numbers in Memory Locations NUM1 (0700H) and NUM2 (0704H). The function of this subroutine, which is named SBADD, is to add these numbers and place their sum into Memory Location SUM (0710H).

A suitable strategy or *algorithm* for this subroutine should be apparent. Since the 8085 instruction set has an ADD instruction that adds the 8-bit contents of a general-purpose register to Register A, all that is necessary is to move one of the numbers to be added into a general-purpose register other than Register A; move the other number into Register A; perform the ADD; and then store the resulting sum into Memory Location SUM. In flowchart form this algorithm is as shown in Fig. 9.16(a). The corresponding 8085 assembly language program is shown in Fig. 9.16(b). Note that the transfer (Reg. B) $\leftarrow$ (NUM1) requires two instructions (Statements 5 and 6).

Figure 9.16(c) illustrates an execution example of Subroutine SBADD for 42H + 1AH = 5CH. The subroutine instructions are executed sequentially, one at a time, starting with Statement 5, which is the first executable instruction. Recall that Statements 1 through 4 are pseudo-instructions that provide information about the program, but they are not executed by the microprocessor. Figure 9.16(c) provides a ''snapshot'' of the contents of the affected registers and memory locations after the execution of each instruction. The reader is urged to trace through the execution to gain a good understanding of the execution of an assembly program. Note that the only instruction that affects the flags (Register F) is the ADD instruction of Statement 8. ■ ■

Example 9.2 Double-Byte Subtract Next, consider writing a subroutine program, to be called DBSUB, for subtracting two 2-byte (16-byte) numbers. As illustrated in Fig. 9.17, the calling program will store the least significant byte of the 2-byte minuend into Location MINLB (0700H) and the most significant byte into Location MINLB + 1 (0701H). The least significant byte of the 2-byte subtrahend will be stored into Location SUBLB (0710H) and the most significant byte into Location SUBLB + 1 (0711H). The function of Subroutine DBSUB is to subtract the 16-bit subtrahend from the 16-bit minuend and place the resulting 16-bit difference into Register Pair BC.

Unfortunately, the 8085 instruction set has no 16-bit subtraction instruction. But, it has the SUB instruction that subtracts an 8-bit number in a general-purpose register from another 8-bit number in Register A, and places the 8-bit difference into Register A. Consequently, we need to determine a way to obtain a 16-bit subtraction with this instruction that performs only an 8-bit subtraction.

To gain some insight into obtaining an algorithm for this subroutine, consider the subtraction example of two 16-bit numbers given in Fig. 9.18(a). From Fig. 9.18(a), observe that the 16-bit resulting DIFFERENCE can be represented as a 2-byte number: DIFF_{MSB} and DIFF_{LSB}. Also, the following appears to be true:

$$\text{DIFF}_{\text{LSB}} = \text{MINUEND}_{\text{LSB}} - \text{SUBTRAHEND}_{\text{LSB}} = 37\text{H} - 1\text{AH} = 1\text{DH}$$
$$\text{DIFF}_{\text{MSB}} = \text{MINUEND}_{\text{MSB}} - \text{SUBTRAHEND}_{\text{MSB}} = 4\text{EH} - 13\text{H} = 3\text{BH}$$

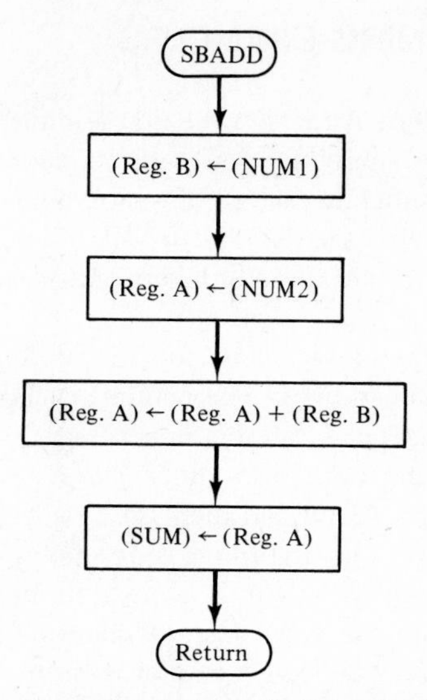

(a) Algorithm in flowchart form for Subroutine SBADD.

```
Statement
  1    NUM1     EQU   0700H ;  One of the operands for the addition is
                            ;  stored in Memory Location 0700H.
  2    NUM2     EQU   0704H ;  The other operand is stored in 0704H.
  3    SUM      EQU   0710H ;  The resulting sum will be stored in 0710H.
  4             ORG   0800H ;  This subroutine program will be stored
                            ;  beginning in Memory Location 0800H.
  5    SBADD:   LDA   NUM1  ;  Moves operand NUM1 into Register A.
  6             MOV   B, A  ;  Transfers NUM1 into Register B.
  7             LDA   NUM2  ;  Moves operand NUM2 into Register A.
  8             ADD   B     ;  Adds NUM1 to NUM2 with the sum in Register A.
  9             STA   SUM   ;  Stores the sum into Memory Location 0710H.
 10             RET         ;  Returns to the calling program.
```

(b) 8085 assembly program for Subroutine SBADD

Statement number	Reg. A	Reg. F	Reg. B	Memory Locations 0700H	0704H	0710H
Initial values	XX	44	XX	42	1A	XX
5	42	44	XX	42	1A	XX
6	42	44	42	42	1A	XX
7	1A	44	42	42	1A	XX
8	5C	04	42	42	1A	XX
9	5C	04	42	42	1A	5C
10	5C	04	42	42	1A	5C

(c) Execution for initial values of NUM1 = 42H, NUM2 = 1AH, and Register F = 44H; all values are in hexadecimal, and XX is don't care

Figure 9.16 Single-byte addition subroutine and execution example.

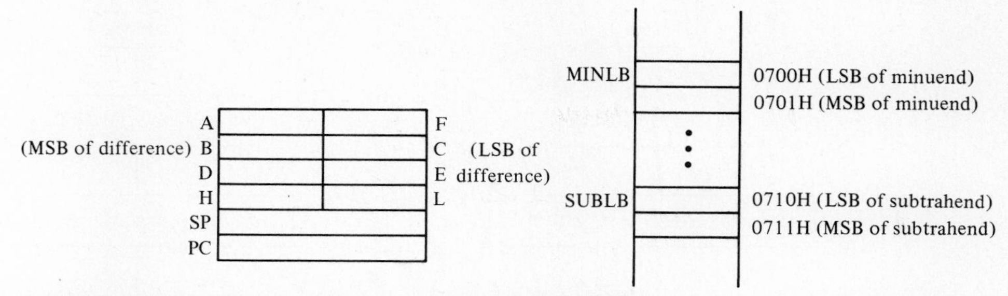

Figure 9.17 Problem specification for Subroutine DBSUB.

This observation is the basis for the preliminary algorithm given in Fig. 9.19(a). To verify this algorithm, we will apply it to another example of 16-bit subtraction, that of 4F27H − 1890H:

$$\text{DIFF}_{\text{LSB}} = \text{MINUEND}_{\text{LSB}} - \text{SUBTRAHEND}_{\text{LSB}} = 27\text{H} - 90\text{H} = 97\text{H} \text{ (with a borrow)}$$

$$\text{DIFF}_{\text{MSB}} = \text{MINUEND}_{\text{MSB}} - \text{SUBTRAHEND}_{\text{MSB}} = 4\text{FH} - 18\text{H} = 37\text{H}$$

Unfortunately, it appears that the preliminary algorithm does not work in this case because the algorithm gives an answer of 3797H, but the correct difference is 3697H, as shown in Fig. 9.18(b). A close inspection shows that this error is caused by the failure of the algorithm to take into account the borrow from bit 0 of $\text{MINUEND}_{\text{MSB}}$ to bit 7 of $\text{MINUEND}_{\text{LSB}}$. We can easily correct this error by using a suitable instruction from the Subtract-with-borrow group. Each instruction of this group utilizes the Carry Flag, which is 1 whenever there is a borrow. Figure 9.19(b) shows the revised algorithm.

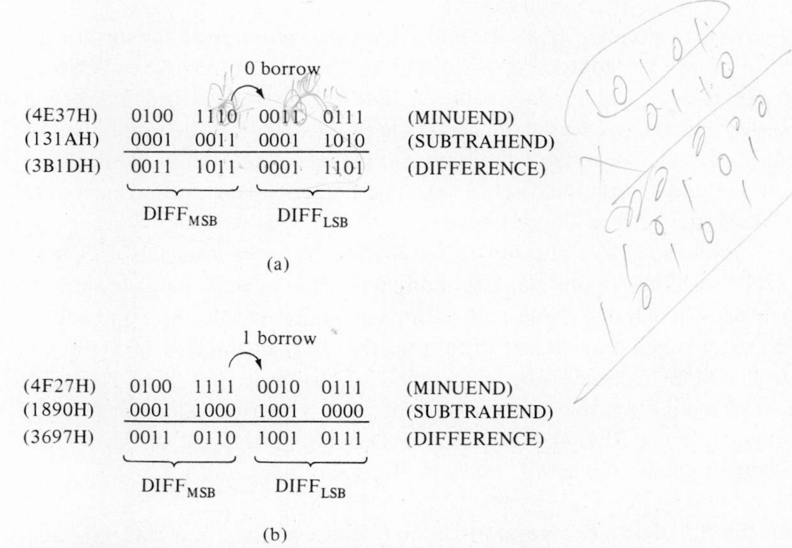

Figure 9.18 Examples of double-byte subtraction.

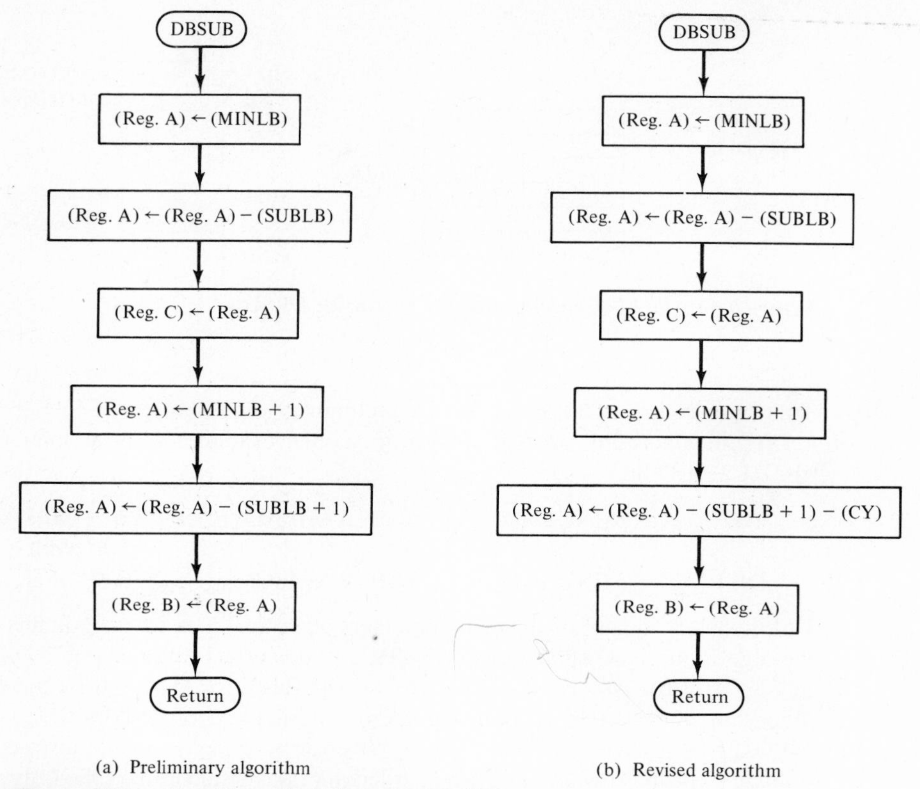

Figure 9.19 Algorithms for Subroutine DBSUB.

A corresponding 8085 assembly language program is shown in Fig. 9.20(a). Note the use of indirect addressing with Register Pair DE in moving each byte of the minuend into Register A (Statements 4, 6, 9, and 11). Also, we use indirect addressing with Register Pair HL in subtracting each byte of the subtrahend, stored in a memory location, from the respective byte of the minuend stored in Register A (Statements 5, 7, 10, and 12). Also, the instruction SBB M (Statement 12) accounts for the borrow (if any) between the MSB and LSB of the minuend.

Figure 9.20(b) illustrates the execution of Subroutine DBSUB for the subtraction 4F27H − 1890H = 3697H. Beginning with Statement 4, the subroutine instructions are executed sequentially. Note that during the entire execution of the subroutine, the Flag Register changes only twice: once after the SUB instruction in Statement 7, and again after the SBB instruction in Statement 12. Also note that the CY Flag is set to 1 after the SUB instruction in Statement 7, indicating a borrow. This borrow (CY = 1) is then available for the SBB instruction in Statement 12 because the intervening instructions (Statements 8, 9, 10, and 11) do not affect any flags. ■ ■

Example 9.3 Code Conversion Using Table Lookup For this example we will write a subroutine program for producing the bit patterns required for displaying two hex digits

```
Statement

  1    MINLB    EQU    0700H     ; The LSB of the minuend is stored in Memory Location
                                 ; 0700H (the MSB is stored in 0701H).
  2    SUBLB    EQU    0710H     ; The LSB of the subtrahend is stored in Memory Location
                                 ; 0710H (the MSB is stored in 0711H).
  3             ORG    0800H     ; The subrouting program is stored beginning in Memory
                                 ; Location 0800H.
  4    DBSUB:   LXI    D,MINLB   ; Initializes the DE register pair with the address of the LSB of
                                 ; the minuend in preparation for using DE for indirect
                                 ; addressing.
  5             LXI    H,SUBLB   ; Initializes the HL register pair with the address of the LSB of
                                 ; the subtrahend in preparation for using HL for indirect
                                   addressing.
  6             LDAX   D         ; Using indirect addressing, loads the minuend LSB into
                                 ; Register A.
  7             SUB    M         ; Using indirect addressing, subtracts the subtrahend LSB from
                                 ; the minuend LSB. The resulting difference is in Register A.
                                 ; Note that if the subtraction requires a borrow, then CY = 1;
                                 ; otherwise, CY = 0.
  8             MOV    C,A       ; Moves the difference LSB into Register C.
  9             INX    D         ; Increments DE to the address of the minuend MSB.
 10             INX    H         ; Increments HL to the address of the subtrahend MSB.
 11             LDAX   D         ; Loads the minuend MSB into Register A.
 12             SBB    M         ; Using indirect addressing, subtracts the subtrahend MSB from
                                 ; the minuend MSB. Also subtracts from the minuend any
                                 ; borrow from the LSB subtraction.
 13             MOV    B,A       ; Moves the difference MSB into Register B.
 14             RET              ; Returns to the calling program with the 16-bit difference in
                                 ; Register Pair BC.
```

(a) 8085 assembly program for Subroutine DBSUB

Statement	Registers								Memory locations			
number	A	F	B	C	D	E	H	L	0700H	0701H	0710H	0711H
Initial values	XX	00	XX	XX	XX	XX	XX	XX	27	4F	90	18
4					07	00	↓	↓				
5	↓						07	10				
6	27	↓										
7	97	81		↓								
8				97	↓	↓						
9					07	01	↓	↓				
10	↓						07	11				
11	4F	↓										
12	36	04	↓									
13	↓	↓	36	↓	↓	↓	↓	↓	↓	↓	↓	↓
14	36	04	36	97	07	01	07	11	27	4F	90	18

Notes:
(1) All values are in hexadecimal.
(2) The arrows imply no change in values.
(3) XX is don't care.

(b) Example execution of Subroutine Program DBSUB

Figure 9.20 Double-byte subtraction subroutine with execution example.

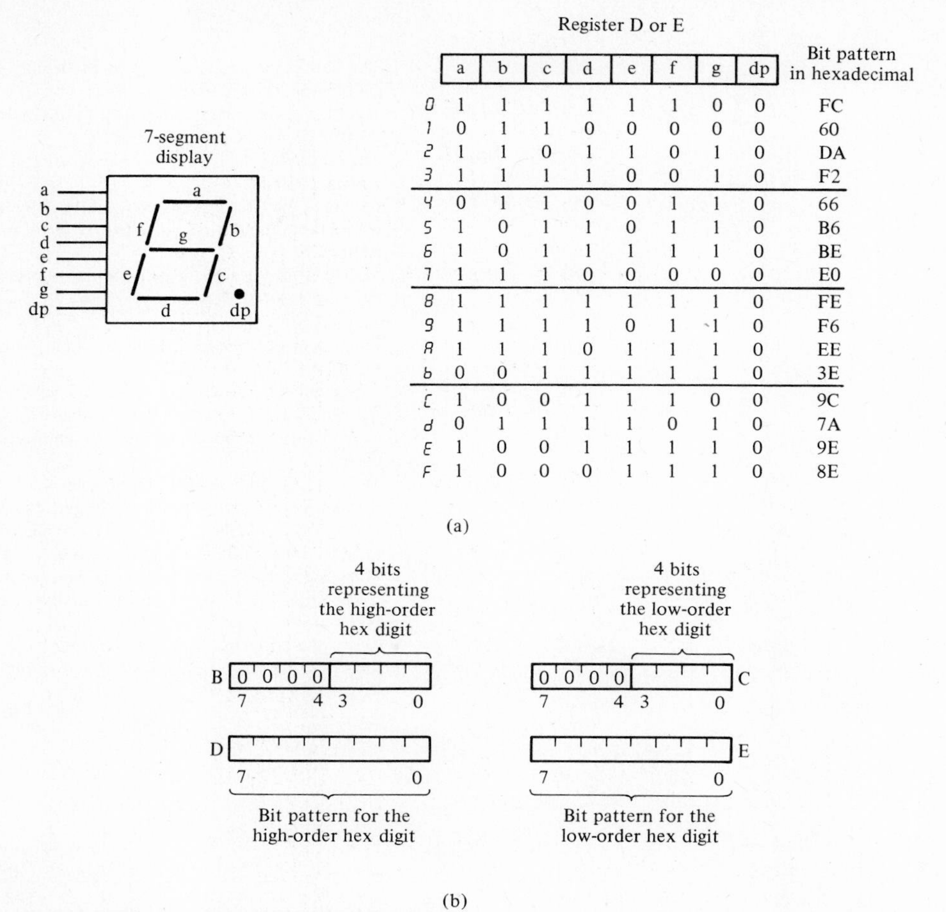

	a	b	c	d	e	f	g	dp	Bit pattern in hexadecimal
0	1	1	1	1	1	1	0	0	FC
1	0	1	1	0	0	0	0	0	60
2	1	1	0	1	1	0	1	0	DA
3	1	1	1	1	0	0	1	0	F2
4	0	1	1	0	0	1	1	0	66
5	1	0	1	1	0	1	1	0	B6
6	1	0	1	1	1	1	1	0	BE
7	1	1	1	0	0	0	0	0	E0
8	1	1	1	1	1	1	1	0	FE
9	1	1	1	1	0	1	1	0	F6
A	1	1	1	0	1	1	1	0	EE
b	0	0	1	1	1	1	1	0	3E
C	1	0	0	1	1	1	0	0	9C
d	0	1	1	1	1	0	1	0	7A
E	1	0	0	1	1	1	1	0	9E
F	1	0	0	0	1	1	1	0	8E

Figure 9.21 Illustration for Example 9.3.

on two 7-segment displays. Recall from Chapter 4 that the display of a 0 on a 7-segment display, such as the one shown in Fig. 9.21(a), requires the following signals:

$$a = 1 \text{ (true)},\ b = 1,\ c = 1,\ d = 1,\ e = 1,\ f = 1,\ g = 0,\ dp = 0$$

In other words, the bit pattern for a "0" on a 7-segment display is 11111100. Recall further that the bit pattern for the display of a "1" is 01100000, and so forth. Figure 9.21(a) gives the bit patterns for the displays of all the hex digits.

For this example, assume that the bits for the two hex digits to be displayed are stored in Registers B and C, as indicated in Fig. 9.21(b). The function of the subroutine is to convert these two hex digits into two bit patterns through a table-lookup technique and to store these bit patterns in Registers D and E, as shown in Fig. 9.21(b). The problem specification of this example is summarized in more detail in the initial comment section of the assembly program given in Fig. 9.22(a).

```
            ; Subroutine Name: CONV7
            ; Input to subroutine:
            ;       High-order hex digit stored in (B3–B0) of Register B,
            ;           (B7–B4) should be 0.
            ;       Low-order hex digit stored in (C3–C0) of Register C,
            ;           (C7–C4) should be 0.
            ;
            ; Output from the subroutine:
            ;       Bit pattern for the 7-segment display for high-order hex digit
            ;           in Register D.
            ;       Bit pattern for the 7-segment display for low-order hex digit
            ;           in Register E.
            ;
            ; Function:
            ;       This subroutine will perform a table lookup to obtain the
            ;       corresponding bit pattern for the 7-segment display for the
            ;       two hex digits stored in Registers B and C. The actual value
            ;       of the hex digit is used as an offset into the table.
            ;
            ; Registers that will be changed:
            ;       D, E, A, H, L
            ;
            ORG    0700H     ; The lookup table is stored beginning at 0700H.
 1  TABL:   DB     0FCH      ; Bit pattern for displaying 0.
 2          DB     60H       ; for 1
 3          DB     0DAH      ; for 2
 4          DB     0F2H      ; for 3
 5          DB     66H       ; for 4
 6          DB     0B6H      ; for 5
 7          DB     0BEH      ; for 6
 8          DB     0E0H      ; for 7
 9          DB     0FEH      ; for 8
10          DB     0F6H      ; for 9
11          DB     0EEH      ; for A
12          DB     3EH       ; for B
13          DB     9CH       ; for C
14          DB     7AH       ; for D
15          DB     9EH       ; for E
16          DB     8EH       ; for F
17          ORG    2000H     ; The subroutine is stored beginning at 2000H.
18  CONV7:  LXI    H,TABL    ; Move the address of TABL (0700H) into HL.
19          MOV    A,L       ; Move the low-order byte of the address of TABL (00H)
                             ; into Register A.
20          ADD    B         ; Add the offset, which is the actual value of the high-order
                             ; hex digit.
21          MOV    L,A       ; Now HL contains 0700H + XXH, where XXH is the
                             ; value of the hex digit.
22          MOV    D,M       ; With indirect addressing, move into Register D the bit
                             ; pattern corresponding to the high-order hex digit.
23          LXI    H,TABL    ; Reinitialize the address back to 0700H.
24          MOV    A,L       ; Set up to add offset again for low-order digit.
25          ADD    C         ; Add low-order digit value as an offset.
26          MOV    L,A       ; Now HL points to corresponding entry for low-order digit.
27          MOV    E,M       ; Bit pattern for low-order digit moves into Register E.
28          RET              ; Return with bit patterns in Registers D and E.
```

(a) Assembly program for Example 9.3

Figure 9.22 Assembly program and sample execution for Example 9.3.

Statement number	Reg. A	Reg. F	Reg. B	Reg. C	Reg. D	Reg. E	Reg. H	Reg. L
Initial values	XX	XX	07	0D	XX	XX	XX	XX
18	↓						07	00
19	00	↓						
20	07	00					↓	↓
21					↓		07	07
22					E0		↓	↓
23	↓						07	00
24	00	↓						
25	0D	00					↓	↓
26						↓	07	0D
27	↓	↓	↓	↓	↓	7A	↓	↓
28	0D	00	07	0D	E0	7A	07	0D

(b) Execution for displaying 7DH. All values are in hexadecimal. Also, XX is don't care.

Figure 9.22 (*cont.*)

In the program of Fig. 9.22(a), note the use of the pseudo-instruction DB (Define Byte). Remember that a pseudo-instruction just provides additional information about the program but is not executed by the microprocessor. The DB pseudo-instruction instructs the *assembler* to put the specified byte value into the specified memory location. For this program, the assembler puts the byte value FCH into Location 0700H, the byte value 60H into Location 0701H, and so forth, and last the byte value 8EH into Location 070FH. This is done at *program assembly time* by the assembler and not at execution time by the microprocessor. We will consider the concept of assembler and program assembly in detail in Chapter 10. At this point of our discussion, let us view a program assembly simply as a step that is required prior to the execution step in order to get the program into a form suitable for execution by the microprocessor.

In the specifications of the byte values, some numbers have leading zeros. Each of these numbers would otherwise have a letter of the alphabet for the first digit. Most commercial assemblers will not recognize a number that does not have a leading decimal digit. We get around this problem by using leading zeros. They do not, of course, alter the values of the numbers.

Subroutine CONV7 of Fig. 9.22(a) illustrates a useful programming technique for code conversion. The lookup table TABL is arranged such that the actual hex digit value is suitable for the offset for entry into the table. Specifically, the 7-segment bit pattern for the hex digit 0 is stored at Memory Location TABL + 0, that for hex digit 1 at TABL + 1, and so forth, with hex digit F stored at TABL + F(hex). Consequently, we can calculate the memory address for the required table entry by adding the value of the hex digit (as an offset) to TABL. Then, using indirect addressing, we can easily obtain the required bit pattern, as illustrated in Fig. 9.22(b) by the sample execution for displaying 7DH.

For the foregoing program, TABL is Memory Location 0700H. Note that the program in Fig. 9.22(a) will not work if TABL is defined to be a memory location such as 06FFH. Why? If TABL is defined as 06FFH, what changes are required for the program? (See Problem 9.14 at the end of this chapter.) ■■

Example 9.4 Introduction to Program Looping For the final example of this section we will write a subroutine program, called INSUB, for inputting 10 bytes of data from Input Port 07H, one at a time, and storing them in ten consecutive memory locations. The first byte is to be stored in a memory location that is EQUated to TABL, the second byte is to be stored in Location TABL + 1, and so forth, with the tenth byte to be stored in Location TABL + 9. We will assume that TABL is already EQUated to a memory location in the calling program, and that a new byte of data is available in Input Port 07H whenever an IN instruction is executed.

A straightforward but undesirable solution to this problem using indirect addressing is outlined in Fig. 9.23. The program is self-explanatory.

A more elegant solution is shown in Fig. 9.24. This program has a loop structure that causes a set of instructions to be executed repeatedly. With this looping, this one set of instructions serves the same function as the ten duplicated sets in the program of Fig. 9.23. A suitable algorithm for this program-loop approach is represented by the flowchart of Fig. 9.24(a), and the corresponding assembly language program is in Fig. 9.24(b). The key instruction is the JNZ (Jump Not Zero) instruction of Statement 7. Execution of this JNZ instruction causes either a jump back to Statement 3 or a continuation on to Statement 8, depending on the value of the Z flag. If the Z Flag is false (0), the instruction of Statement 3 is executed next. But if the Z Flag is true (1), then the instruction of Statement 8 is executed next. The Z Flag value depends on the counter

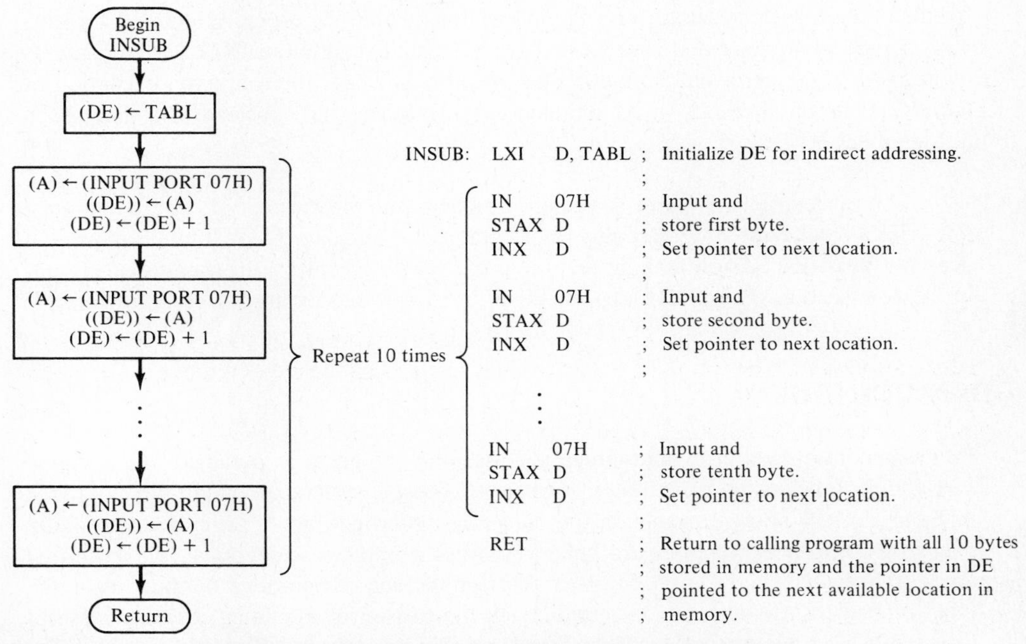

Figure 9.23 A straightforward but undesirable solution for Subroutine INSUB.

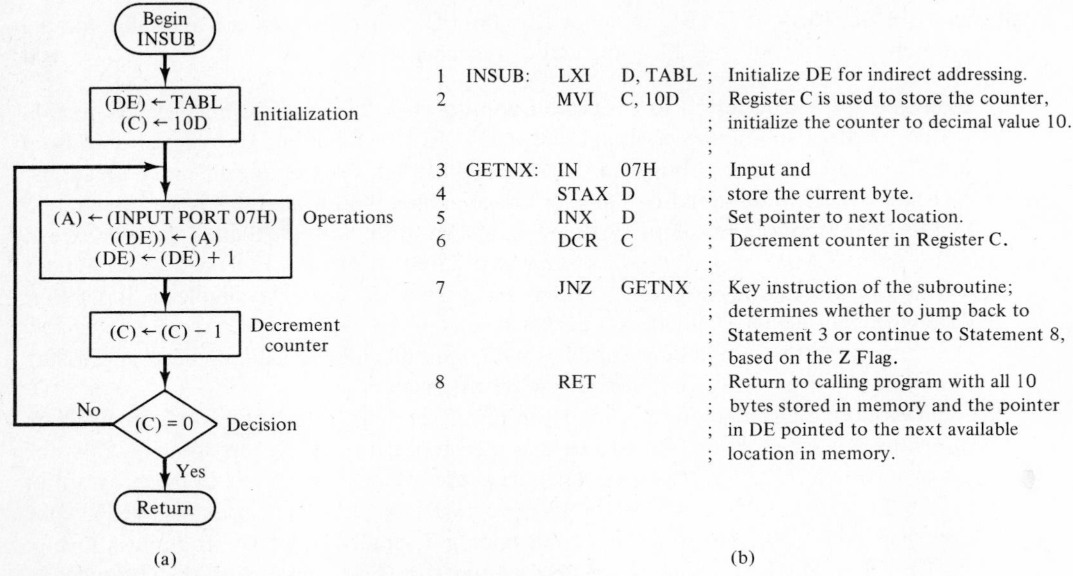

Figure 9.24 Use of a program loop for Subroutine INSUB.

stored in Register C. This counter is set to 10D initially (Statement 2) and then decremented each time instruction DCR C of Statement 6 is executed, which is the instruction immediately preceding the JNZ instruction. Since ten executions of DCR C are required for the counter to decrease to zero, the instructions within the program loop (Statements 3, 4, 5, and 6) are executed ten times. After the tenth execution of DCR C, the contents of Register C are zero, and so the Z Flag is set to 1. Then, the following execution of the JNZ instruction results in a continuation on to Statement 8 instead of a jump back to Statement 3. ■ ■

The preceding example was a brief introduction to the concept of *program control* in assembly language programming. A programming language is of limited value if it does not provide decision making and the corresponding control of program execution. In the next section we will consider the 8085 instructions necessary for program control.

9.6 PROGRAM CONTROL

Program control within a microprocessor assembly program is obtained with *branch instructions* (also called branching instructions). These instructions, one of which is JNZ of Example 9.4, allow a change in the sequential order in which a program is normally executed. Branch instructions are generally divided into two categories: *unconditional* and *conditional*. This section contains descriptions and illustrations of the use of the 8085 branch instructions. Also presented are comparison instructions, which are useful for establishing the desired conditions for controlling conditional branching.

9.6.1 Branch Instructions

With an unconditional branch, the program control is transferred to the instruction whose address is specified in the branch instruction. In other words, the normal sequence of program execution is interrupted and program control is transferred to another part of the program. In the 8085 the JMP (jump) instruction gives an unconditional branch. This instruction is useful for ''jumping over'' program segments that are not to be executed at that time. Figure 9.25 contains the general format, description, and an example of the JMP instruction.

In the ''nonsense'' program segment of Fig. 9.25(c), what happens is that first Register A is initialized to FFH. Next, the value in Register A is outputted to Output Port 03H. Then, this Register A value is decremented by 1. The outputting, decrementing, and jumping occur indefinitely. Statement 5 will never be executed because each time the JMP instruction of Statement 4 is executed, the program control transfers back to Statement 2.

With a conditional branch instruction, the program control may or may not be transferred to the instruction whose address is specified in the branch instruction. The transfer decision depends on the current value of one of the conditional flags; this value, in turn, depends on the result of a previous instruction. Branching occurs only if the condition specified by the branch instruction is true. If the branch condition is false, the microprocessor executes the next instruction in sequence.

In the 8085, the four flags that can be tested by a conditional branch instruction are the Zero (Z) Flag, the Carry (CY) Flag, the Sign (S) Flag, and the Parity (P) Flag. Since each of these flags can be either true or false, there can be (and there are) eight conditional branch or jump instructions, as illustrated in Fig. 9.26. Note that the conditions specified are for prior arithmetic or other data transformation results and not flag values—although, of course, the flag values depend upon such results. In particular, with the execution of JZ there is a jump if the Zero Flag is 1 (and not 0). The ''zero'' in the

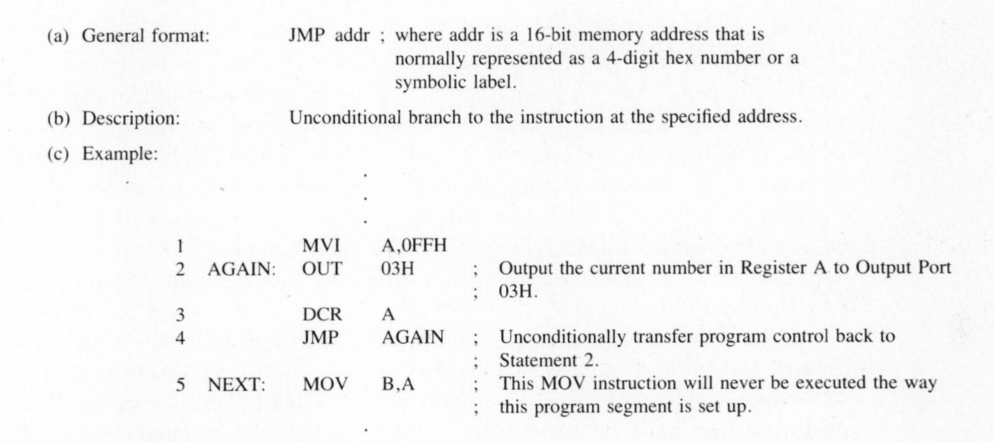

(a) General format: JMP addr ; where addr is a 16-bit memory address that is normally represented as a 4-digit hex number or a symbolic label.

(b) Description: Unconditional branch to the instruction at the specified address.

(c) Example:

```
                  .
                  .
                  .
1          MVI   A,0FFH
2 AGAIN:   OUT   03H      ; Output the current number in Register A to Output Port
                          ; 03H.
3          DCR   A
4          JMP   AGAIN    ; Unconditionally transfer program control back to
                          ; Statement 2.
5 NEXT:    MOV   B,A      ; This MOV instruction will never be executed the way
                          ; this program segment is set up.
                  .
                  .
                  .
```

Figure 9.25 JMP instruction.

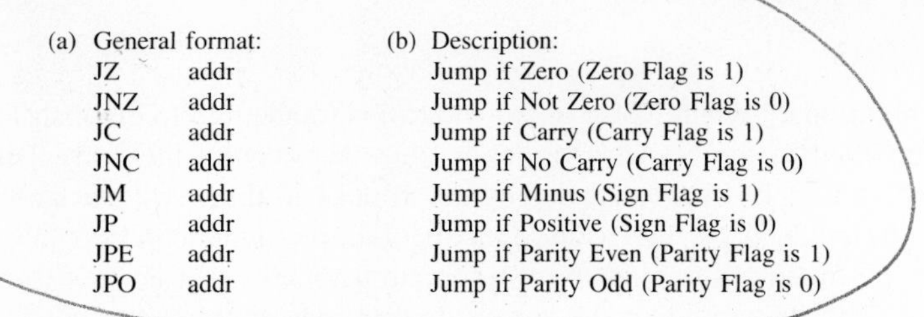

(a) General format:

JZ	addr
JNZ	addr
JC	addr
JNC	addr
JM	addr
JP	addr
JPE	addr
JPO	addr

(b) Description:

Jump if Zero (Zero Flag is 1)
Jump if Not Zero (Zero Flag is 0)
Jump if Carry (Carry Flag is 1)
Jump if No Carry (Carry Flag is 0)
Jump if Minus (Sign Flag is 1)
Jump if Positive (Sign Flag is 0)
Jump if Parity Even (Parity Flag is 1)
Jump if Parity Odd (Parity Flag is 0)

Where addr is a 16-bit memory address that is normally represented as a 4-digit hex number or a symbolic label.

(c) Example:

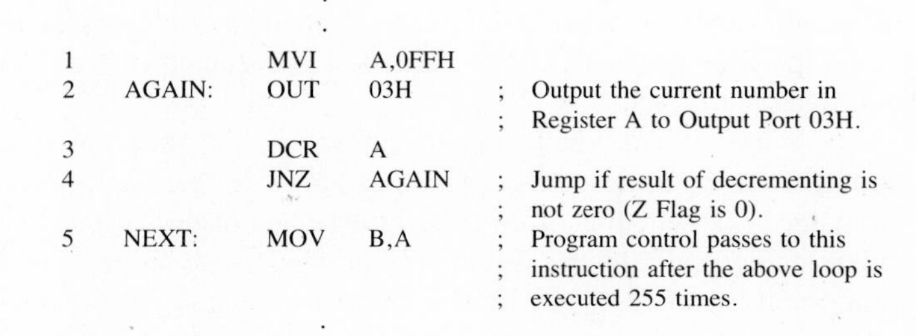

```
                 .
                 .
                 .
1            MVI   A,0FFH
2  AGAIN:    OUT   03H       ; Output the current number in
                             ; Register A to Output Port 03H.
3            DCR   A
4            JNZ   AGAIN     ; Jump if result of decrementing is
                             ; not zero (Z Flag is 0).
5  NEXT:     MOV   B,A       ; Program control passes to this
                             ; instruction after the above loop is
                             ; executed 255 times.
                 .
                 .
                 .
```

Figure 9.26 Conditional jump instructions.

JZ description refers to a prior data transformation result that can cause the Zero Flag to set to 1. Similarly, with JNZ there is a jump if the result is nonzero (Z Flag is 0).

We can correct the programming example of Fig. 9.25 by replacing the unconditional jump instruction (JMP) with a conditional jump instruction (JNZ), as shown in Fig. 9.26(c). In the execution of this program segment, Register A is initialized to FFH (255 decimal). Next the contents of Register A are outputted to Output Port 03H. Then, the contents of Register A are decremented by 1. As long as the contents are not zero (and consequently the Z Flag is not 1), the JNZ instruction of Statement 4 transfers program control back to Statement 2. Unlike the example of Fig. 9.25(c), however, this program will eventually proceed to execute Statement 5. Observe that with each pass through the program loop the execution of Statement 3 decrements by 1 the value in Register A. After the 255th time that this happens, the contents of Register A become zero, and consequently the Z Flag is set to 1. Then when the JNZ instruction of Statement 4 is executed, program control does *not* transfer to Statement 2, but instead proceeds sequentially to Statement 5.

The other jump instructions in Fig. 9.26 can be used in a similar manner. At the time of execution of a jump instruction the decision to jump or not to jump depends on the value of the corresponding condition flag. For the decision to be meaningful the condition flag must be set or cleared by a previously executed instruction that was intentionally included to control the flag value, and that has the capability of doing this. As we have seen, not all 8085 instructions affect the condition flags. Consequently, at the time of the branching decision the instruction that determines the value of the corresponding flag may or may not be the instruction that immediately precedes the conditional jump instruction. Thus a programmer must take care with the selection and placement of instructions that precede conditional jump instructions. Some of the common pitfalls in the use of the 8085 conditional jump instructions are given in Fig. 9.27.

(a)

```
                  .
                  .
                  .
1             MVI A,01H
2             MVI C,01H
3             DCR A          ; Contents of A = 00H, Z Flag = 1.
4             INR C          ; Contents of C = 02H, Z Flag = 0.
5             JZ   LABLX     ; At the execution of this instruction, Z Flag is 0.
6    LABLY:
                  .
                  .
                  .
```

(b)

```
                  .
                  .
                  .
1             MVI A,07H
2             DCR A          ; Contents of A = 06H, Z Flag = 0.
3             MVI A,00H      ; Contents of A = 00H, no effect on flags.
4             JZ   LABLX     ; At the execution of this instruction, Z Flag is 0.
5    LABLY:
                  .
                  .
                  .
```

(c)

```
                  .
                  .
                  .
              ; Assume CY Flag is 0
              at this point.
              ;
1             MVI B,0FFH
2             INR B          ; Contents of B = 00H, but no effect on CY Flag.
3             JC   LABLX     ; At the execution of this instruction, CY was 0 since INR
                             ; instruction had no effect on the CY Flag.
4    LABLY:
                  .
                  .
                  .
```

(d)

```
                  .
                  .
                  .
1             MVI D,07H
2             DCR D          ; Contents of D = 06H, and Z Flag = 0.
3             LXI  B,0001H   ;
4             DCX B          ; Contents of BC become 0000H, but no effect on Z Flag.
5             JZ   LABLX     ; At the execution of this instruction, Z Flag = 0.
6    LABLY:
                  .
                  .
                  .
```

Figure 9.27 Common pitfalls in the use of the 8085 branch instructions.

The program segment of Fig. 9.27(a) relates to a common misbelief that the JZ instruction is controlled by the contents of Register A instead of the value of the Z Flag. With this belief, one would erroneously conclude that the execution of the JZ instruction of Statement 5 causes a jump to LABLX (not shown) since the contents of Register A are zero then. It is true, of course, that these contents are zero as a result of the execution of the DCR instruction of Statement 3. However, the execution of the next instruction INR C of Statement 4 causes the Z Flag to change from 1 to 0, although it does not affect the contents of Register A. Consequently, when Statement 5 is executed, there is no jump, and the program control continues sequentially to LABLY.

The pitfall of the program segment of Fig. 9.27(b) is the failure to remember which instructions affect the Z Flag. Here, presumably, a jump is desired when the JZ instruction of Statement 4 is executed. And although the execution of the MVI instruction of the preceding Statement 3 does put a zero in Register A, this instruction does not affect any flag. The most recently executed instruction *that affects the Z Flag* is DCR of Statement 2, which causes the Z Flag to become false (0). Therefore, although the contents of Register A are made zero immediately before the JZ instruction is executed, the Z Flag is still false (0), so there is no jump, and program control continues sequentially to LABLY.

The pitfall of the program segment of Fig. 9.27(c) is the failure to recall which flags are affected by the execution of instructions that affect the flags. Here, a jump is desired when the JC instruction is executed. Note that relative to the JC instruction, the most recently executed instruction that affects the flags is the INR instruction of Statement 2. However, as specified in Table 9.2, this instruction affects all the condition flags *except* the CY Flag. Therefore, even though the result of the execution of the INR instruction appears to cause an overflow into the CY Flag, it actually does not. Since the CY Flag remains 0, there is no jump, and the program control continues to LABLY.

The pitfall of the program segment of Fig. 9.27(d) centers around the difference in the effect on the flags that the two types of decrement instructions have. Here a jump is desired when the JZ instruction of Statement 5 is executed. And since the DCX instruction of Statement 4 causes the contents of Register Pair BC to become 0, it would appear that the Z Flag will be set to 1 with the execution of Statement 4. But as shown in Table 9.2, the DCX instruction does not affect any condition flags, although the DCR decrement instruction does. Therefore, the most recently executed instruction that affects the Z Flag is the Statement 2 DCR instruction, which causes the Z Flag to clear to 0. Therefore, upon execution of the JZ instruction there is no jump, and program control just continues sequentially on to LABLY.

9.6.2 Comparison Instructions

The examples of Fig. 9.27 illustrate the point that to use the conditional branch instructions a programmer must know precisely the effect that each instruction has on the condition flags. One group of instructions that are used exclusively for affecting the condition flags are the comparison instructions. The general format, description, and examples of these comparison instructions are given in Fig. 9.28. Generally speaking, the CMP and CPI comparison instructions have exactly the same effect on the condition flags as the corresponding subtraction instructions. Specifically, in the execution of a

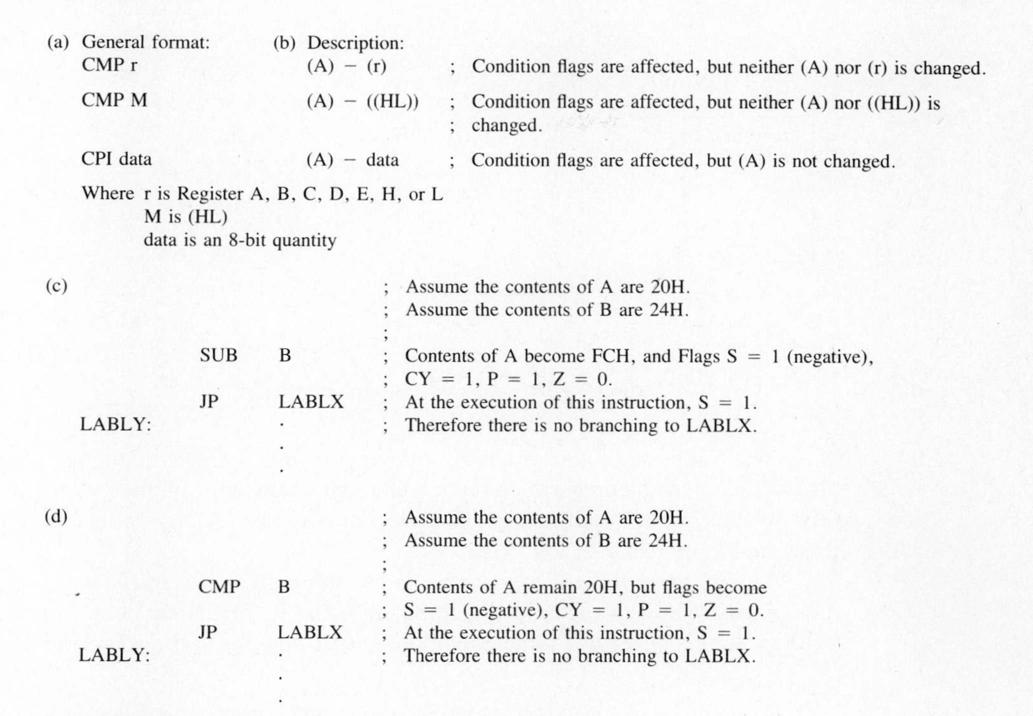

Figure 9.28 Comparison instructions CMP and CPI.

CMP or CPI instruction, all flag register bits are set exactly as if a SUB or SBI instruction had been executed. But this is the only effect, and the result of the subtraction is not stored anywhere. Therefore, unlike the case of the execution of the SUB instruction, there is no change in any register contents; specifically, the contents of Register A remain unchanged.

The purpose of using CMP for comparing two numbers is usually for setting the condition flags such that a jump will or will not occur, depending on which number is larger when a following conditional jump instruction is executed. Preceding each comparison instruction, one of the numbers must be, of course, in Register A, perhaps as a result of an arithmetic operation. The other number can be the contents of any other register or of any memory location, if the CMP instruction is used. Or, the second number can be carried by the CPI instruction. We will consider these instructions more in the next section.

9.7 PROGRAM STRUCTURES

The instructions presented in the preceding section are essential for the construction of the *program structures* that are the foundation of every computer program. Program structures can be classified into three general structural forms: (1) Sequence, (2) Selection, and (3) Repetition.

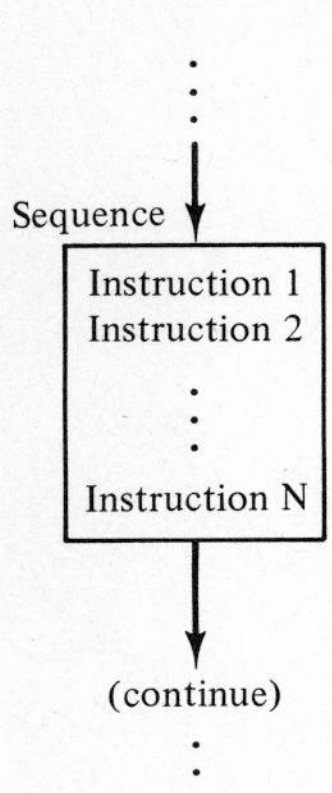

Figure 9.29 Sequence program structure.

The simplest of the program structures is the *Sequence* structure, which is the unconditional, sequential execution of a block of instruction. In other words, the program flow of execution within a Sequence program structure is sequential. This structure is illustrated graphically in Fig. 9.29.

The program flow of execution can be altered by using a *Selection* program structure. A basic form of the Selection structure is the IF-THEN structure illustrated in Fig. 9.30(a). In this structure, *if* the condition is true, *then* the specified instructions in block

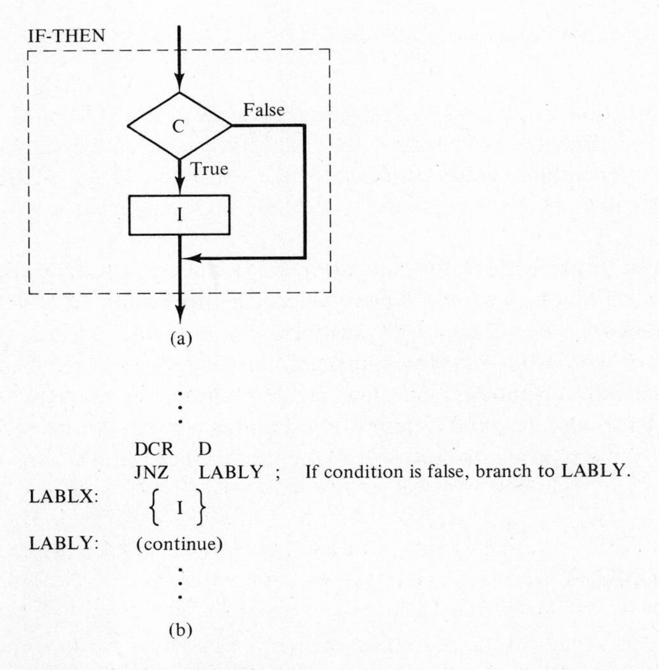

Where C is the test condition
I is a block of instructions

Figure 9.30 IF-THEN Selection program structure.

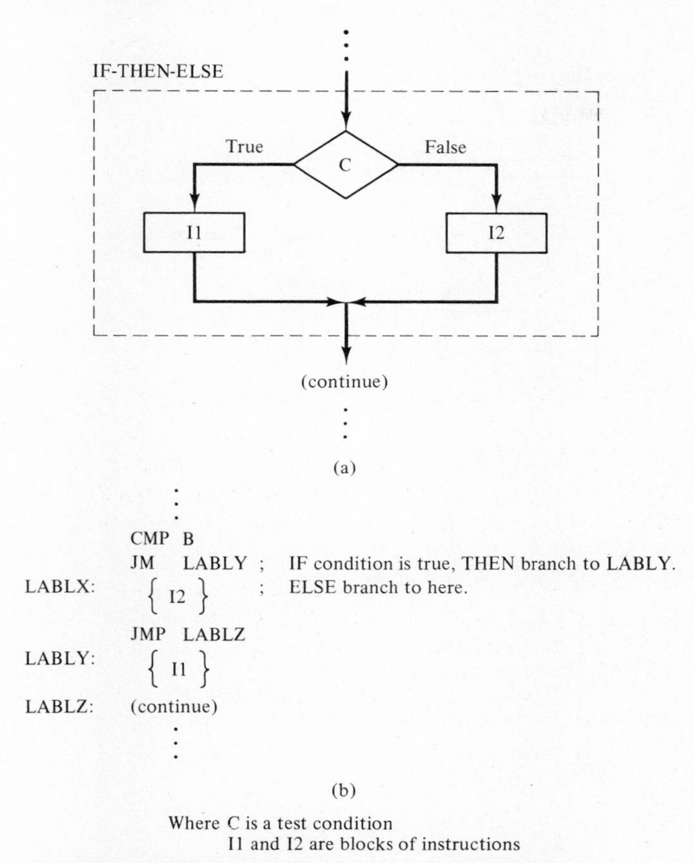

Figure 9.31 IF-THEN-ELSE Selection program structure.

I are executed. But *if* the condition is false, *then* the specified instructions in block I are skipped. Figure 9.30(b) illustrates the typical pattern of instructions required to realize the IF-THEN structure with 8085 assembly language instructions. In this example, the IF-THEN structure is obtained through the use of the Z Flag and the JNZ instruction. Of course, any of the four condition flags and eight conditional jump instructions can be used. Furthermore, a ''compound'' condition can be obtained by using several condition flags and conditional jump instructions.

Another form of the Selection program structure is the IF-THEN-ELSE structure illustrated in Fig. 9.31(a). This is a variation of the IF-THEN structure. In the IF-THEN-ELSE structure, *if* the condition is true, *then* the instructions in block I1 are executed. But if the condition is false (*else*), then the instructions in block I2 are executed. A typical pattern of instructions required to realize the IF-THEN-ELSE structure is given in Fig. 9.31(b).

The *Repetition* program structure is a program loop structure in which a block of instructions are repeatedly executed, with the number of repetitions determined by some test condition(s). As illustrated in Figs. 9.32(a) and 9.33(a), the Repetition structure can

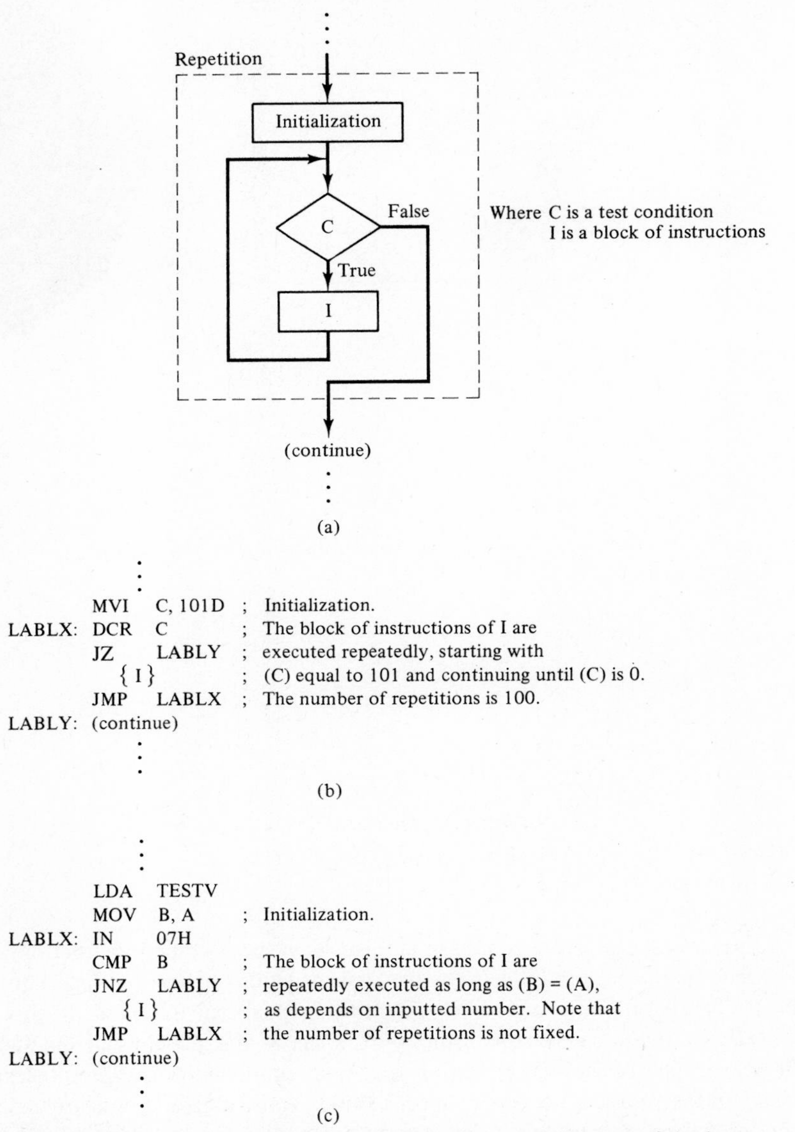

Figure 9.32 Repetition program structure with test at the start of the loop.

be realized in two ways. For the Repetition structure of Fig. 9.32(a), the condition is tested first at the *beginning* of the program loop. If it is true, then the instructions of block I are executed. Otherwise, they are not. As long as the condition is true at the time of each test, these instructions will be repeatedly executed. Observe, though, that the value(s) of the condition flag(s) must be modified by some instruction(s) within the program loop at some time. Otherwise, an infinite loop will result. This Repetition structure is commonly referred to as a ''WHILE'' structure or a ''FOR'' structure. Two

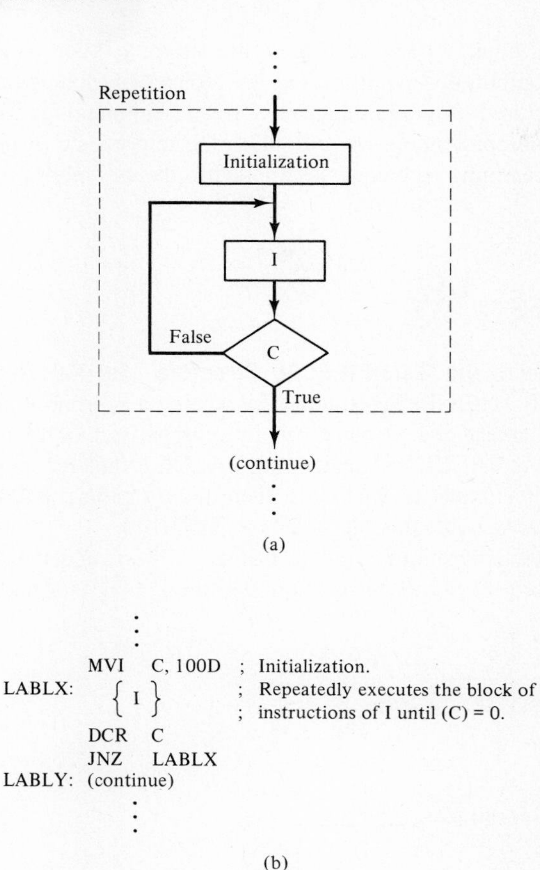

Where C is a test condition
I is a block of instructions

Figure 9.33 Repetition program structure with test at the end of the loop.

typical patterns of instructions that will produce this Repetition structure are given in Figs. 9.32(b) and (c).

In the Repetition structure illustrated in Fig. 9.33(a), the condition is tested at the *end* of the program loop. In other words, the block of instructions are executed first, then a test is performed to determine whether another iteration through the loop is required. This structure is commonly referred to as a ''REPEAT-UNTIL'' structure. A typical pattern of instructions for realizing this Repetition structure is given in Fig. 9.33(b).

The program structures of this section are the fundamental building blocks of a computer program. Every program, however simple or complex, is constructed with some combination and variation of these program structure building blocks. The manner in which these building blocks are put together is the *art of computer programming,* a subject which has been only briefly considered. A rigorous treatment of this subject is beyond the scope of this book. Interested readers are strongly urged to become better educated in the art of programming with the study of topics that include data structures, program structures, structured programming, and software engineering. Throughout the following chapters on microprocessors we will stress the importance of good programming practices and programming techniques as applied to the assembly programming of microprocessors.

9.8 PROGRAMMING EXAMPLES

Example 9.5 Program with the IF-THEN-ELSE Structure In this example we will illustrate the use of the IF-THEN-ELSE structure by writing a subroutine that outputs to Output Port 01H a bit pattern that depends on the comparison of two 8-bit values: VALUE1 and VALUE2. If VALUE1 is greater than VALUE2, then the output bit pattern is 00000001. If VALUE1 is equal to VALUE2, then this bit pattern is 00000010. And if VALUE1 is less than VALUE2, this bit pattern is 00000100.

An algorithm for this subroutine is given in Fig. 9.34(a), and the corresponding assembly language program in Fig. 9.34(b). Note that the Z Flag is tested to determine

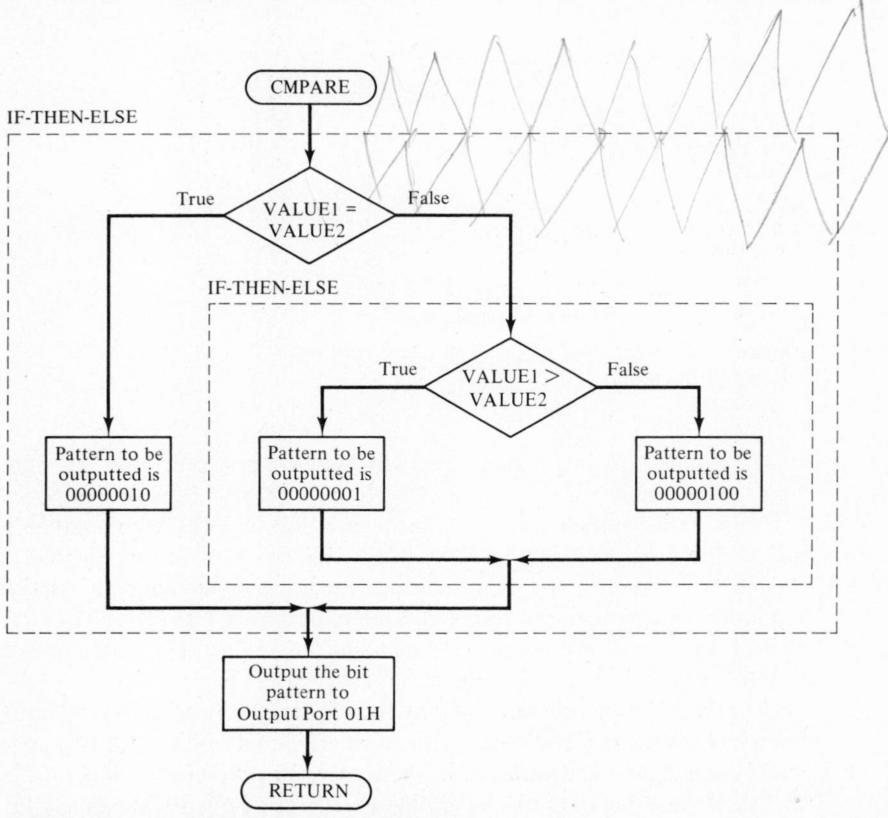

Figure 9.34(a) Algorithm for Example 9.5.

```
                  ; Subroutine Name: CMPARE
                  ; Input to subroutine:
                  ;      Register B - VALUE1, an 8-bit binary number
                  ;      Register C - VALUE2, an 8-bit binary number
                  ;
                  ; Output from subroutine:
                  ;      Status byte (8-bit pattern) to Output Port 01H
                  ;
                  ; Description:
                  ;      This subroutine will compare VALUE1 and VALUE2.
                  ;      If VALUE1 > VALUE2, then bit pattern 00000001 is outputted
                  ;         to Output Port 01H.
                  ;      If VALUE1 = VALUE2, then bit pattern 00000010 is outputted
                  ;         to Output Port 01H.
                  ;      If VALUE1 < VALUE2, then bit pattern 00000100 is outputted
                  ;         to Output Port 01H.
                  ;
                  ; Registers that may be altered
                  ;     A, F
                  ;
1                 ORG   2000H    ; The subroutine is stored beginning at 2000H.
2   CMPARE:       MOV   A, B     ; Move VALUE1 to Register A.
3                 CMP   C        ; Compare VALUE1 to VALUE2; i.e., "VALUE1 - VALUE2."
4                 JZ    EQUAL    ; If Z Flag is true, then VALUE1 = VALUE2.
                                 ; Therefore, branch to EQUAL.
5   NOTEQ:        JC    LESSTH   ; If not equal, then if CY Flag is true, then VALUE1
                                 ; is less than VALUE2. Therefore, branch to LESSTH.
6   GREATR:       MVI   A, 01H   ; At this point we know that VALUE1 > VALUE2. Therefore,
                                 ; the bit pattern to be outputted is 00000001.
7                 JMP   OUTPT    ;
                                 ;
8   LESSTH:       MVI   A, 04H   ; For VALUE1 less than VALUE2, the bit pattern to be
                                 ; outputted is 00000100.
9                 JMP   OUTPT    ;
                                 ;
10  EQUAL:        MVI   A, 02H   ; For VALUE1 = VALUE2, the bit pattern to be outputted
                                 ; is 00000010.
11  OUTPT:        OUT   01H      ;
12                RET            ;
```

(b)

Figure 9.34(b) Assembly program for Example 9.5.

whether VALUE1 is equal to VALUE2 (Statement 4), and the CY Flag is tested to determine whether VALUE1 is less than VALUE2 (Statement 5). Both the Z and CY Flags are set by the same CMP instruction of Statement 3. Note also that an IF-THEN-ELSE structure can be embedded in other program structures. In this manner, an overall program of any complexity can be realized. ■ ■

Example 9.6 Overlapping Program Structures This example illustrates the case in which two program structures overlap. The function of this subroutine is to search a table stored in memory, the search being made with a test data byte previously stored in Register B. If the test value is found in the table, then the CY Flag is set to 1 and the address of the found value is stored in Register Pair DE. If the test value occurs more than once in the table, the subroutine detects only the first one found. If the test value is not found, then the test value is added to the end of the table and the CY Flag is cleared to 0. A more detailed specification of the subroutine is given in the initial comment section of the assembly program of Fig. 9.35(b). The corresponding algorithm is given in Fig. 9.35(a).

Note that in the algorithm of Fig. 9.35(a), the table is searched with a Repetition structure, or program loop, that can be terminated in one of two ways:

1. The end of the table is reached without the test byte having been found. This is the normal terminal of the Repetition structure.
2. The test byte is found to be equal to the current entry from the table. In this case there is branching out of the program loop because of the IF-THEN-ELSE structure.

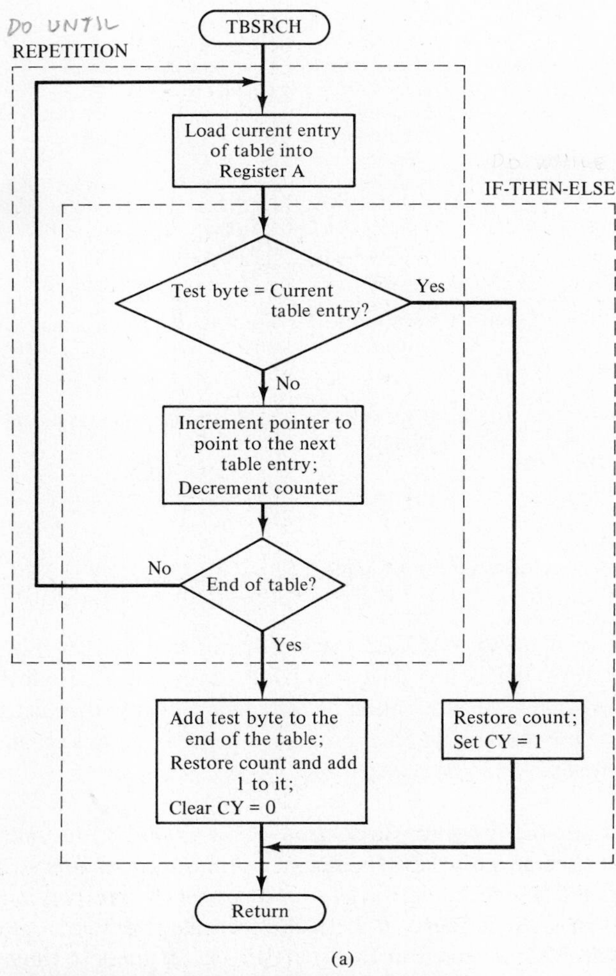

(a)

Figure 9.35(a) Algorithm for Example 9.6.

```
                        ; Subroutine Name: TBSRCH
                        ; Input to subroutine:
                        ;       Register B - the number to be compared: the test data byte.
                        ;       Register Pair DE - the address of the first entry of the table.
                        ;       Register C - the number of entries currently in the table.
                        ;
                        ; Output from subroutine:
                        ;       If the entry is found in the table, set
                        ;          CY to 1.
                        ;          Register Pair DE contains the address of the entry found in
                        ;          the table. Register C remains the same.
                        ;       If the entry is not found in the table, clear
                        ;          CY to 0.
                        ;          Register Pair DE contains the address of the added entry.
                        ;          Register C is equal to the old Register C plus 1.
                        ;
                        ; Description:
                        ;       Given a test data byte in Register B, this subroutine uses
                        ;       this value to search a table. If the test value is found
                        ;       in the table, the CY Flag is set to 1. If the test value is
                        ;       not found in the table, then the test byte is added to the
                        ;       end of table and CY is cleared to 0. Furthermore, the
                        ;       count in Register C is incremented by 1.
                        ;
                        ; Registers that may be altered:
                        ;       A, F, C, D, E, L
                        ;
 1               ORG    2000H    ; The subroutine is stored beginning at 2000H.
 2  TBSRCH:      MOV    L, C     ; Save original count in Register L.
 3  REPEAT:      LDAX   D        ; Load current entry of the table into Register A.
 4               CMP    B        ; Compare with the test data byte.
 5               JZ     FOUND    ; If (A) = (B), then branch to FOUND.
                                 ; Else,
 6  NOTFD:       INX    D        ; branch to here and set up for the next entry in table.
 7               DCR    C        ; Decrement count by 1.
 8               JNZ    REPEAT   ; If not yet processed all the entries in the table,
                                 ; then branch back to REPEAT.
 9  ENDTB:       MOV    A, B     ; Else, we've found the end of the table, therefore
10               STAX   D        ; add test byte to the end of the table.
11               MOV    C, L     ; Restore the original count.
12               INR    C        ; Increment count by 1.
13               STC             ; Set CY flag to 1.
14               CMC             ; Complement the CY Flag (effectively clear CY to 0).
15               JMP    FNSRCH   ;
                                 ;
16  FOUND:       MOV    C, L     ; Restore the original count.
17               STC             ; Set CY Flag to 1.
18  FNSRCH:      RET             ;
```

(b)

Figure 9.35(b) Assembly program for Example 9.6.

Because of the two ways of terminating the program loop, there is an overlapping of the Repetition and the IF-THEN-ELSE structures. Overlapping is generally not considered good programming practice, especially in a high-level programming language such as PASCAL or FORTRAN. It is, however, sometimes necessary in assembly language programming because assembly language instructions do not always support structured programming. Moreover, the forcing of nonoverlapping structures in assembly language programming can sometimes decrease the clarity of a program. Yet, it is still good programming practice to avoid overlapping program structures whenever reasonably possible. ■■

Example 9.7 Nested Program Loops and a Delay Loop This example illustrates a subroutine that has a loop inside a loop (nested loops). Additionally, the sole function of the interior loop is to produce a fixed time delay. The function of the subroutine is to output 80 data bytes from memory to Output Port 03H. The data bytes are outputted one

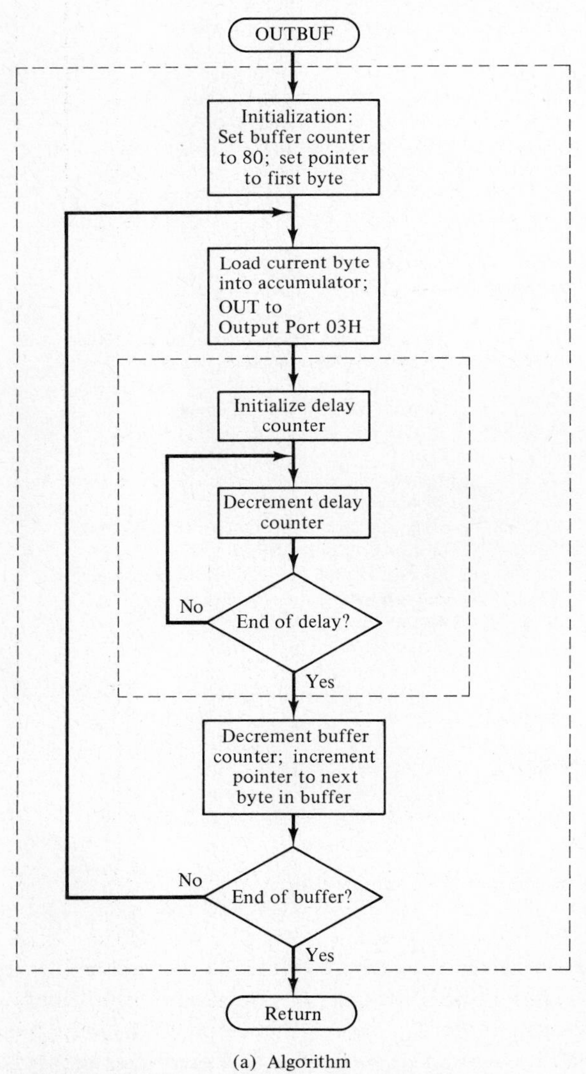

(a) Algorithm

Figure 9.36(a) Algorithm for Example 9.7.

```
                ; Subroutine Name: OUTBUF
                ; Input to subroutine:
                ;       BUFFER has been EQUated in the calling program to the
                ;       location of the beginning of an 80-byte table.
                ;
                ; Output from subroutine:
                ;       The 80 bytes in BUFFER are outputted to Output Port 03H,
                ;       one at a time, at approximately 1-millisecond intervals.
                ;
                ; Description:
                ;       This subroutine will output 80 bytes of data, stored
                ;       beginning at BUFFER, one at a time, to Output Port 03H.
                ;       There will be an approximately 1-millisecond delay between
                ;       each byte output.
                ;
                ; Registers that may be altered:
                ;       A, B, C, D, E, F
                ;
 1              ORG     0700H
 2  OUTBUF:     MVI     C, 80D     ; Initialize buffer counter to 80.
 3              LXI     D, BUFFER  ; Set pointer to the first byte in the buffer.
                                   ;
 4  NXBYTE:     LDAX    D          ; Load current byte into accumulator (Register A).
 5              OUT     03H        ;
                                   ;
 6              MVI     B, 220D    ; Set delay counter (to equivalent to 1 ms).
                                   ; This instruction requires 7 states to execute.
 7  DELAY:      DCR     B          ; Decrement delay counter (requires 4 states).
 8              JNZ     DELAY      ; If not yet 1 ms, then branch back to DELAY
                                   ; (requires 10 states).
 9              DCR     C          ; Decrement buffer counter.
10              INX     D          ; Increment pointer to next byte in BUFFER.
                                   ;
11              JNZ     NXBYTE     ; If not end of buffer, branch back to NXBYTE.
                                   ;
12              RET                ; If end of buffer, return.
```

(b) Assembly program

Figure 9.36(b) Assembly program for Example 9.7.

at a time, with approximately a 1-ms (one-millisecond) delay between successive byte outputs. A more detailed specification of the subroutine is given in the initial comment section of the assembly program of Fig. 9.36(b). The corresponding algorithm is in Fig. 9.36(a).

As stated, this subroutine has a program loop nested within another program loop. The test condition for controlling the outer loop is the buffer counter stored in Register C. The condition for the inner loop is the delay counter stored in Register B. For each iteration of the outer loop, the inner loop is executed 220 times.

This subroutine also illustrates a useful programming technique for producing a time delay with a program loop. This technique exploits the fact that the execution of an instruction requires a finite and known amount of time. To see how the desired 1-ms delay is produced by the program of Fig. 9.36(b), note that between the output of a byte of data (Statement 5) and the output of the next byte, there are the following intervening instructions:

	Instructions		Number of states required in the execution of each instruction
	MVI	B,220D	7
220 times {	DCR	B	4 (each)
	JNZ	DELAY	10 (each)
	DCR	C	4
	INX	D	6
	JNZ	NXBYTE	10
	LDAX	D	7

As shown, … s for its execution. The number of … *CS-80/85 Family User's Manual*. H… …114

Assuming t… …anoseconds), the total elapsed tim… Actually, to program the 1-ms de… …ds to obtain the initial delay count… ■ ■

To un… of states required to execute a particular instruction, we must understand both the stored program concept presented next in Chapter 10 and the hardware aspects of the microprocessor system presented in Chapters 11 and 12. A knowledge of the material of these chapters is essential for acquiring programming skills, for as this last example has shown, programming a microprocessor system is not purely a software endeavor. Also required is an awareness of hardware-software interaction.

SUPPLEMENTARY READING (see Bibliography)

[Andrews 82], [Gorsline 85], [Intel 79], [Intel-A], [Intel-B], [Liu 84], [Muchow 83], [Short 81], [Wiatrowski 80]

PROBLEMS

Note: For all programming problems assigned in this chapter, you should always include a flowchart of the algorithm corresponding to each program. In addition, use only the instructions that are presented in this chapter unless otherwise specified. Finally, each program should be well commented.

9.1. List all the general-purpose registers in the 8085 μP and briefly explain the function of each.

9.2. List all the special-purpose registers in the 8085 μP and briefly explain the function of each.

9.3. The 8085 μP can directly address up to 64K bytes of memory. Is there a reason for this limitation? If so, what is the reason?

9.4. There are three addressing modes for the memory reference instructions of the 8085. Name them and give an example of each.

9.5. The 8085 μP has three different types of load or move instructions for moving a value from a memory location into Register A. Illustrate these three types by specifying instructions for moving the value in Location 2004H into Register A in three different ways.

9.6. Briefly describe the flags that are available in the F register of the 8085. Which ones are accessible to the programmer? Which ones are not?

9.7. Given the following 8085 subroutine:

Statement number			
1		ORG	2000H
2	NUMB1	DB	14H
3	NUMB2	DB	0D9H
4		LDA	NUMB1
5		MOV	C,A
6		LDA	NUMB2
7		ADD	C
8		MVI	B,21H
9		STAX	B
10		RET	

(a) What are the contents of Register A (in hex) and the flags (CY, Z, S, and P) after Statement 7 is executed?

(b) In Statement 9, what is the address of the memory location in which the value is stored?

9.8. Given the following 8085 subroutine:

Statement number			
1		ORG	2000H
2		LXI	B,0102H
3		DCR	B
4		JZ	THERE1
5		DCX	B
6		JZ	THERE2
7	THERE1:	DCX	B
8		JZ	THERE2
9		STA	THERE3
10	THERE2:	RET	
11	THERE3	DB	00H

(a) What are the contents of Registers B and C (in hex) and the contents of the flags (CY, Z, S, and P) after the execution of Statement 2 and also Statement 3. Assume that the initial values of the flags are all zero.

(b) After Statement 4 is executed, does program control jump to THERE1 or proceed to Statement 5?

(c) What are the contents of Registers B and C (in hex) and the contents of the flags (CY, Z, S, and P) after either Statement 5 or Statement 7 is executed?

(d) After Statement 6 or Statement 8 is executed, does program control jump to THERE2 or proceed to the next instruction?

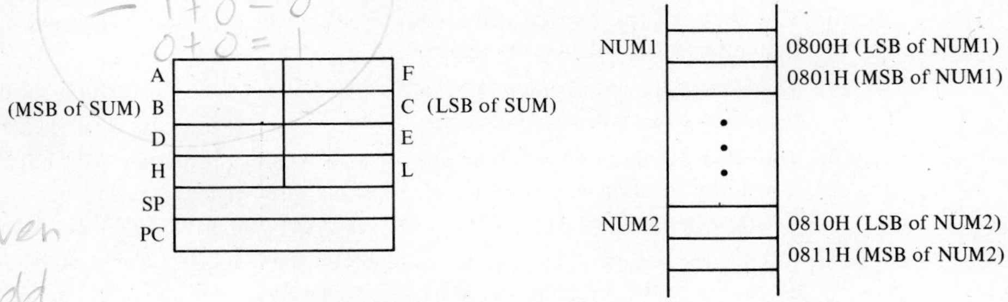

A
F
(MSB of SUM) B
C (LSB of SUM)
D
E
H
L
SP
PC
NUM1
0800H (LSB of NUM1)
0801H (MSB of NUM1)
NUM2
0810H (LSB of NUM2)
0811H (MSB of NUM2)

Figure 9.37 Storage specifications for Problem 9.9.

9.9. Write a subroutine, DBADD, for adding two 2-byte (16-bit) numbers. As illustrated in Fig. 9.37, the two numbers to be added are already stored in memory, and the resulting SUM is to be stored in Register Pair BC.

9.10. Write a subroutine, ADD4, for adding two 4-byte (32-bit) numbers. As illustrated in Fig. 9.38, the two numbers to be added are already stored in memory starting in Memory Locations NUM1 and NUM2, respectively. The resulting sum is to be stored in memory starting in Memory Location SUM.

9.11. Write a subroutine, ZERO16, for determining whether a 16-bit number stored in Register Pair BC is zero. If this number is zero, then clear the Carry Flag (i.e., STC followed by CMC). But if this number is nonzero, then set the Carry Flag (i.e., STC).

9.12. Write a subroutine, PAR16, for determining the parity of a 16-bit word stored in Register Pair BC. If the word has even parity, then set the Carry Flag. But if the word has odd parity, then clear the Carry Flag.

NUM1
0800H (LSB of NUM1)
0801H
0802H
0803H (MSB of NUM1)
NUM2
0810H (LSB of NUM2)
0811H
0812H
0813H (MSB of NUM2)
SUM
0820H (LSB of SUM)
0821H
0822H
0823H (MSB of SUM)

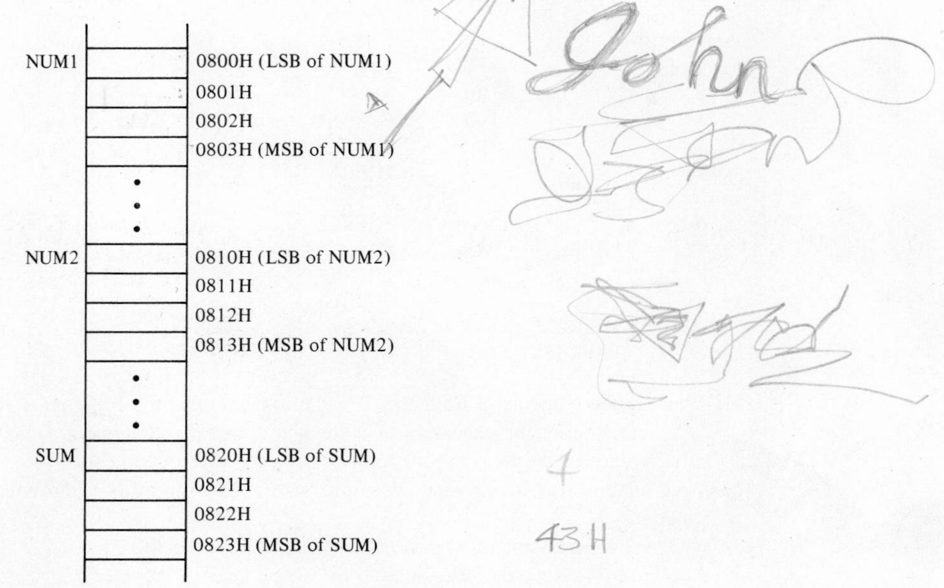

Figure 9.38 Memory specifications for Problem 9.10.

9.13. Write a subroutine, COMP16, for comparing two 16-bit numbers stored in Register Pairs BC and DE. If these two numbers are equal, then clear the Carry Flag. But if these numbers are not equal, then set the Carry Flag.

9.14. For Example 9.3 and Figs. 9.21 and 9.22, assume that the lookup table is stored beginning at 06FFH instead of 0700H (i.e., ORG 06FFH appears before Statement 1). Modify Subroutine CONV7 to accommodate this change.

9.15. Repeat Problem 9.14 if the lookup table is stored beginning at 0710H.

9.16. What are the main differences between the IF-THEN and IF-THEN-ELSE program structures?

9.17. What are the differences among the FOR, WHILE, and REPEAT-UNTIL program structures?

9.18. Write a subroutine for clearing a table (i.e., putting zero in each entry) stored in memory beginning at Memory Location TABLE, as illustrated in Fig. 9.39. Assume that there are 256 entries in this table.

9.19. Write a subroutine for determining the length of a table stored in memory beginning in Memory Location TABLE, as is illustrated in Fig. 9.39. As stated in the following comments, the end of the table is denoted by the first NULL (i.e., 00H) byte encountered.

```
; Input to the subroutine: a table of values beginning at Memory
;      Location TABLE.
; Output from the subroutine:
;      Register A contains the number of entries in TABLE.
; Function:
;      This subroutine will search TABLE for the first occurrence of a
;      NULL (i.e., 00H) byte, which denotes the end of the TABLE. The
;      length of the TABLE (not including the NULL byte) is returned
;      in Register A.
TABLE   EQU 0800H
TABLEN: .
        .
        .
        RET
```

9.20. Write a subroutine for searching TABLE, shown in Fig. 9.39, for the largest value in TABLE. Assume that there are 256 entries in TABLE and that all the values are unsigned integers. This largest value is to be placed in Register A, and its index in TABLE is to be

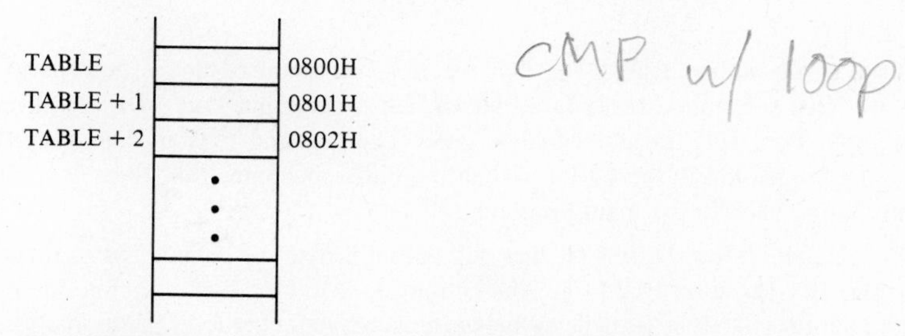

Figure 9.39 Memory locations for Problem 9.18.

placed in Register B. Note that the index of the first entry in TABLE is 0 and that the index of the last entry is 255.

9.21. Write a subroutine, as described by the following comments, that compares two equal-length blocks of memory to determine whether they are identical.

```
;   Inputs:
;       Register Pair HL contains the starting address of block 1.
;       Register Pair DE contains the starting address of block 2.
;       Register C contains the block length.
;   Output:
;       CY Flag.
;   Function:
;       If the blocks are identical, the subroutine returns with CY set.
;       If the blocks are not identical, CY is cleared and the address
;          of the location where they first differ is put into HL.
;
CMPARE:
            .
            .
            .
            RET
```

9.22. In the adding of 2s-complement numbers, if the two numbers to be added are of opposite signs, then there is never an overflow. And, of course, if the two numbers are of the same sign, the result should be of the same sign. Otherwise, it is an overflow. Complete the following subroutine for adding 2s-complement numbers, which includes checking for overflow.

```
;   One number is in Register B.
;   One number is in Register A.
;   The result is placed into Register A.
;   If there is an overflow, set Carry Flag.
;   If there is no overflow, clear Carry Flag.
SGADD:
            .
            .
            .
            RET
```

9.23. Write a subroutine, RDBYTE, that functions as specified in the flowchart shown in Fig. 9.40. This subroutine reads in a byte of data from Input Port 11H whenever the contents of Input Port 10H are not equal to zero. The inputted byte is then stored into the next available location in the buffer. When the buffer becomes full, there is a return from this subroutine back to the main program.

9.24. Write a subroutine, DSPSCN, that will output 8 bytes of data stored in memory beginning at Memory Location BUFFER. The outputting is to be 1 byte at a time to Output Port 00H with approximately a 1-ms delay between each byte outputted. Additionally, corresponding to each byte outputted to Output Port 00H, a different bit of Output Port 01H is set as follows:

Data to be outputted to Output Port 00H	Bit pattern to be outputted to Output Port 01H
BUFFER	00000001
BUFFER + 1	00000010
BUFFER + 2	00000100
BUFFER + 3	00001000
BUFFER + 4	00010000
BUFFER + 5	00100000
BUFFER + 6	01000000
BUFFER + 7	10000000

Assume that the address of BUFFER is already stored in Register Pair HL. [*Hint:* Since the shift instructions have not been presented in this chapter, you cannot use them. (They are presented in Chapter 10.) Instead, use the DB pseudo-instruction to define a table of bit patterns to be outputted to Output Port 01H.]

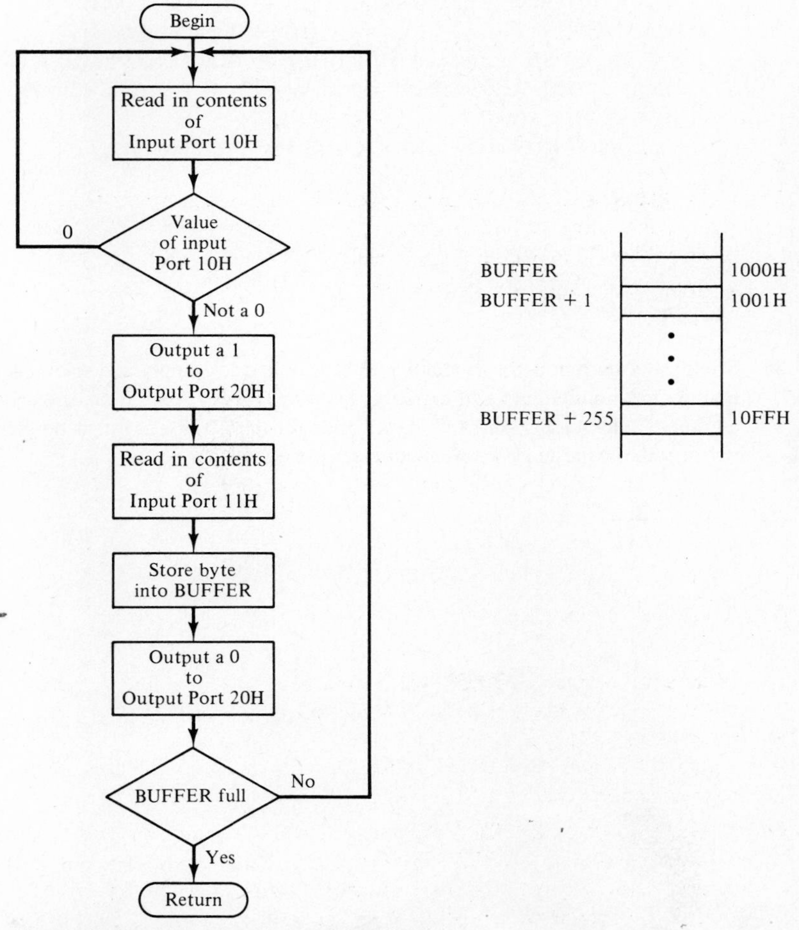

Figure 9.40 Flowchart for Problem 9.23.

14[2] **9.25.** Write a subroutine, described by the following comment statements, that implements a software stack structure.

```
; Inputs:
;       Register A contains an ''opcode''
;       If A = 0 : DEFINE a new stack
;              1 : PUSH
;              2 : POP
;       Register Pair BC contains the 2-byte value to be PUSHed.
;       Input Port 00H contains the low byte of the new stack pointer.
;       Input Port 01H contains the high byte of the new stack pointer.
;
; Output:
;       Register Pair BC contains the 2-byte value that is POPped.
;
; Functions:
;       DEFINE a new stack; i.e., (L) ← (Input Port 00H)
;                                 (H) ← (Input Port 01H)
;       PUSH: decrement (HL); ((HL)) ← (B); decrement (HL); ((HL)) ← (C)
;       POP: (C) ← ((HL)); increment (HL); (B) ← ((HL)); increment (HL)
;
; Don't worry about stack empty or stack overflow.
;
STACK:
          .
          .
          .
          RET
```

9.26. Modify the subroutine of Problem 9.25 to handle stack empty and stack overflow. Assume that the maximum number of entries in the stack is 256. Also, if an error occurs (i.e., POP an empty stack or PUSH a full stack), then set the CY Flag. But if no error occurs, then perform the requested operation and clear the CY Flag.

Chapter 10

Microprocessor Software Concepts II

10.1 INTRODUCTION

Chapter 9 introduced the assembly language instructions and program structures that are required to construct basic assembly language programs. The present chapter continues with additional assembly language instructions and programming concepts. First of all, however, in Secs. 10.2–10.4 there is a discussion of program assembly and machine language for the purpose of providing some insight into computer programming and program execution at the *machine* level. The material in these sections forms the foundation for the discussion of the hardware and interfacing concepts to be presented in Chapters 11 and 12. The rest of Chapter 10 (Secs. 10.5–10.8) has the remaining assembly language instructions and programming concepts. More specifically, Sec. 10.5 presents logic instructions, and also instructions that manipulate data at the bit level. Section 10.6 introduces the concept of the stack. Section 10.7 presents subroutines and the implementation of subroutine linkage using a stack. Finally, for completeness, Sec. 10.8 has the remaining instructions in the 8085 instruction set.

10.2 MACHINE LANGUAGE AND PROGRAM ASSEMBLY

In a paper published in 1946 [Taub 63], A. W. Burks, H. H. Goldstine, and J. von Neumann outlined the organization of an automatic computing machine. The architecture of the machine, commonly called the von Neumann architecture, is the basis for the architecture of modern-day conventional computer systems, including that of a microprocessor system. A key concept introduced by the von Neumann machine is the concept of the *stored program*. With it, program instructions are stored in the memory module along with the data. In the execution of a program, each instruction is fetched from memory by the control unit of the processor (in our case the microprocessor), decoded, and then executed. The fetch-decode-execute cycle continues until the end of the program.

A consequence of the stored-program concept is that the assembly language instructions must be transformed into a form that can be stored in memory locations, and also that can be processed by the microprocessor. In other words, an assembly language program, also called the *source code,* has to be transformed into a *machine language program,* also called the *object code*. This transformation process, called *program assembly,* can be done automatically by a computer program called an *assembler,* as will be explained. The machine language program is simply a coded version of the assembly language program, with each machine language instruction corresponding to an assembly language instruction. Figure 10.1 shows some 8085 assembly language instructions and their corresponding machine language codes. Note from Fig. 10.1(c) that for the 8085 μP, in the machine code of an instruction that has an address, the low-order byte of the address precedes the high-order byte.

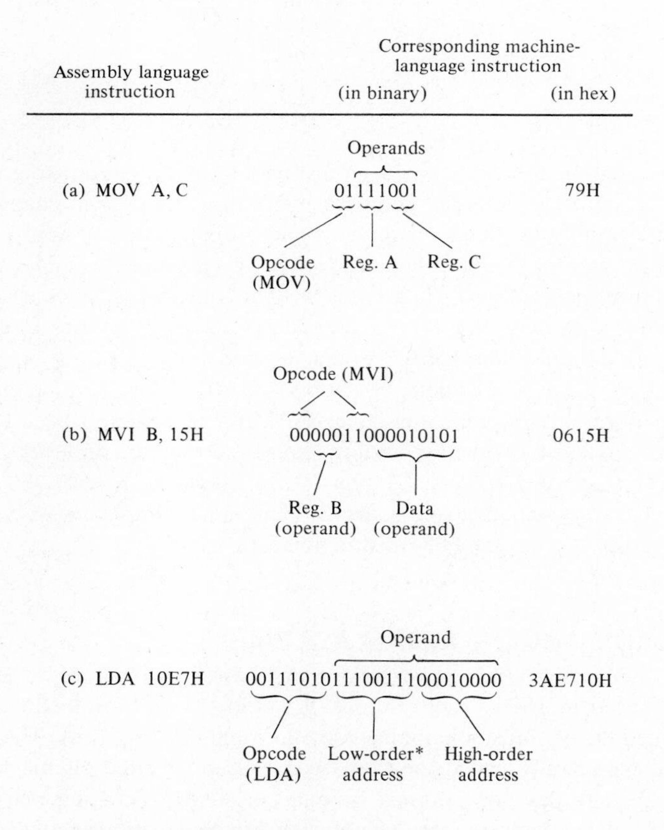

Figure 10.1 Examples of machine language codes.

Generally, a machine language instruction consists of an operation-code (opcode) field and an operand field. The opcode specifies the operation to be performed, such as LDA, MOV, ADD, JNZ, etc. Although each instruction must have an opcode, some instructions do not have an operand. Others, though, have one or even two operands, which is the maximum. Both the opcode and the operand(s) are coded as bit patterns of 0s and 1s. The exact instruction format and bit pattern length depend on the particular instruction. As illustrated in Fig. 10.1, an 8085 machine language instruction can contain either 8 bits, 16 bits, or 24 bits. Since 1 byte is required to store an 8-bit quantity, 2 bytes to store a 16-bit quantity, and 3 bytes to store a 24-bit quantity, the corresponding instructions are called 1-byte, 2-byte, and 3-byte instructions. In specifying machine codes, we will, as always, use the convenient hex notation rather than the cumbersome binary representation. Keep in mind, however, that the hex notation is merely a shorthand representation of the actual binary machine code.

Each instruction in the instruction set of a microprocessor has a unique machine code. The machine code (in hex) for each 8085 instruction is specified in the Appendix. By looking up the corresponding machine code for each 8085 instruction, we can readily assemble an entire assembly language program into a corresponding machine language program in a straightforward manner, as will now be illustrated.

Example 10.1 Hand Assembly of SBADD The first example of hand assembly will be the subroutine program SBADD of Example 9.1 of Sec. 9.5, which is reproduced in Fig. 10.2(a), for convenience. Our task is to assemble the assembly language program into a machine language program and store it into the appropriate locations in memory. Only instructions that are to be executed by the microprocessor need be assembled. For this program, Statements 5–10 are executable statements, but Statements 1–4 are pseudo-instructions. Like all pseudo-instructions, these are directions to the assembler, providing information for the assembly process. Here, since we are ''hand assembling'' the program ourselves, ''we'' are the assembler and the pseudo-instructions are directions to us. In Statement 1 the pseudo-instruction EQU informs us that the symbol NUM1 is equated to 0700H. Consequently, when the symbolic address NUM1 is subsequently encountered in the program, we must replace it with the absolute address 0700H. Similarly, symbolic address NUM2 needs to be replaced with absolute address 0704H, and SUM with 0710H. The pseudo-instruction ORG in Statement 4 specifies to the assembler to initialize a *location counter* to 0800H. In other words, it specifies that the following instruction (Statement 5), after being assembled, will be stored in Memory Location 0800H. The other instructions will be stored in successive locations following 0800H. Since Statement 5 has a label (SBADD), this symbolic label is associated with absolute address 0800H. Generally, the above information is saved for subsequent use in a *symbol table,* as shown in Fig. 10.2(b). With the above information available, we are ready to assemble the program.

In Statement 5 the instruction is LDA NUM1. From the Appendix, we find that the opcode for LDA is 3AH (00111010 in binary). The operand for this LDA instruction is a 16-bit address corresponding to NUM1, which from the symbol table is 0700H. Therefore, the machine code for the LDA NUM1 instruction is 3A0007H. (Remember that in the 8085 μP, the low-order address byte precedes the high-order byte.) Since the location counter is initialized to 0800H, the machine code for this 3-byte instruction is

(a) 8085 assembly language program

```
Statement
 1   NUM1     EQU   0700H   ; One of the operands of the addition is
                            ; stored in Memory Location 0700H.
 2   NUM2     EQU   0704H   ; The other operand is stored in 0704H.
 3   SUM      EQU   0710H   ; The resulting sum will be stored in 0710H.
 4            ORG   0800H   ; This subroutine program will be stored
                            ; beginning in Memory Location 0800H.
 5   SBADD:   LDA   NUM1    ; Moves operand NUM1 into Register A.
 6            MOV   B, A    ; Transfers NUM1 into Register B.
 7            LDA   NUM2    ; Moves operand NUM2 into Register A.
 8            ADD   B       ; Adds NUM1 to NUM2 with the sum in Register A.
 9            STA   SUM     ; Stores the sum into Memory Location 0710H.
10            RET           ; Returns to the calling program.
```

(b) Symbol table

Symbol	Corresponding quantity
NUM1	0700H
NUM2	0704H
SUM	0710H
SBADD	0800H

(c) Assembled program

Memory locations (in hex)	Contents of locations (in hex)	(in binary)	
0800	3A	00111010	LDA NUM1
0801	00	00000000	
0802	07	00000111	
0803	47	01000111	MOV B, A
0804	3A	00111010	LDA NUM2
0805	04	00000100	
0806	07	00000111	
0807	80	10000000	ADD B
0808	32	00110010	STA SUM
0809	10	00010000	
080A	07	00000111	
080B	C9	11001001	RET

Figure 10.2 Illustration for Example 10.1.

stored in Locations 0800H, 0801H, and 0802H, as shown in Fig. 10.2(c). Then, the location counter is incremented to 0803H. The next instruction is MOV B,A. From the Appendix, we find that it is a 1-byte instruction with a machine code of 47H. This is stored in the memory location specified by the location counter: Memory Location 0803H, and the location counter is incremented to 0804H. In the same manner, LDA NUM2 is assembled to 3A0407H and stored in Locations 0804H, 0805H, and 0806H. Also, the

1-byte instruction ADD B is assembled to 80H and stored in Location 0807H. And, the 3-byte instruction STA SUM is assembled to 321007H and stored in Locations 0808H, 0809H, and 080AH. Finally, the 1-byte instruction RET is assembled to C9H and stored in Location 080BH. ■ ■

Example 10.2 Hand Assembly of Code Conversion Subroutine For another illustration of program assembly, let us hand assemble the subroutine program for Example 9.3 of Sec. 9.5. The assembly language program of Fig. 9.22 is reproduced in Fig. 10.3(a). The resulting symbol table is shown in Fig. 10.3(b), and the corresponding machine language program in Fig. 10.3(c).

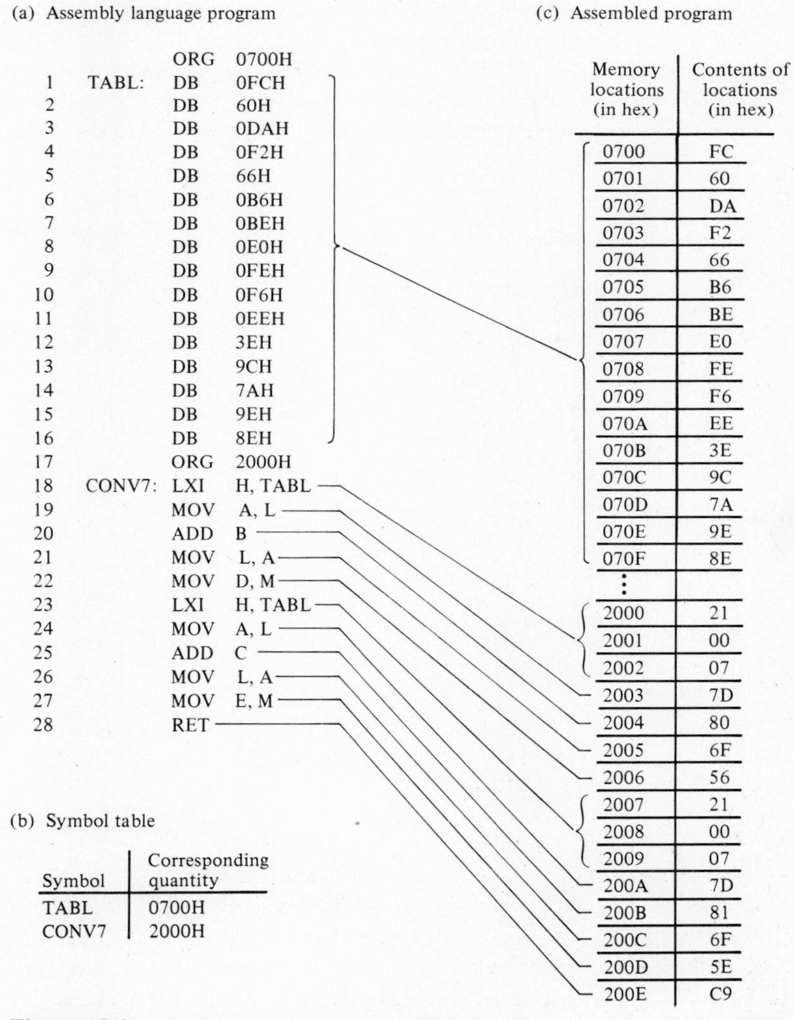

Figure 10.3 Illustration for Example 10.2.

Note that the pseudo-instruction DB (define byte) instructs the assembler to store a specified byte value into a specified memory location. In Statement 1 the specified byte value is FCH and the specified location is the memory location corresponding to TABL. As a result of the preceding ORG pseudo-instruction, TABL corresponds to 0700H, as shown in the symbol table. Therefore, byte FCH is stored in Location 0700H, as illustrated in Fig. 10.3(c). The next byte value 60H (Statement 2) is to be stored in the next available memory location pointed to by the location counter, which is 0701H, and so forth. This is a clear illustration of the stored-program concept, showing that both the data and the program are stored within the memory component of the microprocessor system. ■■

Example 10.3 Hand Assembly of TBSRCH For the final example of hand assembly let us hand assemble the subroutine program for Example 9.6 of Sec. 9.8. The assembly language program of Fig. 9.35(b) is reproduced in Fig. 10.4(a). As we proceed through the assembly process, a problem is encountered at Statement 5 with JZ FOUND. The opcode for JZ can be determined easily enough: it is CAH. However, as illustrated in

(a) Assembly language program

```
 1              ORG    2000H
 2  TBSRCH:     MOV    L, C
 3  REPEAT:     LDAX   D
 4              CMP    B
 5              JZ     FOUND
 6  NOTFD:      INX    D
 7              DCR    C
 8              JNZ    REPEAT
 9  ENDTB:      MOV    A, B
10              STAX   D
11              MOV    C, L
12              INR    C
13              STC
14              CMC
15              JMP    FNSRCH
16  FOUND:      MOV    C, L
17              STC
18  FNSRCH:     RET
```

(b) Partially assembled program and symbol table

Memory locations (in hex)	Contents of locations (in hex)	
2000	69	MOV L, C
2001	1A	LDAX D
2002	B8	CMP B
2003	CA	JZ
2004	?	FOUND
2005	?	

Symbol	Corresponding quantity
TBSRCH	2000H
REPEAT	2001H

Figure 10.4 Illustration for Example 10.3.

(c) Contents of the memory and the symbol table after the first pass

Memory locations (in hex)	Contents of locations (in hex)	
2000	69	MOV L, C
2001	1A	LDAX D
2002	B8	CMP B
2003	CA	JZ
2004	?	FOUND
2005	?	
2006	13	INX D
2007	0D	DCR C
2008	C2	JNZ
2009	01	REPEAT
200A	20	
200B	78	MOV A, B
200C	12	STAX D
200D	4D	MOV C, L
200E	0C	INR C
200F	37	STC
2010	3F	CMC
2011	C3	JMP
2012	?	FNSRCH
2013	?	
2014	4D	MOV C, L
2015	37	STC
2016	C9	RET

Symbol	Corresponding quantity
TBSRCH	2000H
REPEAT	2001H
NOTFD	2006H
ENDTB	200BH
FOUND	2014H
FNSRCH	2016H

Figure 10.4 (*cont.*)

Fig. 10.4(b), the memory address corresponding to the symbol FOUND cannot be found in the symbol table at this point of the assembly process. In fact, the address corresponding to this symbol cannot be determined until Statement 16 has been assembled and the symbol table updated. A similar problem is encountered with Statement 15, which is JMP FNSRCH. In fact, this problem occurs for any instruction that has a *forward* reference to a symbol. Note that this problem does not exist for Statement 8 since the symbol REPEAT is a backward reference and can be found in the symbol table at the time that Statement 8 is being assembled.

To solve the problem of the forward reference of symbols, assembling is done in two passes. During the first pass, the program is assembled as always, except that any symbol that is referenced before it is defined is left unresolved and the corresponding memory location left "blank." For this example, the result after the first pass of the assembly process is shown in Fig. 10.4(c). Since after the first pass, the symbol table is complete, it can be used during the second pass to complete the blanks for the undetermined symbol values. In this example, the contents of Locations 2004H and 2005H become 14H and 20H, respectively, and the contents of Locations 2012H and 2013H become 16H and 20H, respectively. ■■

10.3 ASSEMBLER AND ASSEMBLER DIRECTIVES

Although the hand assembly process is a useful learning tool, seldom in practice is it necessary. As will be explained in Chapter 13, a variety of developmental tools are available to assist the digital designer. Included among these is a software tool called, quite appropriately, an *assembler*. As has been briefly mentioned, an assembler is simply a computer program that will automatically convert an assembly language program into machine codes suitable for execution by the microprocessor for which the assembler is written. Most assemblers generally follow an algorithm similar to the two-pass assembly procedure described in the preceding section. Assemblers allow a digital designer to concentrate on problem solving and the development of the program at the symbolic assembly language level without becoming entangled in machine level details.

The input to an assembler is the assembly program in the form of executable instructions to be assembled, plus pseudo-instructions. These pseudo-instructions, also called *assembler directives,* are directions to the assembler, providing information to it for the assembly process. Although the mnemonics and details of the assembler directives vary among different assemblers, there is a set of common pseudo-instructions that are recognized by most assemblers written for an 8085 assembly language program. These common assembler directives, which have been briefly mentioned in the preceding examples, are described in more detail below.

10.3.1 ORG (Origin) Directive

As a program is being assembled, the assembled code of an instruction or the data value of a memory allocation pseudo-instruction (e.g., DB) is stored in the next available memory location specified by the location counter of the assembler. The ORG directive can be used to alter the location counter by setting it to any location in memory. The format of the ORG directive is given in Fig. 10.5(a). Examples of the ORG directive are given in a program segment in Fig. 10.6. These ORG directives specify the storage of the JMP 2100H instruction beginning in Location 0008H, the JMP 2150H instruction

(a)		ORG	operand	where operand is a 16-bit address or an expression that evaluates to a 16-bit address.
(b)	(label:)	DB	operand	where operand is an 8-bit value, a list of bytes, or an expression that evaluates to an 8-bit value.
(c)	(label:)	DW	operand	where operand is a 16-bit value, a list of words, or an expression that evaluates to a 16-bit value.
(d)	(label:)	DS	operand	where operand is a value or an expression that evaluates to a value.
(e)	Name	EQU	operand	where operand is a value or an expression that evaluates to a value.
(f)	Name	SET	operand	where operand is a value or an expression that evaluates to a value.
(g)		END		

Figure 10.5 Format of the common assembler directives for the 8085.

(a) Program segment

```
LED       EQU   15H
TEMSUM    EQU   C; Register C
SUM       EQU   0710H
MIN       EQU   10D
VALIDN    EQU   100D
MAX       EQU   MIN + VALIDN
          ORG   08H
          JMP   2100H
          ORG   10H
          JMP   2150H
          ORG   1000H
COUNT:    DB    00H
TABLE:    DB    25H, 24H, 92H, 14H
OFFSET:   DB    VALIDN + 5H
BUFF:     DS    10H
TABLE2:   DW    142EH, 7A12H
          DB    'A'; ASCII for A is 41H
          DB    15H
          ORG   2000H
MAIN:     LXI   SP, 27FFH
          .
          .
          .
          MOV   TEMSUM, A
          .
          .
          .
          CPI   MAX
          .
          .
          .
          OUT   LED
          .
          .
          .
          STA   SUM
          .
          .
          .
          END
```

(b) Symbol table

Symbol	Value
LED	15H
TEMSUM	C
SUM	0710H
MIN	0AH
VALIDN	64H
MAX	6EH
COUNT	1000H
TABLE	1001H
OFFSET	1005H
BUFF	1006H
TABLE2	1016H
MAIN	2000H

(c) Contents of memory locations

Memory locations (in hex)	Contents of locations (in hex)	
⋮		
0008	C3	JMP 2100H
0009	00	
000A	21	
⋮		
0010	C3	JMP 2150H
0011	50	
0012	21	
⋮		
1000	00	COUNT
1001	25	TABLE
1002	24	
1003	92	
1004	14	
1005	69	OFFSET
1006	**	BUFF
1007	**	
⋮		
1015	**	
1016	2E	TABLE2
1017	14	
1018	12	
1019	7A	
101A	41	'A'
101B	15	
⋮		
2000	31	LXI SP 27FFH
2001	FF	
2002	27	

Figure 10.6 A nonsense program segment illustrating the use of the common assembler directives of the 8085.

beginning in Location 0010H, and the LXI SP,27FFH instruction beginning in Location 2000H. Furthermore, Location 1000H is to be labeled COUNT, and a value of 00H will be stored in it. For most assemblers, the assembly begins in Memory Location 0000H if no ORG directive is used in the source code.

10.3.2 DB (Define Byte) Directive

The DB directive allocates space in memory and also initializes memory locations to specified values *at the time of assembly*. The format of the DB directive is given in Fig. 10.5(b), and some examples of its use are in Fig. 10.6. Note that the operand of the DB directive can be an 8-bit value, a list of bytes, or an expression that evaluates to an

8-bit value. The label associated with the DB directive is optional, as indicated by the parentheses.

10.3.3 DW (Define Word) Directive

Like the DB directive, the DW directive allocates space in memory and initializes memory locations to specified values at assembly time. The difference is that the DW directive allocates space and specifies values in words (16 bits) instead of bytes (8 bits). The format of the DW directive is given in Fig. 10.5(c), and an example of its use is in Fig. 10.6. As can be noted from Fig. 10.6(c), the assembler stores the least significant byte of a 16-bit value in the first memory location, and the most significant byte in the next memory location. This is true of most assemblers for the 8085.

10.3.4 DS (Define Storage) Directive

The DS directive allocates a block of storage in memory, but it does not initialize the contents of the allocated memory locations. The format of the DS directive is given in Fig. 10.5(d), and an example of its use is in Fig. 10.6. As indicated by the ** notation in Fig. 10.6(c), the contents of the memory locations allocated for BUFF remain undefined.

10.3.5 EQU (Equate) Directive

The format for the EQU directive is given in Fig. 10.5(e). The EQU directive informs the assembler to equate the specified symbol name to the value of the operand. In other words, when the symbolic name is subsequently encountered in the assembly process, the assembler replaces it with the binary value of the corresponding operand. The operand can be either a value or an expression that can be evaluated. The value can be a data value as in the operands of MIN and VALIDN in Fig. 10.6(a), a 16-bit address as in the operand of SUM, a port address such as the operand of LED, or even an 8085 register as in the operand of TEMSUM.

The EQU directive should be used generously to improve the clarity and readability of the assembly program. As an illustration, STA SUM is certainly more meaningful to a programmer than is STA 0710H. Furthermore, if it is necessary to change a value of a data constant (e.g., MIN), then only one instruction needs to be changed if an EQU directive is used. Otherwise, it is necessary to locate and modify every instruction in which that value is used. Similarly, if the port number of, say, an LED display has not been determined at the time that the program is developed, then the symbolic name LED can be used throughout the program, and only one modification to the EQU directive is required later.

10.3.6 SET Directive

The SET directive is similar to the EQU directive but allows more flexibility. For an illustration, in Fig. 10.6 the symbolic name VALIDN is equated to 100D and remains equated to 100D throughout the entire assembly of the program. If, however, the SET directive is used instead, then VALIDN can be set to another value at a later point in the program.

```
        .
        .
        .
VALIDN SET 100D
        .
        .
        .
VALIDN SET 150D
        .
        .
        .
```

Then, when the assembler encounters the symbolic name VALIDN after the second SET directive, it substitutes the second value (150D) instead of the first value (100D).

10.3.7 END Directive

The END directive informs the assembler that the end of the assembly language program has been encountered. The assembler ignores any statement following the END directive.

10.4 PROGRAM EXECUTION

After being assembled and stored in memory, a program is ready for execution. The details and actual timing of the microprocessor signals during the execution of an instruction are presented in detail in Chapters 11 and 12. In this section we will consider the execution of a stored program at the register transfer level.

Conceptually, the execution of a program stored in memory is a repeated sequence of fetch, decode, and execute operations. For a general understanding of these operations refer to Fig. 10.7, which is a block diagram of the 8085 μP system. In it, the program counter (PC) contains the address of the memory location that contains the next instruction to be executed. Under the control of the control unit, a memory read is performed to fetch the instruction from that memory location into the instruction register (IR). Next, the decoding circuitry decodes the instruction in the IR to identify it. The control unit then performs the operations required for the execution of that instruction. Upon the completion of the current execution, the PC is incremented to point to the next instruction. Then, the cycle begins again and the next instruction is fetched into the IR. The sequence of fetch, decode, and execute operations continues until the end of the program.

Example 10.4 Execution of SBADD For an illustration of these concepts the execution of the SBADD subroutine program of Fig. 10.2 will be traced in detail. Here it is assumed that the program is assembled and stored into memory, as shown in Fig. 10.8(a). The

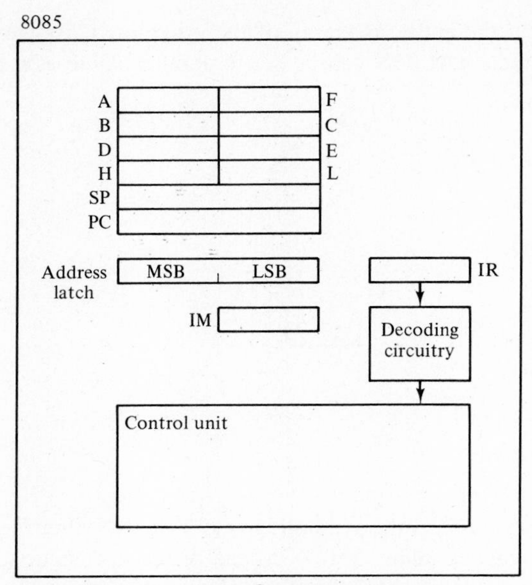

Figure 10.7 Block diagram of the 8085 μP system.

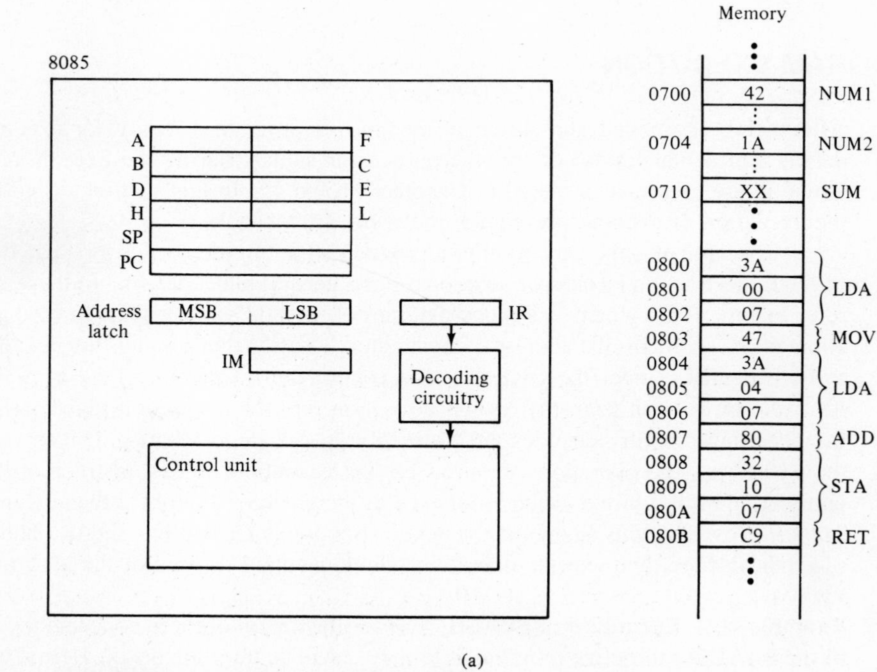

Figure 10.8 Illustration for Example 10.4.

Step	Assembly language statement	IR	Address latch MSB	LSB	Reg. A	Reg. F	Reg. B	PC	Memory locations 0700H	0704H	0710H
	Initial values	XX	XX	XX	XX	44	XX	0800	42	1A	XX
1	LDA NUM1	3A						0801			
2				00				0802			
3			07	00				0803			
4					42						
5	MOV B, A	47					42	0804			
6	LDA NUM2	3A						0805			
7				04				0806			
8			07	04				0807			
9					1A						
10	ADD B	80			5C	04		0808			
11	STA SUM	32						0809			
12				10				080A			
13			07	10				080B			
14											5C
15	RET	C9	07	10	5C	04	42	RT*	42	1A	5C

*All values are in hex;
XX is don't care;
RT is the return address of the calling program

(b)

Figure 10.8 (*cont.*)

execution of this subroutine program, with the detailed contents of the relevant registers, is given in Fig. 10.8(b). Initially, it is assumed that this subroutine is called by a calling program. At that point in time, the contents of the PC are 0800H, and those of Register F, Location 0700H, and Location 0704H are assumed to be 44H, 42H, and 1AH, respectively.

Since at the beginning of the subroutine program execution, the PC contains 0800H, the contents of 0800H (3AH) are first fetched into the IR, as illustrated in step 1 of Fig. 10.8(b). The control unit, through the decoding circuitry, identifies 3A as the opcode for an LDA instruction and so "expects" the operand for the LDA instruction to be stored in the next two locations of memory. Therefore, in step 2, a memory read is performed to read the contents of Location 0801H into the least significant byte of an address latch within the microprocessor. Another memory read is performed in step 3 to read the contents of Location 0802H into the most significant byte of the address latch. Now that the LDA instruction has been fetched and decoded, the LDA instruction can be executed. In this execution, another memory read is required to read into Register A the contents of the memory location whose address (0700H) is stored in the address latch, as illustrated in step 4. Note that four steps, each a "machine cycle," are required to fetch, decode, and execute the LDA instruction. The formal definition of a machine cycle will be given in Chapter 11. At this point, the PC contents have been incremented to 0803H, which is the address of the memory location that contains the next instruction to be executed.

In step 5 the opcode 47H is fetched into the IR. The control unit, through the decoding circuit, identifies it as the opcode for the 1-byte MOV B,A instruction. Since all the required operands are already available, the instruction can be executed without another memory reference. So the fetch, decode, and execute operations are performed within one step, or within a single machine cycle.

The next opcode fetched is again 3AH, another LDA opcode. Consequently, the operations in steps 6, 7, 8, and 9 are similar to those in steps 1, 2, 3, and 4. In step 10, the opcode is 80H, a 1-byte ADD B instruction. As in the MOV instruction, the fetch, decode, and execute operations are performed within one machine cycle. The next opcode, 32H, represents an STA instruction. As for the LDA instruction, two more memory reads are required to determine the operand. Furthermore, a memory *write* is required for storing the contents of Register A into Memory Location 0710H. So, the STA instruction requires four machine cycles. Finally, the last opcode, C9H, is recognized to be the 1-byte RET instruction. In the execution of an RET instruction, the "return address" of the calling program is placed in the PC. The result is that the next instruction executed is the instruction immediately following the calling instruction in the main program. The topic of subroutine linkage is presented in more detail in Sec. 10.7. ■■

As illustrated by Example 10.4, the execution of a stored program is simply a repeated sequence of opcode fetch, decode, and execute operations. Under the control of the control unit, the opcode of the instruction is fetched into the IR, and then decoded. The operations that follow depend on the particular instruction. If additional operands are required, as for a 2- or 3-byte instruction, then additional memory references are required. Otherwise, the execution can be performed within a single machine cycle.

To complete this section on program execution, let us say a few words about the initiation of program execution. Recall that the PC contains the address of the memory location that contains the next instruction to be executed by the microprocessor. Therefore, for the initiation of the execution of a program the address of the memory location that contains the *first* executable instruction of the *main* program must be placed in the PC. For the 8085, by default, the PC contents at power-up are 0000H. If it is desired to begin execution of the main program at power-up, the obvious thing to do is to store the main program beginning in Location 0000H. This is, however, generally not a good programming practice. As we will see later, some of the memory locations in the lower address range (e.g., 0008H) are reserved for special functions. Consequently, the common practice is to place in Location 0000H a JMP instruction that branches to the first instruction of the main program. This first instruction should be stored at a memory location beyond the specially reserved memory locations.

In summary, the intent of Secs. 10.2–10.4 is to provide a foundation for the discussion of the hardware and interfacing concepts presented in Chapters 11 and 12. Furthermore, the material in these three sections provides new insight into assembly language programming. With this new perspective we will now finish this chapter with the remaining assembly language instructions and programming concepts.

10.5 DATA TRANSFORMATION—LOGIC AND BIT MANIPULATION OPERATIONS

Basically, a μP system performs its functions by a series of data transfers and data transformations. Data is transformed through the use of the data transformation instructions of the instruction set. These instructions are generally divided into two categories: arithmetic and logic. In Chapter 9 the instructions for the basic arithmetic operations were presented. In this section we will describe and illustrate the use of the instructions for the logic operations, including instructions that manipulate data at the bit level. The formats and descriptions of these instructions for the 8085 are summarized in Table 10.1. These instructions are divided into three groups: those that perform Boolean operations, those that perform rotate (shift) operations, and those that operate on the CY Flag.

For each Group I instruction except CMA, the Boolean operation is performed *bit by bit* between an 8-bit quantity in Register A and another 8-bit quantity. This other 8-bit quantity may be from another general-purpose register (r), indirectly from a memory

TABLE 10.1 LOGIC INSTRUCTIONS

Instruction	Description	Flags affected	
Group I—Boolean operations			
(a) ANA r	(A)←(A) $\wedge$ (r)	AC = 1	CY = 0 Z, S, P are affected in the normal manner
(b) ANA M	(A)←(A) $\wedge$ ((HL))	AC = 1	
(c) ANI data	(A)←(A) $\wedge$ data	AC = 0	
(d) XRA r	(A)←(A) $\forall$ (r)	AC = 0	
(e) XRA M	(A)←(A) $\forall$ ((HL))	AC = 0	
(f) XRI data	(A)←(A) $\forall$ data	AC = 0	
(g) ORA r	(A)←(A) $\vee$ (r)	AC = 0	
(h) ORA M	(A)←(A) $\vee$ ((HL))	AC = 0	
(i) ORI data	(A)←(A) $\vee$ data	AC = 0	
(j) CMA	(A)←$\overline{(A)}$	No flags are affected	
Group II—Rotate operations			
(k) RLC	[diagram: A, CY]	Only the CY Flag is affected	
(l) RRC	[diagram]		
(m) RAL	[diagram]		
(n) RAR	[diagram]		
III—Operations on CY Flag			
(o) STC	(CY)←1	Only the CY Flag is affected	
(p) CMC	(CY)←$\overline{(CY)}$		

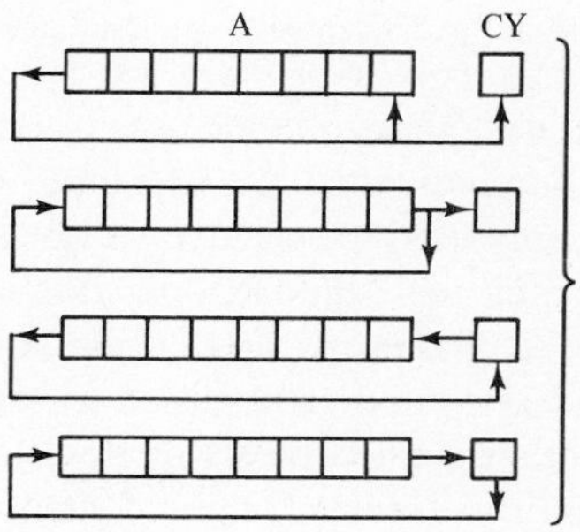

Notes:
r is Register A, B, C, D, E, H, or L.
data is an 8-bit quantity.
$\wedge$ is a logic AND, $\vee$ is logic OR, and $\forall$ is logic Exclusive OR.

location (M), or immediately as a part of the instruction (data). The result of each operation appears in Register A. The condition flags Z, S, and P are affected in the normal manner. The CY Flag, however, is always cleared to 0 after the execution of any of these instructions, regardless of the result of the operation. For the CMA instruction, each bit of Register A is complemented, thereby effectively forming the 1s complement of the original contents of Register A. But, no condition flag is affected.

The 8085 instruction set has four rotate (or shift) instructions. Each rotates the contents of Register A by 1 bit either to the left or right, sometimes through the CY Flag. These instructions are rotate A left (RLC), rotate A right (RRC), rotate A left through the carry (RAL), and rotate A right through the carry (RAR). Table 10.1 clearly shows the operations of these instructions.

Two instructions are available for operations on the Carry Flag: STC and CMC. The STC instruction sets the Carry Flag, and the CMC instruction complements it. Note that there is no instruction for clearing the Carry Flag. For this, two instructions are required:

STC; set CY to 1

CMC; complement CY to 0

Two examples are now presented to illustrate the use of the logic operation instructions.

Example 10.5 Example of the Use of Masks The subroutine program XPAND of this example expands the data byte in Register A into a form suitable for subsequent display by, perhaps, another subroutine. Figure 10.9(a) illustrates the expansion process, and Fig. 10.9(b) shows the corresponding assembly program, which has a detailed specification of the subroutine in the initial comment section.

This example illustrates the use of a Boolean instruction and a *mask* byte for selectively manipulating the individual bits of a data byte. In Statement 5, the mask byte LOMSK (00001111) is used to clear the high nibble of Register A to 0000, since 0 AND any value is 0. Furthermore, since 1 AND any value results in the same value, the low nibble of Register A is unaffected. Similarly, in Statement 8, the mask byte HIMSK (11110000) is used to clear the low nibble of Register A. ■■

Figure 10.10 illustrates other common uses of Boolean instructions and mask bytes for selectively manipulating individual bits of a byte. As illustrated in Fig. 10.10(a), the AND operation is commonly used to selectively clear a bit to 0, as was done in Example 10.5. To selectively set a bit to 1, the OR operation can be used [Fig. 10.10(b)], since ORing any value with 1 results in a 1, and ORing any value with 0 results in the same value. The Exclusive OR operation can be used to selectively complement the value of a bit. For example, if the original contents of Register A are 01011001, then after the instruction in Fig. 10.10(c) is executed, these contents become 01010110.

Boolean instructions and mask bytes can also be used to determine the value of a specific bit. For example, in Fig. 10.10(d), the AND operation is used to mask out all other bits except bit 2. If bit 2 of the original byte was 0, then the resultant byte after the AND operation would be zero. Otherwise, it would be nonzero.

Finally, Figs. 10.10(e) and (f) illustrate two common and advantageous uses of

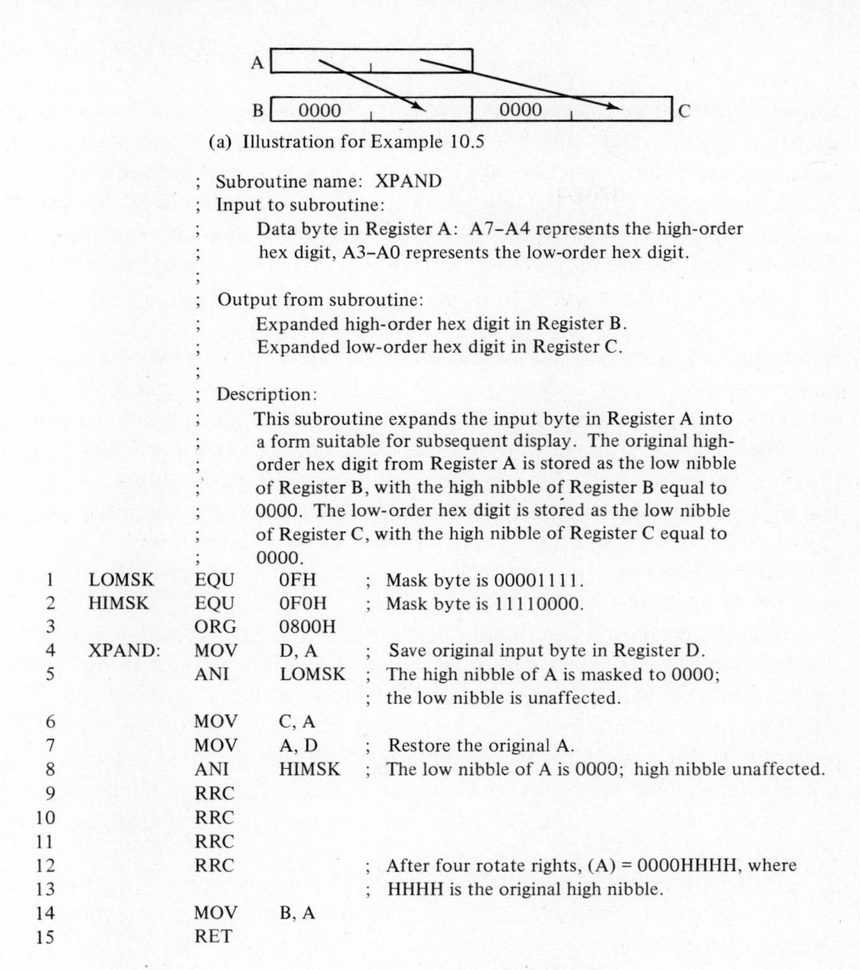

(a) Illustration for Example 10.5

```
                  ; Subroutine name: XPAND
                  ; Input to subroutine:
                  ;      Data byte in Register A: A7–A4 represents the high-order
                  ;      hex digit, A3–A0 represents the low-order hex digit.
                  ;
                  ; Output from subroutine:
                  ;      Expanded high-order hex digit in Register B.
                  ;      Expanded low-order hex digit in Register C.
                  ;
                  ; Description:
                  ;      This subroutine expands the input byte in Register A into
                  ;      a form suitable for subsequent display. The original high-
                  ;      order hex digit from Register A is stored as the low nibble
                  ;      of Register B, with the high nibble of Register B equal to
                  ;      0000. The low-order hex digit is stored as the low nibble
                  ;      of Register C, with the high nibble of Register C equal to
                  ;      0000.
 1   LOMSK    EQU   0FH      ; Mask byte is 00001111.
 2   HIMSK    EQU   0F0H     ; Mask byte is 11110000.
 3            ORG   0800H
 4   XPAND:   MOV   D, A     ; Save original input byte in Register D.
 5            ANI   LOMSK    ; The high nibble of A is masked to 0000;
                             ; the low nibble is unaffected.
 6            MOV   C, A
 7            MOV   A, D     ; Restore the original A.
 8            ANI   HIMSK    ; The low nibble of A is 0000; high nibble unaffected.
 9            RRC
10            RRC
11            RRC
12            RRC            ; After four rotate rights, (A) = 0000HHHH, where
13                           ; HHHH is the original high nibble.
14            MOV   B, A
15            RET
```

(b) Assembly program for Example 10.5

Figure 10.9 Illustration and assembly program for Example 10.5.

```
(a) Example:   Selectively clear bit 3 of Register A to 0.
               ANI    0F7H ;     The mask byte is 11110111.

(b) Example:   Selectively set bits 0, 1, and 5 to 1.
               ORI    23H ;      The mask byte is 00100011.

(c) Example:   Selectively complement the values of bits 0, 1, 2, and 3.
               XRI    0FH ;      The mask byte is 00001111.

(d) Example:   Determine if bit 2 is 0 or 1.
               ANI    04H ;      The mask byte is 00000100.
               JZ     ZERO2

(e) A one-byte instruction to clear Register A to 0 and affect flags.
               XRA    A

(f) Example:   Set condition flags without affecting register values.
               LDA    DBYTE ;  This instruction does not affect flags.
               ORA    A     ;  ANA can be used as well.
               (conditional jump instruction)
```

Figure 10.10 Some common uses of Boolean instructions and mask bytes.

Boolean instructions that are specific to the 8085. The instruction XRA A initializes Register A to 0 while also affecting the flags. Being a 1-byte instruction, it has an advantage over the instruction MVI A,00H which performs the same operation, but is a 2-byte instruction. Either the ORA A or the ANA A instruction of Fig. 10.10(f) sets the condition flags for conditional branching without affecting the values of any registers. These instructions are useful where conditional branching is required after an instruction, such as LDA, that does not affect any flags.

Example 10.6 Software Multiplication of Unsigned Binary Numbers The 8085, like many microprocessors, does not include a multiply or a divide instruction in its instruction set. These operations are usually provided by user-written subroutines. The function of the subroutine program in this example is to multiply two 8-bit unsigned numbers to obtain a 16-bit product. The algorithm for unsigned binary multiplication is essentially the shift-add algorithm used in the design of the hardware multiplication unit of Sec. 7.8.4. A version of the algorithm, modified to be suitable for the architecture of the 8085 microprocessor, is given in Fig. 10.11(b). And, the resultant assembly language program is given in Fig. 10.11(c).

Recall that in the hand multiplication of two unsigned binary numbers, such as is illustrated in Fig. 10.11(a), each bit of the multiplier is examined. If the current multiplier bit is 1, then the shifted multiplicand is added to the partial product. The purpose of the left-shifting of the multiplicand is to account for the weight of the multiplier bit. In the algorithm of Fig. 10.11(b), the left-shifting of the multiplicand is replaced by the functionally equivalent operation of the right-shifting of the partial product. Because of the register structure of the 8085 microprocessor, this right-shifting technique is more efficient.

The corresponding assembly language program is in Fig. 10.11(c). Note that the HL register pair is used for containing the partial product and, eventually, the entire product of the multiplication process. Since there is no 16-bit right-shift instruction in the 8085 instruction set, it is necessary to use Statements 9, 12, 13, 14, 15, and 16 to shift the contents of Register Pair HL one place to the right. Also note that when the partial product is initialized in Statement 2, only Register H is initialized to 0. Register L does not need to be initialized because it is not used until subsequent bit values of Register H are shifted into it. The shifted-in values overwrite the original values, thereby making initialization unnecessary.

Finally, a more elegant, but less clear, program can be written by using Register L for storing both the multiplier and the lower byte of the partial product. Initially, Register L is used to store the multiplier since this register is not used initially to store the LSB of the partial product. As the multiplication process proceeds, the multiplier

```
        1 0 1 0     (Multiplicand)
      × 1 0 1 1     (Multiplier)
-----------------
0 0 0 0 1 0 1 0     Multiplier(0) = 1;  multiplicand is shifted 0 times and added to the partial product.
0 0 0 1 0 1 0 0     Multiplier(1) = 1;  multiplicand is shifted 1 time and added to the partial product.
0 0 0 0 0 0 0 0     Multiplier(2) = 0;  0 is added to the partial product.
0 1 0 1 0 0 0 0     Multiplier(3) = 1;  multiplicand is shifted 3 times and added to the partial product.
-----------------
0 1 1 0 1 1 1 0     (Product)
```

Figure 10.11(a) Illustration of hand multiplication of unsigned binary numbers.

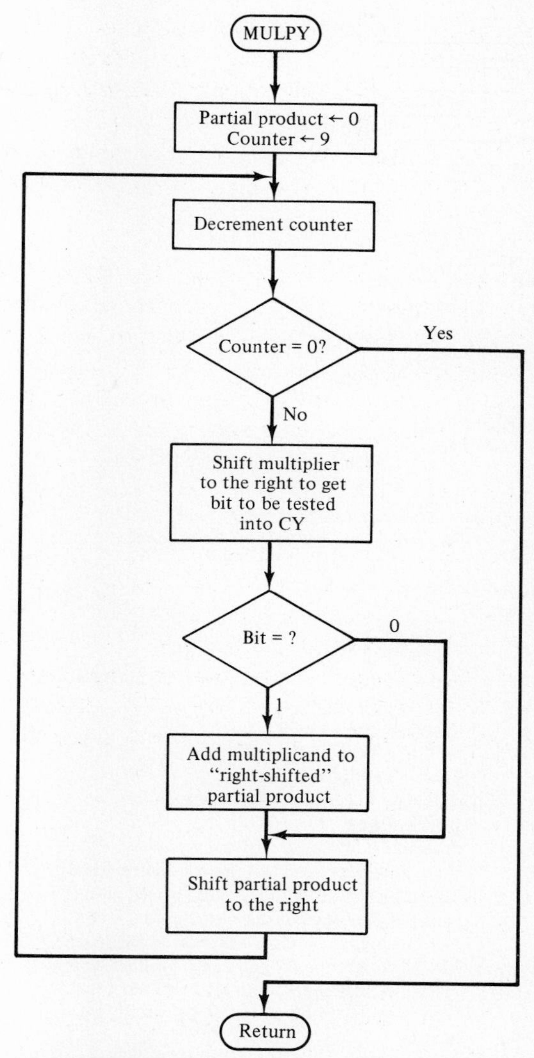

Figure 10.11(b) Algorithm for Example 10.6.

is shifted out of Register L one bit at a time. At the same time, the lower byte of the partial product is shifted into Register H. This approach saves the use of a register and also saves a few instructions. See Problem 10.18 at the end of this chapter. ■■

10.6 STACK AND STACK OPERATIONS

We will introduce the concept of a stack and its associated operations through the study of two instructions in the 8085 instruction set: PUSH and POP. Then we will consider the stack data structure.

A			F
B	Multiplicand	Counter	C
D	Multiplier		E
H	MSB of p.p.	LSB of p.p.	L
SP			
PC			

```
                ; Subroutine name: MULPY
                ; Input to subroutine:
                ;       Register B – 8-bit multiplicand
                ;       Register D – 8-bit multiplier
                ;
                ; Output from subroutine:
                ;       Register pair HL – 16-bit product
                ;
                ; Other register assignment:
                ;       Register C – bit counter
                ;
                ; Description:
                ;       This subroutine will perform unsigned binary multiplication
                ;       between two 8-bit numbers and return a 16-bit product.
                ;
                ; Registers that may be altered:
                ;       A, C, D, H, L, F
                ;
 1              ORG   0800H
 2   MULPY:     MVI   H, 00H   ; Initialized partial product, L needs no initialization.
 3              MVI   C, 09H   ; MLOOP will be performed 8 times.
 4   MLOOP:     DCR   C        ; Decrement bit counter.
 5              JZ    MFIN     ; Return if looped 8 times.
 6              MOV   A, D     ; Bring in multiplier and
 7              RAR            ; shift right to get the bit to be tested into
 8              MOV   D, A     ; the CY Flag.
 9              MOV   A, H     ; Bring in the MSB of the partial product.
10              JNC   SHIFT    ; If the multiplier bit is 0, don't add multiplicand.
11              ADD   B        ; Else if it is 1, add multiplicand to partial product.
12   SHIFT:     RAR            ; Shift right the MSB of the partial product, note
13              MOV   H, A     ; that the original bit 0 of MSB is saved in CY.
14              MOV   A, L     ; Now shift right the LSB of partial product, note
15              RAR            ; bit 0 of MSB is shifted into bit 7 of LSB.
16              MOV   L, A     ; Consequently, 16-bit partial product is shifted
                               ; right by 1.
17              JMP   MLOOP
18   MFIN:      RET
```

Figure 10.11(c) Assembly language program for Example 10.6.

10.6.1 PUSH and POP Instructions

The general formats, descriptions, and examples of the PUSH and POP instructions are given in Figs. 10.12 and 10.13, respectively. We will first consider the PUSH instruction.

The PUSH instruction is used to store the contents of a register pair into two consecutive memory locations. The source register pair can be BC, DE, HL, or PSW (A and F). The destination memory locations in which the 2 bytes of a register pair are stored are dependent on the contents of the stack pointer register (SP). As illustrated in

(a) General format: PUSH rp ;where rp is B (representing BC),
D (representing DE),
H (representing HL),
or PSW (representing A and F)

(b) Description: (SP) ← (SP) − 1
((SP)) ← (rh)
(SP) ← (SP) − 1
((SP)) ← (rl)

(c) Example: PUSH B

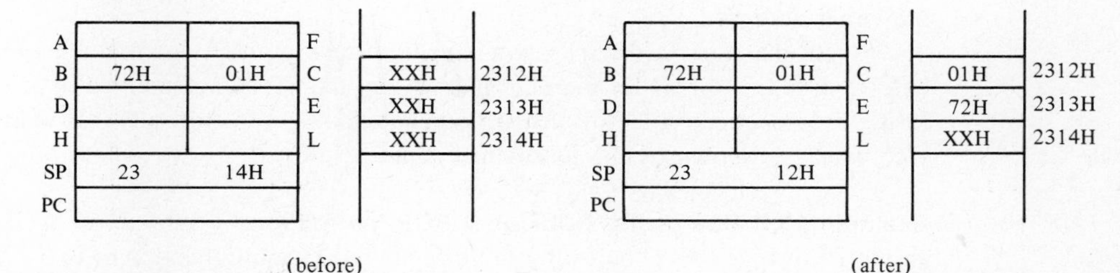

Figure 10.12 PUSH instruction.

Fig. 10.12, the contents of the high-order register of the register pair are moved to the memory location whose address is one less than the original contents of the SP. And, the contents of the low-order register are moved to the memory location whose address is two less than the original contents of the SP. Consequently, the memory location with the larger address receives the contents of the high-order register (rh), and the memory location with the smaller address receives the contents of the low-order register (rl). At

(a) General format: POP rp ;where rp is B (representing BC),
D (representing DE),
H (representing HL),
or PSW (representing A and F)

(b) Description: (rl) ← ((SP))
(SP) ← (SP) + 1
(rh) ← ((SP))
(SP) ← (SP) + 1

(c) Example: POP PSW

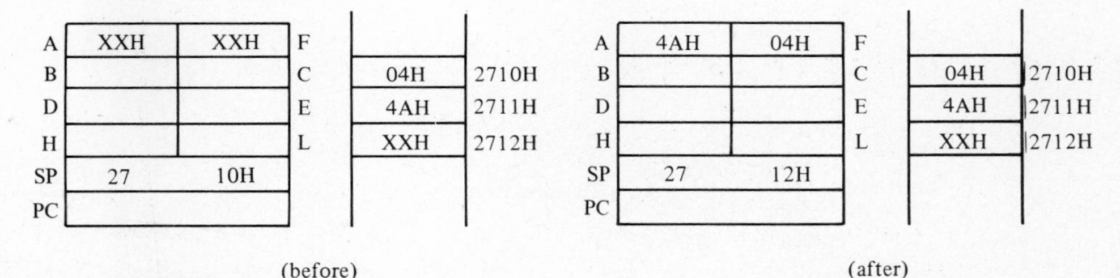

Figure 10.13 POP instruction.

the end of the execution of the PUSH instruction, the SP contains the address of the memory location that contains the contents of the low-order register.

The POP instruction is used to load into a register pair the contents of two consecutive locations in memory. The destination register pair can again be BC, DE, HL, or PSW. Analogous to the PUSH instruction, the addresses of the source memory locations are dependent on the contents of the SP. As illustrated in Fig. 10.13, the high-order register receives the contents of the memory location with the larger address, and the low-order register receives the contents of the memory location with the smaller address. At the end of the execution of the POP instruction, the SP contains one plus the larger address.

Note that for the PUSH instruction, the stack pointer is decremented *before* each byte is stored, whereas for the POP instruction, the SP is incremented *after* each byte is loaded. Consequently, a POP instruction can undo the effect of a PUSH instruction, and vice versa, as shown by the following example.

Example 10.7 Use of the PUSH and POP Instructions Subroutine MULPY of Example 10.6 (Fig. 10.11) performs an unsigned binary multiplication of two 8-bit numbers and returns a 16-bit product into Register Pair HL. Unfortunately, at the end of the execution of the subroutine, the original multiplier in Register D, and the original contents of Registers C, A, and F are destroyed. It is often desirable for a subroutine to return to the calling routine with the original contents of the registers preserved. In this case, the contents of any registers used by the subroutine should be saved at the beginning of the subroutine and restored at the end of the subroutine.

The PUSH and POP instructions provide a convenient way to save and restore the register contents. In Fig. 10.14, Subroutine MULPY is modified to include instructions to preserve the relevant registers. Note the reversal of the order of the PUSH and POP instructions. Figure 10.14(b) provides a graphic illustration of the effects that the PUSH and POP instructions have upon the registers and memory locations. Note that preceding

```
        ORG     0800H
MULPY:  PUSH    PSW      ; Saves Registers A, F
        PUSH    B        ;                 B, C
        PUSH    D        ;                 D, E
        MVI     H, 00H   ; Beginning of normal instructions for MULPY.
        MVI     C, 09H
MLOOP:  DCR     C
        JZ      MFIN
        .
        .
        .
        RAR
        MOV     L, A
        JMP     MLOOP
MFIN:   POP     D        ; Restores Registers D, E
        POP     B        ;                    B, C
        POP     PSW      ;                    A, F
        RET
```

Figure 10.14(a) Modified Subroutine MULPY.

A	32	22	F
B	15	3E	C
D	10	0B	E
H	A2	70	L
SP	23	79	
PC			

	XX
2373	XX
2374	XX
2375	XX
2376	XX
2377	XX
2378	XX
2379	XX

(1) Initial values at the time the subroutine is called

A	32	22	F
B	15	3E	C
D	10	0B	E
H	A2	70	L
SP	23	77	
PC			

	XX
2373	XX
2374	XX
2375	XX
2376	XX
2377	22
2378	32
2379	XX

(2) Values after PUSH PSW

A	32	22	F
B	15	3E	C
D	10	0B	E
H	A2	70	L
SP	23	75	
PC			

	XX
2373	XX
2374	XX
2375	3E
2376	15
2377	22
2378	32
2379	XX

(3) Values after PUSH B

A	32	22	F
B	15	3E	C
D	10	0B	E
H	A2	70	L
SP	23	73	
PC			

	XX
2373	0B
2374	10
2375	3E
2376	15
2377	22
2378	32
2379	XX

(4) Values after PUSH D

A	XX	XX	F
B	XX	XX	C
D	10	0B	E
H	XX	XX	L
SP	23	75	
PC			

	XX
2373	0B
2374	10
2375	3E
2376	15
2377	22
2378	32
2379	XX

(5) Values after POP D

A	XX	XX	F
B	15	3E	C
D	10	0B	E
H	XX	XX	L
SP	23	77	
PC			

	XX
2373	0B
2374	10
2375	3E
2376	15
2377	22
2378	32
2379	XX

(6) Values after POP B

A	32	22	F
B	15	3E	C
D	10	0B	E
H	XX	XX	L
SP	23	79	
PC			

	XX
2373	0B
2374	10
2375	3E
2376	15
2377	22
2378	32
2379	XX

(7) Values after POP PSW

Figure 10.14(b) Effects that the PUSH and POP instructions have upon the registers and memory locations.

any stack operations, the stack pointer SP must be initialized. This is usually done at the beginning of the main program with an LXI instruction. For an illustration,

```
LXI SP,2379H
```

will initialize the stack pointer to point to Memory Location 2379H. ■■

10.6.2 Stack Data Structure

A *stack* is a commonly used data structure in which the *first* element stored (pushed) onto the stack is the *last* element retrieved (popped). An analogous mechanism that is frequently used to describe a stack data structure is the lunchroom tray dispenser found in cafeterias. The only access to a stack of lunchroom trays is at the top of the stack. Therefore, the first tray "pushed" onto the stack of trays is the last one "popped" out.

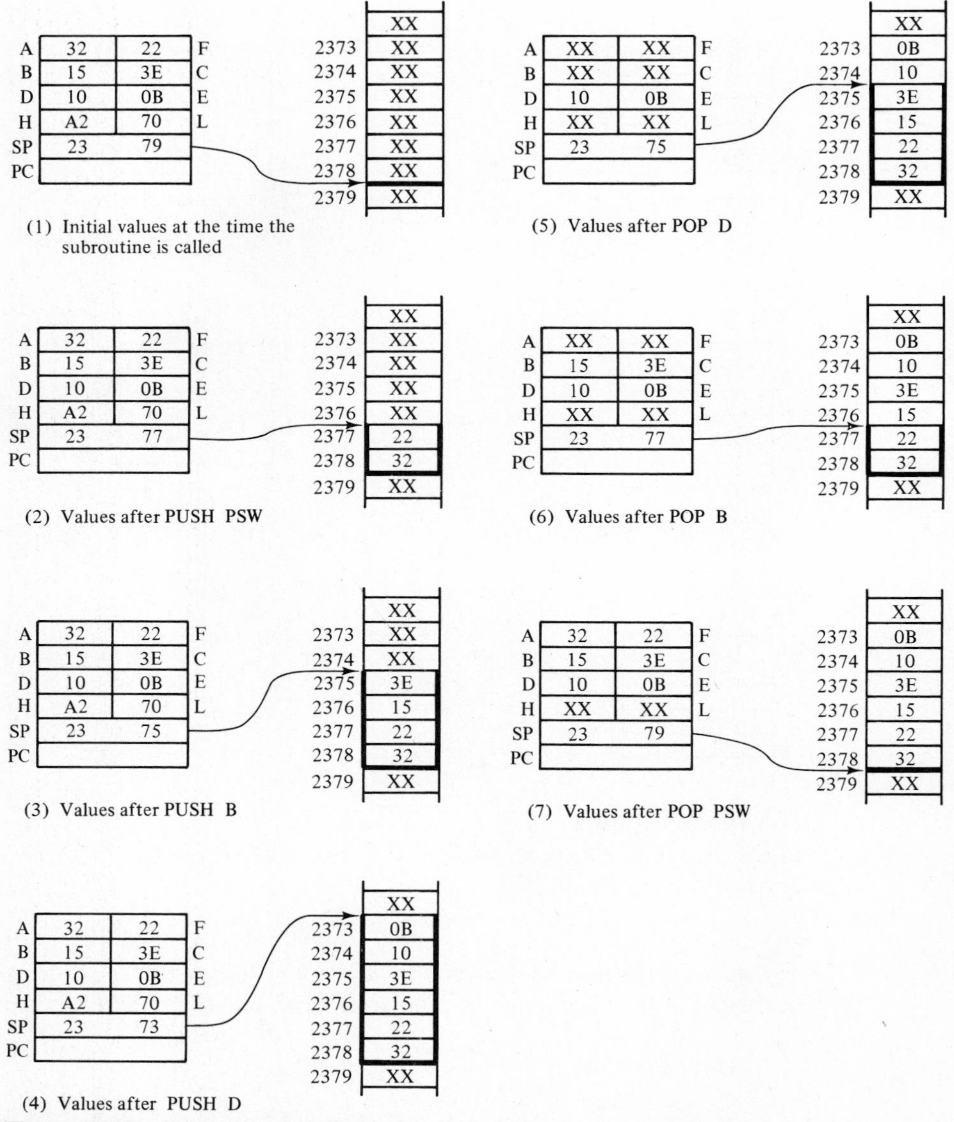

(1) Initial values at the time the subroutine is called

(2) Values after PUSH PSW

(3) Values after PUSH B

(4) Values after PUSH D

(5) Values after POP D

(6) Values after POP B

(7) Values after POP PSW

Figure 10.15 Stack data structure for Example 10.7.

Similarly, for a stack *data structure,* the first *data byte* stored is the last one retrieved. The PUSH and POP instructions of the 8085 are specifically designed to support the stack data structure.

With this understanding of the stack concept, we can view the example of Fig. 10.14(b) in a different perspective. As shown in Fig. 10.15, the bold-lined portion of the memory represents the stack data structure, with the stack pointer pointing to the *top* of the stack. For a stack data structure the only access to the data bytes (like the trays of a lunchroom tray dispenser) is at the top of the stack. Further, the top of the stack, which is specified by the contents of the SP, moves up each time a PUSH instruction is executed and moves down each time a POP instruction is executed. In this way, the stack data structure can store a variable amount of data as it expands or contracts with each PUSH or POP instruction.

10.7 SUBROUTINES

At the assembly language level, the operations of the subroutine call and return are the same as those of a high-level programming language such as PASCAL or FORTRAN. The process is illustrated in Fig. 10.16. In Fig. 10.16(a), the flow of control for the

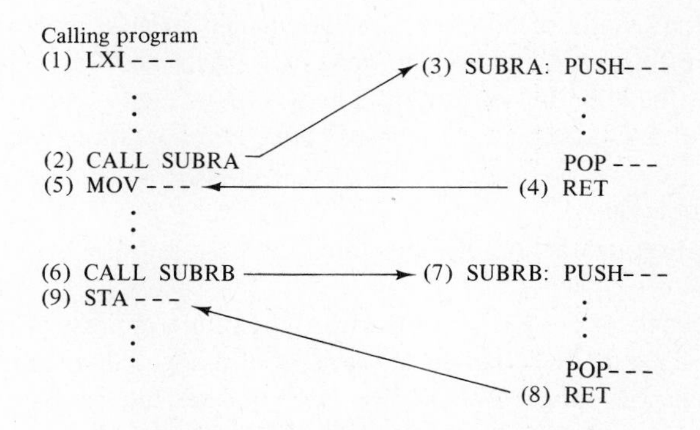

(a) A calling program calling two subroutines

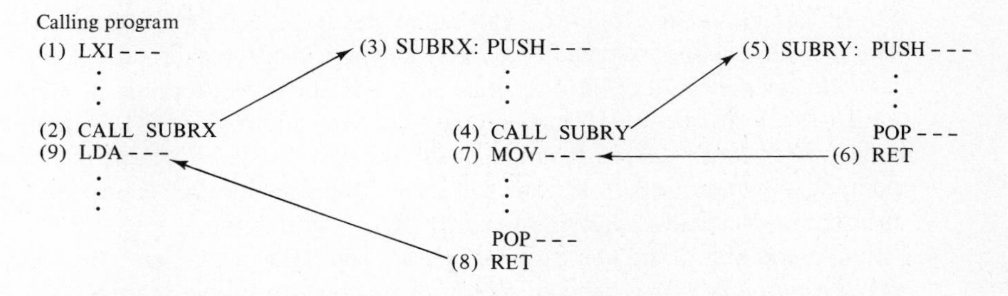

(b) Example of two levels of nested subroutine calls

Figure 10.16 Illustration of subroutine calls at the assembly language level.

program begins at the first instruction of the calling program (1). It proceeds in a normal manner through the execution of the instructions in the calling program. When the subroutine CALL instruction is encountered (2), the control is passed to the first instruction in the subroutine SUBRA. In other words, just as for a jump instruction, a branch is made to the instruction at the specified address (SUBRA) (3). However, unlike for a jump instruction, when a subsequent RET instruction is encountered (4), another branch is automatically made. It is back to the calling program to the next instruction following the CALL instruction (5). To perform this return operation, the microprocessor automatically preserves the *return address* during the execution of the CALL instruction. Continuing with Fig. 10.16(a), a second subroutine SUBRB is called from the calling program. In fact, any number of subroutines can be called from a calling program.

Figure 10.16(b) illustrates the case where a subroutine that is called by a calling program can itself call another subroutine. In other words, the subroutine calls can be *nested*. Figure 10.16(b) illustrates two levels of nested subroutine calls. Except as limited by the hardware of the μP system, subroutine calls can be nested for any number of levels.

10.7.1 Subroutine Instructions for the 8085

Although in writing an assembly language program, a programmer does not usually have to be concerned with the details of how the 8085 implements the subroutine calls and returns, it is instructive to consider this. The general formats and descriptions of the instructions provided by the 8085 for subroutine calls and returns are given in Fig. 10.17.

In the execution of a subroutine CALL instruction, two operations are performed:

1. Save the return address.
2. Branch to the first instruction of the subroutine.

At the time of the execution of the CALL instruction, the return address is stored in the program counter (PC), since the PC contains the address of the would-be next instruction to be executed. To save this return address, the 8085 pushes the contents of the PC (high-byte PCh and low-byte PCl) onto the stack, which is the same stack used by the programmer to store data with the PUSH and POP instructions. Then to branch to the subroutine the 8085 loads the address of the first instruction of the subroutine into the PC, as shown in Fig. 10.17(a). This completes the execution of the CALL instruction. The next instruction executed is the first instruction of the subroutine.

In the subroutine, the execution of the instructions proceeds in a normal manner until a RET instruction is encountered. The execution of a RET instruction causes a branch back to the calling program. As shown in Fig. 10.17(b), this is accomplished by popping the return address from the stack back into the PC. Consequently, the next instruction executed is that following the CALL instruction.

In addition to the unconditional CALL and RET instructions, the 8085 instruction set also has conditional call and return instructions, as shown in Figs. 10.17(c) and (d). For each of these conditional instructions the call or return operation is performed only if the test condition is satisfied. Otherwise the program control continues sequentially to

(a) Unconditional call

General format: CALL addr ; where addr is a 16-bit memory address of the first instruction of the subroutine. It is normally represented by a 4-digit hex number or a symbolic label.

Description:

```
 (SP)  ← (SP) − 1  ⎫
((SP)) ← (PCh)     ⎪  Push the return address
 (SP)  ← (SP) − 1  ⎬  onto the stack.
((SP)) ← (PCl)     ⎭
 (PC)  ← addr         Branch to subroutine.
```

(b) Unconditional return

General format: RET

Description:

```
(PCl) ← ((SP))    ⎫
(SP)  ← (SP) + 1  ⎪  Pop the return address
(PCh) ← ((SP))    ⎬  from the stack into PC.
(SP)  ← (SP) + 1  ⎭
```

(c) Conditional call

General format: Ccondition addr

Description:

```
IF (condition is true)
THEN  (SP)  ← (SP) − 1
     ((SP)) ← (PCh)
      (SP)  ← (SP) − 1
     ((SP)) ← (PCl)
      (PC)  ← addr
ELSE
  control continues sequentially
```

Conditional call instructions:

CZ	addr	Call if Zero
CNZ	addr	Call if Not Zero
CC	addr	Call if Carry
CNC	addr	Call if No Carry
CM	addr	Call if Minus
CP	addr	Call if Positive
CPE	addr	Call if Parity Even
CPO	addr	Call if Parity Odd

(d) Conditional return

General format: Rcondition

Description:

```
IF (condition is true)
THEN
  (PCl) ← ((SP))
  (SP)  ← (SP) + 1
  (PCh) ← ((SP))
  (SP)  ← (SP) + 1
ELSE
  control continues sequentially
```

Conditional return instructions:

RZ	Return if Zero
RNZ	Return if Not Zero
RC	Return if Carry
RNC	Return if No Carry
RM	Return if Minus
RP	Return if Positive
RPE	Return if Parity Even
RPO	Return if Parity Odd

Figure 10.17 8085 CALL and RET instruction formats and descriptions.

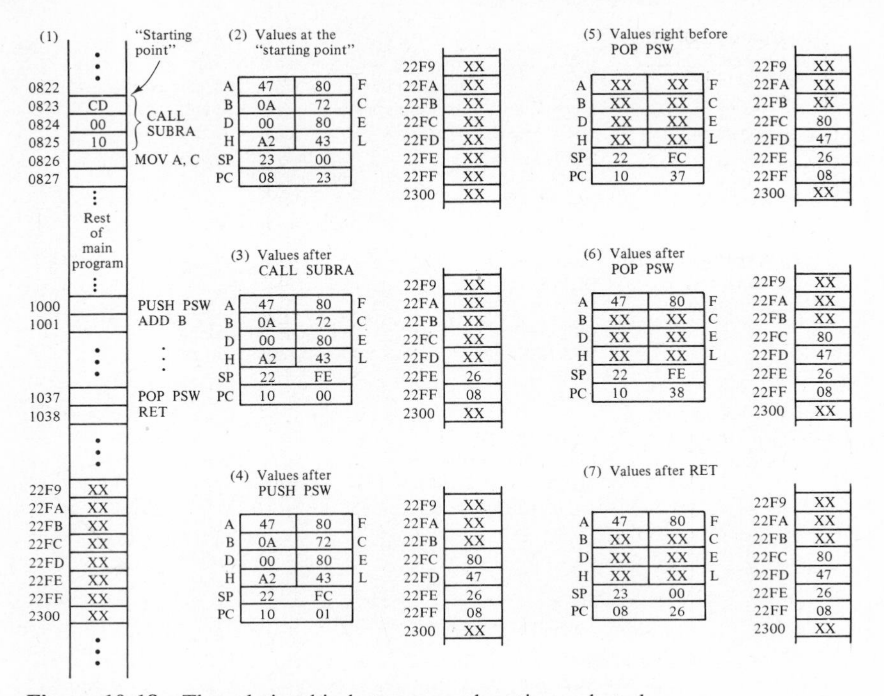

Figure 10.18 The relationship between a subroutine and stack.

the next instruction. Just as for the conditional jump instructions, the test conditions are based on the four condition flags: Zero (Z), Carry (CY), Sign (S), and Parity (P).

To better understand the effects of the CALL and RET instructions, consider the program segment of Fig. 10.18. The starting point of this illustration is in (1) of Fig. 10.18. At this time, the execution of the last regular instruction has just been completed, and we are ready to execute the CALL SUBRA instruction stored beginning in Location 0823H. (The subroutine SUBRA, itself, is stored in Locations 1000H through 1038H.) At this time, and as shown in (2), the program counter (PC) contains 0823H. Contents of the other registers are also shown. The effects of the subsequent CALL, PUSH, POP, and RET instructions are illustrated in (3) through (7).

Several points should be noted from this illustration. First, the number of POP instructions must equal the number of PUSH instructions. To understand this, consider what would happen if the POP PSW instruction were inadvertently omitted. In this case, when the RET instruction was encountered, the SP would be pointing to 22FCH, and so the contents 4780H of the top of the stack would be popped into the PC. As a result, the program control would be returned to the "instruction" in Location 4780H instead of to the instruction following the CALL instruction. On the other hand, if an extra POP instruction was included in the subroutine, then at the time of the execution of the RET instruction, the SP would be pointing to 2300H. Consequently, the RET instruction would return the program control to the "instruction" in the location "XXXXH" specified by the contents of 2300H and 2301H, whatever value that happens to be.

Note also from this illustration that for the 8085 the stack grows upward (to a smaller memory address) as each subroutine CALL or PUSH instruction is executed. Consequently, the stack should be defined at a sufficiently high memory address that it does not "grow into" and destroy any program or data stored at the lower addresses. Note also that the size of the stack is limited only by the size of the memory.

Since stack pushing and popping are inherent in the use of subroutines, the stack pointer (the contents of the SP) must be initialized before any subroutine call or return instruction is executed. As mentioned in the PUSH and POP discussion, this initialization is usually done with an LXI SP instruction at the beginning of the main program. If the SP is not initialized, then when a subroutine is called, the stack will begin at the location corresponding to whatever value that happens to be in the SP at that time. Possibly this location could be in the middle of the main program.

10.8 REMAINING 8085 INSTRUCTIONS

For completeness, the remaining instructions in the 8085 instruction set, not previously presented, will be discussed in this section. The general formats and descriptions of these instructions are given in Table 10.2. Included are the instructions that perform operations on double-byte data, most of which are centered around the Register Pair HL.

The LHLD (Load HL Direct) and SHLD (Store HL Direct) are analogous to the LDA and STA instructions. However, instead of involving Register A and a single memory location, the LHLD and SHLD instructions involve Register Pair HL and two consecutive memory locations. Note that the memory location with the lower address (addr) corresponds to the low-byte register (L), and the memory location with the higher address (addr + 1) corresponds to the high-byte register (H).

The SPHL instruction moves the contents of HL to the stack pointer SP, thereby dynamically establishing a new top of stack pointer. And, the PCHL instruction moves the contents of HL to the program counter. Note that PCHL is the only instruction in the entire 8085 instruction set that explicitly changes the contents of the program counter. The PCHL instruction is, in effect, a "jump indirect" instruction. In other words, when the PCHL instruction is executed, it causes a jump to the instruction whose address is currently stored in HL. However, unlike the jump instructions that we studied in Sec. 9.6, the branch address is not specified as a part of the jump instruction. Rather, it is stored in HL, possibly as a result of a computation or of a table lookup. (See Problem 10.25.)

There are also two exchange instructions that allow a programmer to exchange the contents of two 16-bit registers conveniently. The XCHG instruction exchanges the contents of Register Pairs HL and DE. And, the XTHL instruction exchanges the contents of Register Pair HL with the top of the stack. This top of the stack comprises the contents of two consecutive bytes in memory whose addresses are pointed to by the stack pointer SP. As always with the 8085, the memory location with the lower address (contents of SP) corresponds to the low-byte register (L) and the memory location with the higher address (contents of SP + 1) corresponds to the high-byte register (H).

For double-byte arithmetic, the DAD (Double Add) instruction can be used to add the two-byte contents of a register pair to the contents of HL. This register pair can be BC, DE, or HL.

TABLE 10.2 REMAINING 8085 INSTRUCTIONS

Instructions	Descriptions	Comments
Double-byte data transfer instructions		
LHLD addr	(L)←(addr) (H)←(addr + 1)	Load H and L direct
SHLD addr	(addr)←(L) (addr + 1)←(H)	Store H and L direct
SPHL	(SP)←(HL)	Move HL to SP
PCHL	(PC)←(HL)	Move HL to PC
Exchange instructions		
XCHG	(HL)↔(DE)	Exchange HL and DE
XTHL	((SP))↔(L) ((SP) + 1)↔(H)	Exchange top of stack with HL
Arithmetic instructions		
DAD rp	(HL)←(HL) + (rp)	Add register pair to HL; only the CY Flag is affected
DAA	The 8-bit number in Register A is adjusted to form two 4-bit BCD digits.	
Interrupt-related instructions		
EI	Enable interrupts	
DI	Disable interrupts	
RST	Restart	
SIM	Set interrupt masks	
RIM	Read interrupt masks	
Others		
NOP	No operation	
HLT	Halt	

The next instruction, DAA (Decimal Adjust Accumulator), will not be considered now since it is the subject of the following subsection. Also, the interrupt-related instructions will not be considered now. They will be discussed in detail in Chapter 12 where the interrupt structure of a microprocessor is presented.

The NOP (No Op) instruction performs no operation. The registers and flags are unaffected. It does, however, require four clock cycles for its execution. Consequently, it is useful in implementing software delay loops. Finally, the HLT instruction stops the processor.

10.8.1 DAA (Decimal Adjust Accumulator)

Frequently, for applications such as displaying values on 7-segment displays, it is more convenient to manipulate and store data values as decimal numbers instead of hexa-

Two-digit decimal number	BCD equivalent
00	00000000
01	00000001
02	00000010
03	00000011
04	00000100
05	00000101
06	00000110
07	00000111
08	00001000
09	00001001
10	00010000
11	00010001
12	00010010
13	00010011
14	00010100
.	.
.	.
.	.
98	10011000
99	10011001

Figure 10.19 The BCD equivalents of the 2-digit decimal numbers.

decimal (or binary) numbers. The binary coded decimal (BCD) numbering system is commonly used for this purpose in microprocessor systems. As illustrated in Fig. 10.19, in this numbering system, each decimal digit is coded separately with 4 bits, as follows:

0000 BCD = 0D (Decimal)
0001 BCD = 1D
0010 BCD = 2D
0011 BCD = 3D
0100 BCD = 4D
0101 BCD = 5D
0110 BCD = 6D
0111 BCD = 7D
1000 BCD = 8D
1001 BCD = 9D

Each additional decimal digit requires 4 additional bits. So, the BCD equivalent of 15D is 00010101, the BCD equivalent of 135D is 000100110101, and so forth. Note that, unlike for hexadecimal numbers, the bit patterns of 1010 through 1111 are never used in the BCD numbering system.

The DAA (Decimal Adjust Accumulator) instruction provides a convenient way to manipulate and store BCD numbers in a microprocessor system. Perhaps the function of the DAA instruction is best demonstrated with an illustration. Suppose, as shown in Fig.

(a) Contents of Registers A and C:

00111000	A
00100100	C

(b) Contents after an ADD C instruction:

01011100	A
00100100	C

(c) Contents after a DAA instruction immediately following the ADD C instruction:

01100010	A
00100100	C

Figure 10.20 Illustration of the DAA instruction.

10.20(a), Register A contains 00111000, which represents the BCD equivalent of the decimal number 38D. Also, suppose that Register C contains 00100100, which represents the BCD equivalent of 24D. The sum of 38D and 24D is, of course, 62D. Unfortunately, as illustrated in Fig. 10.20(b), the execution of the 8085 ADD C instruction produces in Register A the sum 01011100, which does not correspond to the desired decimal number 62. If, however, a DAA instruction is executed immediately after the ADD C instruction, the microprocessor automatically adjusts the contents of Register A to the valid BCD number equivalent, as shown in Fig. 10.20(c). Therefore, for the manipulation and storage of data values as decimal (BCD) numbers, a decimal adjust operation is required after every arithmetic operation. There is one word of caution, though. For the 8085, the DAA instruction functions only if used immediately after an add or increment instruction. So, if an arithmetic operation is other than an add or increment instruction, it should be followed by, for example, an ADI 0H instruction before the DAA instruction is used.

The material presented in Chapters 9 and 10 have provided fundamentals of software development for a microprocessor system. We are now ready to begin the exploration of the hardware and interfacing fundamentals that are essential for the full exploitation of the power of a microprocessor system.

SUPPLEMENTARY READING (see Bibliography)

[Andrews 82], [Gorsline 85], [Intel 79], [Intel-A], [Intel-B], [Liu 84], [Muchow 83], [Short 81], [Taub 63], [Wiatrowski 80]

PROBLEMS

Note: For all programming problems assigned in this chapter you should always include a flowchart of the algorithm corresponding to each program. Also, always use the PUSH and POP instructions to preserve the contents of all the registers that are not used to contain subroutine outputs. As always, each program should be well commented.

10.1. Distinguish the following terms: assembly language program, program assembly, assembler, machine language program, source code, and object code.

10.2. In the assembly of an assembly program, what is meant by a forward reference to a symbol? Also, what is a symbol table?

10.3. What is the difference between an instruction in a μP's instruction set and a pseudo-instruction (i.e., assembler directive)?

10.4. Hand assemble Subroutine MULPY of Example 10.6 (Fig. 10.11). Include the symbol table and the assembled code, as is shown in Fig. 10.2.

10.5. Repeat Problem 10.4 for the assembly program shown in Fig. 9.2.

10.6. Repeat Problem 10.4 for the following assembly program:

```
 OUTPA   EQU    01H
  INPA   EQU    00H
         ORG    0H
         JMP    MAIN
         ORG    8H
         JMP    INTRH
         ORG    100H
 TEST1   DB     50H
         ORG    800H
 MAIN:   LXI    SP,17FFH
         EI
         LDA    TEST1
         MOV    B,A
         LDA    TEST2
         MOV    C,A
 LOOP:   IN     INPA
         CMP    B
         JC     LOOP
         CMP    C
         JNC    LOOP
         OUT    OUTPA
         STA    BUFF
         JMP    LOOP
  BUFF   DS     1
 TEST2   DB     7FH
INTRH:   PUSH   PSW
         LDA    BUFF
         OUT    OUTPA
         POP    PSW
         EI
         RET
```

10.7. Disassemble the following 8085 machine code program. In other words, produce the assembly program corresponding to it.

Memory locations	Contents
0700H	21H
0701H	01H
0702H	08H
0703H	01H
0704H	02H
0705H	08H
0706H	3AH
0707H	00H
0708H	08H
0709H	96H
070AH	02H
070BH	C9H

Symbol Table

Symbol	Value
SBSUB	0700H
MINUM	0800H
SUBHAN	0801H
DIFFCE	0802H

10.8. Repeat Problem 10.7 for the following 8085 machine code program.

Memory locations	Contents
1000H	F5H
1001H	11H
1002H	00H
1003H	08H
1004H	21H
1005H	00H
1006H	09H
1007H	0EH
1008H	FFH
1009H	1AH
100AH	77H
100BH	13H
100CH	23H
100DH	0DH
100EH	C2H
100FH	09H
1010H	10H
1011H	F1H
1012H	C9H

Symbol Table

Symbol	Value
BUFF1	0800H
BUFF2	0900H
INBUFF	1000H
LOOP	1009H

10.9. The execution of a stored program is a repeated sequence of three operations. Name them and briefly explain each.

10.10. Analogous to the LDA instruction, the execution of an STA instruction requires four machine cycles. Briefly describe each of them.

10.11. The execution of an MVI instruction requires two machine cycles. Briefly describe each of them.

10.12. Trace through the execution of Subroutine CONV7 of Example 10.2 (Fig. 10.3). In other words, using the assembled machine code shown in Fig. 10.3(c), produce a detailed table, similar to that of Fig. 10.8, tracing the execution of the subroutine. Assume that the initial values for Registers B, C, and F are 03H, 0CH, and 00H, respectively. (*Hints:* The LXI instructions require three machine cycles for execution; the MOV D,M and MOV E,M instructions require two machine cycles. Each of the other instructions requires one machine cycle.)

10.13. Write a subroutine, PAR16, for determining the parity of a 16-bit word stored in Register Pair BC. If this word has even parity, then set the Carry Flag. But if this word has odd parity, then clear the Carry Flag. Note that this is a repeat of Problem 9.12. With the use of logic instructions, however, the number of instructions required is substantially reduced.

10.14. Write a subroutine, SWAP, that will swap the contents of the low-order nibble (i.e., 4 bits) of Register A with the contents of the low-order nibble of Memory Location DATA without affecting the high-order nibbles:

$$DATA_3DATA_2DATA_1DATA_0 \leftrightarrow A_3A_2A_1A_0$$

10.15. Write a subroutine, PACK, that will do the opposite of the XPAND subroutine of Example 10.5 and Fig. 10.9. Specifically, PACK will strip off the higher-order nibble of each byte in Registers B and C and pack the two lower-order nibbles into Register A.

10.16. Write a subroutine, WRBYTE, that will wait in an idle state as long as bit 4 of Input Port 00H is 0. But when this bit becomes 1, then the subroutine will function as specified in the flowchart shown in Fig. 10.21.

10.17. Modify Subroutine MULPY of Example 10.6 to use a REPEAT-UNTIL program structure (i.e., test the condition at the end of the loop) instead of the FOR program structure (which tests at the beginning of the loop) that is used, as is shown in Fig. 10.11.

10.18. Modify Subroutine MULPY of Example 10.6 (Fig. 10.11) to utilize Register L for storing both the multiplier and the lower byte of the partial product, as described at the end of Sec. 10.5.

10.19. Repeat Problem 9.25, which is the implementation of a software stack structure, using the PUSH and POP instructions along with the stack pointer register SP.

10.20. Write a subroutine, STKCTR, that is the calling subroutine that calls the subroutine STACK of Problem 10.19 to implement the software stack structure. The functions of STKCTR are specified in the flowchart of Fig. 10.22.

10.21. Given the following "nonsense" assembly program,

```
Statement
    1                ORG     0600H
    2                LXI     SP,0900H
    3                CALL    SUBA
    4                HLT
    5      SUBA:     PUSH    B
    6                LXI     B,0900H
    7                PUSH    PSW
    8                PUSH    H
    9                POP     B
   10                POP     H
   11                RET
   12                ORG     0800H
   13      SUBB:     LXI     H,1A1AH
   14                RET
```

Initial contents of registers:
A = 42H
Flag Register = B4H
HL = 0000H
BC = 0800H

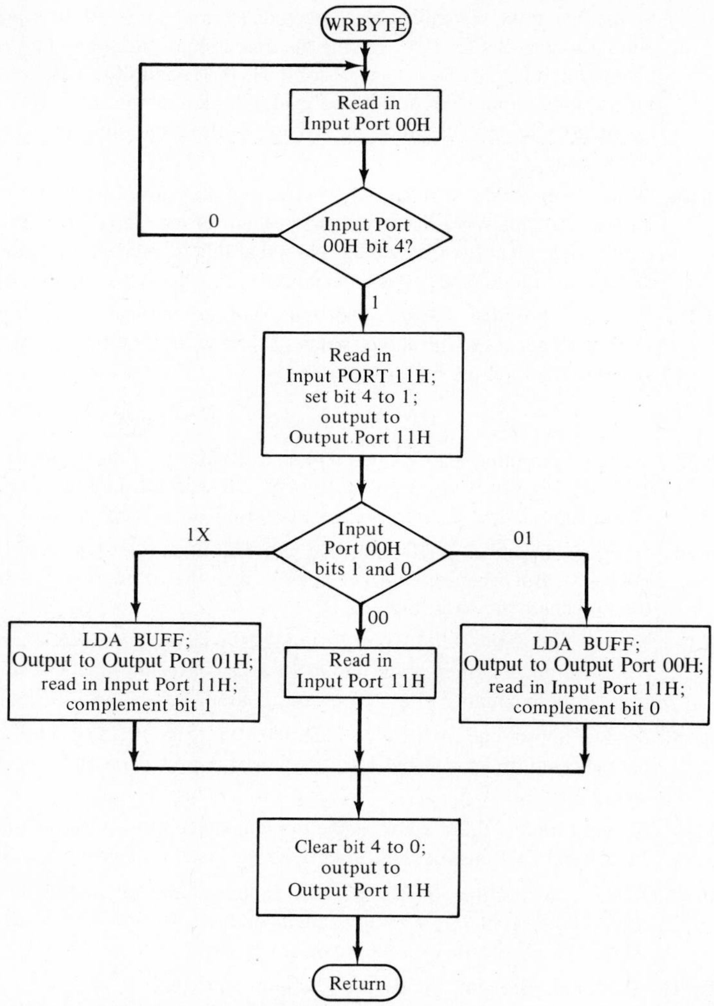

Figure 10.21 Flowchart for Problem 10.16.

(a) Trace the execution of this program. Your answer should be in the form of

Statements 2, 3, 5, . . . , 4

(b) At the end of the execution, what are the values (in hex) of the contents of the following registers: PSW, HL, BC, and SP?

10.22. Given the following "nonsense" assembly program,

Statement		
1	ORG	2000H
2	LXI	SP,2300H
3	CALL	SUBA
4	CALL	SUBB
5	HLT	

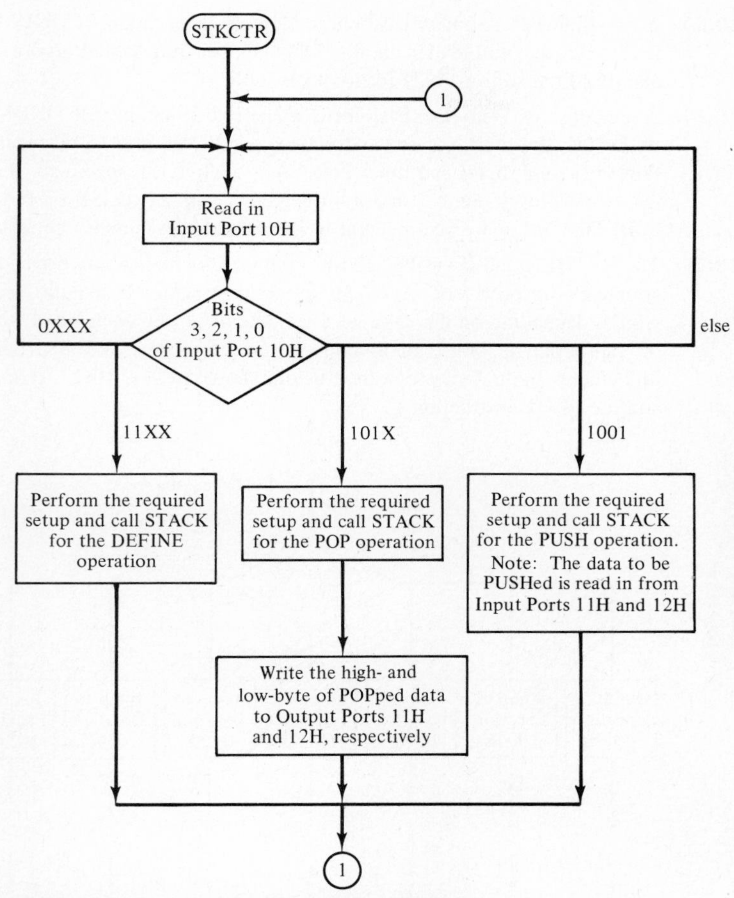

Figure 10.22 Flowchart for Problem 10.20.

```
 6            ORG    2100H
 7   SUBA:    LXI    B,0000H
 8            PUSH   B
 9            PUSH   PSW
10            POP    H
11            MVI    A,22H
12            STA    22FDH
13            RET
14            ORG    2200H
15   SUBB:    INX    B
16            RET
```

(a) Trace the execution of this program. Your answer should be in the form of

Statements 2, 3, 7, . . . , 5

(b) At the end of the execution, what are the values (in hex) of the contents of Registers BC and SP?

BC = 0002 SP = 2300

10.23. A two-digit BCD number is stored in memory in Location BCDNUM. Write a subroutine, BCDAD2, that will read in another BCD number from Input Port 00H, add it to BCDNUM, and place the sum in BCD form into BCDSUM.

10.24. A four-digit BCD number is stored in memory beginning at BCDNUM. Write a subroutine, BCDAD4, that will function as follows: It will read in a 16-bit binary number from Input Port 00H (high byte) and Input Port 01H (low byte), respectively. Further, it will convert the 16-bit binary number into a 4-digit BCD number (using the DAA instruction), add it to BCDNUM, and place the sum in BCD form into memory beginning at BCDSUM.

10.25. The IF-THEN and IF-THEN-ELSE program structures are two-way and three-way branch structures, respectively. A CASE program structure is a multi-way structure. In other words, depending on the value of a variable (e.g., the contents of Register A), the control of a program is branched to one of many locations. Write a part of a program that implements the CASE program structure shown in Fig. 10.23. (*Hint:* Use a lookup table and the PCHL instruction.)

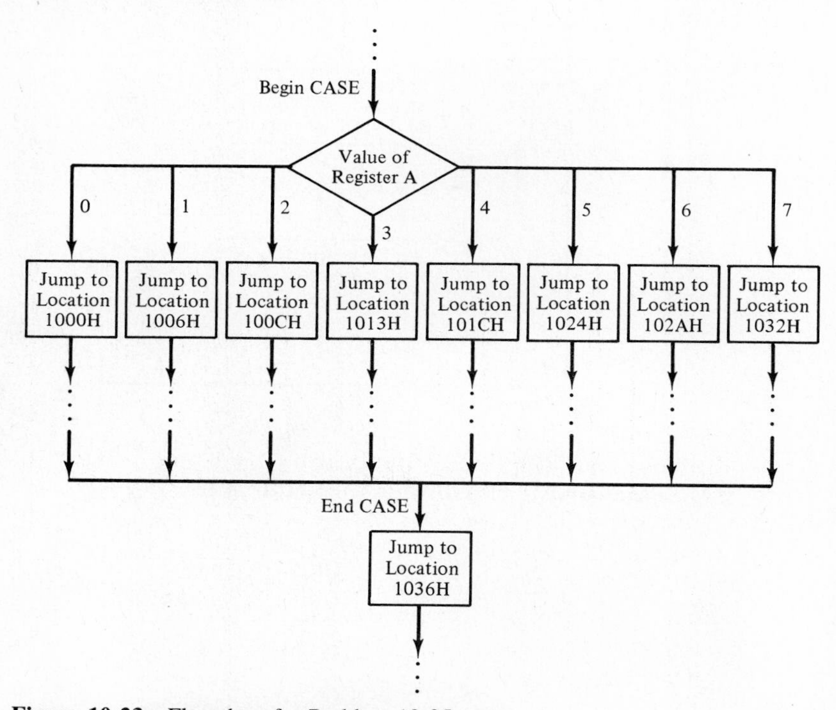

Figure 10.23 Flowchart for Problem 10.25.

Chapter 11

Microprocessor Hardware and Interfacing Concepts I

11.1 INTRODUCTION

In the preceding two chapters we studied the software concepts that are essential for the full exploitation of the power of a μP system. In Chapter 9 the assembly language instructions and program structures required to construct basic assembly language programs were introduced. These were continued in Chapter 10, in which program assembly and machine language concepts were also presented for the gaining of insight into computer programming and the execution of programs at the machine level. The material of Secs. 10.2–10.4 forms the foundation for the discussion of the hardware and interfacing concepts to be presented in this chapter and also in Chapter 12.

In Sec. 11.2 a model of a microprocessor system is presented. Also included are a model of the memory module, a model of a generalized interface module, and a model of an 8-bit microprocessor. These general models provide an overview of the operations of a microprocessor system. For a greater insight into these operations, the operations of a microprocessor system based on the 8085 μP are presented in detail in the remaining sections of this chapter.

11.2 MODEL OF A μP SYSTEM

Figure 11.1 shows a block diagram of a microprocessor system, which is the same as that shown in Fig. 8.2, and is reproduced here for convenience. A μP system is a digital system in which the key processing element is a microprocessor that, like the 8085, contains the arithmetic logic unit, registers, and control logic necessary for performing operations on data and for synchronizing and controlling transfer operations among the μP system components.

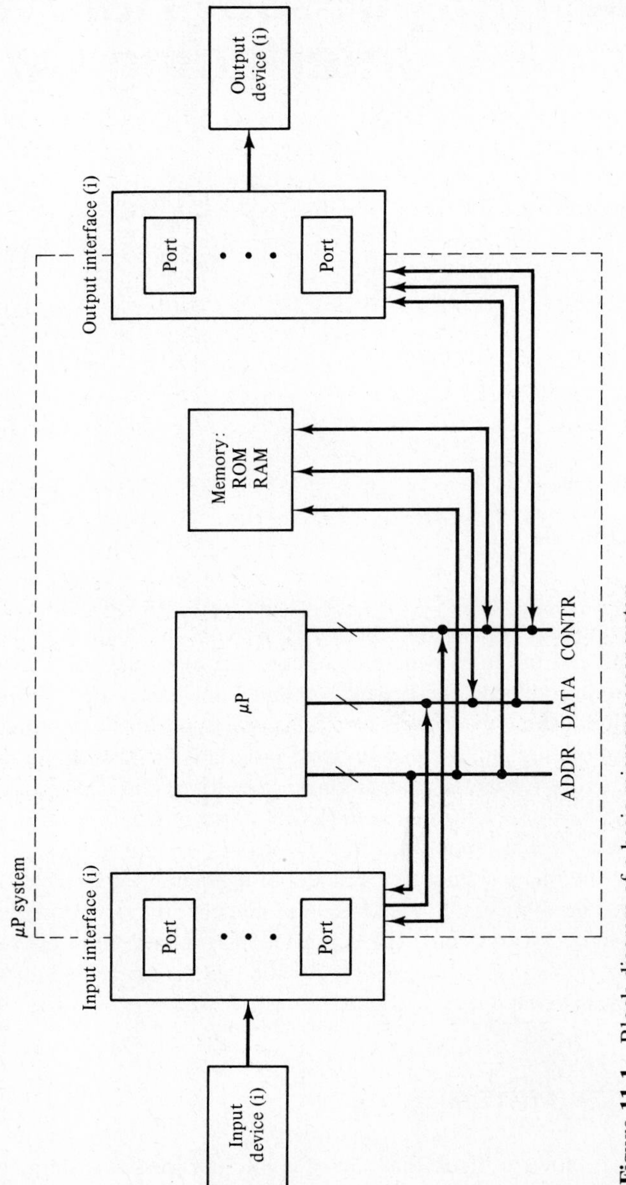

Figure 11.1 Block diagram of a basic microprocessor system.

The function of a microprocessor system is to process the digital data inputted from the input devices, to perform any required transformations, and to transfer the resultant data to output devices. Each input device is connected to the μP system through an input interface circuit that is unique to that device. Through its interface circuit, each input device provides input data to its assigned port(s). At the appropriate time, as synchronized between the microprocessor and the interface circuitry, the data is moved from an input port to a register within the microprocessor. In the 8085, for example, the execution of an IN instruction produces this transfer. Similarly, each output device is connected to the microprocessor system through a unique output interface circuit. Through its output interface circuit, each output device removes from the assigned output port(s) the data placed there by the microprocessor as a result of—for example, for the 8085—the execution of an OUT instruction.

The sequence of operations to be performed by a μP system is directed by a program that has been assembled and stored in the system memory module. As discussed in Sec. 10.4, the execution of a machine code program stored in memory is a repeated sequence of opcode fetch, decode, and execute operations. The tasks carried out by the program are simply a well-synchronized sequence of data transfers and data transformations.

The communication and data transfers among the component blocks of the μP system occur along three busses: the data bus (DATA), the address bus (ADDR), and the control bus (CONTR). The data bus provides a common path for the transfer of data between any two components. More specifically, under the control of the microprocessor, signals from the address bus, along with signals from the control bus, specify the type of data transfer and the source and destination of the data transfer.

11.2.1 A Model of the Memory Module of a μP System

Basic memory concepts were presented in detail in Chapter 6. We will now review those memory concepts that are relevant to the memory module of a μP system. Shown in Fig. 11.2 is a block diagram of a 64K-byte memory module. Conceptually, it is a

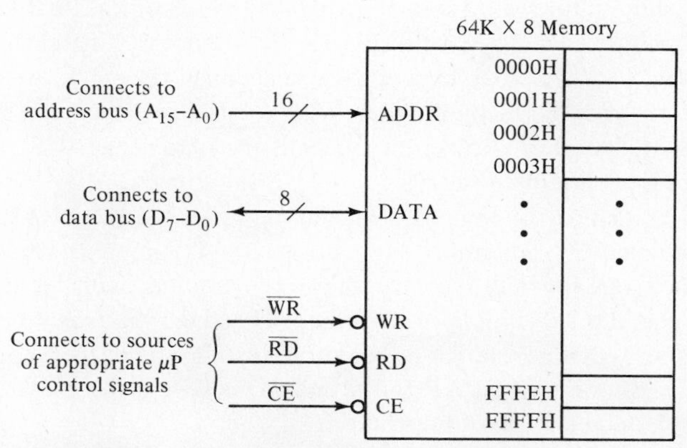

Figure 11.2 Model of a 64K-byte memory module.

collection of 64K (actually $2^{16} = 65{,}536$) addressable storage registers, each called a memory location. Associated with each memory location is a unique address, beginning with Memory Location 0000H and ending with Memory Location FFFFH.

For a read of the contents of a memory location, the corresponding 16-bit address must be stable on the ADDR inputs. Also, the chip-enable (CE) and the read (RD) inputs must be true, and the write (WR) input must be false, which correspond to applied signals of $\overline{CE} = L$, $\overline{RD} = L$, and $\overline{WR} = H$. After a time equal to the access time of the memory, the 8-bit contents from that memory location are available on the DATA outputs. For a write into a memory location, the corresponding 16-bit address must again be stable on the ADDR inputs. Additionally, the data to be written must be stable on the DATA inputs. The CE and WR inputs must be true and the RD input must be false, corresponding to applied signals of $\overline{CE} = L$, $\overline{WR} = L$, and $\overline{RD} = H$. After a time equal to the memory write time, the data on the DATA inputs is transferred into that memory location.

As an illustration, for the μP to read the contents of Memory Location 2300H, which could occur for an opcode fetch with PC = 2300H or with the execution of an LDA 2300H instruction, the μP must provide the following signals at the inputs of the memory module.

$$
\begin{aligned}
A_{15}\text{–}A_0 &= 2300\text{H} = 0010001100000000\text{B} \\
\overline{WR} &= H \\
\overline{RD} &= L \\
\overline{CE} &= L
\end{aligned}
$$

Similarly, for the μP to change the contents of Memory Location 2310H to 47H (i.e., to perform a memory write), the μP must provide the following signals.

$$
\begin{aligned}
A_{15}\text{–}A_0 &= 2310\text{H} = 0010001100010000\text{B} \\
D_7\text{–}D_0 &= 47\text{H} = 01000111\text{B} \\
\overline{WR} &= L \\
\overline{RD} &= H \\
\overline{CE} &= L
\end{aligned}
$$

Figure 11.3 shows the realization of two memory modules. In Fig. 11.3(a), a 1K × 8 memory module is realized with two 1K × 4 memory modules, each having ten address lines (A_9–A_0) and four bidirectional data lines (D_3–D_0). Also, the write-enable (WE) input controls the memory read (WE = false) and memory write (WE = true) operations. And, the chip is functional only if the chip-select (CS) input is true. Otherwise, the data lines are three-stated.

In the 1K × 8 memory of Fig. 11.3(a), the two 1K × 4 modules have identical address inputs. Consequently, when a 10-bit address (A_9–A_0) signal is applied, the same relative memory locations for both memory chips are accessed, with the high nibble in one memory chip and the low nibble in the other. Together, the two nibbles form the 8-bit data for that address. Note that the write-enable input (WE) of the 1K × 4 modules is equal to $WR \cdot \overline{RD}$. So, only when WR is true and RD is false is there a write operation. Otherwise, there is a default to a read operation.

To obtain a 2K × 8 memory module, we simply use two of the 1K × 8 memory modules, as shown in Fig. 11.3(b). Since the data lines of the 1K × 4 modules are

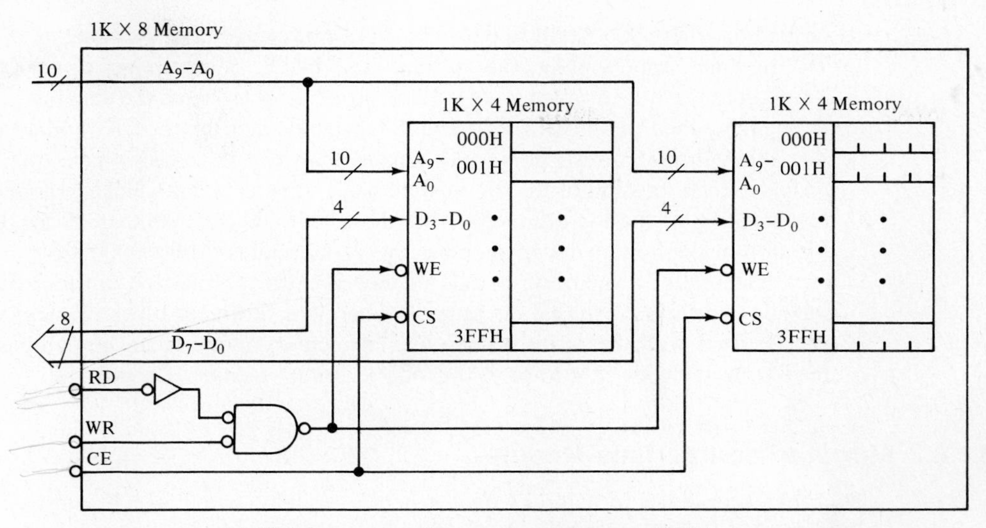

(a) 1K × 8 memory module constructed from two 1K × 4 memory modules

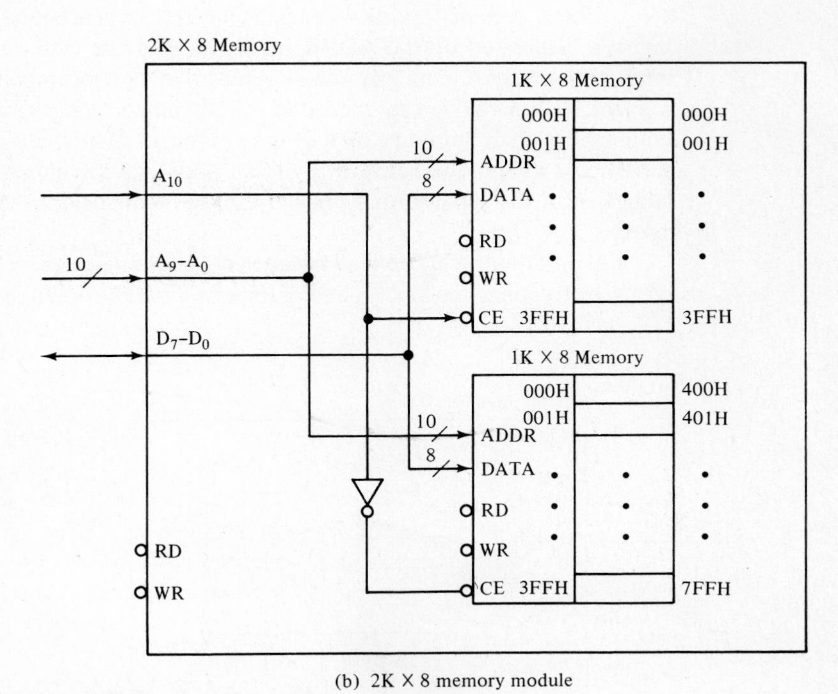

(b) 2K × 8 memory module

Figure 11.3 Realizations of memory modules.

three-state outputs, the 8-bit data lines of the two 1K × 8 memory modules can be connected together, with only the selected outputs connected *electrically* to the data bus. The chip-enable (CE) inputs of the two 1K modules are controlled by the high-order bit A_{10} of the address. If A_{10} is 0, then the top 1K module, representing the first 1K block of memory (00000000000B to 01111111111B), is enabled. At the same time, the bottom 1K module, representing the second 1K block of memory (10000000000B to 11111111111B), is disabled and electrically disconnected from the data bus. Conversely, if A_{10} is 1, then the bottom 1K module is enabled and the top 1K module is disabled. So, although the corresponding locations of the two 1K modules have the same 10-bit address, each location of the 2K module has a unique 11-bit address. For example, the first location of the top 1K module has the 11-bit address 00000000000B, and the first location of the bottom 1K module has the 11-bit address 10000000000B.

Generally, a memory module of any size can be realized with smaller memory modules and some external circuitry. Furthermore, different types of memory modules can be used with the result that an overall memory module can be composed of both read-write memory (RWM or RAM) and read-only memory (ROM).

11.2.2 Model of an Interface Module

A block diagram of a model of a generalized input/output (I/O) interface module is shown in Fig. 11.4. Although this interface module contains four input ports and four output ports, an interface module can contain any number of input and output ports.

Like a memory location, each input port is conceptually an addressable storage register, addressed by the ADDR inputs. Associated with each input port is a unique input-port address. Similarly, each output port is conceptually an addressable storage register, also addressed by the same ADDR inputs. Since both the input ports and the output ports share the same ADDR inputs, the addresses, although unique among themselves, are not unique between the input ports and the output ports. To distinguish the addresses of the input ports from those of the output ports, the RD and WR inputs are used.

In order for the μP to read the contents from an input port by means of, for example, an IN instruction, the corresponding input-port address (either 00, 01, 10, or 11) must

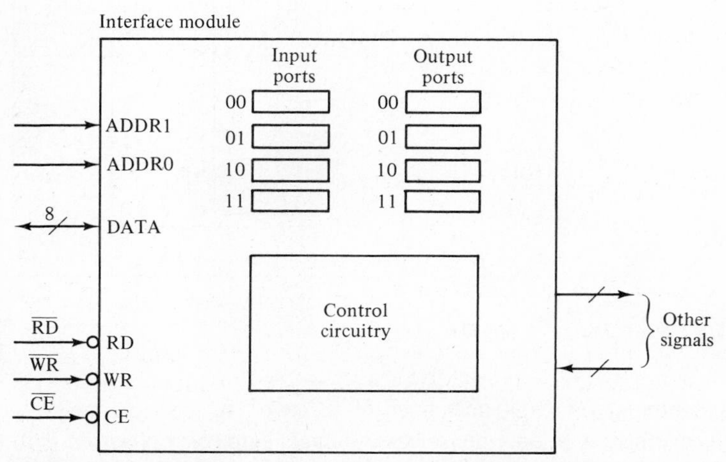

Figure 11.4 Model of a generalized I/O interface module.

be stable on the ADDR inputs. Also, the CE and RD inputs must be true and the WR input must be false. After an appropriate delay, the 8-bit contents of that input port are available on the DATA outputs. In order for the μP to write to an output port, the corresponding address (either 00, 01, 10, or 11) must again be stable on the ADDR inputs. Additionally, the data to be written must be stable on the DATA inputs. And, the CE and WR inputs must be true and the RD input must be false. Then, after an appropriate delay, the data on the DATA inputs is transferred to the output port.

Note that the address space of the input and output ports are separated. In other words, the μP cannot write to an input port or read from an output port. Consequently, with respect to the μP, the input ports are a set of read-only registers and the output ports are a set of write-only registers.

An output port generally serves one of two functions. It can be used as a storage register for the data outputted by the μP to an output device. Then the data is stored there until the output device removes it. Alternatively, the interface module can be designed in such a manner that an output port can be used as a *command register*. For this use, the μP can program the interface module by writing a certain bit pattern into an output port that is designated as a command register.

Similarly, an input port generally serves one of two functions. It can be a storage register in which an input device can deposit data for the μP. Then the data is stored there until removed by the μP as a result of, for example, the execution of an IN instruction. Alternatively, the interface module can be designed such that an input port can be used as a status register. Then, the μP can receive the status of and information about the operational conditions of the interface module by reading the contents of an appropriate input register that has been designated as a status register for that interface module.

The operation of the interface module is controlled by the control circuitry shown in block form in Fig. 11.4. The design of the control circuitry is unique to the function that is performed by the interface module. The control circuitry and additional signals for a programmable communication interface circuit are not the same as those for other types of interface modules such as a CRT (cathode ray tube) controller or a DMA (direct memory access) controller. However, although the functions and control of the different types of interface modules are not the same, the manner in which the interface modules interact and communicate with the μP through the I/O ports are conceptually the same, as will be described in Chapter 12.

11.2.3 Model of a Microprocessor

Figure 11.5 shows a block diagram of a microprocessor that has a data path of 8 bits and a memory space of 64K bytes. Internally, the microprocessor contains a set of general-purpose registers, a set of special-purpose registers (including the program counter PC), the arithmetic/logic unit (ALU), and the control unit (including an instruction register IR and the instruction decoding circuitry). Externally, the microprocessor is connected to the other modules of the μP system through the three busses shown. Also, a set of support pins are connected to circuits that support the operations of the μP system. These circuits include the power supply and clock circuitry, as well as electrical ground.

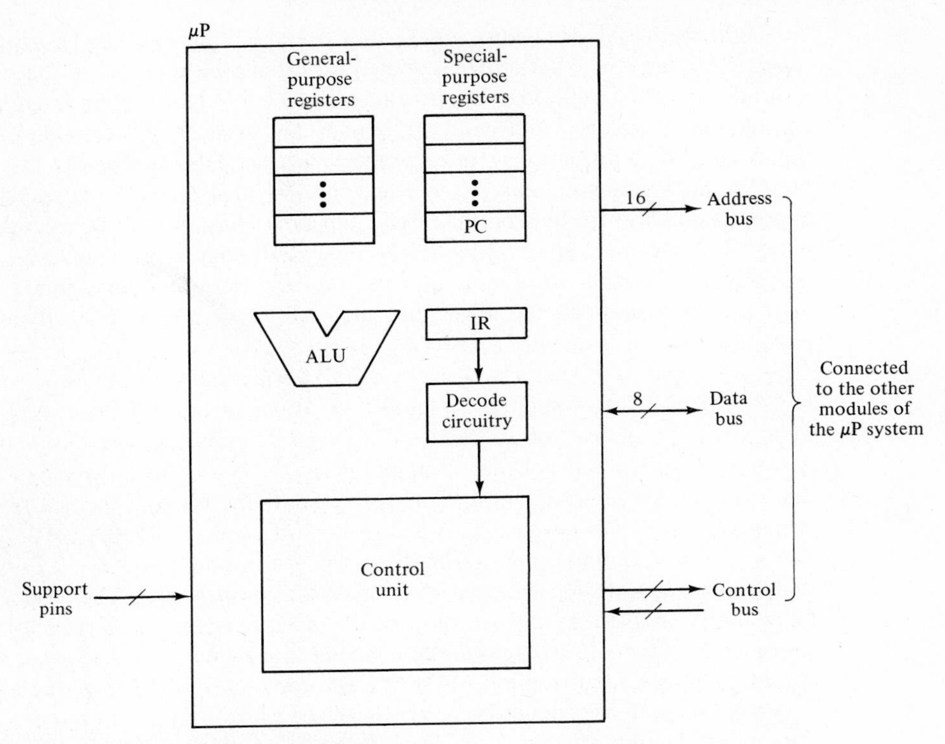

Figure 11.5 Model of a microprocessor.

The key component of the microprocessor is the control unit. The main function of the control unit is to control and synchronize the operations required to fetch, decode, and execute the sequence of instructions of a program that is stored in memory. As has been mentioned, the execution of instructions involves a synchronized sequence of data transfers and data transformations. Under the control of the control unit, data is transformed in the microprocessor by the ALU. The ALU performs the arithmetic and logic operations, as specified by the instructions, on data stored in the general-purpose registers.

Also under the control of the control unit, data can be transferred between registers within the microprocessor and between a register in the microprocessor and a register in an external module such as a memory location or an I/O port. The data transfer, and the handshaking required to synchronize the data transfers with the external modules, are handled through the three external busses: the address bus, the data bus, and the control bus. The general functions of these busses were described in the discussion of Fig. 11.1.

Data transfers between the μP and external modules are accomplished through a sequence of μP machine operations, also called *machine cycles*. The machine cycles for data transfer can be classified into five types: opcode fetch, memory read, memory write, I/O read, and I/O write.

An *opcode-fetch* machine cycle consists of the sequence of primitive operations required to fetch and decode the opcode of an instruction that is to be executed. Of course, immediately before execution begins, the PC contains the address of the memory location in which the instruction is stored. The control unit begins an opcode-fetch machine cycle by placing the required 16-bit memory address (the contents of the PC) onto the 16-bit address bus. Then, the control unit places the required signals on the control bus for enabling the memory module to perform a memory read operation. The memory module responds by placing the contents of the specified memory location onto the data bus. These contents are the 8-bit opcode. After an appropriate delay, the control unit transfers the contents of the data bus into the IR. Finally, the decoding circuitry decodes the opcode to identify the instruction, thereby completing the opcode-fetch machine cycle.

A *memory read* machine cycle is similar to an opcode-fetch memory cycle in that it also fetches an 8-bit quantity from a memory location. However, for a memory read machine cycle, the address placed onto the address bus is not always the contents of the PC. Also, the destination of the data transfer is not the IR but a general-purpose or special-purpose register. For example, a memory read machine cycle is required in the execution of an LDA instruction. The control unit begins a memory read machine cycle by placing the required 16-bit memory address onto the address bus. Next, the control unit places signals on the control bus for enabling the memory module to perform a memory read operation. The memory module responds by placing the contents of the memory location onto the data bus. The control unit then transfers the contents of the data bus into the specified destination general-purpose or special-purpose register, thereby completing the memory read machine cycle.

A *memory write* machine cycle consists of the sequence of operations required for transferring an 8-bit quantity from a register in the microprocessor to a memory location in the memory module, as is needed, for example, in the execution of the STA instruction. The control unit begins a memory write machine cycle by placing the required 16-bit memory address onto the address bus. At the same time, the control unit places on the data bus the data that is to be transferred to the memory location. Next, the control unit places signals on the control bus for enabling the memory module to perform a memory write operation. And, the memory module responds by performing a memory write operation, to complete the memory write machine cycle.

The sequence of operations for an *I/O-read* machine cycle is identical to that of a memory read machine cycle. The source of the data transfer is not, however, a memory location but rather an *input* port in an I/O interface module. Consequently, the control unit places the required 8-bit input-port address onto the address bus, and normally on the low 8-bit A_7–A_0 lines. Also, the control unit places signals on the control bus to enable the required I/O interface module, instead of the memory module, for an I/O read operation.

Similarly, the sequence of operations for an *I/O-write* machine cycle is identical to that of a memory write. The destination of the data is not, however, a memory location, but an *output* port in an I/O interface module. Consequently, the control unit places the required 8-bit output-port address on the low-order 8 bits of the address bus, and the appropriate signals on the control bus, to enable the selected I/O interface module for an I/O write operation.

We have considered a general model of a microprocessor, as well as an overview of its operations. To gain a greater insight into the operations of a microprocessor, we will now study in detail the operations of a specific microprocessor—the 8085 μP.

11.3 THE 8085 MICROPROCESSOR

The 8085, whose instruction set has been presented in detail in Chapters 9 and 10, is a general-purpose microprocessor with a data path of 8 bits. It is capable of directly addressing up to 64K bytes of memory, 256 input ports, and 256 output ports. A pin-out diagram of the 8085 is given in Fig. 11.6(a), and the functional block diagram in Fig. 11.6(b). For the sake of consistency with the literature from the manufacturer (Intel Corporation), all signals labelled on the outside of the block diagram should always be considered to be active-high. Therefore, the notation .H should be assumed unless otherwise indicated.

The 8085 microprocessor is packaged in a 40-pin, dual-in-line package. Four of the pins are support pins: V_{CC}, V_{SS}, X_1, and X_2. Power is supplied by a single 5-V power supply connected across the V_{CC} and V_{SS} pins, positive at V_{CC} and ground at V_{SS}. The synchronization of all the μP operations is based on the frequency of an internal clock. The frequency of this clock is determined by a frequency-selective circuit connected across the X_1 and X_2 inputs. This circuit can be an RC circuit, an LC circuit, or a crystal. Alternatively, an external clock source can be connected to X_1 only. In either case, an internal T flip-flop energized from the X_1 input halves the frequency to make

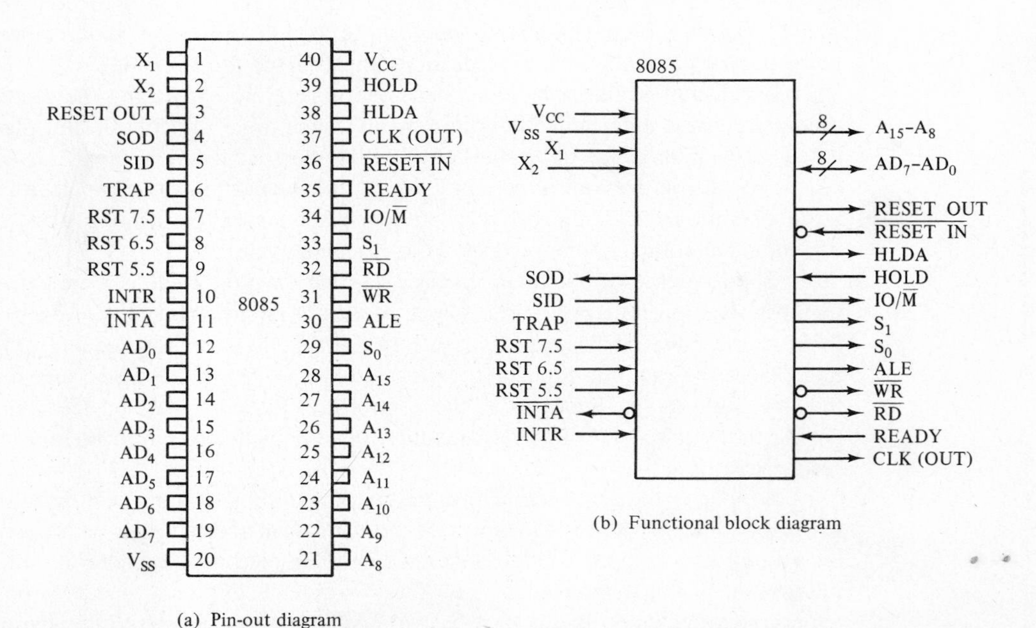

(a) Pin-out diagram

(b) Functional block diagram

Figure 11.6 The 8085 microprocessor.

the internal clock frequency one-half the frequency of the frequency-selective circuit or of the external clock. Therefore, an 8085 μP operated with a 6-MHz crystal will have an internal clock frequency of 3 MHz and a period of approximately 333 ns. The minimum driving frequency for the 8085 is 1 MHz, giving a minimum internal clock frequency of 0.5 MHz. The upper frequency limit depends on the version of the 8085: either the 8085A or the 8085A-2. The maximum driving frequency of the 8085A is 6 MHz, and that of the 8085A-2 is 10 MHz.

Twenty of the forty pins of the 8085 are for control and other miscellaneous functions. The control signals of immediate interest are $\overline{RD}$, $\overline{WR}$, IO/$\overline{M}$, and ALE. The functions of these signals are explained in detail later in the present section. Other control signals, such as INTR, $\overline{INTA}$, RSTs, SID, SOD, and HOLD, will be described in subsequent sections as they are required.

Only 16 pins remain. Unfortunately, this is not enough for the 16-bit address bus and the 8-bit data bus required to realize an 8-bit μP, as shown in Fig. 11.5. To circumvent this problem, the 8085 μP uses a time-multiplexed data bus. More specifically, the eight lines AD_7–AD_0 are used sometimes as the 8-bit data bus and sometimes as the low-order 8 bits of the address bus. Then, together with A_{15}–A_8, they form a 16-bit address bus at the appropriate times. Multiplexing the data bus has the advantage of lowering the pin count, but has the disadvantages of a decrease in speed and an increase in control and circuit complexity.

For the realization of the 8-bit μP of Fig. 11.5 with an 8085 μP, some external circuitry is required, as illustrated in Fig. 11.7. The external circuitry consists simply of an 8-bit latch, which is essentially a set of eight trailing-edge triggered D flip-flops. The

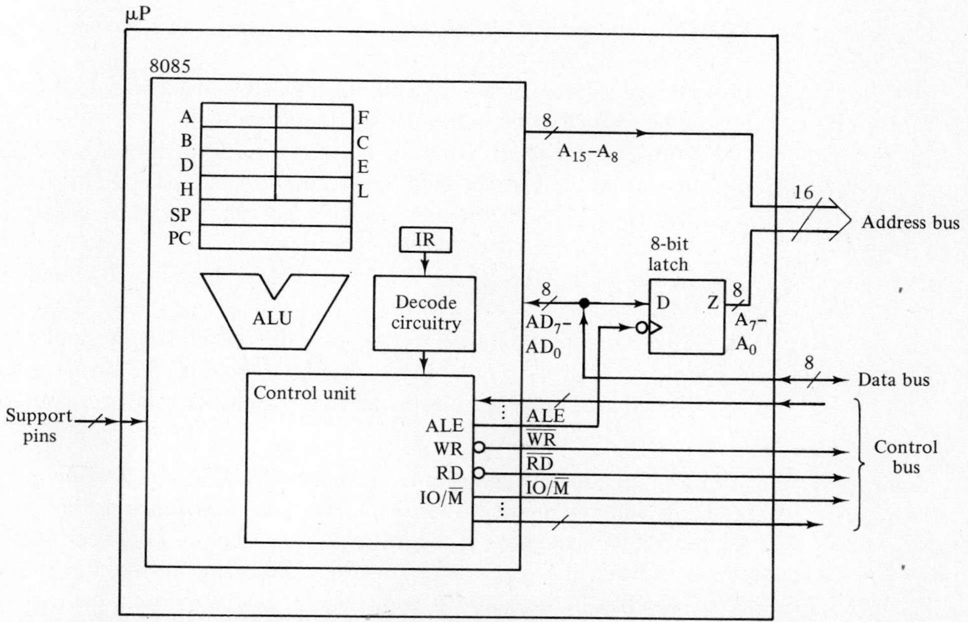

Figure 11.7 An 8085 implementation of an 8-bit microprocessor.

operation of this circuitry can be understood from a detailed explanation of the memory write and memory read operations.

Recall that for a memory write operation (i.e., a memory write machine cycle), the μP places the appropriate 16-bit memory address onto the address bus, and the 8-bit data on the data bus, and then places the appropriate signals on the control bus. For the 8085 these control signals are low for $IO/\overline{M}$ and $\overline{WR}$, and high for $\overline{RD}$. As mentioned, the AD lines have to be time-multiplexed, and this requires the following steps:

1. The high-order 8 bits of the 16-bit address are placed onto A_{15}–A_8 and the low-order 8 bits onto AD_7–AD_0.
2. The signal ALE is pulsed. This pulsing causes the low-order 8 bits of the address to be latched at the trailing edge of this pulse. Consequently, the 16-bit address is now available on the 16-bit address bus.
3. With the low-order 8 bits of the address saved in the 8-bit latch, the AD_7–AD_0 lines are now free to be used as the data bus. Therefore, the μP will place the data to be transferred onto the data bus. Note that even though this 8-bit data appears at the inputs of the 8-bit latch, this data has no effect on the low-order 8 bits of the address since ALE remains low.
4. At this point the signals $IO/\overline{M}$ and $\overline{WR}$ should be low and $\overline{RD}$ should be high.

The AD lines are similarly time-multiplexed for a memory read operation, such as an opcode-fetch or a memory read machine cycle. For this operation the following steps are required.

1. The high-order 8 bits of the address are placed onto A_{15}–A_8, and the low-order 8 bits onto AD_7–AD_0.
2. The signal ALE is pulsed. This pulsing causes the low-order 8 bits of the address to be latched at the trailing edge of this pulse.
3. With the low-order 8 bits of the address saved, the AD_7–AD_0 lines are now free to be used as the data bus. Therefore, the μP relinquishes control of the data bus so that the memory module can place the subsequently retrieved 8-bit data (or opcode) onto the data bus.
4. At this point, the signals $IO/\overline{M}$ and $\overline{RD}$ should be low and $\overline{WR}$ should be high.
5. After an appropriate delay, the data (or opcode) is available on the data bus (and on AD_7–AD_0), having been placed there by the memory module.
6. The control unit then places the data (or opcode) into the appropriate register.

For a more concrete and detailed explanation, let us consider the μP system shown in Fig. 11.8, which comprises the μP of Fig. 11.7 and a memory module connected to it. Assume that the execution of the instruction in Memory Location 0902H has just been completed and that the μP is ready to execute the next instruction (STA 09EFH), the opcode of which is in Memory Location 0903H. In other words, the μP is ready to begin

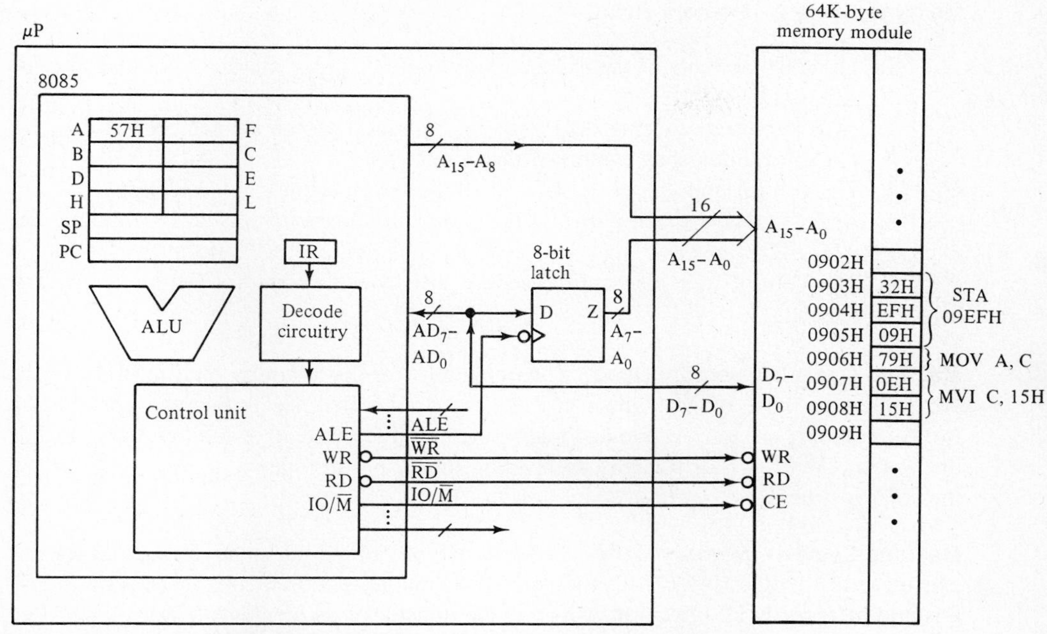

Figure 11.8 A sample μP system configuration.

the fetch-decode-execute cycle for this instruction. At this time, the contents of the PC are 0903H.

The following sequence of operations (machine cycles) are required to fetch, decode, and execute this instruction.

Machine Cycle 1—Opcode Fetch

1. The high-order 8 bits, 09H (00001001B), of the PC are placed onto A_{15}–A_8, and the low-order 8 bits, 03H (00000011B), onto AD_7–AD_0.
2. ALE is pulsed, causing 03H to be latched and become available on A_7–A_0.
3. The μP relinquishes control of the AD lines.
4. The control unit causes $IO/\overline{M}$ and $\overline{RD}$ to become low, and $\overline{WR}$ high.
5. After a delay required by the memory module to output the contents of a memory location, the opcode 32H (from Location 0903H) is available on AD_7–AD_0.
6. The control unit places the opcode 32H into the IR.
7. The instruction is decoded and determined to be an STA instruction. The source register of an STA instruction is Register A, of course, and the destination register is a memory location. The control unit ''knows'' that the address of that destination location is stored in the next two locations in memory following the location containing the STA opcode. Therefore, two additional memory read machine cycles are required. The PC now contains 0904H.

Machine Cycle 2—Memory Read

1. The 09H contents of the PC are placed onto A_{15}–A_8, and the 04H placed onto AD_7–AD_0.
2. ALE is pulsed, causing 04H to be latched and become available on A_7–A_0.
3. The μP relinquishes control of the AD lines.
4. The control unit causes $IO/\overline{M}$ and $\overline{RD}$ to become low, and $\overline{WR}$ high.
5. After a delay, the low-byte EFH of the destination register is available on AD_7–AD_0.
6. The control unit places EFH in the low-byte portion of a 16-bit temporary register.

Machine Cycle 3—Memory Read The operations for this memory read machine cycle are similar to those of Machine Cycle 2. The only difference is that the contents of Location 0905H are retrieved and stored in the high-byte instead of the low-byte portion of the temporary register, thereby completing the 16-bit destination address (09EFH) of the STA instruction.

Machine Cycle 4—Memory Write Now, the μP has all the information required for the execution of the STA 09EFH instruction. The actual execution of this instruction is just a single memory write operation, which is the transfer of the contents (57H) of Register A to Memory Location 09EFH. This requires the following sequence of operations:

1. The 09H contents of the temporary register are placed onto A_{15}–A_8, and the EFH contents are placed onto AD_7–AD_0.
2. ALE is pulsed, causing EFH to be latched and become available on A_7–A_0.
3. The contents (57H) of Register A are placed onto AD_7–AD_0, thereby making them available on D_7–D_0.
4. The control unit causes $IO/\overline{M}$ and $\overline{WR}$ to become low, and $\overline{RD}$ high.
5. After a delay required by the memory module to place data into a memory location, the contents of Memory Location 09EFH are changed to 57H, thereby completing the memory write machine cycle and the execution of the STA instruction.

In summary, the execution of a stored program is simply a synchronized sequence of machine cycles. The execution of any instruction begins with an opcode-fetch machine cycle. Under the control of the control unit, the opcode of the instruction is fetched into the IR and decoded. The actual sequence of operations that follows depends on the particular instruction. For some instructions, such as STA, two additional memory reads are required to obtain the 16-bit destination address stored in the next two memory locations, and a memory write is required to perform the store. For a single-byte instruction such as the MOV A,C instruction (79H) shown in Location 0906H of Fig. 11.8, only the opcode-fetch machine cycle is required. No other machine cycle is necessary because the actual moving of data between registers requires no additional external memory reference. For a 2-byte instruction such as the MVI 15H instruction in Locations 0907H and 0908H, only one additional memory read machine cycle is required along with the opcode-fetch machine cycle. Why?

11.4 INSTRUCTION CYCLE, MACHINE CYCLES, AND T STATES

An *instruction cycle* is defined to be all the machine operations that are required to fetch, decode, and execute an instruction. An instruction cycle consists of a sequence of *machine cycles*. A machine cycle corresponds to a single machine operation on the external busses. For the instructions of the 8085 μP, an instruction cycle is divided into one to five machine cycles, each of which is one of the following seven types:

1. Opcode fetch (OF)
2. Memory read (MR)
3. Memory write (MW)
4. I/O read (IOR)
5. I/O write (IOW)
6. Interrupt acknowledge (INA)
7. Bus idle (BI)

At the beginning of each machine cycle, three control signals, $IO/\overline{M}$, S_1, and S_0, are generated to identify the type of machine cycle. The values of these signals corresponding to each type of machine cycle are shown in Table 11.1.

Under normal circumstances, each of the above machine cycles, with the exception of the opcode fetch, requires three (internal) clock cycles, also called *T states*. In other words, each of the machine cycles (except opcode fetch) is divided into the three T states. An opcode-fetch machine cycle is essentially a special memory read machine cycle that has additional T states for the decoding and, sometimes, execution of an instruction. The number of additional T states ranges from one to three, depending on the instruction. Therefore, an opcode-fetch machine cycle is generally divided into four to six T states.

Figure 11.9(a) contains an illustration of an instruction cycle. Shown are the machine cycles and T states for an STA instruction. In machine cycle M1, the opcode (32H) is fetched from a memory location and decoded in four T states. Two additional memory read machine cycles, M2 and M3, are required for this STA instruction to obtain the 16-bit destination address stored in the next two locations in memory. Finally, machine

TABLE 11.1 8085 MACHINE CYCLE CHART

Machine cycle	$IO/\overline{M}$	S_1	S_0	$\overline{RD}$	$\overline{WR}$	$\overline{INTA}$
Opcode fetch (OF)	0	1	1	0	1	1
Memory read (MR)	0	1	0	0	1	1
Memory write (MW)	0	0	1	1	0	1
I/O read (IOR)	1	1	0	0	1	1
I/O write (IOW)	1	0	1	1	0	1
Interrupt acknowledge (INA)	1	1	1	1	1	0
Bus idle (BI): DAD	0	1	0	1	1	1
INA(RST/TRAP)	1	1	1	1	1	1
HALT	TS	0	0	TS	TS	1

0 = low, 1 = high, TS = three-state.

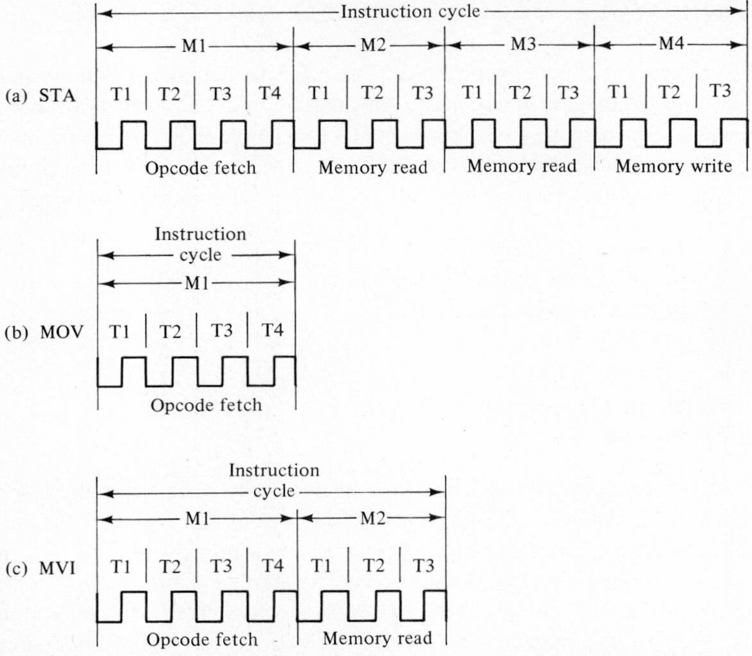

Figure 11.9 Instruction cycles, machine cycles, and T states for STA, MOV, and MVI instructions.

cycle M4 is required for the memory write for the actual store operation. In Figs. 11.9(b) and (c), two other instruction cycles are shown, those for the MOV and MVI instructions, respectively.

Note that for the STA instruction, the opcode-fetch machine cycle requires four T states, but each of the other three machine cycles requires just three T states each. Consequently, 13 T states are needed for the execution of an STA instruction. The corresponding time depends on the frequency of the internal clock. If the 8085 μP is operating with a 6-MHz crystal, then each T state is approximately 0.333 μs in duration, and the time required to execute an STA instruction is $13 \times 0.333 = 4.33$ μs.

Figure 11.10 shows a timing diagram of an opcode-fetch machine cycle for the STA instruction. It details the relative timing of the signals that were discussed in the STA 09EFH instruction execution of the preceding section. At the beginning of the opcode-fetch machine cycle, in T state T1, the $IO/\overline{M}$ is made low to designate that this machine cycle requires a memory reference, as opposed to an I/O reference. Also in T1, the high-order 8 bits (PCh) of the PC are placed onto A_{15}–A_8, as indicated graphically by the crossed lines in Fig. 11.10. PCh remains there throughout the first three T states. Finally during T1, the low-order 8 bits (PCl) of the PC are placed onto AD_7–AD_0, and signal ALE is pulsed. This pulsing causes PCl to be latched at the trailing edge of the pulse, and so become available on A_7–A_0 (not shown).

During the second T state, T2, the μP relinquishes control of the AD lines, as indicated by the dashed lines in Fig. 11.10. Also in T2, the signal $\overline{RD}$ is made low to

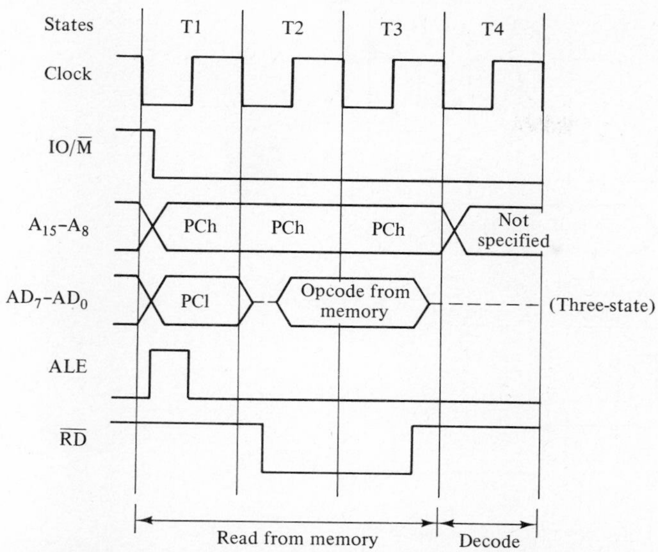

Figure 11.10 Opcode-fetch machine cycle for the STA instruction

enable the memory module for a read operation. The memory module has until almost the end of T3 to place the opcode onto AD_7–AD_0. Near the end of T3, the signal $\overline{RD}$ is made high, which results in the transfer of the opcode into the instruction register IR, and the termination of the read operation for the memory module. During T4, the instruction is decoded to determine whether additional T states are required. For most 8085 instructions, including STA, no additional T states are required for the opcode-fetch machine cycle.

Figure 11.11(a) shows a timing diagram for a memory read machine cycle, and Fig. 11.11(b) shows a timing diagram for a memory write machine cycle. The memory read machine cycle is similar to the first three T states for an opcode-fetch machine cycle. The only differences are that the address placed onto the address bus is not always the contents of the PC, and the destination of the data transfer is not the IR but a general-purpose or special-purpose register. Likewise, a memory write machine cycle is similar to a memory read machine cycle. The only differences are that for a memory write machine cycle, the signal $\overline{WR}$ instead of $\overline{RD}$ is made low, and the transfer of data on the AD lines is in the opposite direction.

Figure 11.12 shows a memory read machine cycle with a TWAIT state inserted. A TWAIT state is used to "stretch" a machine cycle to accommodate slow memory modules that cannot return the data within the time required by the 8085 μP. As shown in Fig. 11.12, during T state T2, the μP examines the value of the READY signal, which is an input signal from the slow memory module being read from. If READY is high, the μP proceeds to T3. But if READY is low, indicating that the memory is not ready, then a TWAIT state is inserted. Additional TWAIT states can be added indefinitely until READY becomes high. When the READY signal goes high, the μP will exit TWAIT and enter T3 to complete the machine cycle. TWAIT states can also be inserted for the opcode-fetch and memory write machine cycles.

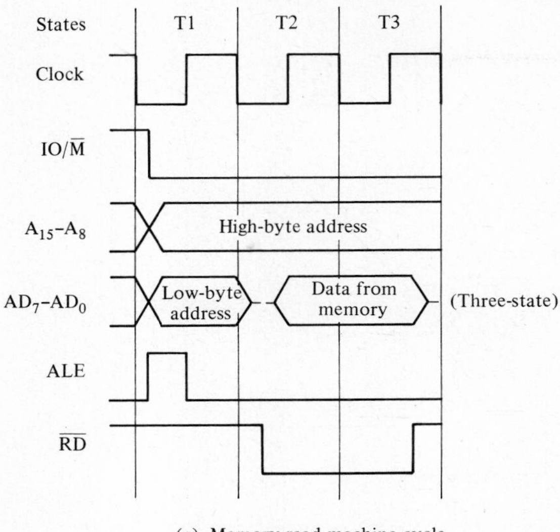

(a) Memory-read machine cycle

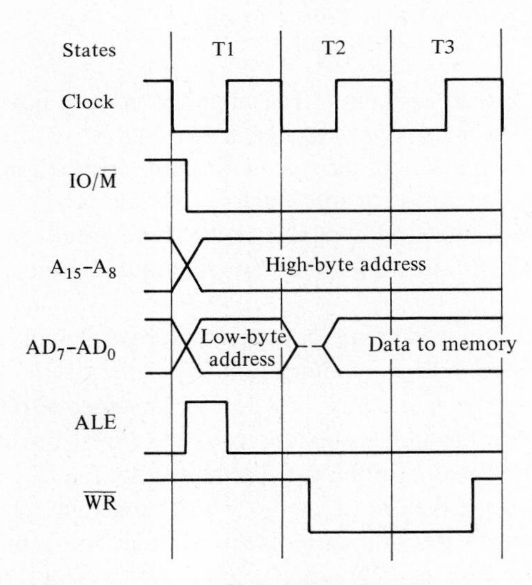

(b) Memory-write machine cycle

Figure 11.11 Memory read and write machine cycles.

11.5 INTERFACING THE MEMORY MODULE

The previously considered μP system of Fig. 11.8 has a 64K-byte memory module connected to the μP. Consequently, each of the 64K memory locations has a unique 16-bit address, from 0000H to FFFFH. However, the memory modules of most smaller μP systems are often less than 64K bytes, and so require fewer than 16 address bits. As an illustration, Fig. 11.13(a) shows a μP system configuration with a 2K memory module.

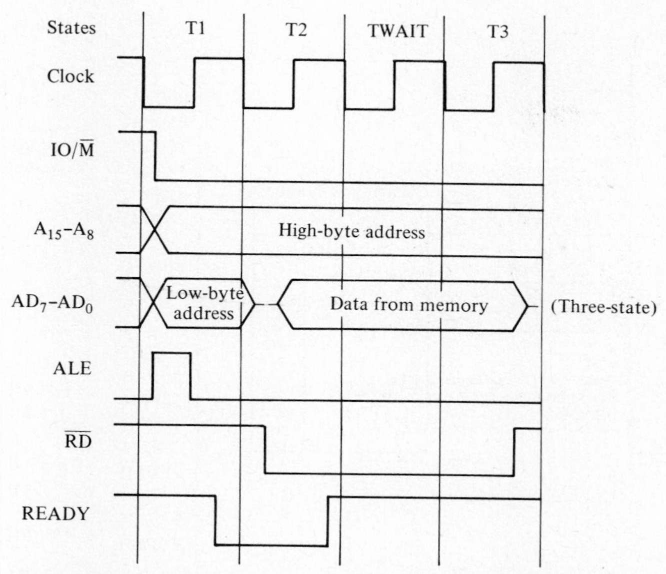

Figure 11.12 Memory read machine cycle with TWAIT state.

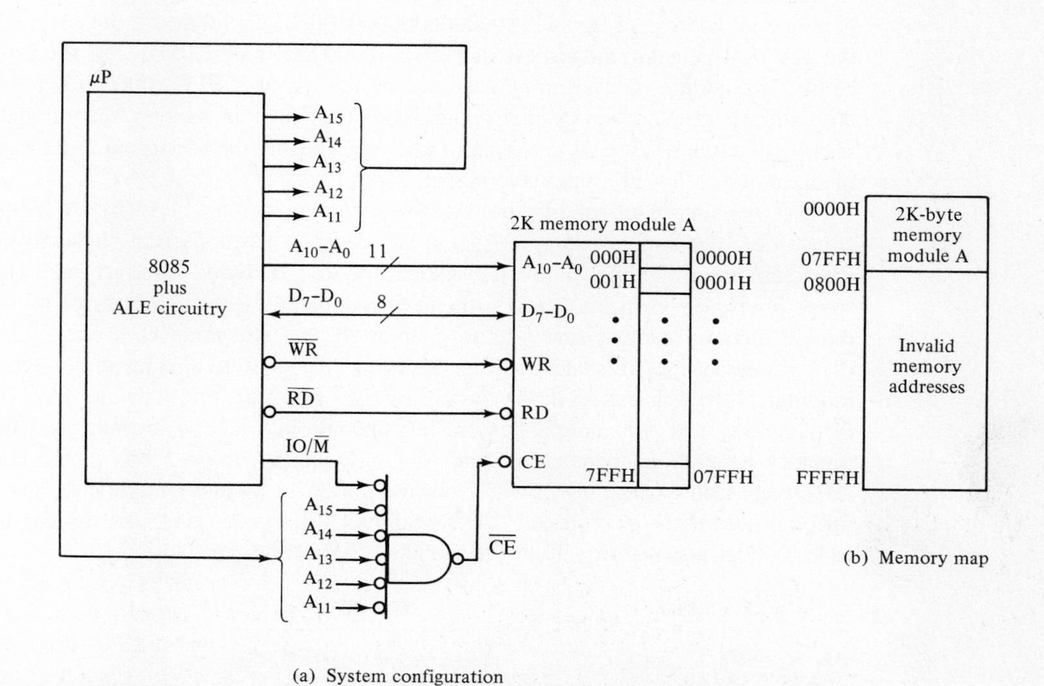

(a) System configuration

Figure 11.13 A fully decoded system with a 2K-byte memory module.

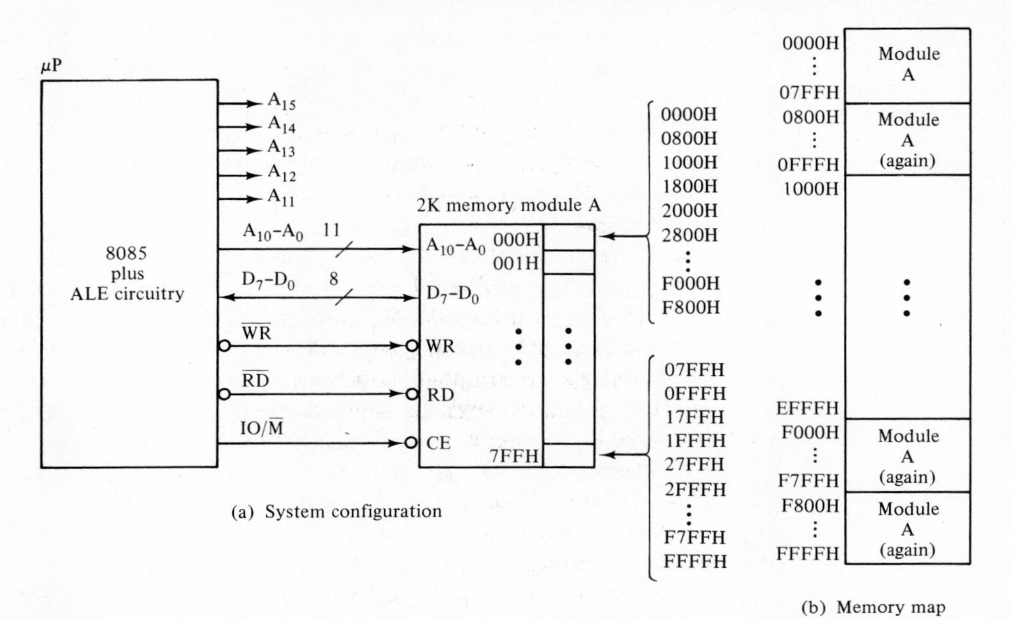

Figure 11.14 A partially decoded system with a 2K-byte memory module.

In this case, to reference the first location of the 2K-byte memory module, the μP must generate an address of 0000H (0000000000000000B). To reference the second location, the μP must generate an address of 0001H (0000000000000001B). For the last location of the 2K module, the μP must generate an address of 07FFH (0000011111111111B). Any address greater than 07FFH is not valid since such an address has a nonzero value for one or more of A_{15}, A_{14}, A_{13}, A_{12}, and A_{11}, causing the $\overline{CE}$ signal to be high instead of the required low for an enable signal.

A *memory map* for this μP system is shown in Fig. 11.13(b). A memory map shows how the logical addresses of the μP memory address space physically map into the actual memory units. In this μP system, the first 2K bytes of the μP memory address space physically map into the 2K memory module A. But the addresses of the rest of the μP memory address space do not map at all, and so cannot be used.

Consider the μP system of Fig. 11.14(a). This system also has a 2K-byte memory module. However, unlike in the system of Fig. 11.13(a), the high bits (A_{15}, A_{14}, A_{13}, A_{12}, and A_{11}) of the address bus are left unconnected and so are not used to select a memory location in memory module A. Consequently, these 5 bits of the address are effectively don't cares. As a result, each location in memory module A has $2^5 = 32$ different addresses or "aliases." For example, the signals generated by the μP for an LDA 0000H instruction and an LDA 0800H instruction are

LDA 0000H	LDA 0800H
A_{15}–A_0 = 0000000000000000	A_{15}–A_0 = 0000100000000000
$\overline{WR}$ = H	$\overline{WR}$ = H
$\overline{RD}$ = L	$\overline{RD}$ = L
IO/$\overline{M}$ = L	IO/$\overline{M}$ = L

Since A_{15}–A_{11} have no effect on memory module A, the executions of these two instructions produce the same result on memory module A—the retrieval of the contents of the first memory location. The same is true for instructions LDA 1000H, LDA 1800H, . . . , LDA 0F800H. A memory map for this μP system is shown in Fig. 11.14(b). As shown, all 64K memory addresses of the μP memory address space are valid addresses. However, since there are only 2K *physical* memory locations, each physical memory location is shared by 32 addresses.

The system configuration of Fig. 11.13(a) illustrates a μP system in which the addresses are *fully decoded,* and that of Fig. 11.14(a) a μP system in which the addresses are *partially decoded.* In a μP system in which the addresses are fully decoded, every bit of the address is used to select a physical memory location. Consequently, each physical memory location corresponds to a *unique* address in the μP memory address space. But in a μP system in which the addresses are only partially decoded, not all bits of the address are used to select a physical memory location. Consequently, a don't-care situation exists for some address bits, and so each physical memory location corresponds to more than one address of the μP memory address space.

A partially decoded system has the advantage of being simple and inexpensive since no additional circuitry is required. But it has the disadvantage that since each memory location has more than one address, care must be taken to ensure that the addresses are referenced in some systematic manner in the software programs. Otherwise, difficulties may arise in the maintenance of the software and hardware of the μP system. For example, it may be difficult to expand the memory module at some later date. This is not a problem for a fully decoded system, which has the advantage that each memory location has just one address. But a fully decoded system has the disadvantage of requiring additional address decoding circuitry, which may be substantial, especially if the memory module is complex.

A final illustration of the memory concepts of this section is presented in Fig. 11.15. The desired memory map is shown in Fig. 11.15(a) and the corresponding system configuration in Fig. 11.15(b). Since the 8085 μP has been shown in several preceding examples, it has been omitted for convenience. Note that the system memory module comprises three memory units: two 1K-byte RAMs and a single 2K-byte ROM. As is evident, a mixture of fully decoded and partially decoded schemes is used. Note that for a memory reference by the μP, the signal $IO/\overline{M}$ must be low to enable the 2-to-4 decoder, which in turn controls the enabling of the individual memory units.

For memory unit ROM-B, every bit of the 16-bit address is used to specify a location, specifically,

$$\left.\begin{aligned} A_{15} &= 0 \\ A_{14} &= 0 \\ A_{13} &= 0 \end{aligned}\right\} \text{ to enable the decoder}$$

$$\left.\begin{aligned} A_{12} &= 1 \\ A_{11} &= 0 \end{aligned}\right\} \begin{aligned} &\text{to activate the appropriate decoder output (2)} \\ &\text{to enable the ROM-B chip-enable (CE) input} \end{aligned}$$

$$A_{10}\text{–}A_0 = \text{the actual address within ROM-B}$$

Therefore, the address range for ROM-B is from $\underline{0001}000000000000$B to $\underline{00010}11111111111$B (1000H to 17FFH). Since the addresses for ROM-B are fully decoded, each location in ROM-B has a unique address. Also, since a ROM is read-only,

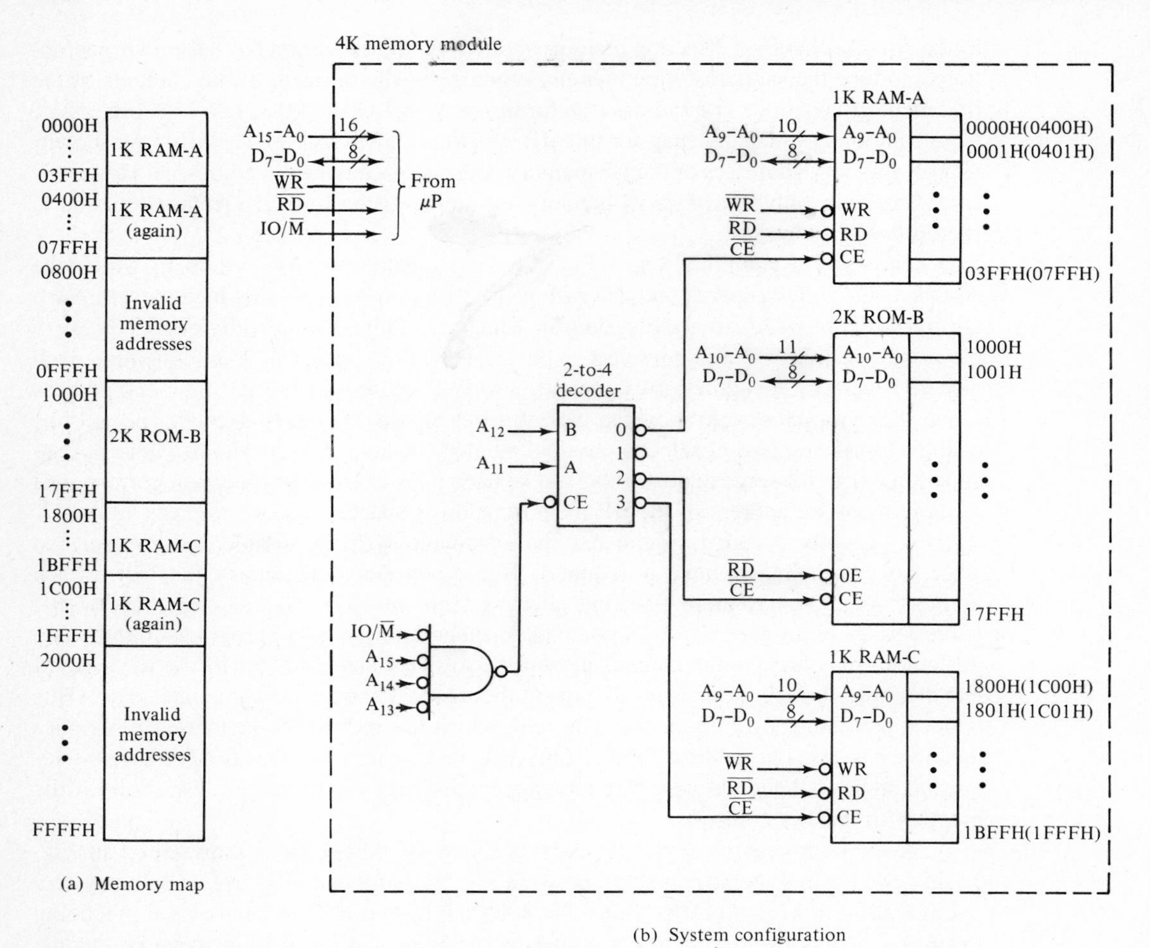

Figure 11.15 A μP system with a 4K-byte memory module.

just the $\overline{RD}$ signal from the μP control bus is required to be connected to the output-enable (OE) input of the ROM.

For memory unit RAM-A, every bit of the 16-bit address, except A_{10}, is used to specify a location. Specifically,

$$\left.\begin{aligned} A_{15} &= 0 \\ A_{14} &= 0 \\ A_{13} &= 0 \end{aligned}\right\} \text{ to enable the decoder}$$

$$\left.\begin{aligned} A_{12} &= 0 \\ A_{11} &= 0 \end{aligned}\right\} \begin{aligned} &\text{to activate the appropriate decoder output (0)} \\ &\text{to enable the RAM-A chip-enable (CE) input} \end{aligned}$$

$$\begin{aligned} A_{10} &= \text{don't care} \\ A_9\text{–}A_0 &= \text{the actual address within RAM-A} \end{aligned}$$

Since the addresses for RAM-A are partially decoded, each location has two different addresses. The addresses range from 0000H to 03FFH, and repeat from 0400H to 07FFH.

The operation of RAM-C is similar to that of RAM-A. Each location in RAM-C also has two different addresses. These addresses range from 1800H to 1BFFH, and repeat from 1C00H to 1FFFH. This repeating is a result of A_{10} being a don't care. Also, address bits A_{11} and A_{12} must both be 1 to activate the appropriate decoder output (3) to enable the RAM-C chip-enable (CE) input.

Note that the addresses in the range 0800H to 0FFFH are not valid since the corresponding decoder output (1) does not enable any memory unit. Furthermore, any address greater than 1FFFH is invalid since no memory unit is enabled. More specifically, for such an address at least one of A_{15}, A_{14}, and A_{13} is nonzero, and any nonzero value of these address bits prevents the enabling of the 2-to-4 decoder.

Now consider what would happen if the CE input of ROM-B was connected to decoder output 1 instead of to decoder output 2. Obviously, the memory map for the μP would change. Then the address range for ROM-B would become 0800H to 0FFFH, and the addresses in the address range of 1000H to 17FFH would become invalid. Conceptually, the memory module can be configured and the addresses can be decoded in any manner desired by the μP system designer. However, the addresses referenced in the software programs must be consistent with the hardware configuration. In other words, the data must be where the program "thinks" it is. The key point here is that the μP system designer must be aware of the hardware/software interaction that is inherent in the design of a μP system.

SUPPLEMENTARY READING (see Bibliography)

[Clements 82], [Conffron 83], [Gorsline 85], [Intel 79], [Intel-A], [Intel-B], [Liu 84], [Monolithic], [Muchow 83], [Wiatrowski 80]

PROBLEMS

11.1. In the model of a microprocessor system shown in Fig. 11.1, explain the directions of the arrows of the three busses. For example, why is there a bidirectional data bus between the memory module and the μP but only a unidirectional data bus between the input and output interface modules and the μP?

11.2. Draw block diagrams for the following RAM modules. Specify the number of address lines and data lines.

(a) 128×4 bits

(b) 8192×8 bits

(c) $128\text{K} \times 8$ bits

11.3. Realize the $2\text{K} \times 8$ RAM module of Fig. 11.16 with two $1\text{K} \times 8$ memory modules of the type shown in Fig. 11.3(a). Note that unlike the $2\text{K} \times 8$ memory module of Fig. 11.3(b), this memory module has a chip-enable (CE) input.

11.4. Realize the $8\text{K} \times 8$ RAM module shown in Fig. 11.17 by using four $2\text{K} \times 8$ RAMs (Fig. 11.16) and a 2-to-4 decoder.

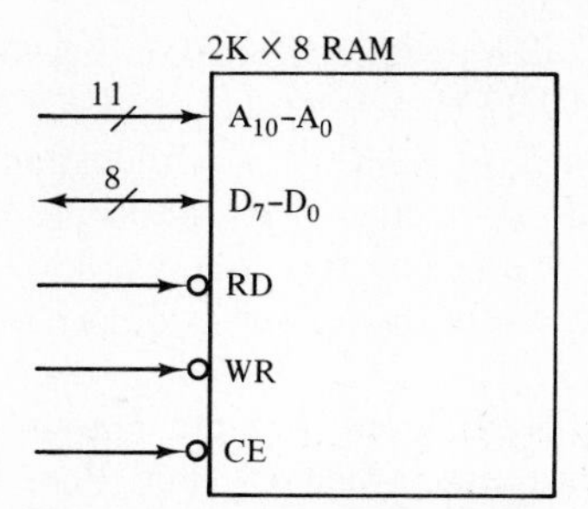

Figure 11.16 2K × 8 RAM module for Problem 11.3.

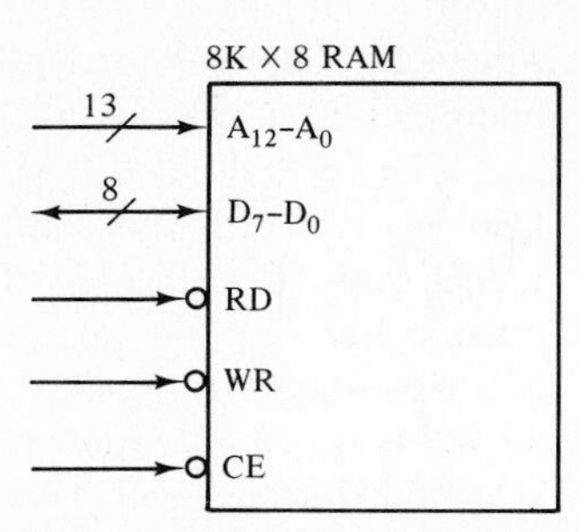

Figure 11.17 8K × 8 RAM module for Problem 11.4.

11.5. Construct the memory module shown in Fig. 11.18, which provides 6K × 8 bits of EPROM and 2K × 8 bits of RAM. *Hint:* Use three 2716 EPROMs (Sec. 6.5.2 and Fig. 6.29) and two 1K × 8 RAM modules [Fig. 11.3(a)], a 2-to-4 decoder, and any additional logic that may be required.

11.6. What are some of the functions of the input ports of an interface module?

11.7. What are some of the functions of the output ports of an interface module?

11.8. The address spaces of the input ports and the output ports of the 8085 μP are independent of each other. Explain what that means.

11.9. For the realization of an 8-bit μP with an 8085 μP, the AD lines of the 8085 are time-multiplexed to obtain a 16-bit address bus and an 8-bit data bus. Explain in words how this is accomplished through the use of the 8-bit latch shown in Fig. 11.7.

11.10. Define the following terms: instruction cycle, machine cycle, and T state.

11.11. Why is the memory read machine cycle M2 necessary in Fig. 11.9(c)?

11.12. How many instruction cycle(s), machine cycle(s), and T state(s) are required to execute each of the following instructions?

(a) ADI 01H **(b)** INR B **(c)** LXI SP,07FFH
(d) MOV M,B **(e)** JMP 1000H **(f)** POP PSW

11.13. Draw, in the form of Fig. 11.9, the instruction cycle(s), machine cycle(s), and T state(s) that are required to execute each of the instructions of Problem 11.12.

11.14. If an 8085 μP is operating with a 6-MHz crystal, what is the time required (in microseconds) to execute each of the instructions of Problem 11.12?

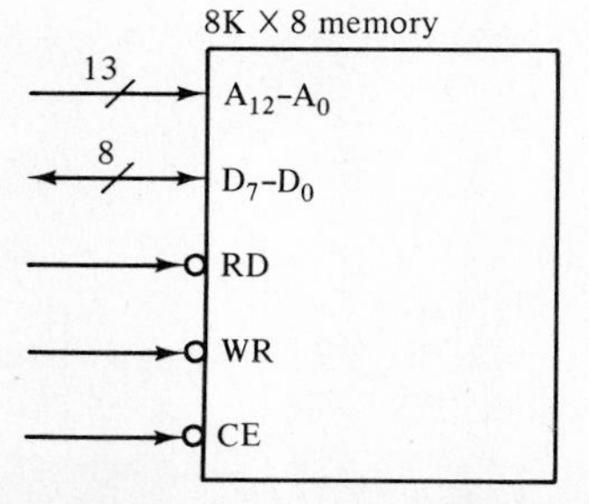

0000H–07FFH: 2K of EPROM locations
0800H–0FFFH: 2K of RAM locations
1000H–1FFFH: 4K of EPROM locations

Figure 11.18 Memory module for Problem 11.5.

11.15. If an 8085 μP is operating with a 4-MHz crystal, what time delay is realized through the execution of the following subroutine?

```
DELAY:  MVI C,100D
 LOOP:  DCR C
        JNZ LOOP
        RET
```

11.16. If an 8085 μP is operating with a 6-MHz crystal, complete the following DELAY subroutine for obtaining a time delay of approximately $\frac{1}{3}$ second.

```
DELAY:  LXI B,     D    ; Answer should be in decimal.
 LOOP:  ORA A           ; Does nothing but "kill time."
        DCX B           ; Decrement 16-bit counter.
        MOV A,B
        ORA C           ; Set flags.
        JNZ LOOP
        RET
```

11.17. Consider the μP system configuration of Fig. 11.8 and the following memory contents:

Memory location	Contents
.	
.	
.	
0812H	7BH
0813H	E6H
0814H	15H
0815H	07H
.	
.	
.	

Assume that the system is executing a program stored in memory and has just completed the execution of the instruction in Location 0812H. The system is now ready to begin the opcode-fetch machine cycle of the next instruction, which is in Location 0813H.

(a) The detailed timing diagram (without labels) for the opcode-fetch machine cycle is given in Fig. 11.19. What are the values for the following pins at the indicated times of t_0, t_1, and t_2?

	t_0	t_1	t_2
A_{15}–A_8 =			
AD_7–AD_0 =			
A_7–A_0 =			
$IO/\overline{M}$ =			
ALE =			
$\overline{RD}$ =			

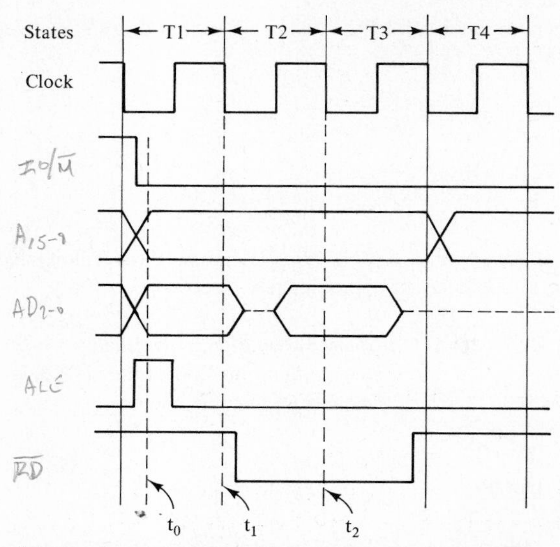

Figure 11.19 Timing diagram for Problem 11.17.

(b) Draw detailed timing diagrams for any additional machine cycle(s) that are required to complete the execution of this instruction. Your timing diagrams should be in the form of Figs. 11.10 and 11.11, except that you are to use actual values for the high-byte address, the low-byte address, and the data.

11.18. Repeat Problem 11.17 for the following memory contents, assuming that the system has just finished the execution of the instruction in Location 0926H and is ready to begin the opcode-fetch machine cycle of the next instruction, which is in Location 0927H.

Memory location	Contents
.	
.	
.	
0926H	5FH
0927H	21H
0928H	00H
0929H	20H
.	
.	
.	

11.19. For the detailed timing diagram of the opcode-fetch machine cycle given in Fig. 11.19 assume that at the indicated times the values for the pins are as specified below. Then, from this diagram and these values, deduce the instruction that is being executed and also the memory location in which it is stored.

	t_0	t_1	t_2
A_{15}–A_8 =	07H	07H	07H
AD_7–AD_0 =	14H	14H	7AH
A_7–A_0 =	13H	14H	14H
$IO/\overline{M}$ =	0	0	0
ALE =	1	0	0
$\overline{RD}$ =	1	1	0

714 7A MOV A, D

11.20. The logic that is connected to the CE input of the memory module A (CE-A) in Fig. 11.13 is equal to

$$\overline{IO/\overline{M}} \cdot \overline{A_{15}} \cdot \overline{A_{14}} \cdot \overline{A_{13}} \cdot \overline{A_{12}} \cdot \overline{A_{11}}$$

Suppose that the logic connected to CE-A is changed to

18 - 1F
58 - 5F
98 - 9F
D8 - DF

$$\overline{IO/\overline{M}} \cdot \overline{A_{13}} \cdot A_{12} \cdot A_{11}$$

in which case the addresses become partially decoded.

(a) How many addresses or aliases does each memory location have?

(b) Specify the memory map for this new configuration.

11.21. The memory module in Fig. 11.14 has addresses that are partially decoded. Reconfigure the system in a way such that the addresses for the 2K memory module A are fully decoded and are mapped into the addresses F000H–F7FFH.

11.22. Consider the following changes to the system configuration of Fig. 11.15.

The CE input of the 1K RAM-A is connected to decoder output 1 (DC_1).

The CE input of the 2K ROM-B is connected to decoder output 3 (DC_3).

The CE input of the 1K RAM-C is connected to decoder output 0 ORed with decoder output 2 ($DC_0 + DC_2$).

Specify the memory map for the new configuration.

11.23. Consider the following changes to the system configuration of Fig. 11.15:

The CE input of the 1K RAM-A is connected to $DC_0 \cdot \overline{A_{10}}$.
The CE input of the 2K ROM-B is connected to DC_2.
The CE input of the 1K RAM-C is connected to $DC_3 \cdot \overline{A_{15}} \cdot \overline{A_{10}}$.

The CE input of the decoder is connected to $\overline{IO/\overline{M}} \cdot A_{14} \cdot \overline{A_{13}}$.

Note that DC_i above is decoder output i. Specify the memory map for the new configuration.

0 3
14

11.24. Given the following memory map, reconfigure the system of Fig. 11.15 to realize it.

RAM-A = 2000H–23FFH
A000H–A3FFH (repeats)

ROM-B = 2800H–2FFFH

RAM-C = 3000H–33FFH
3400H–37FFH
B000H–B3FFH
B400H–B7FFH

All other addresses are invalid.

Chapter 12

Microprocessor Hardware and Interfacing Concepts II

12.1 INTRODUCTION

To review, an instruction cycle is defined to be all the machine operations required to fetch, decode, and execute an instruction. And, an instruction cycle consists of a sequence of machine cycles. For the 8085 μP, there are seven types of machine cycles:

1. Opcode fetch (OF)
2. Memory read (MR)
3. Memory write (MW)
4. I/O read (IOR)
5. I/O write (IOW)
6. Interrupt acknowledge (INA)
7. Bus idle (BI)

The opcode-fetch, memory read, and memory write machine cycles were presented in detail in Chapter 11 and were used as the basis for the discussion of the interfacing concepts relating to the memory module of a μP system.

In this chapter the I/O-read and I/O-write machine cycles are presented in Sec. 12.2. They, in turn, are used as the basis for the discussion of the interfacing concepts relating to the I/O interface modules of a μP system. In Sec. 12.3 is described the interfacing of an I/O module using fully decoded and partially decoded schemes. Also in it, I/O-mapped and memory-mapped I/O are distinguished. In Sec. 12.4 is presented the synchronization of I/O operations using the polling method and the interrupt-driven method. Additionally, the interrupt structure of the 8085 μP is considered in some detail. Finally, in Sec. 12.5 some commercially available I/O interface modules are surveyed. One final introductory point: Although Sec. 12.5 is the only section specifically directed to the 8085 μP, the terminology of the 8085 is used, for convenience, in the other sections as well. The concepts, however, are not limited to the 8085 μP but generally apply to other microprocessors as well.

12.2 I/O-READ AND I/O-WRITE MACHINE CYCLES

An I/O instruction is a data transfer instruction that moves data between a general-purpose register within a μP and an input or output port. As we know, the 8085 μP has two I/O instructions:

IN port# ; where port# is an 8-bit input-port address
OUT port# ; where port# is an 8-bit output-port address

For the IN instruction, the destination register is Register A and the source register is the input port specified by the port# specification in the IN instruction. For the OUT instruction, the source register is Register A and the destination register is the output port specified by the port# specification in the OUT instruction.

The instruction cycle, machine cycles, and T states for an IN instruction are shown in Fig. 12.1(a) and for an OUT instruction are shown in Fig. 12.1(b). As with any instruction, the first machine cycle for an IN instruction is an opcode fetch. After the instruction is decoded and determined to be an IN instruction, an additional memory read machine cycle is required to read in the second byte of this 2-byte instruction. This second byte is, of course, the 8-bit input-port address that is stored in the next location in memory. Finally, to perform the execution of an IN instruction, an I/O-read machine cycle is required to transfer the data from the specified input port into Register A. As shown in Fig. 12.1(b), the instruction cycle for an OUT instruction is similar to that for

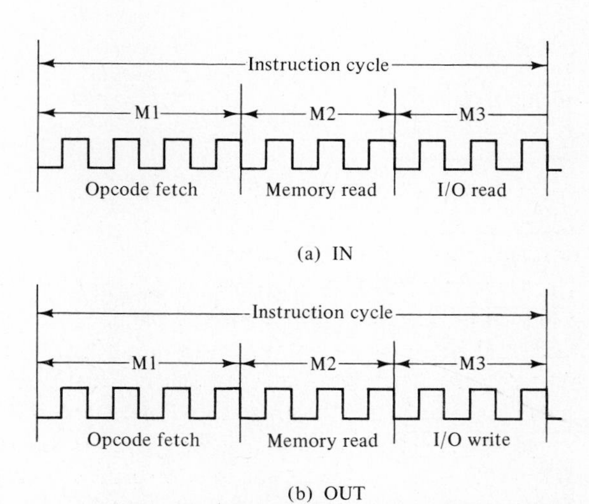

Figure 12.1 Instruction cycles for the IN and OUT instructions.

an IN instruction. The only difference is that in machine cycle M3, an I/O-write machine cycle is required to transfer the data from Register A to the specified output port, while for an IN instruction the transfer of data is in the opposite direction.

The timing of the opcode-fetch and memory read machine cycles were presented in detail in Chapter 11. In this section we are interested in the detailed timing of the I/O-read and I/O-write memory cycles.

Figure 12.2 shows a timing diagram of an I/O-read machine cycle. It details the relative timing of the pertinent μP signals. At the beginning of the I/O-read machine cycle, in T state T1, the $IO/\overline{M}$ signal is made high to indicate that this machine cycle requires an I/O reference, as opposed to a memory reference. Also in T1, the 8-bit input-port address is placed onto AD_7–AD_0 and *also* onto A_{15}–A_8. Further, the signal ALE is pulsed, causing the 8-bit input-port address to be latched at the trailing edge of ALE. If a configuration such as the one shown in Fig. 11.7 is used, then the 8-bit port address is available on both A_7–A_0 and A_{15}–A_8 at the same time.

In the second T state T2, the μP relinquishes control of the AD lines, as indicated by the dashed lines in Fig. 12.2. Also in T2, the signal $\overline{RD}$ is made low. Together with $IO/\overline{M}$, it enables the selected I/O interface module for an I/O read operation. The 8085 μP allows the I/O interface module until near the end of T3 to place the data onto AD_7–AD_0. Near the end of T3, the signal $\overline{RD}$ is made high to transfer the data into Register A and to terminate the I/O read operation for the I/O interface module. As can be seen, the timing of an I/O-read machine cycle is similar to that for the memory read machine cycle shown in Fig. 11.11(a). The principal difference is that $IO/\overline{M}$ is made high for an I/O-read, but low for a memory read machine cycle.

Figure 12.3 shows a timing diagram for an I/O-write machine cycle. The timing for an I/O-write machine cycle is similar to that for an I/O-read machine cycle. One difference, though, is that for a I/O-write machine cycle, the signal $\overline{WR}$, instead of the

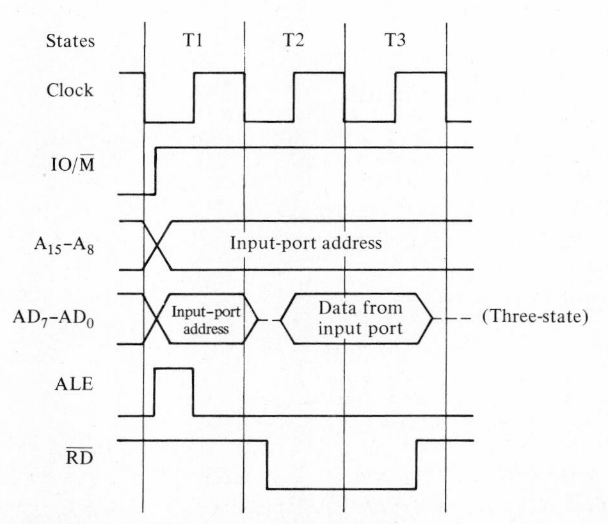

Figure 12.2 I/O-read machine cycle.

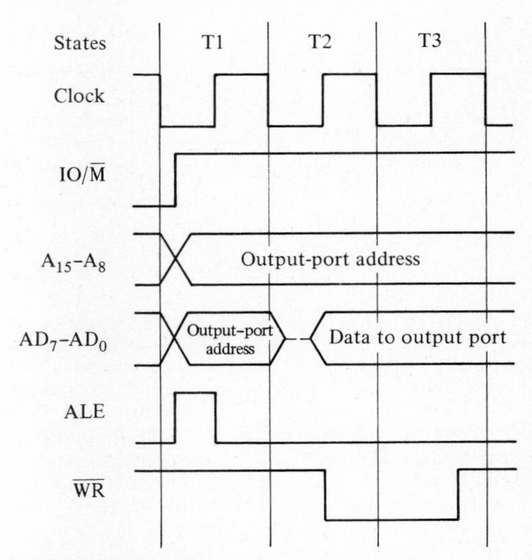

Figure 12.3 I/O-write machine cycle.

signal $\overline{RD}$, is made low. Also, the direction of the data transfer on the AD lines is opposite.

12.3 INTERFACING AN I/O INTERFACE MODULE

The memory and I/O address spaces of the 8085 μP are illustrated in Fig. 12.4. The 8085 μP can directly address up to 64K bytes (0000H to FFFFH) of memory in response to any of the memory reference instructions shown in Fig. 12.4(a). When any of these instructions is executed, the μP will generate the following signals:

$IO/\overline{M} = L$

$\overline{RD}$ or $\overline{WR} = L$ (e.g., $\overline{RD} = L$ for an LDA instruction
$\overline{WR} = L$ for an STA instruction)

And depending on the A_{15}–A_0 values, one of the 64K memory locations will be accessed.

The 8085 μP can also directly address up to 256 (00H to FFH) input ports in response to the IN instruction. When an IN instruction is executed, the μP will generate the following signals:

$$IO/\overline{M} = H$$
$$\overline{RD} = L$$
$$\overline{WR} = H$$

And, depending on the values of A_7–A_0, one of the 256 input ports will be accessed. Similarly, the 8085 μP can directly address up to 256 (00H to FFH) output ports in response to the OUT instruction. When an OUT instruction is executed, the μP will generate the following signals:

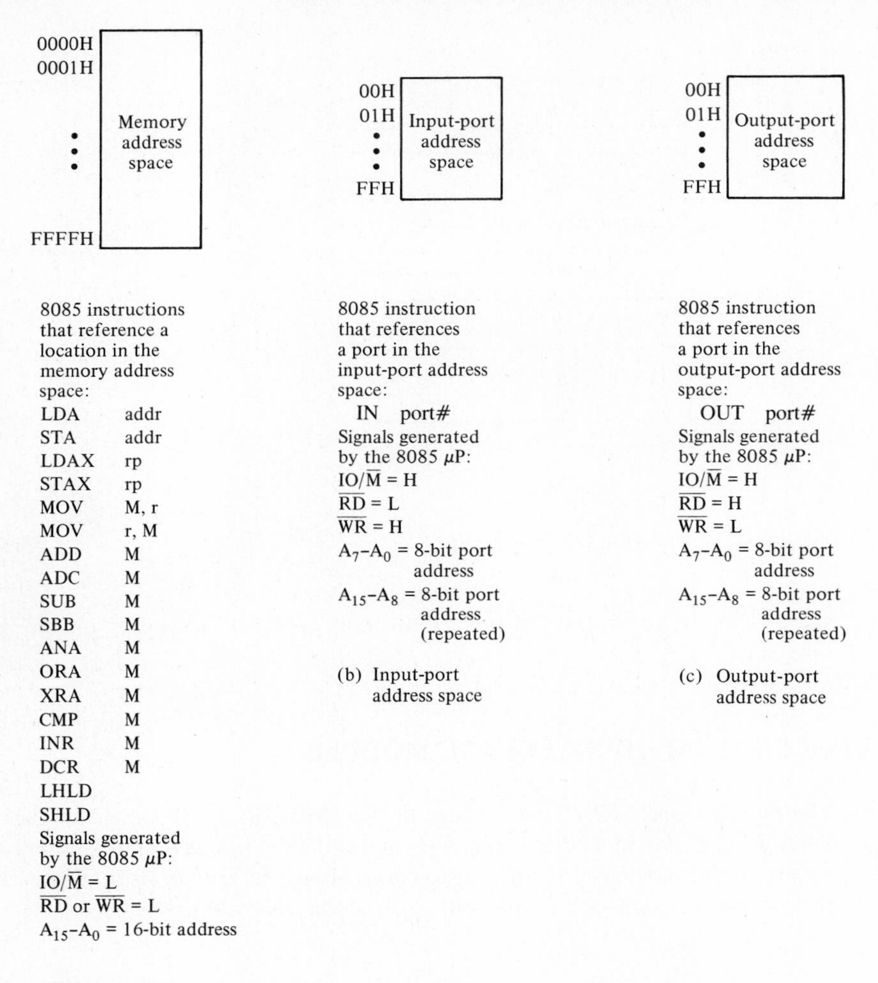

Figure 12.4 The address spaces of the 8085 μP.

$$IO/\overline{M} = H$$
$$\overline{RD} = H$$
$$\overline{WR} = L$$

Then depending on the values of A_7–A_0, one of the 256 output ports will be accessed.

As described in Sec. 11.5, the memory modules required for most μP systems are often less than 64K bytes each. Similarly, only a small portion of the input-port space and of the output-port space are required for most μP systems. Consider Fig. 12.5(a) in which the shown μP system configuration has two interface modules. Within these modules are a total of 11 input ports and 11 output ports. The memory map and *I/O*-

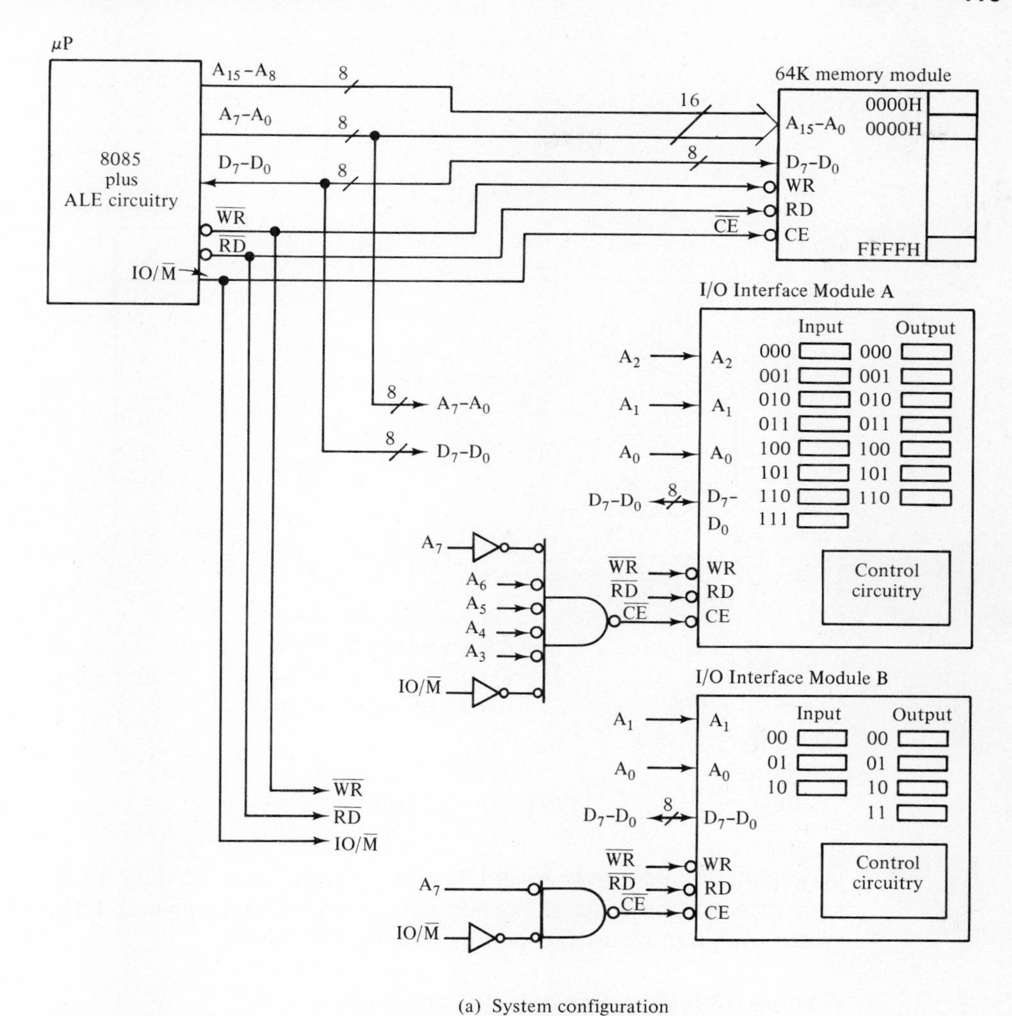

(a) System configuration

Figure 12.5 A μP system with memory and I/O interface modules.

port maps for this μP system are shown in Fig. 12.5(b). Similar to a memory map, the I/O-port maps show how the logical addresses of the I/O address spaces of the μP are mapped physically into the actual I/O ports. In other words, the I/O maps provide the I/O-port address assignment(s) for each I/O port in the μP system. For clarity, the input-port and output-port maps are presented in the form of tables, as shown in Fig. 12.5(b).

For this μP system, note that the port addresses are *fully decoded* for the ports in Interface Module A. In other words, all 8 bits of the port address, A_7–A_0, are used to select a port in this interface module. Consequently, each I/O port within Interface Module A has a unique port address. On the other hand, for the ports in Interface Module B, the port addresses are *partially decoded* since only A_7, A_1, and A_0 are used to select

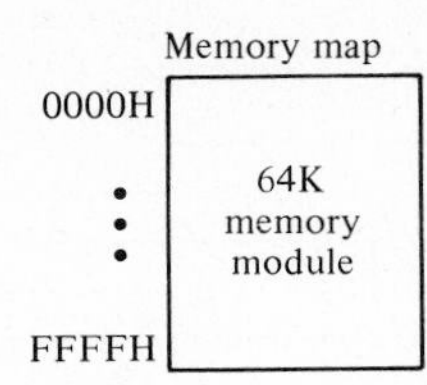

Input-port map

Input port	Input-port address
A-000	80H
A-001	81H
A-010	82H
A-011	83H
A-100	84H
A-101	85H
A-110	86H
A-111	87H
B-00	0XXXXX00B
B-01	0XXXXX01B
B-10	0XXXXX10B

Output-port map

Output port	Output-port address
A-000	80H
A-001	81H
A-010	82H
A-011	83H
A-100	84H
A-101	85H
A-110	86H
B-00	0XXXXX00B
B-01	0XXXXX01B
B-10	0XXXXX10B
B-11	0XXXXX11B

Notes: (1) All other I/O-port addresses (e.g., 90H) are invalid in this μP system.
(2) A-000 designates the port 000 in I/O Interface Module A.
(3) B-00 designates the port 00 in I/O Interface Module B.

(b) Memory and I/O-port maps

Figure 12.5 (*cont.*)

these ports. Consequently, each I/O port within Interface Module B is associated with more than one port address. The port addresses associated with each I/O port of this μP system are given by the I/O-port maps in Fig. 12.5(b).

Example 12.1 Execution of an IN Instruction

Consider the execution of an IN 85H instruction, which transfers into Register A the 8-bit data from Input Port 101 in Interface Module A. Let us trace the instruction cycle of that instruction with respect to the circuit shown in Fig. 12.5(a). Of course, the first two machine cycles are the opcode fetch and the memory read, which fetch the opcode and the input-port addresses, respectively. The third machine cycle is an I/O-read machine cycle in which the contents of Input Port 101 in Interface Module A are transferred into Register A.

At the beginning of this I/O-read machine cycle, in T state T1, the IO/$\overline{\text{M}}$ signal is made high. Further, the 8-bit port address 85H (10000101B) is placed onto AD_7–AD_0 and also onto A_{15}–A_8. The signal ALE is pulsed, with the result that the value of 85H becomes available on A_7–A_0.

In T-state T2 the μP relinquishes control of the AD lines. Also, the signal $\overline{\text{RD}}$ is made low and $\overline{\text{WR}}$ is made high. At this point in time, the status of the relevant signals at the memory and I/O interface modules are as shown in Fig. 12.6. We see that the

Memory Module	Interface Module A	Interface Module B
A_{15}–A_0 = 8585H	A_2 = 1 (H)	A_1 = 0
$\overline{WR}$ = H	A_1 = 0 (L)	A_0 = 1
$\overline{RD}$ = L	A_0 = 1 (H)	$\overline{WR}$ = H
$\overline{CE}$ = IO/$\overline{M}$ = H	$\overline{WR}$ = H	$\overline{RD}$ = L
	$\overline{RD}$ = L	$\overline{CE}$ = H
	$\overline{CE}$ = L	

Figure 12.6 Signal status for Example 12.1.

only module that is enabled is the I/O Interface Module A because of the status of the following signals.

$$
\begin{aligned}
A_7 &= H \\
A_6 &= L \\
A_5 &= L \\
A_4 &= L \\
A_3 &= L \\
IO/\overline{M} &= H
\end{aligned}
$$

which activate only the Module-A CE input. Also, the selected port within Interface Module A is Input Port 101 and the operation is a read because of the status of the following signals.

$$
\begin{aligned}
A_2 &= H \\
A_1 &= L \\
A_0 &= H \\
\overline{RD} &= L
\end{aligned}
$$

As a result, Interface Module A responds by placing the contents of Input Port 101 onto D_7–D_0. Near the end of T3, the signal $\overline{RD}$ is made high by the μP to transfer the data into Register A and to terminate the I/O read operation. ■■

Example 12.2 Execution of an OUT Instruction

Consider the execution of OUT 7EH. Referring to the output-port map in Fig. 12.5(b), and the second row from the bottom, we see that this instruction will transfer the contents of Register A to Output Port 10 in Interface Module B, since 7EH produces A_7, A_1, and A_0 bits of 0, 1, 0, respectively. Let us trace the instruction cycle of this instruction with respect to the system shown in Fig. 12.5(a).

Again, the first two machine cycles are the opcode fetch and memory read, which fetch the opcode and the output-port address, respectively. However, the third machine cycle is an I/O-*write* machine cycle. At the beginning of the I/O-write machine cycle, in T-state T1, the IO/$\overline{M}$ signal is made high. Further, the 8-bit port address 7EH (01111110B) is placed onto AD_7–AD_0 and A_{15}–A_8. Moreover, the signal ALE is pulsed, with the result that 7EH becomes available on A_7–A_0.

During T-state T2, the μP places the contents of Register A onto the AD_7–AD_0 lines, which are also the D_7–D_0 lines. At this point in time, the status of the relevant signals at the memory and I/O modules are as shown in Fig. 12.7. We see that the only

Memory Module	Interface Module A	Interface Module B
A_{15}–A_0 = 7E7EH	A_2 = 1	A_1 = 1
$\overline{WR}$ = L	A_1 = 1	A_0 = 0
$\overline{RD}$ = H	A_0 = 0	$\overline{WR}$ = L
$\overline{CE}$ = IO/$\overline{M}$ = H	$\overline{WR}$ = L	$\overline{RD}$ = H
	$\overline{RD}$ = H	$\overline{CE}$ = L
	$\overline{CE}$ = H	

Figure 12.7 Signal status for Example 12.2.

module that is enabled is I/O Interface Module B, the selected port within it is Output Port 10, and the operation is an I/O write. The Interface Module B will respond by placing the contents of D_7–D_0 into Output Port 10. ■ ■

In summary, the following points should be noted from Examples 12.1 and 12.2. The address, data, and the various control signals are available to all the modules that are connected to the system busses. However, only the one module whose $\overline{CE}$ input is enabled will respond to the signals. With respect to the μP, the signal IO/$\overline{M}$ is used to distinguish a memory reference (IO/$\overline{M}$ = L) from an I/O reference (IO/$\overline{M}$ = H). The address, fully or partially decoded, is used to select the module and a particular memory location or I/O port within the selected module. Finally, the $\overline{RD}$ and $\overline{WR}$ signals are used to specify a read or a write operation.

12.3.1 I/O-Mapped I/O and Memory-Mapped I/O

In a microprocessor such as the 8085 μP, the memory address space is distinct from the input-port address space and the output-port address space, as shown in Fig. 12.4. In other words, the instructions used for memory reference are different from those used for references of input ports and output ports, and the values of the signals generated by each group of these instructions (IO/$\overline{M}$, $\overline{RD}$, and $\overline{WR}$) are different. We can take advantage of this fact to increase the flexibility in configuring the I/O interface modules in a μP system.

In the design of a μP system, I/O ports can be configured in the system in one of two structures:

1. I/O-mapped I/O structure
2. Memory-mapped I/O structure

In an *I/O-mapped I/O structure,* an I/O port is assigned an address in the μP I/O-port address space. In this case, the only way in which such an I/O port can be accessed is with an IN or OUT instruction. But in a *memory-mapped I/O structure,* an I/O port is assigned an address in the μP *memory* address space. Consequently, in this case such an I/O port *cannot* be accessed with an IN or OUT instruction, but instead must be accessed with *any* of the memory reference instructions shown in Fig. 12.4(a). In a memory-mapped I/O structure, an I/O port is simply treated by the μP as a memory location. From an activation point of view, an I/O structure is I/O-mapped if the signal IO/$\overline{M}$ = H causes the CE input to be true, but is memory-mapped if IO/$\overline{M}$ = L does this.

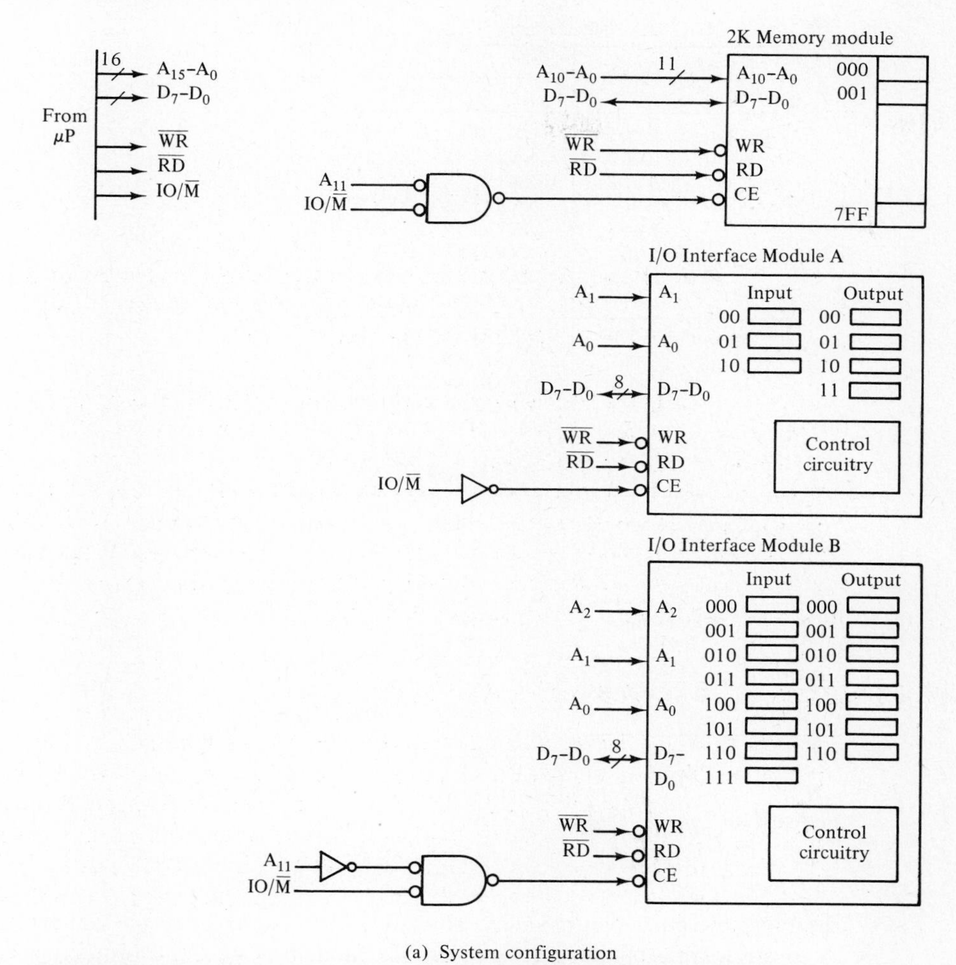

(a) System configuration

Figure 12.8 An illustration of memory-mapped I/O and I/O-mapped I/O.

For an illustration, consider the μP system shown in Fig. 12.8(a). The memory map and the I/O-port maps for this μP system are given in Fig. 12.8(b). Again, for clarity, all three maps are given in the form of tables. There is a memory map for the 2K memory module and Module B, and an input-port map and an output-port map for Module A.

The 2K memory module shown in Fig. 12.8(a) is conventional and also is not pertinent to this discussion. Hence it will not be considered except to note that it will be enabled (CE = true) if A_{11} = L and IO/$\overline{M}$ = L.

The I/O ports within Interface Module A are configured in an *I/O-mapped* I/O structure, because for the transfer of the contents of any of the input ports within Interface Module A to Register A an IN instruction is required. When an IN instruction is executed, the signal IO/$\overline{M}$ will be made high by the μP, thereby enabling Interface Module A and

Memory map

"Memory" modules	Address
2K memory module	X000H–X7FFH
Input ports: B-000	XXXX1XXXXXXXX000B
B-001	XXXX1XXXXXXXX001B
B-010	XXXX1XXXXXXXX010B
B-011	XXXX1XXXXXXXX011B
B-100	XXXX1XXXXXXXX100B
B-101	XXXX1XXXXXXXX101B
B-110	XXXX1XXXXXXXX110B
B-111	XXXX1XXXXXXXX111B
Output ports: B-000	XXXX1XXXXXXXX000B
B-001	XXXX1XXXXXXXX001B
B-010	XXXX1XXXXXXXX010B
B-011	XXXX1XXXXXXXX011B
B-100	XXXX1XXXXXXXX100B
B-101	XXXX1XXXXXXXX101B
B-110	XXXX1XXXXXXXX110B

Note: All other memory addresses are invalid for the μP system.

Input port map

Input port	Input-port address
A-00	XXXXXX00B
A-01	XXXXXX01B
A-10	XXXXXX10B

Output-port map

Output port	Output-port address
A-00	XXXXXX00B
A-01	XXXXXX01B
A-10	XXXXXX10B
A-11	XXXXXX11B

(b) Memory and I/O-port maps

Figure 12.8 (*cont.*)

disabling the other two modules. Similarly, for the transfer of the contents of Register A to any of the output ports in Interface Module A an OUT instruction is required. Observe that since only the values of A_1 and A_0 select the ports, the other address bits (A_7–A_2) are don't cares, as shown in Fig. 12.8(b).

On the other hand, the I/O ports within Interface Module B are configured in a *memory-mapped* I/O structure, as is evident, because only if $IO/\overline{M}$ = L can the CE input possibly be true. For this structure the transfer of the contents of any input port in Interface Module B into Register A requires a memory reference instruction such as an LDA instruction. For example, for the transfer of the contents of Input Port B-000 to Register A the instruction LDA 0800H can be used. (Note from Fig. 12.8 that because of the address don't cares, many other address are equally suitable, such as 0808H, 7948H, etc.). When this LDA instruction is executed, the μP generates the following signals.

$$IO/\overline{M} = L$$
$$\overline{RD} = L$$
$$\overline{WR} = H$$
$$A_{15}\text{–}A_0 = 0800H$$

Consequently, Interface Module B is enabled ($IO/\overline{M}$ = L, A_{11} = H) and Input Port B-000 is selected (A_2, A_1, A_0 = LLL).

Observe that the execution of the LDA instruction is ''intended'' by the μP to access a memory location that is assigned to that address in the memory address space. In reality, though, an input port has been assigned to it and therefore is treated by the μP simply as another memory location. Because of this fact, the following instructions can also be used to perform the same data transfer as LDA 0800H:

```
(a) LXI  H,0800H     (b) LXI  B,0800H     (c) LXI  D,0800H
    MOV  A,M             LDAX B               LDAX D
```

Similarly, to transfer the contents of Register A to an output port in Interface Module B, such as B-110, any of the following instructions can be used:

```
(a) STA 0806H  (b) LXI H,0806H  (c) LXI  B,0806H  (d) LXI  D,0806H
                   MOV M,A          STAX B            STAX D
```

Furthermore, any of the data transformation instructions shown in Fig. 12.4(a) can be used to perform arithmetic and logic operations directly with the contents of the input ports. For example, the following instructions will add the contents of Input Port B-000 to the contents of Register A:

```
LXI  H,0800H
ADD  M
```

Consider the instructions that would be required to perform the equivalent operation with an I/O-mapped input port such as A-00. (See Problem 12.12 at the end of this chapter.)

As we have seen, the main advantage of memory-mapped I/O is the flexibility it affords the μP system designer. Many of the more powerful memory reference instructions can be used with I/O ports. But, there are disadvantages. One minor disadvantage is that the μP system cannot have a full 64K-byte memory module. More important is the fact that the memory reference instructions require a 2-byte (16-bit) address as opposed to a 1-byte address for an IN or OUT instruction. The implication is that more storage and more execution time are required to process the extra byte. Finally, unless the instructions are well commented, it would not be easy to distinguish a memory reference instruction from an I/O-reference instruction, which is desirable in terms of the debugging and maintenance of the program.

12.4 SYNCHRONIZATION OF I/O OPERATIONS

I/O operations are operations that facilitate the transfer of data between the μP and I/O devices, via I/O interface modules. A data transfer from an input device to the μP is called an input operation. And, a data transfer from the μP to an output device is called an output operation. An input device can be simply a set of toggle switches, a keyboard, an analog-to-digital (A/D) converter, or even an I/O interface module of another μP system. An output device can be simply a set of LED displays, a cathode ray tube (CRT) display, a printer, a digital-to-analog (D/A) converter, or again even an I/O interface module of another μP system. Other devices such as a tape drive or a disk drive can function as an input or an output device. I/O devices are generally connected to the μP through an I/O interface module.

For an input operation it is the responsibility of the input device to place the data into a designated input port at the "appropriate" time. And, it is the responsibility of the μP to read the data from the input port (e.g., with an IN instruction) at the "appropriate" time. The obvious question is how do the μP and the input device determine when is the "appropriate" time? Similarly, for an output operation the μP needs to determine the appropriate time to place the data into a designated output port so that an output device can read it at the appropriate time. Since the I/O devices operate asynchronously with respect to the μP system, there needs to be a mechanism to synchronize the transfer of data between the μP and the I/O devices. We will now consider a typical procedure for the synchronization of I/O operations between the μP and I/O devices.

Consider Fig. 12.9(a), in which an interface module connects Input Device A and Output Device B to a μP. There are four input ports and four output ports in this interface module. The input ports of immediate interest are Input Port 00H [labeled PA in Fig. 12.9(a)] and Input Port 03H (labeled Status). The output port of interest is Output Port 00H (labeled PB).

Typically, the procedure for the synchronization between the μP system and an input device for an input operation can be outlined as follows:

Responsibility of the Input Device

1. It is the responsibility of Input Device A to determine whether the designated input port (Input Port PA) is "empty." In other words, Input Device A needs to sense the PA_EMPTY signal to determine whether the previous byte of data has been read by the μP (i.e., determine if PA_EMPTY = true). If PA_EMPTY is not true, then Input Device A waits until it is true.
2. IF Input Device A determines that the designated input port is empty (PA_EMPTY = true), then Input Device A can place the next byte of data into the input port. This is done by placing the data on the PA_DATA lines and strobing the LD_PA signal.

Responsibility of the Interface Module

1. When the LD_PA signal is strobed, the data on the PA_DATA lines is loaded into Input Port PA.
2. Additionally, the controller of the interface module produces the following output signals:
   (a) PA_EMPTY is made false. (PA is now full.)
   (b) PA_INTR is made true (to be discussed later).
   (c) In the Status port, which is shown in Fig. 12.9(b), the PA_EMPTY bit (bit 2) is set to 0, and the PA_INTR bit (bit 3) is set to 1.

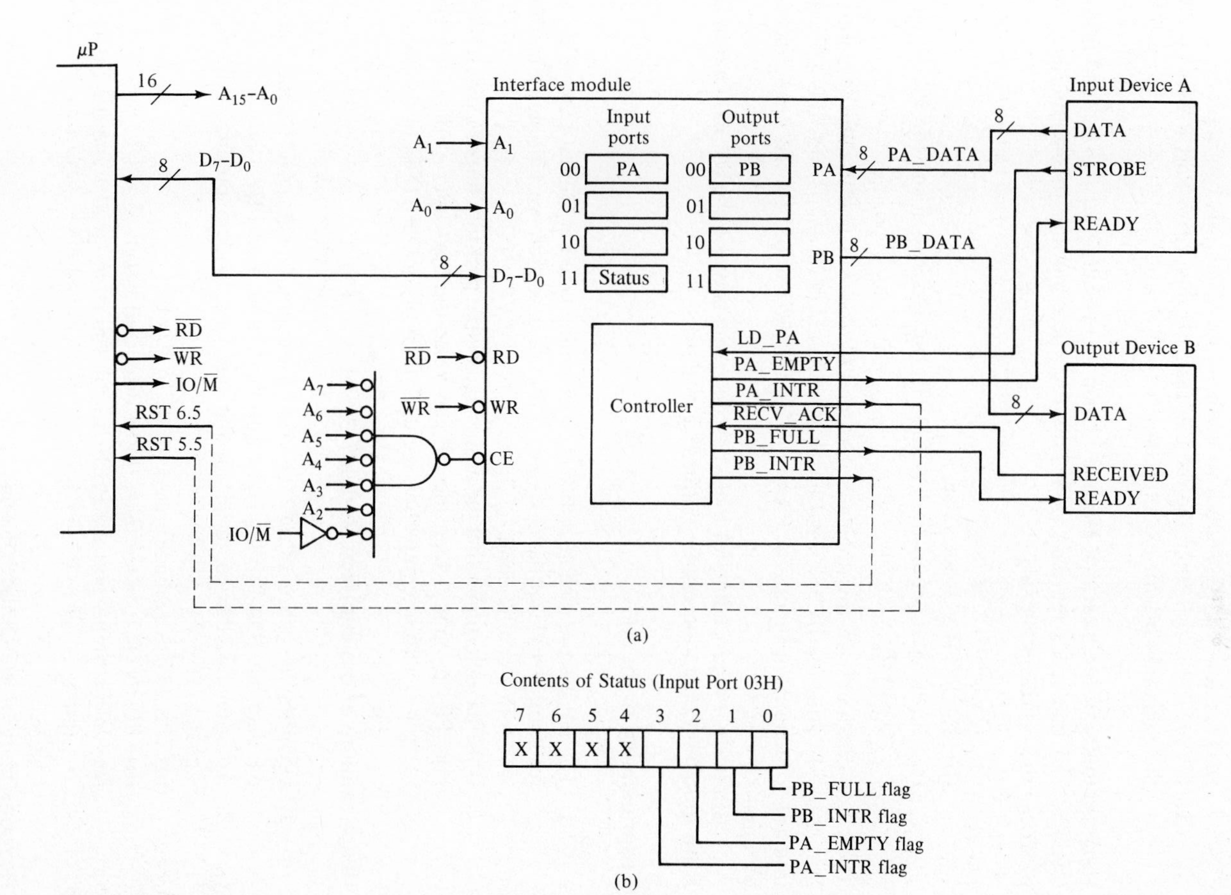

Figure 12.9 Synchronization between the μP and I/O devices.

Responsibility of the μP

1. The μP needs to determine whether the designated input port is "full." In the next section we will study two methods by which this can be done.
2. If Port PA is full, then the μP can read the data from Port PA by completing the execution of, for example, an IN 00H instruction.

Responsibility of the Interface Module Upon the execution of an I/O read instruction (such as IN) that reads the data from Port PA, the controller of the interface module produces the following signals:

- **(a)** PA_EMPTY is made true.
- **(b)** PA_INTR is made false.
- **(c)** In the Status port the PA_EMPTY bit is set to 1 and the PA_INTR bit is set to 0.

The cycle continues. In this manner, the input device will load a data byte into Input Port PA only after the μP has read in the previous byte of data. And, the μP will read in a byte of data from Input Port PA only after a new byte of data has been placed there by Input Device A.

Similarly, the procedure for the synchronization between the μP and Output Device B for an output operation can be outlined as follows:

Responsibility of the μP

1. The μP has the responsibility of determining whether Output Device B is ready for another byte of data. Again, in the next section we will study methods by which this can be done.
2. If the μP determines that the designated output port for Output Device B (Output Port PB) is empty, then the μP can place the next byte of data into Port PB by, for example, completing the execution of an OUT 00H instruction.

Responsibility of the Interface Module

1. Upon the execution of an I/O write instruction (such as OUT), the data on the μP data bus (D_7–D_0) is placed into the Output Port PB.
2. Additionally, the controller of the interface module produces the following signals:
   - **(a)** PB_FULL is made true.
   - **(b)** PB_INTR is made false (to be discussed later).
   - **(c)** In the Status port the PB_FULL bit (bit 0) is set to 1 and the PB_INTR bit (bit 1) is set to 0.

Responsibility of the Output Device

1. Output Device B has the responsibility of determining whether the designated output port (Output Port PB) is full (i.e., whether PB_FULL is true).

2. If PB_FULL is true, then the output device has the responsibility of reading the data from the output port and signaling to the μP that it has done so by making the signal RECV_ACK (receive acknowledge) true.

Responsibility of the Interface Module When the RECV_ACK signal becomes true, the controller of the interface module produces the following signals:

(a) PB_FULL is made false. (PB is no longer full.)
(b) PB_INTR is made true.
(c) In the Status port the PB_FULL bit is set to 0 and the PB_INTR bit is set to 1.

The cycle again continues. In this manner, the μP will write to Output Port PB only after Output Device B has read the previous byte of data. Also, Output Device B will read from Output Port PB only after a new byte of data has been placed there by the μP.

In the design of a μP system there are two principal methods by which the synchronization, often called *handshaking*, between the μP and an I/O device can be programmatically controlled by the μP:

1. Polling method
2. Interrupt-driven method

An example of the polling method of handshaking is given in the next section (Sec. 12.4.1), and an example of the interrupt-driven method of handshaking is given in Sec. 12.4.2. Additionally, the interrupt structures of the 8085 μP are presented in some detail in Sec. 12.5.

12.4.1 Polling Method of I/O Synchronization

In the polling method of handshaking, the μP ''polls'' the I/O devices to determine the readiness of each. In this method an input device that has placed the next byte of data into the designated input port, or an output device that is ready for the next byte of data, indicates so by setting a status flag. For the μP system shown in Fig. 12.9(a) this setting is done in the interface module by setting a designated bit in the Status port. The μP will test each status flag in turn and service those devices that require servicing.

For the μP system of Fig. 12.9(a), an 8085 assembly program can be written to use the polling method to control the synchronization of the I/O operations. A portion of such a program is shown in Fig. 12.10. In this program the status of Input Device A is checked first (Statements 4, 6, and 7). If Device A is ready (i.e., PORTA is full), then the data byte for PORTA is read (Statement 8) and processed (Statement 9 and Subroutine PROCA). It is assumed that Subroutine PROCA processes the data in a manner that is required for this particular application.

After Device A is serviced, or if Device A was not ready in the first place, then the status of Output Device B is checked (Statements 10, 11, 12). If Device B is ready for another byte of data, then Subroutine PROCB is called (Statement 13) to perform

```
 1     PORTA    EQU    00H
 2     PORTB    EQU    00H
 3     STATUS   EQU    03H
 .               .
 .               .
 .               .
 4     LOOP:    IN     STATUS   ; Read in the status word. It contains the status of Input Device
                                ; A (bit 2) and Output Device B (bit 0).
 5              MOV    B,A      ; Save original STATUS in Register B.
 6     CHECKA:  ANI    04H      ; Use mask (00000100B) to isolate bit 2 (PA_EMPTY) of the
                                ; status word.
 7              JNZ    CHECKB   ; If PA_EMPTY is 1 (i.e., still empty), then branch to check
                                ; the status of Device B.
 8              IN     PORTA    ; Else, a new data byte is in PORTA, therefore IN PORTA.
 9              CALL   PROCA    ; Call Subroutine PROCA and process the data byte from
                                ; PORTA.
10     CHECKB:  MOV    A,B      ; Restore the original status word.
11              ANI    01H      ; Use mask (00000001B) to isolate bit 0 (PB_FULL) of the
                                ; status word.
12              JNZ    NORMPR   ; If PB_FULL is 1 (i.e., Device B not ready), then branch to
                                ; normal processing.
13              CALL   PROCB    ; Else, call PROCB to place a new output data byte into
                                ; Register A.
14              OUT    PORTB    ;
15     NORMPR:
 .               .
 .               .         (Normal processing)
 .               .
16              JMP    LOOP     ; Finished normal processing, go check status.
17     PROCA:   PUSHs           ; PROCA is a subroutine that processes the inputted data byte in
 .               .              ; a manner that is required for this particular application.
 .               .              ;
 .               .              ;
18              POPs            ;
19              RET             ;
                                ;
20     PROCB:   PUSHs           ; PROCB is a subroutine that performs the processing required
 .               .              ; to place a new output data byte into Register A.
 .               .              ;
 .               .              ;
21              POPs            ;
22              RET             ;
```

Figure 12.10 An example of the polling method of I/O synchronization.

any processing required to place a new output data byte into Register A. The OUT instruction (Statement 14) outputs the data byte to PORTB. After Device B is serviced, or if Device B was not ready in the first place, then the "normal processing" will resume.

In this example, the sequence of check/process Device A, check/process Device B, and normal processing continues indefinitely. Note that if both Device A and Device B require service at the same time, Device A is serviced first. Note also that normal processing is suspended while the statuses of Devices A and B are checked, even if neither device is requesting service. Suspending normal processing unnecessarily may not be desirable in some applications. Furthermore, if Device A or Device B requires servicing during normal processing, it will not be serviced until normal processing is finished and the program control branches back to LOOP. This may not be desirable in an application in which the normal processing portion of the program is long and the response to I/O devices is time critical. In the next section, an interrupt-driven method of I/O synchronization is illustrated that will eliminate these problems of the polling method of synchronization.

12.4.2 Interrupt-Driven Method of I/O Synchronization

In the interrupt-driven method of handshaking, an I/O device that is ready for service indicates so by "interrupting" the μP. If the device is allowed to interrupt, then the μP will complete the execution of the current instruction, save the return address onto the stack, and branch to a special location to execute a subroutine that will service the interrupting device. At the completion of the interrupt-service subroutine, the program control is returned to the instruction at the return address. So the μP resumes its normal processing. In this manner, normal processing is interrupted only if an I/O device requires servicing. In other words, normal processing will not be suspended simply to check on the status of the I/O devices. Further, unlike the polling method of I/O synchronization where the I/O devices are serviced only *after* normal processing is completed, the I/O devices can interrupt and receive service "immediately," even during the middle of normal processing.

To illustrate the interrupt-driven method of I/O synchronization, we will again use the μP system of Fig. 12.9(a) and will again use 8085 terminology although the concepts we will consider apply to microprocessors in general. For this illustration, we will see the purpose of the dashed-line connections between PA_INTR and PB_INTR of the interface module and the interrupt request inputs (RST5.5 and RST6.5) of the μP. In Fig. 12.11 there is a portion of a program for controlling the I/O synchronization using the interrupt-driven method. Let us now trace through the execution of that program.

Upon power-up, the contents of the program counter PC are assumed to be 0000H, as they are for the 8085 μP. As a result, for this program the first instruction to be executed after power-up is JMP INIT (Statement 4), which will transfer program control to the instruction in Location 0800H (Statement 10). What follows are instructions for performing the necessary initialization for this program, including the initialization of the stack pointer (Statement 11) and the enabling of the interrupt system (Statement 12).

At this point normal processing begins (at Statement 13). As long as there is no request from Device A or Device B for service, the normal processing part of this program is performed in an uninterrupted manner, with continuous looping. However, if, for example, Device A requires servicing, then the interface module will, among other actions, make the signal PA_INTR (and RST5.5) true. When the RST5.5 interrupt request input of the μP becomes true, the μP responds as follows:

1. It completes the execution of the current instruction. For example, if the current instruction being executed happens to be Statement 14, then the execution of the MOV A,B instruction is completed before the μP responds to the interrupt request.
2. The μP then saves the return address. In this case the address corresponding to Statement 15 is saved onto the stack.

```
1          PORTA   EQU    00H
2          PORTB   EQU    00H
.                  .
.                  .
.                  .
3                  ORG    0000H     ; At power-up, the first instruction executed is this instruction in
4                  JMP    INIT      ; Memory Location 0000H.
                                    ;
5                  ORG    002CH     ; Special location assigned to RST5.5 interrupt.
6                  JMP    SERVA     ; Jump to interrupt service routine for Device A.
                                    ;
7                  ORG    0034H     ; Special location assigned to RST6.5 interrupt.
8                  JMP    SERVB     ; Jump to interrupt service routine for Device B.
                                    ;
9                  ORG    0800H     ; The main program starts in Location 0800H.
10        INIT:    .                ;
                   .                ; Normal initializations for this program.
                   .                ;
11                 LXI    SP,2FFFH  ; Initialize the stack pointer.
12                 EI               ; Enable the interrupt system.
                                    ;
13       NORMPR:                    ; Beginning of the normal processing part of the program.
                   .                ; Either Device A or Device B can interrupt during the
                   .                ; execution of any instruction within this block of instructions.
                   .                ;
                                    ;
14                 MOV    A,B       ; These two instructions are just illustrative instructions within
15                 ANA    C         ; this block of instructions.
                   .                ;
                   .                ;
                   .                ;
16                 JMP    NORMPR    ; End of the normal processing loop.
                                    ;
17        SERVA:   PUSHs            ; Save the appropriate registers.
18                 IN     PORTA     ;
                   .                ; Process the inputted data byte in a manner required for this
                   .                ; particular application.
                   .                ;
19                 POPs             ;
20                 EI               ; The interrupt system was automatically disabled. This will
                                    ; reenable it.
21                 RET              ;
                                    ;
22        SERVB:   PUSHs            ;
                   .                ; Performs the processing required to place a new output data
                   .                ; byte into Register A.
                   .                ;
23                 OUT    PORTB     ;
24                 POPs             ;
25                 EI               ;
26                 RET              ;
```

Figure 12.11 An illustration of the interrupt-driven method of I/O synchronization.

3. Program control is transferred to a special location designated for that particular interrupt request input. For the RST5.5 interrupt request input, the designated address is assumed to be 002CH, as it is for the 8085. Consequently, the next instruction executed is JMP SERVA (Statement 6).
4. Program execution proceeds in a normal manner (Statements 6, 17, 18, etc.)

until a RET instruction is encountered. In this case the entire subroutine SERVA is executed. Presumably, SERVA is a subroutine that inputs data from Input Port PA and processes the inputted data in a manner that is required for this particular application.

5. When the RET instruction is executed (Statement 21), the return address is popped from the stack and program control transfers back to the instruction stored in that return address. In this case the next instruction to be executed is Statement 15. In this way, normal processing resumes at the point at which it was interrupted.

An interrupt request from Device B produces a similar result. In this case the interface module makes the signal PB_INTR (and RST6.5) true. The designated special location for the RST6.5 interrupt request input is assumed to be Memory Location 0034H, as it is for the 8085. Consequently, the subroutine SERVB is executed to service an interrupt from Device B.

In general, there must be a priority arrangement for interrupt requests to resolve the situation when several interrupt requests occur at the same time. Here, suppose that, as is the case for the 8085, the RST6.5 interrupt request input has a higher priority than does the RST5.5 interrupt request input. Then, if both Device A and Device B require service, and both RST5.5 and RST6.5 are true, the RST6.5 interrupt request is handled first.

Another general point is that the μP will respond to an interrupt request only if the interrupt system is enabled (e.g., with an EI instruction). The interrupt system may be disabled at power-up, as it is for the 8085 μP. In this case, an EI instruction is required to explicitly enable the interrupt system (Statement 12). Otherwise, the μP will *not* respond to any interrupt requests from any of the I/O devices. Also, in order that the subroutine that is servicing an interrupt request (e.g., SERVA or SERVB) will not itself be unintentionally interrupted, the interrupt system is automatically disabled upon a response to an interrupt request. Consequently, the last instruction of an interrupt service routine is normally an EI instruction (Statements 20 and 25) to enable the interrupt system again.

In summary, an interrupt request from an external device functions, in effect, like a subroutine call. But there is the difference that the subroutine call is initiated by a CALL instruction of a program stored *within* the μP system, while an interrupt is initiated by a device *external* to the μP system.

12.4.3 The Interrupt Facilities of the 8085 μP

For the 8085 μP, five interrupt request inputs and one output are provided for interrupts, as shown in Fig. 12.12(a). The RST5.5 and RST6.5 interrupt operations were explained in the preceding section. The operation of the RST7.5 interrupt is similar. However, RST7.5 is edge-sensitive whereas RST6.5 and RST5.5 are level-sensitive. In other words, for RST6.5 or RST5.5, the input must be held high by the requesting device until the μP has responded. On the other hand, for RST7.5 there is an internal flip-flop that will store the RST7.5 interrupt request at the rising edge of the request signal. Therefore,

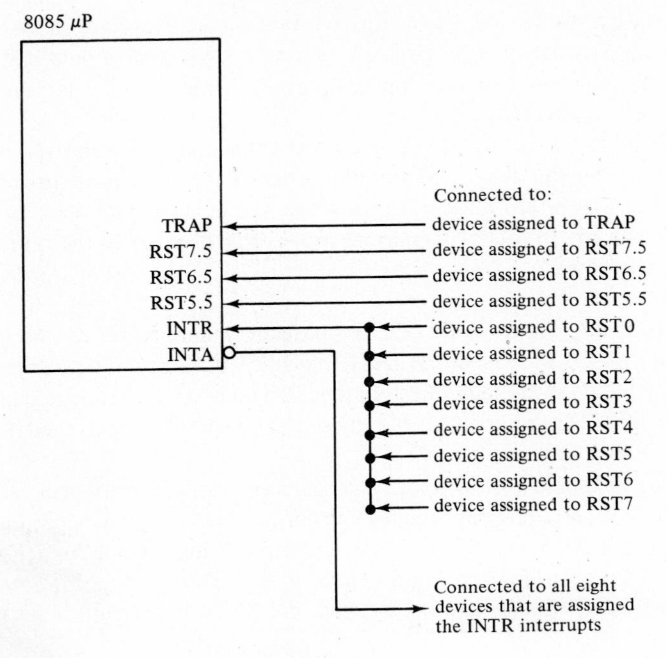

(a) 8085 interrupts

Instruction	Description
EI	Enable Interrupts
DI	Disable Interrupts
SIM	Set Interrupt Masks
RIM	Read Interrupt Masks

(b) 8085 instructions used for interrupts

Figure 12.12 The interrupt facilities of the 8085 μP.

the RST7.5 input need not be held high by the requesting device. The designated address for the RST7.5 interrupt is 3CH, as shown in Fig. 12.12(c).

In conjunction with the SIM (Set Interrupt Mask) instruction, any one or all three of these interrupts (RST7.5, RST6.5, and RST5.5) can be enabled or disabled by using the EI or DI instruction, respectively. More specifically, through the setting or clearing of the designated bits in the interrupt mask with the SIM instruction, each of the three interrupts can be selectively enabled or disabled when the EI or DI instruction is executed. Furthermore, an RIM (Read Interrupt Mask) instruction can be used to determine the current status (enabled, disabled, pending interrupts, etc.) of the three interrupt inputs.

The TRAP interrupt request, the input for which is shown in Fig. 12.12(a), is different from the RST7.5, RST6.5, and RST5.5 interrupts in that the TRAP interrupt is nonmaskable. In other words, TRAP cannot be disabled. The SIM and DI instructions have no effect on the TRAP interrupt. This interrupt is normally used for catastrophic situations such as power failure detection or any situation in which an interrupt must be

Priority	Interrupt	Designated Address
1	TRAP	24H
2	RST7.5	3CH
3	RST6.5	34H
4	RST5.5	2CH
5	INTR–RST0	00H
	RST1	08H
	RST2	10H
	RST3	18H
	RST4	20H
	RST5	28H
	RST6	30H
	RST7	38H

(c) Designated addresses for the 8085 interrupts

Interrupt	Opcode
RST0	11000111
RST1	11001111
RST2	11010111
RST3	11011111
RST4	11100111
RST5	11101111
RST6	11110111
RST7	11111111

(d) Opcodes corresponding to the INTR interrupts

Figure 12.12 (*cont.*)

handled immediately. As shown in Fig. 12.12(c), the designated address for the TRAP interrupt is 24H.

The INTR interrupt request input and the INTA (Interrupt Acknowledge) output can accommodate up to eight additional interrupts directly, corresponding to RST0, RST1, RST2, RST3, RST4, RST5, RST6, and RST7. As shown in Fig. 12.12(a), each of the eight devices sharing the INTR inputs is assigned uniquely to one of these interrupts. The designated address for each of the eight interrupts is given in Fig. 12.12(c).

When INTR is made true, one of these eight interrupts needs to be serviced. The μP cannot, however, determine directly which of the eight is requesting an interrupt since they all share the same INTR interrupt request input. Consequently, when INTR is made true and the interrupt system is enabled, the μP will respond as follows:

1. As before, the execution of the current instruction is completed before the μP responds to the INTR request.
2. The μP enters into an *Interrupt Acknowledge* machine cycle.
   (a) The INTA signal is made true (low). Note from Fig. 12.12(a) that all eight devices will receive the INTA signal.
   (b) However, only the interrupting device will respond by placing (normally) the opcode of the designated RST instruction onto the μP system data bus. For example, if the device that is assigned RST3 is the interrupting device,

then it has the responsibility of detecting the INTA signal and placing the RST3 opcode (11011111B) onto the data bus. The opcode (in binary) corresponding to each of the INTR interrupts is given in Fig. 12.12(d).

(c) After receiving the opcode, the μP decodes the RST instruction and branches to the corresponding designated location. In this case, for RST3, the designated location is 18H, as shown in Fig. 12.12(c).

3. Subsequently, the interrupt is serviced like the other interrupts that have been discussed.

The INTR interrupts are enabled and disabled collectively by the EI and DI instructions, respectively. They cannot be selectively enabled and disabled by using the SIM instruction, as can RST7.5, RST6.5, and RST5.5.

As shown in Fig. 12.12(c), the priority of the 8085 interrupts, from the highest to the lowest, is as follows: TRAP, RST7.5, RST6.5, RST5.5, and the INTR interrupts. The eight interrupts that share the INTR inputs have the same priority. Therefore, if more than one of the eight devices sharing INTR are interrupting at the same time, there needs to be some *external* logic for resolving the priority and allowing only the device with the highest priority to respond to the INTA signal.

The 8085 interrupt facilities described in this section illustrate only some of the concepts of interrupts. For additional information concerning the specifics of the 8085 interrupts, refer to Intel's *MCS-80/85 Family User's Manual.*

12.5 EXAMPLES OF COMMERCIALLY AVAILABLE I/O INTERFACE MODULES

In this section, we will briefly consider examples of some commercially available I/O interface modules that integrate some of the more common digital functions onto a single chip.

12.5.1 Intel 8155

The Intel 8155 is a multifunctional chip that is a combination of a memory module containing 256 bytes of RAM, an I/O interface module with three programmable I/O ports, a 14-bit binary counter/timer, and circuitry for demultiplexing the AD bus of the 8085 μP. The functional block diagram and the pin-out diagram for the 8155 are shown in Figs. 12.13(a) and (b), respectively. As usual, the chip is enabled when the CE input is made true. Also, the IO/$\overline{\text{M}}$ input must be false for a memory access and true for an I/O access. And, the RD and WR inputs control whether the access is a read or a write operation. The RESET input resets the chip into an initial state.

8155 RAM Functions

For the access of one of the 256 memory locations on board the 8155, the following signals must have the indicated values:

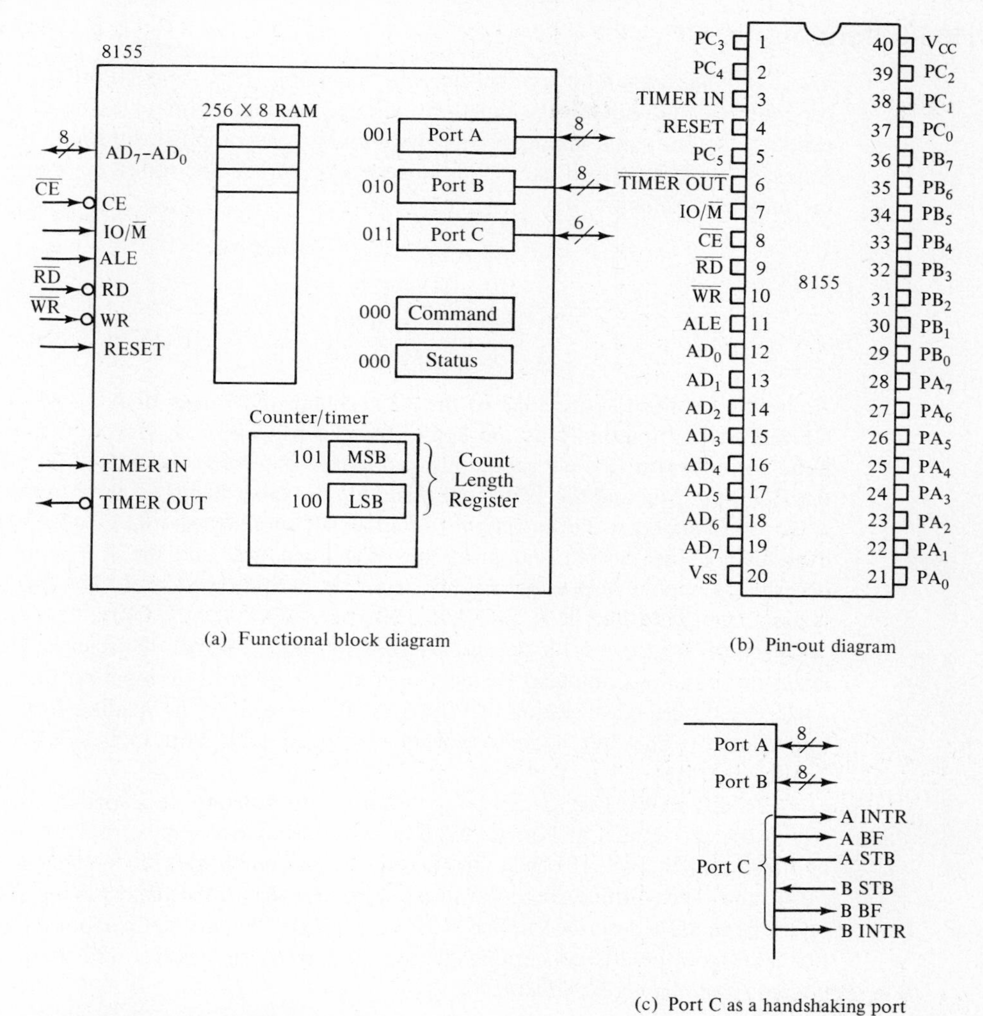

Figure 12.13 The Intel 8155.

(a) $AD_7–AD_0$ = 8-bit address
(b) CE = true
(c) $IO/\overline{M}$ = false

At the trailing edge of the ALE signal, the values of $AD_7–AD_0$, CE, and $IO/\overline{M}$ are latched internally by the 8155. For a memory read operation, the μP then relinquishes control of the $AD_7–AD_0$ bus and makes the RD input true and the WR input false. After a time equal to the access time of the memory module, the data is available on the AD bus. On the other hand, for a memory write operation, the μP places onto $AD_7–AD_0$ the 8-bit data to be written and makes the WR input true and the RD input false.

8155 I/O Ports

The 8155 has two 8-bit general-purpose I/O ports: Port A and Port B. Each can be programmed individually to function either as an input port or as an output port, as needed by a particular application. Programming is accomplished through the Command Register (to be described later). For an access of Port A, the following inputs must have the indicated values:

(a) AD_7–AD_3 = don't care
(b) AD_2–AD_0 = 001
(c) CE = true
(d) $IO/\overline{M}$ = true

As before, at the trailing edge of the ALE signal, the values of AD_7–AD_0, $IO/\overline{M}$, and CE are latched internally by the 8155. To read from Port A, if Port A is programmed to be an input port, the μP should now relinquish control of the AD_7–AD_0 bus and make the RD input true and the WR input false. On the other hand, to write to Port A, if Port A is programmed to be an output port, the μP should now place onto AD_7–AD_0 the 8-bit data to be written and make the WR input true and the RD input false. This description applies also to the reading from and writing to Port B. The I/O address that is associated with Port B is XXXXX010B (i.e., AD_2–AD_0 = 010).

There is also a 6-bit general-purpose I/O port, Port C, that can be programmed, again through the Command Register, to function either as an input port or as an output port. The described operation of Port A applies as well to the reading from and writing into Port C. The I/O address that is associated with Port C is XXXXX011 (i.e., A_2–A_0 = 011).

Additionally, Port C can be programmed to function as a special-purpose port in which the six signals of Port C function as handshaking signals for Port A and Port B, as illustrated in Fig. 12.13(c). In this mode the signals INTR (Interrupt), BF (Buffer Full), and STB (Strobe) function in a manner similar to the handshaking signals of the interface module described in Sec. 12.4 (Fig. 12.9). We will not belabor the details here. For additional details concerning the specifics of these signals, refer to Intel's *Microsystem Components Handbook.*

8155 Programmable Counter/Timer

The 8155 also has a 14-bit programmable counter/timer that counts the number of incoming pulses on the TIMER IN input. An initial count length, up to 14 bits long, and a timer output mode (2 bits) are loaded into a 2-byte Count Length Register, the format of which is shown in Fig. 12.14(a). The least-significant byte (LSB) of the Count Length Register contains the low-order 8 bits of the 14-bit count length. The most significant byte (MSB) of the Count Length Register contains the high-order 6 bits of the 14-bit count length in bits 5 to 0, and a 2-bit timer output mode in bits 7 and 6. The LSB portion of the Count Length Register has an I/O address of XXXXX100, and the MSB portion has an I/O address of XXXXX101.

With the initial count length stored in it, the Count Length Register is decremented each time a pulse is detected on the TIMER IN input. When the terminal count (TC) is

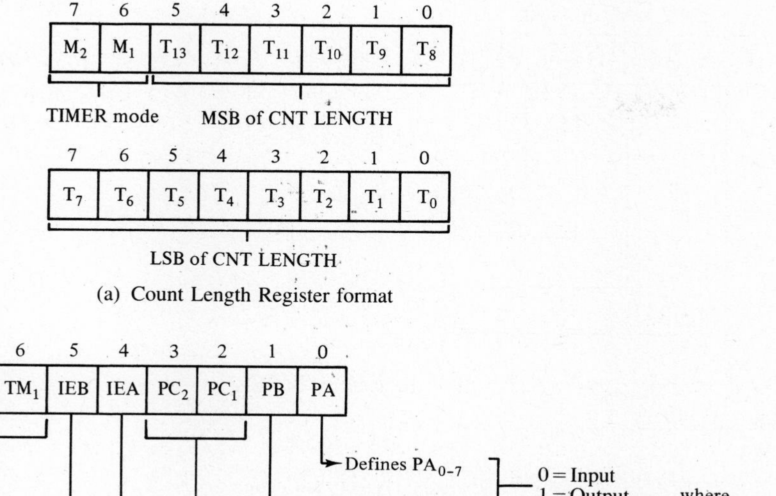

(a) Count Length Register format

(b) Command Register bit assignment

Figure 12.14 Register definition of the 8155. (Courtesy of Intel Corporation.)

reached, then the TIMER OUT output is activated. The type of this output depends on the 2-bit timer output mode stored in bits 15 and 14 of the Count Length Register. The output at TIMER OUT can be a single pulse (at TC) or a single square wave with a period equal to the count length. Or, it can be a pulse train or a square-wave train that has a period equal to the count length. In other words, for continuous operation, the given count length is repeatedly counted down.

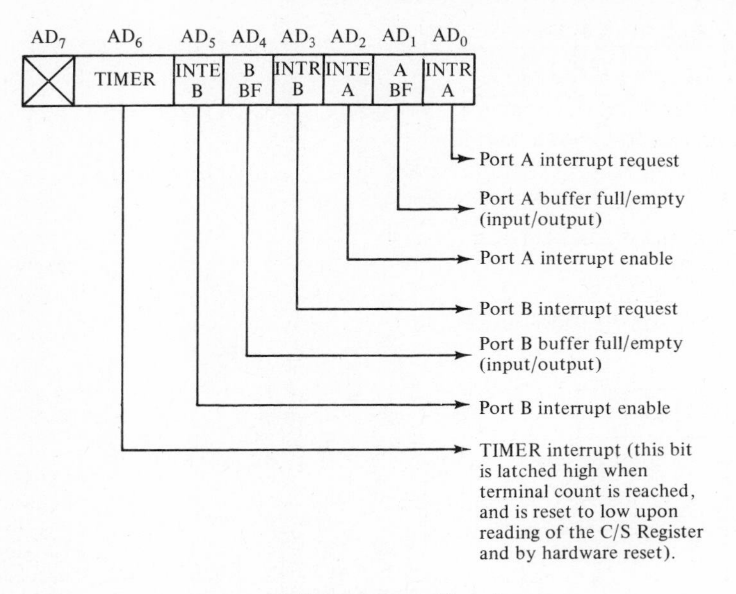

(c) Status Register bit assignment

Figure 12.14 (*cont.*)

For an illustration, the following 8085 instructions will initialize the count length to 380H with a 14-bit pattern of 00001110000000, and will produce a 2-bit timer mode of 10 that will result in a single pulse out when the terminal count is reached. So, these instructions cause the contents of the Count Length Register to be

$$1000001110000000\text{B} = 8380\text{H}$$

Only four instructions are required:

```
MVI  A,80H  ; The low-order byte of the count length is 10000000.
OUT  04H    ; Store in the low-order byte of the Count Length
            ; Register.
MVI  A,83H  ; The timer output mode is 10 for a single pulse at
            ; TC. The high-order 6 bits of the count length
            ; are 000011.
OUT  05H    ; Store in the high-order byte of the Count Length
            ; Register.
```

Programmed in this manner, the Count Length Register will be decremented each time a pulse is detected at the TIMER IN input. Then, when the terminal count of 380H = 896D is reached, and so the count length is reduced to zero, a single pulse is outputted at TIMER OUT. This TIMER OUT pulse might be used, for example, as an interrupt request signal for energizing the RST7.5 interrupt request input of the 8085.

8155 Command Register and Status Register

The programming of most 8155 functions requires the use of the Command Register, which is assigned the *output* (i.e., write-only) port address of XXXXX000. The Command Register has the bit format shown in Fig. 12.14(b). The specified functions are obtained by setting the designated bits within the Command Register to a 0 or a 1. As an illustration, the following instructions will produce a bit pattern in the Command Register that will result in Port A being an input port, Port B being an output port, and the signals of Port C functioning as handshaking signals. Also, it disables the interrupt function of both Port A and Port B. Note, however, that it does not affect the counter/timer operation. Only two instructions are necessary:

```
MVI  A,0AH ; The bit pattern is 00001010.
OUT  00H   ; Store the bit pattern into the Command Register.
```

The Status Register, which is assigned the *input* (i.e., read only) port address of XXXXX000, can be used to poll the status of Port A, Port B, or the counter/timer. The format of the Status Register is shown in Fig. 12.14(c). By the use of an IN instruction, the contents of the Status Register can be brought into Register A of the μP and tested. Then, it can be determined whether Port A is full (bit 1), whether Port A is allowed to interrupt (bit 2), or whether Port A has interrupted (bit 0). In a similar fashion, the status of Port B can be determined (bits 4, 5, and 3). Also, if the counter/timer is in operation, then it can be determined whether the terminal count (TC) has been reached (bit 6).

In summary, the 8155 is a combination of three functional modules: a memory module, an I/O interface module, and a counter/timer. The memory module contains 256 bytes of memory locations. The I/O interface module contains three programmable I/O ports: Port A, Port B, and Port C. All three ports can be programmed individually as an input port or an output port, as required by the particular application. Additionally, Port C can be programmed to support the interrupt method of handshaking for Port A and Port B. The counter/timer module of the 8155 counts down from a given count length of up to 14 bits. When the terminal count is reached, the TIMER OUT output is activated and the TIMER bit of the Status Register is set. Programming of the 8155 functions is generally done through the use of the Command Register. The Status Register contains the statuses of Port A, Port B, and the counter/timer and is generally used to support the polling method of handshaking for Port A, Port B, and the counter/timer.

12.5.2 Intel 8355/8755

The Intel 8355 is another example of a commercially available I/O interface module. Rather than become involved with the specific details, we will simply describe the major functions of the 8355. It is also a multifunctional chip, being a combination of a memory module containing 2K bytes of ROM, an interface module with two programmable I/O ports, and the circuitry for demultiplexing the AD bus of the 8085 μP. Another Intel device, the Intel 8755, is functionally the same as the 8355. The only difference is that the memory module of the 8755 contains 2K bytes of erasable and electrically reprogrammable EPROM instead of ROM. The functional block diagram and the pin-out

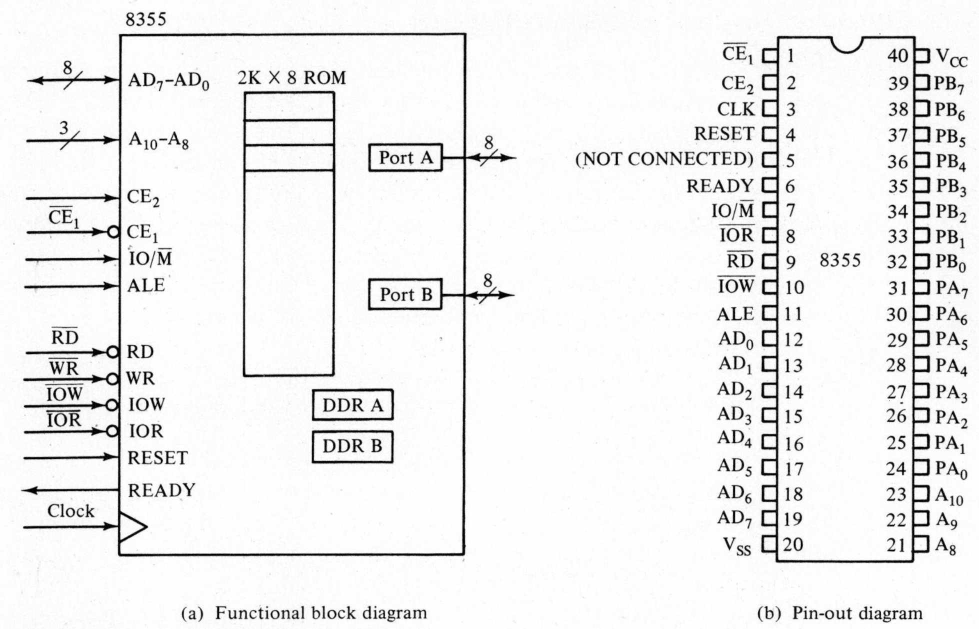

(a) Functional block diagram (b) Pin-out diagram

Figure 12.15 Intel 8355.

diagram of the 8355 are shown in Figs. 12.15(a) and (b), respectively. The chip is enabled when both CE_1 and CE_2 are made true. The RESET input resets the chip into an initial state.

For a read of one of the 2K memory locations of the ROM, the following inputs must have the indicated values:

(a) A_{10}–A_8, AD_7–AD_0 = 11-bit address
(b) CE_2 = true, CE_1 = true
(c) $IO/\overline{M}$ = false
(d) RD = true

Of course, a memory write operation is not possible since the memory module is read-only (ROM).

The 8355 has two 8-bit general-purpose I/O ports: Port A and Port B, that are *bit* programmable. Specifically, each bit of each port can be *individually* programmed to be in the input mode or the output mode. Programming is done through the two corresponding Data Direction Registers (DDR A and DDR B). A 0 in a particular bit position of a DDR specifies that the corresponding I/O port bit is in the input mode, while a 1 specifies an output mode. The I/O port addresses are XXXXXXXXX00B for Port A and XXXXXXXXX10B for Port B.

As is evident, the 8355 is a simpler device than the 8155 described in the preceding section. It is a combination of two functional modules: a memory module that provides

2K bytes of ROM and an I/O interface module that provides two 8-bit bit-programmable I/O ports.

12.5.3 An 8085 Minimum System Configuration

Figure 12.16(a) shows the schematic diagram of a basic microprocessor system that is built around the 8085 μP. This three-chip μP system contains

1. A CPU (8085)
2. A memory module containing 2K bytes of ROM (for storing programs and fixed-data tables) and 256 bytes of RAM (for temporary data storage)
3. An I/O interface module containing 38 I/O lines, providing a variety of parallel I/O ports, a handshaking port, and bit-programmable I/O ports
4. A 14-bit programmable timer/counter

The memory map and the I/O maps corresponding to this μP system are shown in Fig. 12.16(b). These maps are based on standard I/O-mapped I/O.

The purpose of Fig. 12.16(a) is to illustrate what a ''real'' schematic of a μP looks like, complete with terminals (V_{CC} and Gnd) for connection to a power supply. Although some of the details of the schematic should be apparent from the material of the preceding sections, there are other details that are explained only in the manufacturer's data book. There is, however, one detail that we will consider. Note that address bit A_{11} is used for chip enabling, with a 0 enabling the 8155 chip and a 1 enabling the 8355 chip, as is apparent from Fig. 12.16(a). This bit also appears in the I/O port maps of Fig. 12.16(b). There it is the A_3 bit, which is the alias of the A_{11} bit because, as stated in Sec. 12.2, the 8-bit port address is available on both A_7–A_0 and A_{15}–A_8 at the same time. This is the reason why in Fig. 12.16(b) the A_3 bit is shown as 0 for the 8155 and as 1 for the 8355.

This basic μP system can be easily expanded by adding additional memory modules and/or additional I/O interface modules that can interface a variety of I/O devices to the μP system. A browse through a manufacturer's peripheral-device data books reveals that the interfacing functions for most common I/O devices are integrated into single-chip I/O interface modules. Furthermore, these I/O interface modules are designed in such a manner that they can be easily interfaced to the microprocessor family for which they were designed. The study of the specific details of these I/O interface modules is beyond the scope of this text and is dependent on the individual areas of interest. However, although the functions and details of the different types of interface modules are different, the manner in which they interact and communicate with the μP through the I/O ports are conceptually the same as the concepts presented in this chapter.

SUPPLEMENTARY READING (see Bibliography)

[Andrews 82], [Clements 82], [Conffron 83], [Gorsline 85], [Intel 79], [Intel-B], [Liu, 84], [Muchow 83]

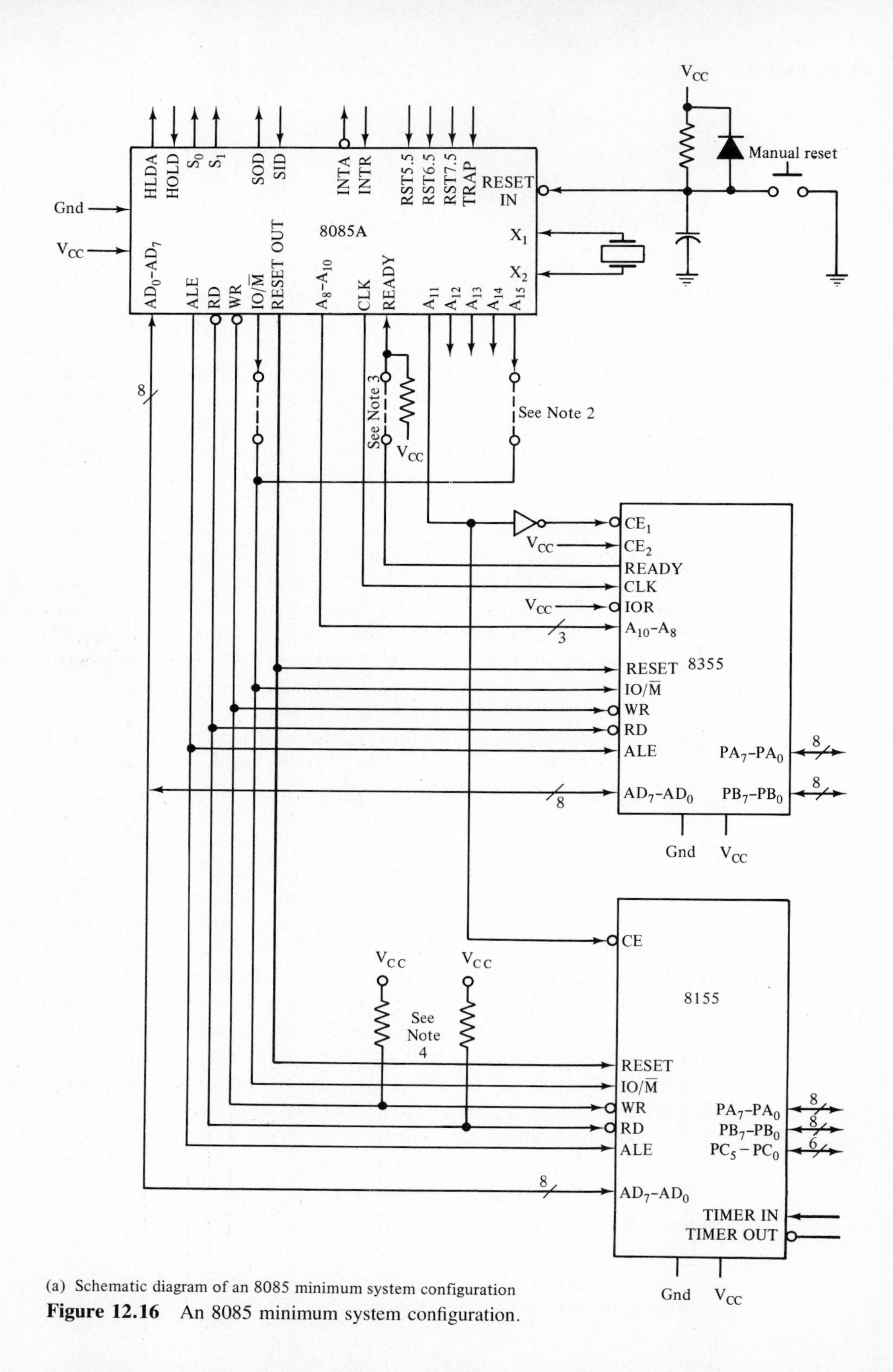

(a) Schematic diagram of an 8085 minimum system configuration

Figure 12.16 An 8085 minimum system configuration.

Memory map

Memory modules	Address	
8155 RAM (256 bytes)	0000H–00FFH	and other aliases (address bits 15, 14, 13, 12, 10, 9, and 8 are don't cares)
8355 ROM (2K bytes)	0800H–0FFFH	and other aliases (address bits 15, 14, 13, and 12 are don't cares)

Input-port map

Input port	Input-port address
8155--	
Port A	XXXX0001B
Port B	XXXX0010B
Port C	XXXX0011B
Status	XXXX0000B
Count Length Reg. (low byte)	XXXX0100B
Count Length Reg. (high byte)	XXXX0101B
8355--	
Port A	XXXX1X00B
Port B	XXXX1X01B
DDR A	XXXX1X10B
DDR B	XXXX1X11B

Output-port map

Output port	Output-port address
8155--	
Port A	XXXX0001B
Port B	XXXX0010B
Port C	XXXX0011B
Command	XXXX0000B
Count Length Reg. (low byte)	XXXX0100B
Count Length Reg. (high byte)	XXXX0101B
8355--	
Port A	XXXX1X00B
Port B	XXXX1X01B
DDR A	XXXX1X10B
DDR B	XXXX1X11B

(b) Memory and I/O-port maps

Notes for Figure 12.16(a)

Note 1: TRAP, $\overline{\text{INTR}}$, and HOLD must be grounded if they aren't used

Note 2: Use IO/$\overline{\text{M}}$ for standard I/O mapping, use A15 for memory mapped I/O

Note 3: Connection is necessary only if one TWAIT state is desired

Note 4: Pull-up resistors recommended to avoid spurious selection when $\overline{\text{RD}}$ and $\overline{\text{WR}}$ are 3-stated

Figure 12.16 (*cont.*)

PROBLEMS

12.1. Repeat Problem 11.17 for the following memory contents. Assume that the system has just finished the execution of the instruction in Location 0702H and is ready to begin the opcode-fetch machine cycle of the next instruction, which is in Location 0703H.

Memory location	Contents
.	
.	
.	
0702H	6AH
0703H	DBH

0704H	15H
0705H	32H
.	
.	
.	

12.2. Repeat Problem 11.17 for the following memory contents. Assume that the system has just finished the execution of the instruction in Location 0719H and is ready to begin the opcode-fetch machine cycle of the next instruction, which is in Location 071AH.

Memory location	Contents
.	
.	
.	
0719H	7AH
071AH	D3H
071BH	07H
071CH	01H
.	
.	
.	

12.3. There are three separate address spaces for the 8085 μP.
(a) Name them.
(b) How are they distinguished in terms of the control signals generated by the 8085 μP?

12.4. Are the I/O ports within I/O Interface Module A in Fig. 12.5(a) configured in an I/O-mapped I/O structure or a memory-mapped I/O structure? Explain.

12.5. What are the advantages and disadvantages of I/O-mapped I/O and memory-mapped I/O?

12.6. The logic that is connected to the CE input (CE-A) of I/O Interface Module A in Fig. 12.5(a) is equal to $A_7 \cdot \overline{A_6} \cdot \overline{A_5} \cdot \overline{A_4} \cdot \overline{A_3} \cdot IO/\overline{M}$. What are the resultant problems (if any) if the logic connected to CE-A is changed to $A_7 \cdot \overline{A_6} \cdot \overline{A_5} \cdot \overline{A_4} \cdot \overline{A_3} \cdot \overline{IO/\overline{M}}$?

12.7. Are the port addresses for the ports of Interface Module A in Fig. 12.8(a) fully decoded or partially decoded? Explain.

12.8. Suppose that the following changes are made to the system configuration in Fig. 12.8(a).

	Existing logic for CEs	New logic for CEs
2K memory module	$\overline{A_{11}} \cdot \overline{IO/\overline{M}}$	$\overline{A_{12}} \cdot \overline{A_{11}} \cdot \overline{IO/\overline{M}}$
I/O Interface Module A	$IO/\overline{M}$	$A_{12} \cdot A_{11} \cdot \overline{IO/\overline{M}}$
I/O Interface Module B	$A_{11} \cdot \overline{IO/\overline{M}}$	$A_{12} \cdot A_{11} \cdot IO/\overline{M}$

(a) Are the ports within I/O Interface Module A now I/O-mapped or memory-mapped?
(b) Are the ports within I/O Interface Module B now I/O-mapped or memory-mapped?
(c) Specify the memory map and I/O maps for the new configuration.

12.9. Repeat Problem 12.8 for the following changes.

	Existing logic for CEs	New logic for CEs
2K memory module	$\overline{A_{11}} \cdot \overline{IO/\overline{M}}$	$\overline{A_{12}} \cdot \overline{A_{11}} \cdot \overline{IO/\overline{M}}$
I/O Interface Module A	$IO/\overline{M}$	$\overline{A_{12}} \cdot A_{11} \cdot \overline{IO/\overline{M}}$
I/O Interface Module B	$A_{11} \cdot \overline{IO/\overline{M}}$	$A_{12} \cdot \overline{A_{11}} \cdot \overline{IO/\overline{M}}$

12.10. Given the following memory map and I/O maps, reconfigure the system in Fig. 12.8(a) to realize them.

Memory map	Address
2K memory module	0000H–07FFH or 4000H–47FFH

(All other addresses are invalid.)

Input-port map	Address
Input ports: A-00	70H
A-01	71H
A-10	72H
B-000	80H
B-001	81H
B-010	82H
B-011	83H
B-100	84H
B-101	85H
B-110	86H
B-111	87H

Output-port map	Address
Output ports: A-00	70H
A-01	71H
A-10	72H
A-11	73H
B-000	80H
B-001	81H
B-010	82H
B-011	83H
B-100	84H
B-101	85H
B-110	86H

12.11. Repeat Problem 12.10 for the following memory map and I/O maps.

Memory map	Address
2K memory module	X800H-XFFFH

(All other addresses are invalid.)

Input-port map	Address
Input ports: A-00	X1H
A-01	X3H
A-10	X5H
B-000	X0H
B-001	X8H
B-010	X4H
B-011	XCH
B-100	X2H
B-101	XAH
B-110	X6H
B-111	XEH

Output-port map	Address
Output ports: A-00	X1H
A-01	X3H
A-10	X5H
A-11	X7H
B-000	X0H
B-001	X8H
B-010	X4H
B-011	XCH
B-100	X2H
B-101	XAH
B-110	X6H

12.12. The following instructions will add the contents of Input Port B-000 in Fig. 12.8(a) to the contents of Register A:

```
LXI   H,0800H
ADD   M
```

Give the instructions required to perform the equivalent operation of adding the contents of Input Port A-00 in Fig. 12.8(a) to the contents of Register A.

12.13. Given the system configuration shown in Fig. 12.8(a), what is the source of the data transfer for each of the following instructions? Each answer should be in the form of "the contents of Memory Location 0501H" or "the contents of Input Port A-01," or the like.
(a) LDA 0001H **(b)** LDA 0801H **(c)** LDA 0807H
(d) LDA 0FFFFH **(e)** IN 01H **(f)** IN 03H
(g) IN 04H

12.14. Given the system configuration shown in Fig. 12.8(a), what is the destination of the data transfer for each of the following instructions?
(a) STA 0001H **(b)** STA 0801H **(c)** STA 0807H
(d) STA 0FFFFH **(e)** OUT 01H **(f)** OUT 03H
(g) OUT 04H

12.15. Basing your work on the system configuration shown in Fig. 12.9, modify the program of Fig. 12.10 to implement the flowchart of Fig. 12.17.

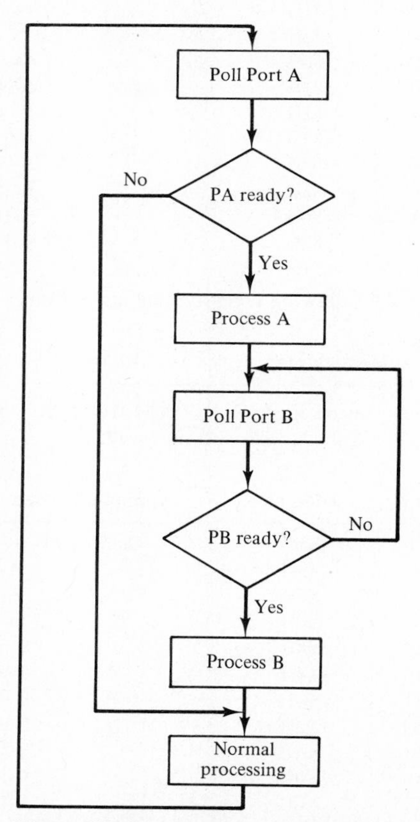

Figure 12.17 Flowchart for Problem 12.15.

12.16. Modify the system configuration shown in Fig. 12.9 and the program given in Fig. 12.11 to function as follows:

Input Device A is assigned to RST7.5.

Output Device B is assigned to RST5.5.

If Device A interrupts, then the corresponding service routine should input a byte from Port PA and store it into a memory location designated as IBUFF.

If Device B interrupts, then the corresponding service routine should output to Port PB the byte from a memory location OBUFF.

12.17. Repeat Problem 12.16 for the following specifications:

Input Device A is assigned to RST6.5.

Output Device B is assigned to RST7.5.

When Input Device A interrupts, a counter (Memory Location COUNT) is incremented by 1.

When Output Device B interrupts,
if COUNT $<$ 50D, then output the value of COUNT to Port PB.
if COUNT $\geqq$ 50D, then output the value FFH to Port PB and reset COUNT to 0.

12.18. Modify the system configuration of Fig. 12.9 and the program of Fig. 12.11 to function as follows: Synchronization with Input Device A (assigned to RST5.5) is interrupt-driven. However, synchronization with Output Device B is obtained by polling the status of Port PB after each iteration of the normal processing.

12.19. Figure 12.18(a) shows a skeleton of a program, and Fig. 12.18(b) shows the contents of some of the memory locations. Assume that the main program is looping indefinitely (JMP

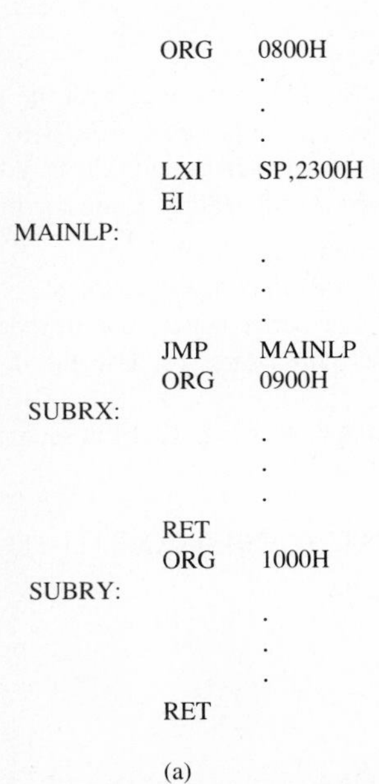
```
        ORG   0800H
              .
              .
              .
        LXI   SP,2300H
        EI
MAINLP:
              .
              .
              .
        JMP   MAINLP
        ORG   0900H
SUBRX:
              .
              .
              .
        RET
        ORG   1000H
SUBRY:
              .
              .
              .
        RET
```

(a)

Figure 12.18 Program skeleton and some memory location contents for Problem 12.19.

Address	Contents
0032H	10H
0033H	00H
0034H	C3H
0035H	00H
0036H	10H
0037H	00H
0038H	10H
0039H	C3H
003AH	C2H
003BH	3AH
003CH	C3H
003DH	00H
003EH	09H
003FH	CAH
0040H	3CH

(b)

Figure 12.18 *(cont.)*

MAINLP) and that at some point in time an RST7.5 interrupt occurs. Explain in words the sequence of events that will occur from the time that the 8085 μP detects the interrupt until the time that normal processing resumes. Be as detailed as you can. Refer to specific locations and subroutine names when possible.

12.20. What are some of the advantages and disadvantages of the polling method and interrupt-driven method of I/O synchronization?

12.21. What are the similarities and differences between a subroutine call and an interrupt request from an external device?

12.22. Replace the I/O interface module in Fig. 12.9(a) with an 8155 multifunctional chip. More specifically,

(a) Draw the new system configuration.

(b) Write a subroutine to program the 8155 to be used with the program of Fig. 12.11. In doing this, program Port A to be an input port, Port B to be an output port, and Port C to be the handshaking signals for Port A and Port B. Also, enable the interrupt functions for both Port A and Port B and specify no-op for the counter operation.

12.23. Given the system configuration shown in Fig. 12.19,

(a) What are the addresses for the following components of the 8355: 2K ROM, Port A, Port B, DDR A, and DDR B? For convenience, use hexadecimal for the memory addresses and binary for the I/O port addresses. Use the don't-care designation X when necessary.

(b) Make connections in the circuit diagram of Fig. 12.19 in such a way that the addresses of the 8155 are as follows:

RAM: 1001 0XXX 0000 0000B – 1001 0XXX 1111 1111B
Command/Status: 90H
Port A: 91H
Port B: 92H
Port C: 93H
LSB Timer: 94H
MSB Timer: 95H

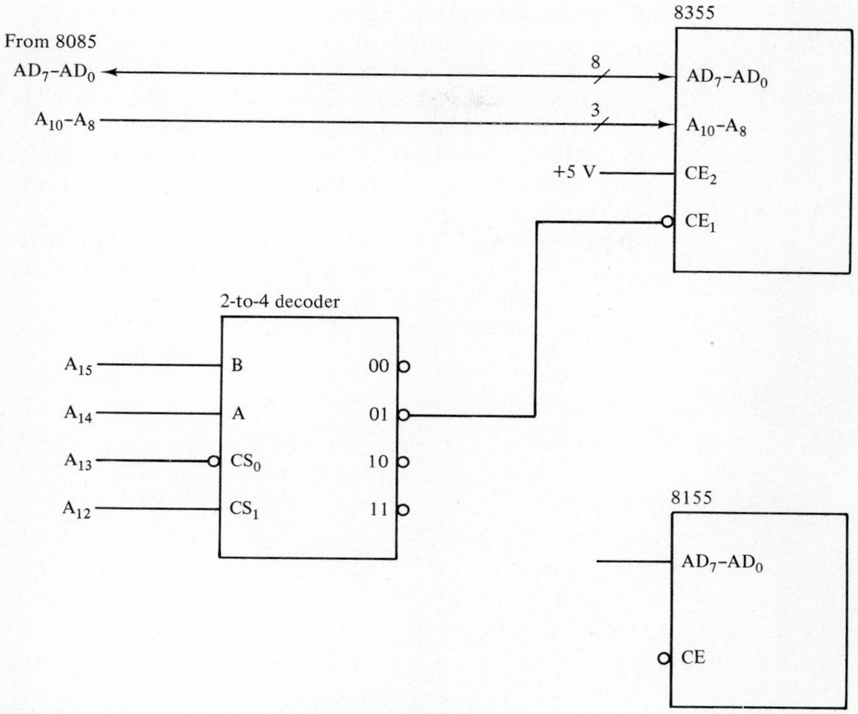

Figure 12.19 System configuration for Problem 12.23.

12.24. For the system components shown in Fig. 12.20, complete all the relevant connections between the 8085 μP and the 8155 based on the information contained in the following subroutine:

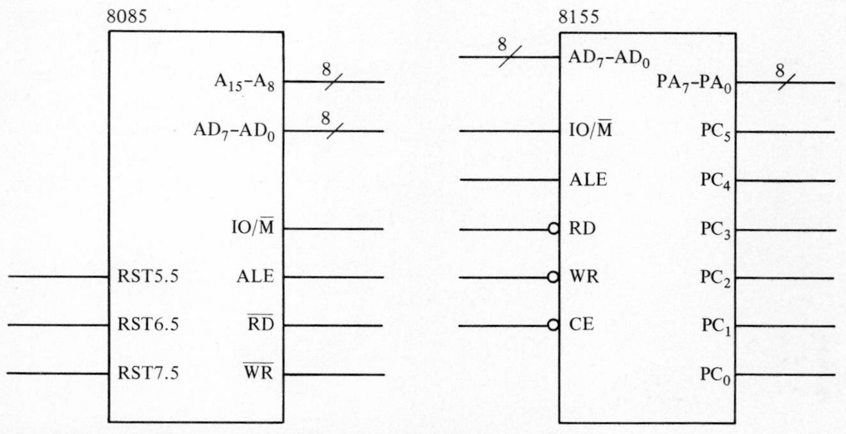

Figure 12.20 System components for Problem 12.24.

```
;  For the 8085 system, when a certain interrupt is requested,
;     the program control will be transferred to Memory Location
;     3CH in which a JMP SENDDT instruction is stored.
;  Assume that the following instructions have been executed in
;     the main program for initialization purposes:
;     MVI A, 00010101B (To program the Command Register of the 8155.)
;     OUT 10H           (10H is the only address of the Command Register.)
;
SENDDT:  OUT  11H  ; Output a data byte from Register A to 8155's Port A.
                   ; Note that 11H is the only address of Port A.
         EI
         RET
```

Chapter 13

Other Microprocessor Topics

13.1 INTRODUCTION

The second half of this book has been devoted to the fundamentals of microprocessors and of microprocessor-based design. Throughout Chapters 9–12 we used a specific microprocessor, the Intel 8085, and its supporting chips to illustrate the various μP concepts. As stated earlier in Chapter 8, the 8085 is not, of course, representable of the state of the art in μP technology. But this is a problem with any microprocessor. Even the most current of the 16-bit or 32-bit microprocessors will be out of date in a relatively short time. It has not been our intention in this book to study the specific details of a current microprocessor but rather to use a microprocessor that is excellent for illustrating those concepts and fundamentals that will remain current. For this purpose an 8-bit microprocessor, such as the 8085, is ideal for an introductory text. The 8085 is powerful enough to illustrate the important μP concepts, but simple enough to avoid obscuring these concepts with complex component details. Finally, it has been our experience that once a specific microprocessor and its applications have been mastered, then it is straightforward, with some training, to become proficient in the use of other microprocessors and their applications.

To complete our introduction to microprocessors and microprocessor-based designs, we will, in the next section, trace the growth and development of the microprocessor. Then, in Sec. 13.3, we will discuss the important criteria for the selection of a microprocessor for a microprocessor-based product. Next, in Sec. 13.4, we will revisit the microprocessor-based design procedure that was introduced in Chapter 8 and will discuss the steps in the development procedure in more detail. Additionally, we will survey the μP system development tools, both software and hardware, that are available for helping the digital designer in each phase of the design procedure.

13.2 THE DEVELOPMENT OF THE MICROPROCESSOR

The first microprocessor was introduced in 1971 by the Intel Corporation in the form of a 4-bit microprocessor, the 4004, which was intended to be used as the processing element of a calculator. Although it was not intended as a general-purpose processor, the 4004 (and its successor, the 4040) found success in the general-purpose market. Even today, 4-bit microprocessors are found in abundance in calculators, video games, automotive controls, appliance controls, and in any applications where there is no need for the additional power (and cost) of the more powerful microprocessors.

The first 8-bit general-purpose microprocessor was introduced in 1972, also by Intel, in the form of the 8008. It was quickly replaced by the more powerful 8080, the predecessor to the 8085. With the success of the 8080 and the potential of the microprocessor apparent, other companies quickly entered the μP race. Within 2 years, various 8-bit general-purpose microprocessors were introduced into the market. The most successful of these were the Z-80 from Zilog, the 6800 family from Motorola, and the 6205 from MOS Technology (now Commodore Semiconductor Group). Intel, itself, upgraded the 8080 with the 8085.

Zilog is a company that was started by a group of designers who left Intel to form their own company. Their Z-80 was designed to compete directly with Intel's 8080 and 8085. Though not pin-for-pin compatible with the 8080 or the 8085, the Z-80 is architecturally similar to the 8080/8085, with enhancements. It has two identical banks of "8085-type" registers so that bank switching can be employed in response to interrupts or for other uses. Additionally, it has two index registers that are used to support the index addressing mode. The instruction set of the Z-80 is a superset of the 8080/8085 instruction set and is object-code upward compatible. In other words, an assembled 8080/8085 program can be executed without changes in a Z-80 system. Further, the Z-80 has 50 additional instructions, including advanced block move and block search instructions, and instructions that use index addressing and relative addressing, and also more extensive data movement and data manipulation instructions. In general, the Z-80 can do essentially anything that the 8085 can do, and more. The Z-80 microprocessor has been a highly successful product.

The Motorola 6800 was also designed to compete with the 8080. It has two accumulators but no other general-purpose registers. It also has an index register for the index addressing mode. The instruction set is patterned after the successful PDP-11 minicomputer. Since it does not have vectored interrupts, it has to achieve the interrupt function with software. The 6800 is the original in an extensive family of microprocessors produced by Motorola. Included in this family is the 6809, its top-of-the-line 8-bit microprocessor. The 6809 is an enhanced version of the 6800 with more registers, having another 16-bit index register and a second stack pointer. It is upward compatible with the 6800, but only at the source-code level. In other words, existing 6800 programs have to be reassembled before they can be executed in a 6809 system. The 6809 is much more powerful than the 6800, deriving its extra power from the addition of powerful instructions to its instruction set. Included are instructions that employ 16-bit index addressing, some 16-bit data manipulation instructions, and an 8 × 8 bit multiply instruction with a 16-bit product.

The 6502 was introduced by MOS Technology to compete with the 8080 and the

6800. Since MOS Technology was founded in part by a group of designers who helped in the design of the 6800, it is not surprising that the 6502 has many similarities to the 6800. The designer's goal was to achieve as much PDP-11 style addressing as possible onto a single chip. The 6502 has an unusual common-memory architecture in which the instructions, data, and I/O all share the same 64K memory space. The main attribute of the 6502 is its low cost. The 6502 has found success in low-cost applications such as the video game market. In 1982 the 6502 had the highest volume of sales of any 8-bit microprocessor.

Since the introduction of the 8-bit general-purpose microprocessors in the early and mid-1970s, advances in microelectronics and in IC fabrication technology have made it possible to integrate more and more logic functions onto a single chip. Fueled by this increase in integration density, the development of the microprocessor has progressed in two general directions:

1. Increasing the amount of functionalities integrated onto a single chip: single-chip microcomputers
2. Increasing the power and performance of the microprocessor: 16- and 32-bit μPs

13.2.1 Single-Chip Microcomputers

For a microprocessor system (or microcomputer), such as the minimally configured μP system shown in Fig. 12.16, three or more chips are necessary to contain the various components. The microprocessor is contained on one chip, and the memory (ROM and RAM) and some I/O (ports, counters, etc.) are contained on others. In a single-chip microcomputer, however, all three components of a μP system (μP, memory, and I/O) are integrated onto a single chip (they are *on-chip*). Since single-chip microcomputers are generally used in some type of control application, the single-chip microcomputer is commonly called a *microcontroller* (μC).

In 1976, Intel introduced the first popular microcontrollers: the 8048 family. On-chip of each of the various versions of the 8048 μCs are, along with the microprocessor, 1K to 4K bytes of ROM (or EPROM) and 64 to 256 locations of RAM. The ROM or EPROM is used for the storage of the program and fixed data tables. The RAM contains two banks of eight working registers, an eight-level stack, and general RAM locations. Unlike microprocessors, such as the 8085, where all memory is treated the same, the ROM (or EPROM) in the 8048 (called program memory) is treated differently from the RAM (called data memory). In other words, different instructions are used to access the two different memory spaces. For this reason a computer such as the 8048 is said to have a split-memory architecture. Also on-chip are I/O ports and an 8-bit timer/counter. The 8048 I/O has its own I/O space and also instructions to operate directly on the I/O ports. Additionally, the 8048 can accommodate the 8080/8085 peripheral chips, and thereby vastly increase its flexibility. But, the instruction set of the 8048 is not as extensive as that of a general-purpose μP such as the 8085 or the Z-80. For example, it does not have a PUSH or POP instruction or a compare instruction. It does, however, have additional instructions that make use of the additional on-chip functions. The 8048 family has been a very successful product line. In its prime, the 8048 achieved a leading position among the 8-bit single-chip μCs.

Zilog's entry in the 8-bit single-chip μC market is the Z-8 family. It was introduced after the 8048 and therefore is, as to be expected, more sophisticated. It has a unique architecture of three memory spaces: program, data, and register file. In the register file it has 124 general-purpose working registers that can be used as either accumulators or index registers. On-chip, it has 2K to 4K bytes of ROM and 256 bytes of RAM for the register file. Also on-chip, it has two 8-bit timers (each with a 6-bit prescaler) and a built-in duplex UART (Universal Asynchronous Receiver/Transmitter). Its instruction set includes eight addressing modes, block transfer instructions, and sophisticated data and program manipulation instructions. All in all, the Z-8 is a large and ambitious chip.

Intel's upgrade of the 8048 is the 8051 family. It is an enhancement of the 8048 and has everything that the 8048 has plus more of everything: 4K to 8K bytes of ROM on-chip, 128 to 256 bytes of RAM, up to three 16-bit timers/counters, more interrupts, increased stack depth, and new instructions such as multiply, divide, and compare. Also added is a full-duplex hardware UART and A/D (Analog/Digital) converter.

One of Motorola's entries in the 8-bit single-chip μC market is the 6801. The 6801 is an expandable one-chip version of the 6800. It was designed to compete with the Z-8 and 8051. It has all the instructions of the 6800 plus ten more, including an 8×8 multiply instruction. Integrated on-chip are a multifunctional 16-bit timer, a UART, and an 8-bit A/D converter.

Other companies that have made a significant impact on the 8-bit single-chip μC market are Mostek, NEC Electronics, Rockwell, and Texas Instruments.

The next phase of the μC development began with the introduction of the powerful 16-bit single-chip microcontrollers. In 1984, Intel introduced the 8096, a 16-bit single-chip microcontroller. On-chip are 8K bytes of ROM and 232 bytes of register file RAM. I/O capabilities include an A/D converter, a full-duplex UART, a pulse-width-modulated output, high-speed pulsed I/O, four 16-bit timers and a watchdog timer. It has a 16×16 multiply and a 32/16-bit divide instruction. Applications for such a high-performance μC include the sophisticated and real-time control requirements found in robotics, telecommunications, and instrumentation. Other high-performance single-chip microcontrollers followed, including entries from National Semiconductor, NEC Electronics, and Digital Equipment Corporation.

13.2.2 High-Performance Microprocessors

Another result of the increase in integration density has been the development of powerful, high-performance general-purpose microprocessors. In 1978, Intel introduced the 8086. It was the first of the "second-generation" 16-bit microprocessors. Although not fully compatible with the 8085, the 8086 is architecturally very similar and contains all the instructions of the 8085 plus additional instructions such as for multiply, divide, and block operations. It can address up to 1M bytes (1 megabyte $\approx 1 \times 10^6$ bytes) of memory using "segmentation" techniques. The performance can be enhanced by using the 8086 in a coprocessor configuration with the 8087 and/or 8089 coprocessors. The 8087 is a numeric coprocessor used to perform arithmetic, logic, and transcendental operations in parallel with the main 8086 CPU. The result is an increase in speed of 100 times for those operations. Also, an 8089 I/O coprocessor can be used with the 8086 to perform I/O and memory operations in parallel to increase the throughput in an I/O-intensive application. The performance of the 8086 system matches the performance of the midrange minicomputers of its time. As a result, this class of microprocessors is sometimes called the ''micro-mini'' processors.

Included in the same class of processors as the 8086 are Zilog's Z-8000 and Motorola's 68000. The Z-8000 is generally viewed as being more powerful than the 8086, and the 68000 is the most powerful of the three, with its 32-bit wide internal registers and data paths of 16 bits.

Intel's enhancement of the 8086 include the 80186 and the 80286. The 80186 has an enhanced CPU with added clock generation circuitry, a two-channel DMA (Direct Memory Access) controller, an interrupt controller, and three 16-bit timers, all integrated on-chip. The 80286 also has the enhanced CPU. However, instead of containing the other 80186 functions on-chip, the 80286 is designed to have hardware support for memory management that allows the 80286 to address up to 1G bytes (1 gigabyte $\approx 1 \times 10^9$ bytes) of ''virtual memory.'' It also has hardware support for memory protection functions.

The next enhancement from Intel was the 80386, a 32-bit extension of the 80286, with 32-bit data paths and 32-bit addressing. In addition to the 80286-type memory management, the 80386 has a ''program cache'' on-chip to increase the speed of program execution. The 80386 and other 32-bit microprocessors have moved from the world of micro-mini to the world of ''micro-mainframe,'' having the power of mainframe computers. Other 32-bit microprocessors included in this class are Zilog's Z-80000, Motorola's 68020, and National Semiconductor's 32032. As the advances in microelectronics and in IC fabrication technology continue, increasingly powerful processors continue to become available on the market.

Our intention in this book has been to provide an introduction to microprocessor-based design in which the microprocessor is the controlling circuit element in a digital circuit. But, as you can see from the above discussion on the micro-mini and micro-mainframe processors, microprocessor development is moving away from the concept of microprocessor-based circuit design into the realm of full-scale computer systems, dealing with such concepts as virtual memory, program cache, and multiuser and multitasking environment, among others. These subjects are important to those readers who intend to continue on in the field of computer engineering. They are, however, beyond the scope of this book and so will not be considered further here. They have been mentioned here simply for completeness and to provide the reader with a feel for what lies ahead in the world of computer engineering.

13.3 MICROPROCESSOR SELECTION

The first important task in the design of a microprocessor system is the selection of the microprocessor to be used in the system. In this section we will outline some of the important criteria that can be used to select an appropriate microprocessor for a μP system from the variety of microprocessors that are available on the market today.

13.3.1 Performance of the Microprocessor

The most important criterion in the selection of a microprocessor is, of course, for the selected microprocessor to be powerful enough to satisfy the performance requirements of the μP system. An evaluation of the power of a microprocessor can be based on many criteria. Some of the important ones are outlined in the following paragraphs.

Instruction Set The power and flexibility of a microprocessor is generally constrained by the power and flexibility of its instruction set. The instruction set of a candidate microprocessor can be evaluated in terms of its data manipulation instructions, data movement instructions, and program control instructions. For example, what are the addressing modes supported by the instruction set? Are there block-move instructions and multiply/divide instructions? More importantly, the instruction set of the selected microprocessor should be powerful in the area that needs to be powerful for the target μP-based product. For example, for a computationally intensive application it is desirable for the instruction set to have powerful and flexible arithmetic instructions. Then, powerful and flexible I/O instructions or data movement instructions may be of secondary importance.

Instruction-Cycle Time The instruction-cycle time provides the most important indicator of the execution speed of a microprocessor. Particular attention should be paid to the instruction-cycle times of the instructions that will be most used by the target application. Related to the instruction-cycle time is the rate at which the clock source can be run.

Capacity of the Address Space The address capacity of a microprocessor provides an indication of the complexity of the μP system that it can support. A limited memory space limits the program size and complexity. Also, a limited I/O space may hamper an I/O-intensive application.

Word Size The word size of a microprocessor is determined by the width of its internal registers and the width of its data paths. Certainly, a 16-bit microprocessor is inherently more powerful than an equivalent 8-bit microprocessor.

I/O Facilities For applications that are I/O-intensive, important considerations include the I/O expansion capability and the number of I/O lines that are provided by the candidate microprocessor. Also, for applications that require fast response times, it is important to consider the number and types of interrupts that are provided by the microprocessor.

Power Consumption For applications that have special power requirements, the power consumption of the candidate microprocessor may be an important factor.

Package Size For applications that have special packaging requirements, the package size of the candidate microprocessor may be an important factor.

13.3.2 Other Criteria for Microprocessor Selection

In many cases there will be several microprocessors available that satisfy the performance requirements of the μP system to approximately the same extent. Then, additional criteria must be used to further evaluate these microprocessors. Some important ones are outlined in the following paragraphs.

Cost There are many factors that contribute to the total cost of a μP system. Generally, the total cost can be divided into the cost of the initial development and the cost of production and maintenance.

The *development cost* includes the dollar cost of the development and diagnostic support tools required by the candidate microprocessor. More important, it also includes the effort (i.e., the labor-hours/years, which also translates into dollar costs) required to develop and debug the μP-based product. The effort required is, in turn, dependent on some of the other criteria such as the power and flexibility of the instruction set, and the availability of good documentation and of development and diagnostic support tools.

The *production cost* is the cost required to produce the μP-based product in a production environment, including the cost of the parts and the cost of assembling the parts into a μP system. Obviously, the cost of the parts includes the purchase cost of the microprocessor. Additionally, it includes the cost of all the supporting hardware required to construct the μP system with the candidate microprocessor, including memory and I/O chips, power supplies, and so forth. The other production costs are directly proportional to the number of IC packages required to implement the μP-based product.

Development and Diagnostic Support As mentioned above, the development and debugging of a μP system constitute a major portion of its total cost. Therefore, a major selection criterion is the availability of the development and diagnostic support tools for the candidate microprocessor, provided either by the vendor or a third-party supplier. The development and diagnostic support tools, both hardware and software, that are presently available will be surveyed in Sec. 13.4.

Completeness of the μP Family Although a microprocessor can be interfaced with support chips from other μP families, it is generally easier and more compatible to use the support chips designed specifically for that family. Therefore the availability of the support chips (pertinent to the requirements of the target application) for the candidate μP family should be considered, including memory and I/O support chips, coprocessors, etc.

Availability of Parts Given that a candidate microprocessor satisfies all technical and cost criteria, then the availability of the microprocessor and its supporting peripheral chips, in the quantity and time frame required, is a further consideration.

Reputation of the Vendor In the selection of a microprocessor, it is important to investigate the reputation of the vendor supplying the parts regarding the vendor's support for its past and current products. How does the vendor rate in terms of providing development and diagnostic tools, good documentation and applications notes, field technical support, and training courses? Also, how long is the vendor expected to support the candidate microprocessor? If the current microprocessor is replaced by a more powerful version, can we expect this version to be upward compatible?

Second Sources Popular microprocessors are usually second-sourced. In other words, manufacturers other than the original vendor of the microprocessor also supply the product. If a microprocessor is well second-sourced, then the availability of the microprocessor is guaranteed.

13.3.3 Characteristics of the Target Application

Criteria for evaluating a microprocessor have been outlined in the preceding sections. In this section we will outline the characteristics of the μP-based product of the target application and describe how they relate to the μP selection criteria.

Speed Requirement If the μP system has a high speed requirement, then the performance criteria discussed in Sec. 13.3.1 have the utmost importance, particularly the instruction-cycle time. If the application is computation-intensive, then the performance of the arithmetic instructions is important. If the application is I/O-intensive, then the performance of the I/O instructions is important. Furthermore, it is important to determine the availability of the support chips and/or coprocessors that can improve the performance of the μP system. If the μP system does not have a high speed requirement, then these performance criteria are of less importance and so the other selection criteria discussed in Sec. 13.3.2 play a more important role.

Number of Units to Be Produced If the μP system is to be produced in very large quantities, then the production cost becomes more important than the development cost since the initial development cost is averaged over many units. In this case it is important to select a microprocessor and support chips that have a low unit cost and for which a low number of IC packages is required. On the other hand, if the production volume is relatively low, then the dominant cost of the total cost of a μP-based product is the development cost, particularly the development cost of the software. In this case it is important to select a microprocessor that is well supported in terms of development and diagnostic support tools, that has an instruction set that facilitates software development, and that has good documentation.

Life Expectancy of the Product If the μP-based product is expected to have a long life expectancy, say more than 5 to 10 years, then it is important to consider the features of a μP family that promote that longevity: How long can it be expected that the vendor will support the μP family? Will the required parts be available in the long run? Is the μP family well second-sourced? On the other hand, if the life expectancy of the product is relatively short, then the foregoing considerations are less important.

Stability of the Design of the Product If the design of the μP-based product is expected to be frequently changed and/or expanded, it is important to select a μP family that allows such changes and expansions to be easily made. Furthermore, if the vendor replaces the current microprocessor with a more powerful version, it is desirable to have the new version be upward compatible so that a redevelopment effort is not necessary.

Schedule Urgency If the μP-based product is in a competitive marketplace where a 3- to 6-month delay can mean failure, then the ease of development is a key factor in the selection of the microprocessor. In this case it is again necessary to select a microprocessor that is well supported in terms of development and diagnostic tools, that has an instruction set that facilitates software development, and that has good documentation. Furthermore, it is desirable, if possible, to select a μP family that is familiar to the design staff so that the development time is minimized. It is also necessary to be certain that the required parts can be made available by the vendor in a timely manner.

13.4 MICROPROCESSOR SYSTEM DEVELOPMENT—REVISITED

13.4.1 Introduction

The general procedure for the development of a μP system was introduced in Chapter 8 and summarized in the flowchart of Fig. 8.3. In this section we will discuss the steps in the development procedure, doing so in more detail and with a new perspective based on the μP knowledge gained since Chapter 8. Additionally, we will survey the development tools that are available to help the designer at each step of the development process. The block diagram for the procedure for the development of a μP system is reproduced in more detail in Fig. 13.1, along with the development tools that are available to the designer at each step.

13.4.2 System Design

With the functional specification for a μP system given, the first step in the development procedure is to perform a *system design*. The result of the μP system design is the design specification for the μP system hardware and also the design specification for the μP system software. Recall that the *functional specification* specifies *what* the system does, and the *design specification* specifies *how* the system is to be implemented to satisfy the functional specification. In terms of the hardware, the design specification consists of the specification of the hardware modules, including the selection of the μP and the supporting modules as well as the configuration of these modules. In terms of the software, the design specification consists of the selection of the programming language to be used, the specification of the program modules required to control the hardware modules, and the specifications of the data structures that are necessary to support the program modules. Some key tasks of the μP system design are the selection of the appropriate μP family (Sec. 13.3), as well as the determination of the hardware/software trade-off and the selection of the programming language (see the following subsections).

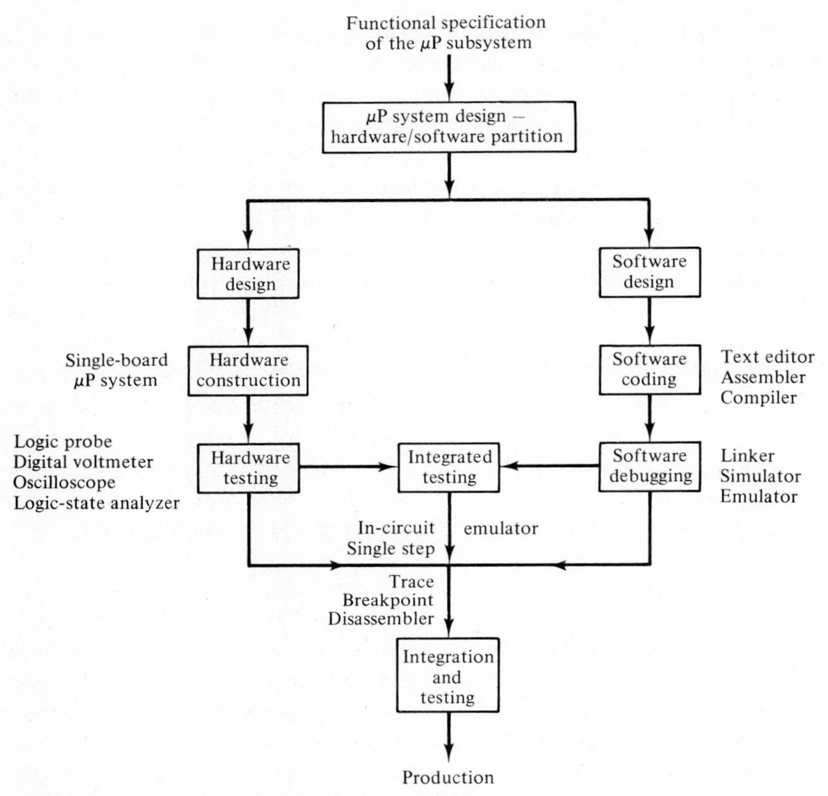

Figure 13.1 Development procedure for a μP system.

Hardware/Software Trade-Off

A key task in μP system design is to determine which functions are best performed by the μP system hardware and which by the μP system software. Generally speaking, a given function can be implemented either by hardware or software. For example, a time delay can be implemented by a program delay loop within a subroutine or by a hardware timer such as the one in the Intel 8155 I/O interface module. For another example, parallel/serial communication can be performed either by a hardware USART or by some fairly complex software subroutines.

The purpose of the hardware/software trade-off determination is to minimize the cost of the μP system for the specified performance requirement. In general, a hardware implementation of a function with available LSI components is faster but is more costly in terms of production cost (cost of the parts, etc.). On the other hand, a software implementation of a function is slower and more costly in terms of development cost but is more flexible in terms of modifications to the design. Consequently, for a μP-based product with a projected large production volume, hardware components should be kept at a minimum, and functions should be implemented with software modules whenever performance requirements permit. The increase in the initial (software) development cost will be averaged over the large number of units produced. Also, for products whose design is not stable, with anticipated changes and enhancements, the flexibility of software implementations of functions is more attractive. On the other hand, if the production volume will be small, then the initial (software) development cost dominates the production cost. In this case, hardware implementations of functions can be more cost-effective. Also, if the μP-based product is in a competitive marketplace where an early completion date is crucial, then for a minimization of development effort and time, hardware implementations are probably preferable. Of course, performance requirements can also dictate the use of hardware implementations.

Selection of the Programming Language

In this book, μP system programs have been exclusively written in an assembly language. One reason for this is that a large amount of programming for μP systems is done in assembly languages. Also, from an educational point of view, programming in an assembly language requires a greater understanding of the important microprocessor concepts, knowlege of which is necessary even for using a high-level programming language to program a μP system.

In practice, though, it is possible and often desirable to program a μP system in a high-level programming language such as PL/M, PASCAL, or C. PL/M, a high-level language developed by Intel for μP systems, has much of the high-level programming constructs of the more commonly known PL/I programming language, including DO and IF-THEN-ELSE statements, arrays, and so on. Additionally, PL/M has a number of low-level features that permit a programmer to have access to μP hardware-level constructs, such as input to or output from I/O ports, stacks, interrupt facilities, etc. PASCAL is a "block-structured" programming language that supports and encourages good programming habits in the writing of structured programs. The various versions of the PASCAL language that are used for programming μP systems are generally extended to include low-level facilities to access μP hardware-level constructs or to allow assembly language routines to be linked to it. The C programming language is another high-level programming language that is gaining popularity for programming μP systems. It has the block structure features of PASCAL and some low-level features that are similar to those of PL/M. The close ties that the C language has with the UNIX operating system has also increased its popularity since the UNIX environment itself is gaining popularity as the operating system for engineering work stations and development systems.

A program written in a high-level language is *compiled* into machine code, just as a program written in an assembly language is *assembled* into machine code. In other words, the end product of a program, whether it is written in a high-level language or an assembly language, is the machine code for the μP system. The differences are the ease in which the program can be developed, the execution and storage efficiency of the resulting machine code, and the effect on the overall "cost" of the μP system.

Again, the choice of the appropriate programming language is a function of the characteristics of the μP-based product that is being developed, including the production volume, schedule urgency, and others. In general, a high-level language has the advantage of ease of software development and so should be used in applications that require low development cost. With a high-level language, a programmer can implement algorithms using high-level programming constructs such as DO loops, IF-THEN-ELSE statements, built-in functions, and recursion, as well as high-level data structures such as arrays, pointers, and complex data types. The programmer no longer has to be concerned explicitly with registers, condition flags, and the like. The disadvantage of using a high-level programming language is inefficiency, both in execution and storage. The object code from a compiled program is far less efficient than an equivalent one from an assembled program.

In most current applications, the software development is a bottleneck in the development of a μP system project, as is evidenced in the μP system development procedure of Fig. 13.1 by the great amount of software development tools that are available. For this reason a good rule of thumb is to use a high-level language whenever performance requirements and memory space permit. Then, if necessary, program other parts (especially the time-critical parts) in assembly language.

Hardware/Software Development

As is shown in Fig. 13.1, after the design specifications for the μP hardware and the μP software have been determined, the development of them can proceed in parallel. Further, the guiding principles for the development of the hardware and software are the same and can be summarized as follows:

Top-Down Design In a top-down design process, the design of the hardware or software system module is partitioned into more detailed designs of less complex modules, which are in turn partitioned, until eventually the resultant modules at the ''bottom level'' can be readily constructed with available components (hardware) or can be readily coded as subroutines (software). In this manner each phase of the design remains manageable. For example, low-level details such as voltage polarity for an output or the number of machine cycles required to implement a 2-ms delay will not interfere with the high-level system/subsystem design. Similarly, although the detailed design of a low-level module depends upon low-level details, it is restricted to only a very small portion of the overall design and therefore remains manageable.

Bottom-Up Implementation/Testing Although the design process is top-down, the implementation and testing process is bottom-up. Individual modules (hardware circuits or software subroutines) are implemented first and tested. Then they are integrated into subsystem modules and tested. The process is continued until the entire μP system module is synthesized and tested to verify that the system requirement specifications have been satisfied.

Modularity Associated with the top-down approach of design and the bottom-up approach of implementation and testing is the notion of modularity. With the design at every level being divided into well-defined modules and the interrelationships of these modules, the design and the implementation/testing of the μP system are performed in manageable bite-size parts. At each phase of the development process, we need only satisfy the input, the output, and the functional and performance requirements of the module under development, with no concern for the other parts of the μP system.

Minimization of the Redesign Loop The development of a module in a μP system is an iterative process consisting of the following steps: design/redesign, implementation, and testing. If errors are detected in the testing step, a redesign of the module is required and the cycle is repeated. In the development of a μP system, it is important to make the redesign loops as short as possible. More specifically, it is desirable to detect errors as early as possible so that the errors are not propagated to the following phases of the development process. The farther an error is propagated, the more work that has to be redone.

Two design concepts are important for the shortening of the redesign loops of a μP system: modularity of design and early integration of hardware and software. Modularity of design permits the errors to be isolated within a module and not passed on to the next module. Early integration of hardware and software permits the early detection of interfacing and timing errors that can be detected only during integrated testing. But if the hardware and software are developed in isolation, then these errors are propagated until the integrated testing step, with the results of a lengthening of the redesign cycle and an increase in the amount of the work that has to be redone.

13.4.3 Hardware Design, Construction, and Testing

In general, the development of the μP hardware is straightforward, as is evidenced by the small amount of hardware development tools available as compared to the amount of software development tools (see Fig. 13.1). Since most commonly required digital functions are integrated as MSI and LSI components, the development of μP hardware involves primarily the selection and configuration of the microprocessor and its supporting chips. If an unusual interface component is required that is not readily available as an LSI chip, then, of course, the hardware development of that module is required in accordance with the digital circuit development procedure of Chapter 7.

Hardware development can be made even simpler if a *single-board μP system* can be suitably used as the core of the μP hardware. This system is a ready-made, minimally configured μP system that can be purchased as a single unit from a manufacturer. It generally consists of a microprocessor, some memory modules (RAM and ROM), some limited I/O facilities, and any required support circuitry such as a power supply, clock circuitry, keypad, and LEDs. Also, some facilities are available for building additional interface circuitry to customize the single-board μP system for a particular application. The main advantage in using a single-board μP system as a core for the μP hardware is that it eliminates the need for constructing the μP hardware from scratch with chips and PC boards; the single-board μP system is already constructed and tested. A μP system constructed from such a board is usually more reliable than one constructed from scratch, and also is a great deal less costly in terms of development time. It is a very attractive alternative for low-volume production.

In the development of the μP hardware, most of the effort is in the testing and debugging of the hardware modules. Testing of the μP hardware consists of both *stand-alone* hardware testing and *software-driven* hardware testing. Most of the necessary tests of the μP are software-driven hardware tests performed during the integrated testing, as will be discussed in Sec. 13.4.5. However, to minimize the redesign loop it is important to perform as much stand-alone testing of the μP hardware as is possible to detect as many errors as possible so that they are not propagated to the integration phase.

Stand-alone testing of the μP hardware module begins with the testing of the power supply voltage levels and the voltage levels of each IC in the module to determine whether they are within the required limits. If a clock signal is required within a hardware module, then its frequency, amplitude, and its rise and fall times need to be checked. Other stand-alone tests include the tests for correct operation of the chip-select logic and the handshaking logic, and any tests that can be performed independently of the software. For this stand-alone testing, various diagnostic tools are available, as will now be considered.

Logic Probe A logic probe is a simple probing device with an LED display that indicates the voltage state (high or low) of any point in the circuit with which the probe is in contact. It can be used in this manner to easily determine the voltage level of an input or output signal in a digital circuit.

Digital Voltmeter A digital voltmeter can be used to detect the actual voltage level of a signal, to determine whether it is within the specified voltage limits for its high or low state. Failure to be in the specified limits can be the result of a faulty power supply or an excessive load on the signal.

Oscilloscope An oscilloscope is used for measuring the dc and ac characteristics of logic signals. One can be used, for example, to measure precisely the rise time and fall time of a signal or the duration of a pulse or to determine whether the setup time of a component has been violated. Some sophisticated oscilloscopes have *multitrace* capabilities for displaying the timing relationships among several signals. *Storage oscilloscopes* can even store, for a given period of time, information about the signals that are under observation. The amount of information stored is determined by the capacity of the oscilloscope memory. A *trigger* switch or signal is used to begin the storage of the information. The stored information can then be recalled for display and analysis.

Logic State Analyzer Similar to a storage oscilloscope, a logic state analyzer (LSA) can store and display the logic values for a set of signals over a period of time. As is illustrated in Fig. 13.2(a), the logic values for M (e.g., $M = 32$) inputs/outputs can be recorded by connecting each input/output to one of the leads of the LSA. If the capacity of the LSA memory is N words $\times$ M bits (e.g., $N = 1\text{K}$), then the logic values for the M inputs/outputs can be recorded for N clock cycles. The stored information can then be recalled for display and analysis.

In the operation of an LSA, a *trigger word* can be used to begin the storage of the data. In this mode of operation, the LSA will not store any data until it detects the specified bit pattern of the trigger word. If, for example, the low-order 16 leads of an LSA are connected to the 16-bit address bus of the microprocessor, and if we want to start the storing of data when the address 060AH is detected, then the trigger word should be specified as XXXX060AH. After the LSA detects the trigger word, it stores this word and then the data for the next $N - 1$ clock cycles, as illustrated in Fig. 13.2(b). Note that the rate at which the data is sampled and stored is a function of the frequency of the clock signal. For most LSAs the user can specify either an internal clock source

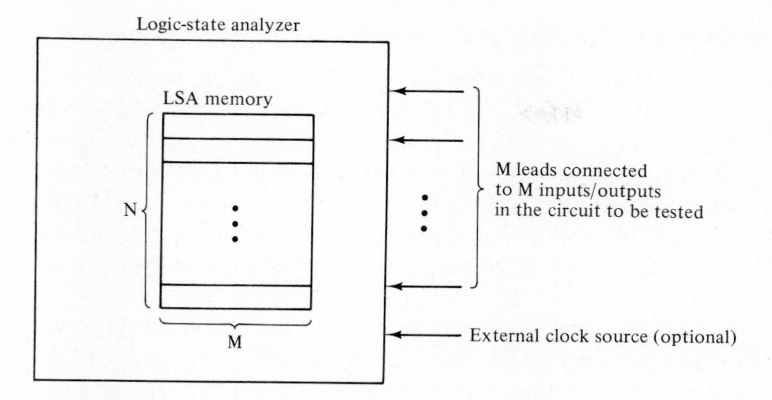

(a) Functional block diagram of the LSA

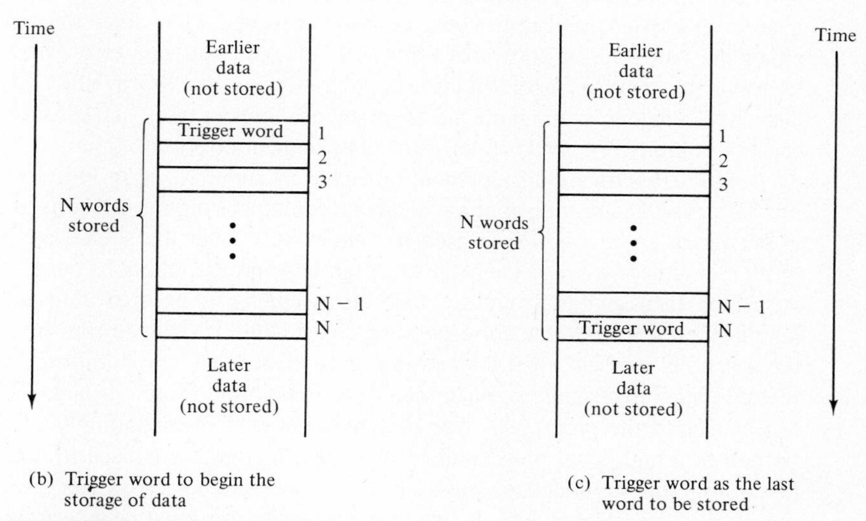

(b) Trigger word to begin the storage of data

(c) Trigger word as the last word to be stored

Figure 13.2 Logic state analyzer.

supplied by the LSA or an external clock source. The external clock source is usually the system clock used for the μP hardware.

Alternatively, the trigger word can indicate the *last* word to be stored, as is illustrated in Fig. 13.2(c). In this case the LSA stores the logic values for the M inputs/outputs for the $N - 1$ clock cycles preceding the clock cycle in which it detects the trigger word pattern. Then the trigger word is the Nth word stored.

Another LSA feature is the *delay* trigger. With this feature activated, the LSA will not begin the recording of data until either a delay of a specified number of clock cycles, or a delay of a specified number of clock cycles after detecting the trigger word. With this and its many other features, the logic state analyzer is a very flexible and powerful diagnostic tool for testing μP hardware.

13.4.4 Software Design, Coding, and Testing

Except for the case of very-high-volume production, the dominant cost of a μP system is generally the initial development cost, particularly the *software* development cost. Consequently, in the development of μP software, it is particularly important to follow the previously discussed design principles of top-down design, bottom-up implementation/testing, modularity, and minimization of the redesign loop. In this section we will discuss the μP software development procedure and survey some important μP software development tools that are available for supporting the software development of a μP system.

If no development tools were available, the software developer would have to write the programs in assembly language, hand assemble them into binary machine code, and then program the machine code into a PROM or an EPROM. In such an environment the programs could not be tested until after the μP hardware had been constructed and the PROM/EPROM containing the programs had been placed in the μP system. Then, if an error existed, its cause would be difficult to determine since the error could be in either the hardware or software. Further, if it was a software error, then it could be a logical error, a syntax error, an error in programming the PROM/EPROM, or something else. As is evident, without the aid of some development tools, μP software development can be prohibitively costly in terms of development effort and time.

The μP software development procedure is summarized in Fig. 13.3. Also shown are the development tools that are available for supporting the μP software development at each step. A *text editor* is used to interactively enter the source program, which is written in either assembly language or a high-level programming language. If the program is written in assembly language, then an *assembler* is used to translate the assembly language program into the corresponding object code. If syntax errors are detected during the assembly process, then the cause of each error must be determined, after which the text editor is used again to make the corrections to the source program. This process is repeated until the program is free of syntax errors. Correspondingly, if the program is written in a high-level programming language, a compiler is used to translate the high-level programming language program into the object code. If syntax errors are detected during the compilation process, the cause of each error must be determined and the text editor used again to make the corrections to the source program. Then the process is repeated.

After the program is free of syntax errors, the resulting machine language program is linked, loaded, and executed to test for logical errors. The test execution of the software can be performed without the full construction of the μP hardware by using *simulators* or *emulators*. If logical errors are detected during the testing of the program, the cause of each error must be determined and the text editor used again to make the corrections to the source program. The entire process is repeated until the program is free of logical errors. Finally, the error-free machine code is programmed into a PROM or an EPROM, ready for the final integrated testing.

The various development tools shown in Fig. 13.3 can be implemented on either a general-purpose computer system or a *microcomputer development system*. The latter is a special-purpose computer system that is tailored for the efficient development of a μP system. It contains tools such as the text editor, assemblers, compilers, and linker

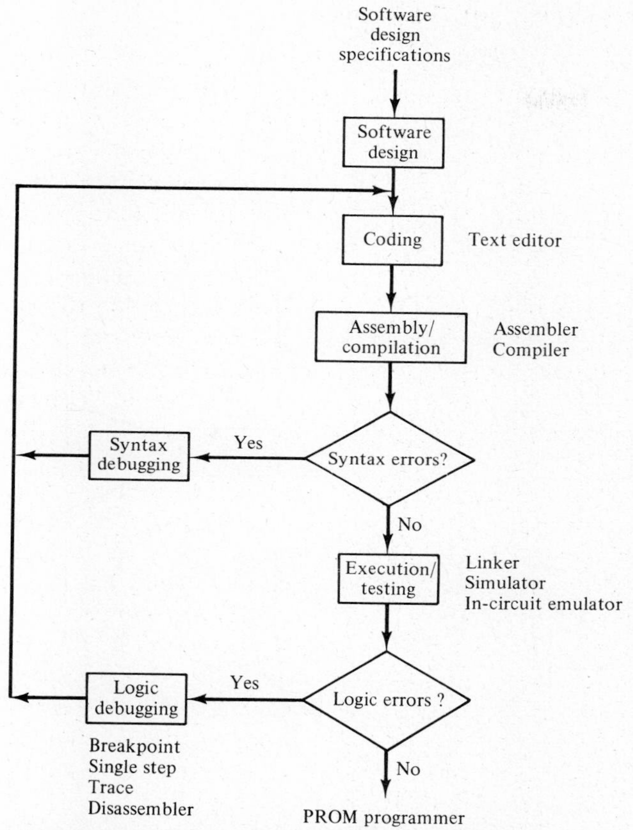

Figure 13.3 Development procedure for μP software.

for supporting software development. It also contains tools such as in-circuit emulators (to be described later) and other analysis tools for supporting the testing of software in an integrated environment *without* the need of the fully constructed μP hardware.

Development systems can be classified into *dedicated* development systems and *universal* development systems. A dedicated development system is designed to support the μP system development for one specific μP family and is generally provided by the manufacturer of that μP family. Since it is designed for the development of only one μP family, the machine code generated by its compilers is highly efficient. For the same reason, however, it cannot be used for the development of a new or different μP family. In contrast, a universal development system is designed to support the μP system development for a variety of μP families from different manufacturers. For this reason, the development of a new μP family can be handled easily by the addition of new software (assembler, compilers) and new emulation hardware. However, the machine code generated by its compilers is generally less efficient than that of a dedicated development system.

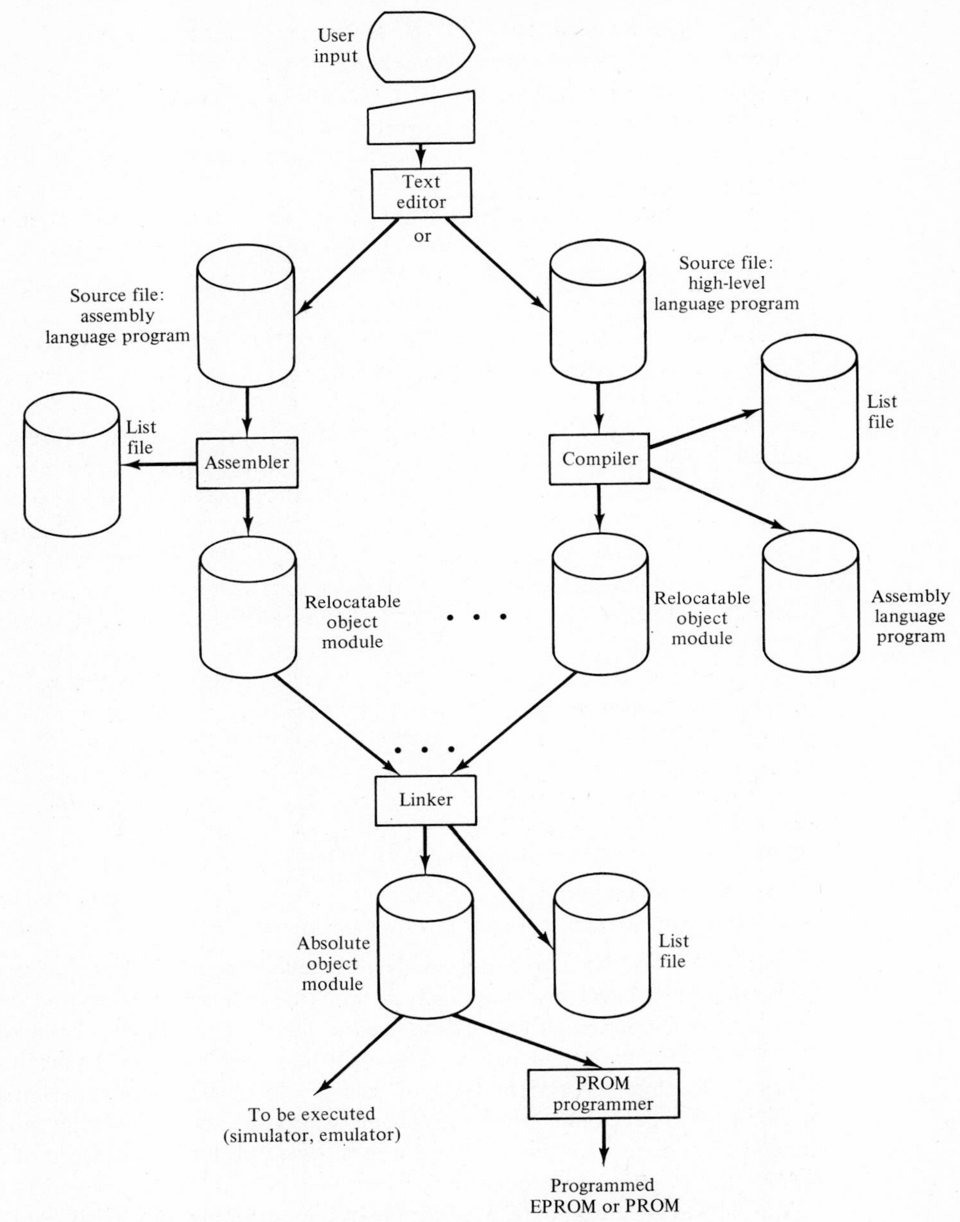

Figure 13.4 μP software development tools and key files generated.

The software development tools that are found in these microcomputer development systems are described in the paragraphs that follow. Shown in Fig. 13.4 are the relationships among these tools and the key files that are produced through the use of them.

Text Editor A text editor is a system-provided program for supporting the editing of a text file. Once invoked, an editor provides all the functions required for helping a user to interactively create a new file or to edit an existing one. The functions that are generally provided by an editor include creating a file, retrieving an existing file into the workspace, inserting new text into the file, deleting text from the file, modifying the text, moving text within the file, finding text in the file, and so forth. Since most readers are probably users of computer systems and as such are probably already familiar with the functions of a text editor, we will not belabor the point here. In the use of a text editor for μP software development, the output of the editor is a file containing the source program, written either in assembly language or a high-level programming language.

Assembler An assembler, another system-provided program, facilitates the translation of a source program written in assembly language into the corresponding machine code for a (micro)processor. It performs essentially the same functions as the hand assembly process discussed in Chapter 10. An assembler, however, performs them faster and more reliably. Furthermore, an assembler can detect syntax errors, and can often point out the causes of the syntax errors in the *assembler list file,* which is an output of the assembler. Also, most assemblers provide a cross-reference file, therein listing all program symbols (i.e., variable names, statement labels, etc.) alphabetically, along with the line number in which each symbol is defined, as well as the line numbers in which it is referenced. The cross-reference file is often helpful in the debugging of subtle syntax or logical errors.

The input to the assembler is the source program file produced by the text editor. The most important output from the assembler is the *relocatable object module,* which is a file containing the resulting machine code in a relocatable form. It is "relocatable" because the absolute addresses for the program have not been assigned and therefore the program can be placed anywhere in memory. The absolute addresses are assigned in the link step by the linker, which we will consider after the compiler.

Compiler A compiler is a system-provided program that facilitates the translation of a source program written in a high-level language into a corresponding assembly language program and/or machine-code program for a (micro)processor. In translating to the machine code, a compiler performs essentially the same functions as the assembler described above and produces essentially the same files as outputs, including the list file, cross-reference file, and relocatable object module. The input to a compiler is the source program written in a high-level language, and produced by using the text editor. A different compiler is required for each language. For example, if the source program is written in PASCAL, then a PASCAL compiler is required. If it is written in the C language, then a C compiler is required, and so forth.

A compiler is generally a more complex program than an assembler. Unlike in the assembly process where each assembly language statement translates into one machine code statement, a high-level language statement bears no resemblance to an assembly language statement or a machine code statement and does not reference the components of the μP architecture (registers, stack, etc.). Consequently, the translation of a high-level language is much more complex. Furthermore, high-level data structures like arrays and pointers require sophisticated memory management by the corresponding machine code program.

Linker The linker is a system-provided program that is used to link several relocatable object modules into a single absolute object module by assigning absolute addresses to the modules. The resultant absolute object module is ready to be loaded into memory for execution.

Perhaps it is not apparent that having a linker is desirable. After all, in a simple μP system where a single program is sufficient for controlling the μP system, we need only assemble (or compile) the source program into object code, which is then ready for execution without the use of a linker. In the general case, however, several program modules are required for controlling a μP system. Then, the assemble-link or compile-link feature is extremely convenient for facilitating the parallel and independent development of the different program modules (perhaps by different programmers). As stated, the output of an assembler or a compiler is the relocatable object module, which is a machine code program that is ready for execution except for lacking absolute addresses. And, these addresses are assigned by the linker in the link step. With the use of a linker, different program modules of a μP system can be assembled (or compiled), linked (assigned absolute addresses), and executed separately. Then, when each individual program module has been successfully debugged, the linker can be invoked again, this time to reassign the absolute addresses, readjusting them if necessary. It also relinks the individual program modules into a single absolute object module that is ready for execution.

13.4.5 Integrated Testing

Because of the hardware/software interactive nature of a μP system, the most important tests are performed during the integrated testing step. In this step, the testing consists of software-driven hardware testing and hardware-driven software testing. The key development tool for this integrated testing is the *in-circuit emulator (ICE).*

In-Circuit Emulation

In the development of a μP system, in-circuit emulation is the process in which the microprocessor and/or the memory modules of the prototype μP system are replaced by the in-circuit emulator, which appears to the remainder of the μP system to be exactly the same in all respects as the original components it replaced. In-circuit emulation is used for the following reasons:

1. Along with the other development tools in the development system, the ICE provides a powerful set of tools for facilitating the integrated debugging of the μP system, both the hardware and the software.
2. The ICE facilitates an early integrated testing of the μP system. In particular, integrated testing of the software can be performed without having to wait for the full construction of the μP hardware. Also, program modules can be added stepwise to the PROM/EPROM in the μP system as each program module is debugged.

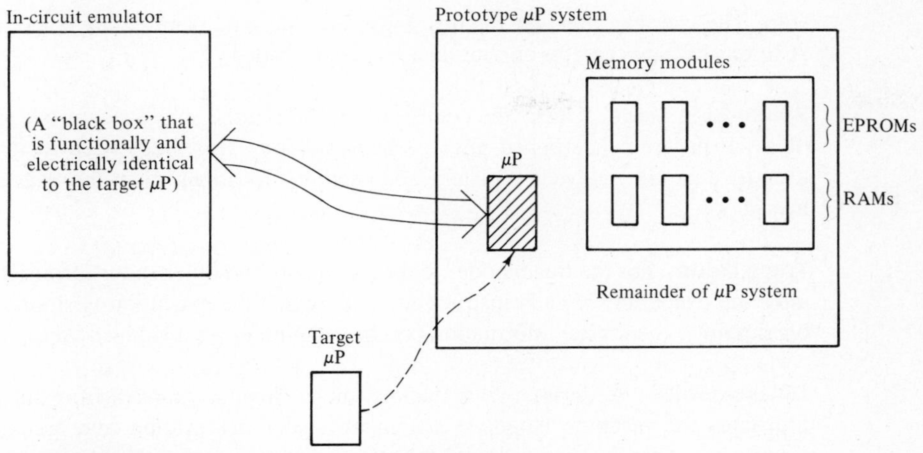

Figure 13.5 Simple use of the in-circuit emulator.

The in-circuit emulator is a very powerful and flexible tool that can be used in various ways in the integrated testing of a μP system. To introduce some of the features of an ICE, let us consider the case in which the ICE is used simply for its debugging capabilities. In this case the μP hardware of the prototype μP system is fully constructed and the μP software has been developed and programmed into the EPROM modules of the μP system. To use the ICE we remove the target microprocessor of the μP system from its socket and, as shown in Fig. 13.5, replace it with the "emulator microprocessor," which has connections to the emulator controller in the development system. To the prototype μP system the ICE is simply another "microprocessor" that is functionally and electrically identical to the removed target microprocessor. Consequently, the μP software stored in the EPROM module of the μP system can be executed as before. The difference is that now the development system has full control over the execution of the "emulation microprocessor." As a result, the user, through the tools provided by the development system, can control the execution of the μP software in such a way that errors in the μP software and/or hardware can be systematically debugged. Some of the debugging functions that are supplied by the development system and the ICE will now be considered.

Breakpoint In a normal mode of operation a μP program begins execution at the instruction initially pointed to by the program counter, and the execution continues until a HALT instruction is encountered. In this case, if the μP system is not functioning properly, it is difficult to identify the cause of the problem. The breakpoint facility alleviates this difficulty. It allows the user to "freeze" the execution of the μP program at a specified point of the execution so that the contents of the relevant registers and memory locations can be examined. For example, if a user wants to be certain that the μP hardware is working properly in loading "PORTA" with the appropriate data, the user can set a breakpoint immediately after the "IN PORTA" instruction. When the breakpoint is encountered during the execution of the μP program, the execution will stop. The user can then use the development system to examine the contents of Register A to ensure that the appropriate data has been loaded.

Single-Step Mode Under the control of the ICE in the single-step mode, the execution of a μP program is stopped after each instruction. This single-step facility allows the user to examine relevant registers and memory locations after the execution of *each* instruction.

Trace Mode For the trace mode, register contents and other status information are saved after the execution of each instruction. Then when the execution is stopped (e.g., via a breakpoint), the stored information can be recalled to be displayed and analyzed.

Disassembler A disassembler operates in an inverse manner from an assembler. It translates the machine language statements under observation into assembly language statements. This function is useful during the debugging of a μP (machine code) program since it allows the user to debug at the symbolic level (i.e., the assembly program level) rather than at the machine code level.

Figure 13.6 shows a more general use of an in-circuit emulator. Note that the target microprocessor is again removed from its socket and replaced by the ICE. For purposes of explanation, let us assume that some of the μP software is still under development and that the corresponding EPROMs (illustrated by the shaded boxes) have not yet been placed on-board the prototype μP system. Also, one of the RAM modules in the prototype μP is not ready for use. Then, in the integrated testing, the physically absent EPROMs

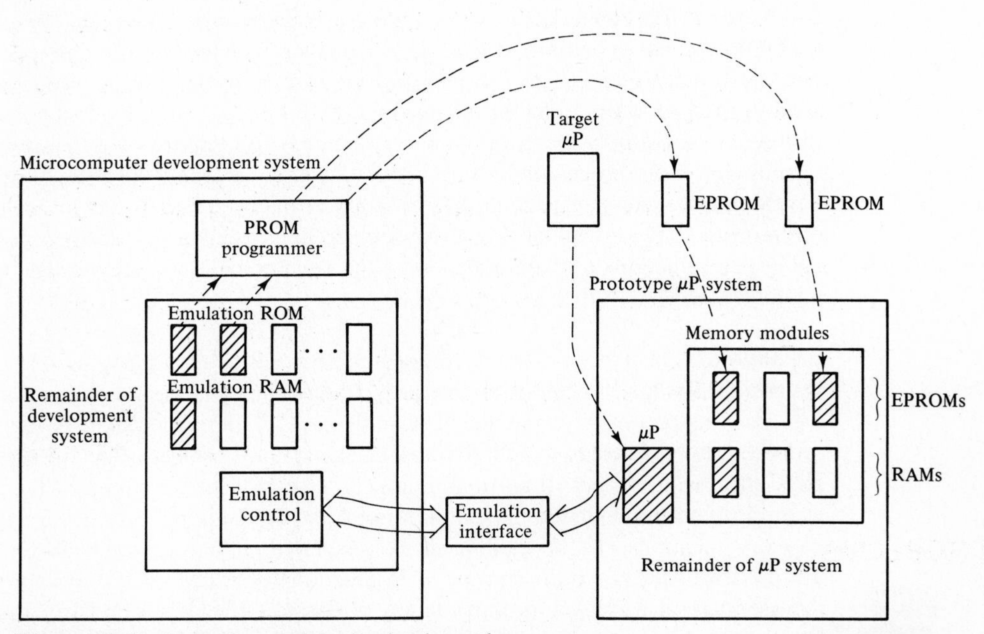

Figure 13.6 General use of the in-circuit emulator.

and RAM will be emulated by the emulation ROM and emulation RAM, respectively. In other words, if during the execution of the μP software, a memory location is to be accessed whose address is a physically present EPROM or RAM, then the "emulation microprocessor" will access the memory location there. If, however, the memory location to be accessed is in a physically absent EPROM or RAM, then the "emulation microprocessor" will access the memory location in the emulation memory. Effectively, during the emulation run, the μP appears to have a full complement of EPROMs and RAMs even though some of the EPROMs and/or RAMs are not physically present on-board the prototype μP system.

If during the integrated testing, the program under development (which is stored in the emulation ROM) does not function properly, then all the development tools that are available in the development system (breakpoint, single-step, trace, disassembler, editor, assembler, compiler, linker, etc.) can be used to detect and correct the errors. When the program under development is finally determined to be free of errors, then the PROM programmer can be used to program the EPROM. The EPROM can then be physically placed on-board the μP system. As a result, for the next emulation run, more EPROMs will be specified to be physical EPROMs and fewer EPROMs will be emulated. Similarly, if more RAMs are put on-board the μP system, then fewer will be specified as emulation RAMs.

If, during the integrated testing, it is determined that an error is caused by a hardware malfunction, then the ICE can be used in conjunction with the hardware diagnostic tools described in Sec. 13.4.3 (logic probe, digital voltmeter, oscilloscope, and logic state analyzer) to debug the hardware malfunction.

In the manner explained, integrated testing can be performed at various stages of completion of the μP hardware and software. Initially, all the μP hardware (microprocessor, all EPROMs and RAMs, I/O) can be emulated while the software is being tested. But as more of the μP hardware is constructed and more of the μP software is debugged, and both are gradually moved on-board the μP system, less hardware and software have to be emulated. The final integration and testing is made after the μP hardware is completely constructed and the μP software completely developed. At this point, most of the errors should have been corrected. However, a final set of predesigned tests should be performed to ensure that the μP system satisfies the specified system requirements. Again, the development system and the in-circuit emulator can be used to verify the final testing. Used in this manner, the development system and the in-circuit emulator provide a set of powerful and flexible μP development tools.

SUPPLEMENTARY READING (see Bibliography)

[Cushman 86], [Freedman 83], [Kline 83], [Peatman 77], [Rafiquzzaman 84], [Roberts 82], [Short 81]

PROBLEMS

13.1. When a microprocessor is classified as a 4-bit, 8-bit, 16-bit, or 32-bit microprocessor, on what general criteria is this classification based?

13.2. What does it mean to say that microprocessor X is object-code compatible with another microprocessor Y? Give some examples of microprocessors that are object-code compatible.

13.3. What does it mean to say that microprocessor X is source-code compatible with another microprocessor Y? Give some examples of microprocessors that are source-code compatible.

13.4. What are the main μP system functions that are generally integrated onto a single-chip microcomputer?

13.5. What is the difference between a microprocessor with a split-memory architecture and one with a common-memory architecture? Give an example of each.

13.6. List the criteria that can be used to evaluate a microprocessor's performance. Briefly explain each.

13.7. List some of the criteria that can be used to evaluate a microprocessor other than by its performance. Briefly explain each.

13.8. What does it mean to have a microprocessor second-sourced?

13.9. The total cost of a μP system can be divided into two main components. Name and briefly explain each.

13.10. Which is more important in the cost of a μP-based product, production cost or development cost? Explain.

13.11. List the characteristics of the target application that have an effect on the selection of the microprocessor used for the μP-based product. Briefly explain each.

13.12. Explain the hardware/software trade-off issues in the development of a μP system. In other words, what determines whether a μP function is to be implemented in hardware or software?

13.13. Compare the hardware (LSI) implementation versus the software implementation of a μP function in terms of the following items. Briefly explain each answer.

(a) Development cost; which method is higher?
(b) Production cost; which method is higher?
(c) High-volume production; which method is preferable?
(d) Low-volume production; which method is preferable?
(e) Early completion of project; which method is preferable?

13.14. Discuss the advantages and disadvantages of using an assembly language or a high-level programming language in programming a μP system. Under what circumstances should each be used?

13.15. A high-level programming language is usually preferred over an assembly language in programming a μP system whenever performance and space requirements permit. True or false? Explain.

13.16. Identify and explain the following key phrases relating to the development of a μP system: top-down, bottom-up, modularity, and redesign loop.

13.17. Briefly explain why ''minimization of the redesign loop'' is important in the development cycle of a μP-based product.

13.18. Identify and briefly explain the following development tools used in the development of μP hardware: single-board μP system, logic probe, digital voltmeter, oscilloscope, and logic state analyzer.

13.19. Explain the μP software development procedure illustrated by the flowchart of Fig. 13.3.

13.20. List the development tools that are available for supporting the development of μP system software. Briefly explain each.

13.21. What is the difference between a dedicated μP development system and a universal μP development system? Give the advantages and disadvantages of each.

13.22. Explain how the following debugging tools, provided by a μP development system, can be used with an in-circuit emulator (ICE) to facilitate the integrated testing of a μP system: breakpoint, single-step mode, trace mode, and disassembler.

Appendix

8085 Opcode and Machine Code Correspondence

DATA TRANSFER GROUP

Move

Instruction	Operands	Hex
MOV	A,A	7F
MOV	A,B	78
MOV	A,C	79
MOV	A,D	7A
MOV	A,E	7B
MOV	A,H	7C
MOV	A,L	7D
MOV	A,M	7E
MOV	B,A	47
MOV	B,B	40
MOV	B,C	41
MOV	B,D	42
MOV	B,E	43
MOV	B,H	44
MOV	B,L	45
MOV	B,M	46
MOV	C,A	4F
MOV	C,B	48
MOV	C,C	49
MOV	C,D	4A
MOV	C,E	4B
MOV	C,H	4C
MOV	C,L	4D
MOV	C,M	4E
MOV	D,A	57
MOV	D,B	50
MOV	D,C	51
MOV	D,D	52
MOV	D,E	53
MOV	D,H	54
MOV	D,L	55
MOV	D,M	56

Move (cont)

Instruction	Operands	Hex
MOV	E,A	5F
MOV	E,B	58
MOV	E,C	59
MOV	E,D	5A
MOV	E,E	5B
MOV	E,H	5C
MOV	E,L	5D
MOV	E,M	5E
MOV	H,A	67
MOV	H,B	60
MOV	H,C	61
MOV	H,D	62
MOV	H,E	63
MOV	H,H	64
MOV	H,L	65
MOV	H,M	66
MOV	L,A	6F
MOV	L,B	68
MOV	L,C	69
MOV	L,D	6A
MOV	L,E	6B
MOV	L,H	6C
MOV	L,L	6D
MOV	L,M	6E
MOV	M,A	77
MOV	M,B	70
MOV	M,C	71
MOV	M,D	72
MOV	M,E	73
MOV	M,H	74
MOV	M,L	75
XCHG		EB

Move Immediate

Instruction	Operands	Hex
MVI	A, byte	3E
MVI	B, byte	06
MVI	C, byte	0E
MVI	D, byte	16
MVI	E, byte	1E
MVI	H, byte	26
MVI	L, byte	2E
MVI	M, byte	36

Load Immediate

Instruction	Operands	Hex
LXI	B, dble	01
LXI	D, dble	11
LXI	H, dble	21
LXI	SP, dble	31

Load/Store

Instruction	Hex
LDAX B	0A
LDAX D	1A
LHLD adr	2A
LDA adr	3A
STAX B	02
STAX D	12
SHLD adr	22
STA adr	32

byte = constant, or logical/arithmetic expression that evaluates to an 8-bit data quantity. (Second byte of 2-byte instructions).

dble = constant, or logical/arithmetic expression that evaluates to a 16-bit data quantity. (Second and Third bytes of 3-byte instructions).

adr = 16-bit address (Second and Third bytes of 3-byte instructions).

* = all flags (C, Z, S, P, AC) affected.

* * = all flags except CARRY affected; (exception: INX and DCX affect no flags).

† = only CARRY affected.

ARITHMETIC AND LOGICAL GROUP

Add*

Instruction	Operand	Hex
ADD	A	87
ADD	B	80
ADD	C	81
ADD	D	82
ADD	E	83
ADD	H	84
ADD	L	85
ADD	M	86
ADC	A	8F
ADC	B	88
ADC	C	89
ADC	D	8A
ADC	E	8B
ADC	H	8C
ADC	L	8D
ADC	M	8E

Subtract*

Instruction	Operand	Hex
SUB	A	97
SUB	B	90
SUB	C	91
SUB	D	92
SUB	E	93
SUB	H	94
SUB	L	95
SUB	M	96
SBB	A	9F
SBB	B	98
SBB	C	99
SBB	D	9A
SBB	E	9B
SBB	H	9C
SBB	L	9D
SBB	M	9E

Double Add †

Instruction	Operand	Hex
DAD	B	09
DAD	D	19
DAD	H	29
DAD	SP	39

Increment**

Instruction	Operand	Hex
INR	A	3C
INR	B	04
INR	C	0C
INR	D	14
INR	E	1C
INR	H	24
INR	L	2C
INR	M	34
INX	B	03
INX	D	13
INX	H	23
INX	SP	33

Decrement**

Instruction	Operand	Hex
DCR	A	3D
DCR	B	05
DCR	C	0D
DCR	D	15
DCR	E	1D
DCR	H	25
DCR	L	2D
DCR	M	35
DCX	B	0B
DCX	D	1B
DCX	H	2B
DCX	SP	3B

Specials

Instruction	Hex
DAA*	27
CMA	2F
STC†	37
CMC†	3F

Rotate †

Instruction	Hex
RLC	07
RRC	0F
RAL	17
RAR	1F

Logical*

Instruction	Operand	Hex
ANA	A	A7
ANA	B	A0
ANA	C	A1
ANA	D	A2
ANA	E	A3
ANA	H	A4
ANA	L	A5
ANA	M	A6
XRA	A	AF
XRA	B	A8
XRA	C	A9
XRA	D	AA
XRA	E	AB
XRA	H	AC
XRA	L	AD
XRA	M	AE
ORA	A	B7
ORA	B	B0
ORA	C	B1
ORA	D	B2
ORA	E	B3
ORA	H	B4
ORA	L	B5
ORA	M	B6
CMP	A	BF
CMP	B	B8
CMP	C	B9
CMP	D	BA
CMP	E	BB
CMP	H	BC
CMP	L	BD
CMP	M	BE

Arith & Logical Immediate

Instruction	Hex
ADI byte	C6
ACI byte	CE
SUI byte	D6
SBI byte	DE
ANI byte	E6
XRI byte	EE
ORI byte	F6
CPI byte	FE

BRANCH CONTROL GROUP

Jump

Instruction	Hex
JMP adr	C3
JNZ adr	C2
JZ adr	CA
JNC adr	D2
JC adr	DA
JPO adr	E2
JPE adr	EA
JP adr	F2
JM adr	FA
PCHL	E9

Call

Instruction	Hex
CALL adr	CD
CNZ adr	C4
CZ adr	CC
CNC adr	D4
CC adr	DC
CPO adr	E4
CPE adr	EC
CP adr	F4
CM adr	FC

Return

Instruction	Hex
RET	C9
RNZ	C0
RZ	C8
RNC	D0
RC	D8
RPO	E0
RPE	E8
RP	F0
RM	F8

Restart

Instruction	Operand	Hex
RST	0	C7
RST	1	CF
RST	2	D7
RST	3	DF
RST	4	E7
RST	5	EF
RST	6	F7
RST	7	FF

I/O AND MACHINE CONTROL

Stack Ops

Instruction	Operand	Hex
PUSH	B	C5
PUSH	D	D5
PUSH	H	E5
PUSH	PSW	F5
POP	B	C1
POP	D	D1
POP	H	E1
POP	PSW*	F1
XTHL		E3
SPHL		F9

Input/Output

Instruction	Hex
OUT byte	D3
IN byte	DB

Control

Instruction	Hex
DI	F3
EI	FB
NOP	00
HLT	76
RIM	20
SIM	30

Hex	Mnemonic	Operand	Hex	Mnemonic	Operand	Hex	Mnemonic	Operand	Hex	Mnemonic	Operand	Hex	Mnemonic	Operand	Hex	Mnemonic	Operand
00	NOP		2B	DCX	H	56	MOV	D,M	81	ADD	C	AC	XRA	H	D7	RST	2
01	LXI	B,dble	2C	INR	L	57	MOV	D,A	82	ADD	D	AD	XRA	L	D8	RC	
02	STAX	B	2D	DCR	L	58	MOV	E,B	83	ADD	E	AE	XRA	M	D9	---	
03	INX	B	2E	MVI	L,byte	59	MOV	E,C	84	ADD	H	AF	XRA	A	DA	JC	adr
04	INR	B	2F	CMA		5A	MOV	E,D	85	ADD	L	B0	ORA	B	DB	IN	byte
05	DCR	B	30	SIM		5B	MOV	E,E	86	ADD	M	B1	ORA	C	DC	CC	adr
06	MVI	B,byte	31	LXI	SP,dble	5C	MOV	E,H	87	ADD	A	B2	ORA	D	DD	---	
07	RLC		32	STA	adr	5D	MOV	E,L	88	ADC	B	B3	ORA	E	DE	SBI	byte
08	---		33	INX	SP	5E	MOV	E,M	89	ADC	C	B4	ORA	H	DF	RST	3
09	DAD	B	34	INR	M	5F	MOV	E,A	8A	ADC	D	B5	ORA	L	E0	RPO	
0A	LDAX	B	35	DCR	M	60	MOV	H,B	8B	ADC	E	B6	ORA	M	E1	POP	H
0B	DCX	B	36	MVI	M,byte	61	MOV	H,C	8C	ADC	H	B7	ORA	A	E2	JPO	adr
0C	INR	C	37	STC		62	MOV	H,D	8D	ADC	L	B8	CMP	B	E3	XTHL	
0D	DCR	C	38	---		63	MOV	H,E	8E	ADC	M	B9	CMP	C	E4	CPO	adr
0E	MVI	C,byte	39	DAD	SP	64	MOV	H,H	8F	ADC	A	BA	CMP	D	E5	PUSH	H
0F	RRC		3A	LDA	adr	65	MOV	H,L	90	SUB	B	BB	CMP	E	E6	ANI	byte
10	---		3B	DCX	SP	66	MOV	H,M	91	SUB	C	BC	CMP	H	E7	RST	4
11	LXI	D,dble	3C	INR	A	67	MOV	H,A	92	SUB	D	BD	CMP	L	E8	RPE	
12	STAX	D	3D	DCR	A	68	MOV	L,B	93	SUB	E	BE	CMP	M	E9	PCHL	
13	INX	D	3E	MVI	A,byte	69	MOV	L,C	94	SUB	H	BF	CMP	A	EA	JPE	adr
14	INR	D	3F	CMC		6A	MOV	L,D	95	SUB	L	C0	RNZ		EB	XCHG	
15	DCR	D	40	MOV	B,B	6B	MOV	L,E	96	SUB	M	C1	POP	B	EC	CPE	adr
16	MVI	D,byte	41	MOV	B,C	6C	MOV	L,H	97	SUB	A	C2	JNZ	adr	ED	---	
17	RAL		42	MOV	B,D	6D	MOV	L,L	98	SBB	B	C3	JMP	adr	EE	XRI	byte
18	---		43	MOV	B,E	6E	MOV	L,M	99	SBB	C	C4	CNZ	adr	EF	RST	5
19	DAD	D	44	MOV	B,H	6F	MOV	L,A	9A	SBB	D	C5	PUSH	B	F0	RP	
1A	LDAX	D	45	MOV	B,L	70	MOV	M,B	9B	SBB	E	C6	ADI	byte	F1	POP	PSW
1B	DCX	D	46	MOV	B,M	71	MOV	M,C	9C	SBB	H	C7	RST	0	F2	JP	adr
1C	INR	E	47	MOV	B,A	72	MOV	M,D	9D	SBB	L	C8	RZ		F3	DI	
1D	DCR	E	48	MOV	C,B	73	MOV	M,E	9E	SBB	M	C9	RET		F4	CP	adr
1E	MVI	E,byte	49	MOV	C,C	74	MOV	M,H	9F	SBB	A	CA	JZ	adr	F5	PUSH	PSW
1F	RAR		4A	MOV	C,D	75	MOV	M,L	A0	ANA	B	CB	---		F6	ORI	byte
20	RIM		4B	MOV	C,E	76	HLT		A1	ANA	C	CC	CZ	adr	F7	RST	6
21	LXI	H,dble	4C	MOV	C,H	77	MOV	M,A	A2	ANA	D	CD	CALL	adr	F8	RM	
22	SHLD	adr	4D	MOV	C,L	78	MOV	A,B	A3	ANA	E	CE	ACI	byte	F9	SPHL	
23	INX	H	4E	MOV	C,M	79	MOV	A,C	A4	ANA	H	CF	RST	1	FA	JM	adr
24	INR	H	4F	MOV	C,A	7A	MOV	A,D	A5	ANA	L	D0	RNC		FB	EI	
25	DCR	H	50	MOV	D,B	7B	MOV	A,E	A6	ANA	M	D1	POP	D	FC	CM	adr
26	MVI	H,byte	51	MOV	D,C	7C	MOV	A,H	A7	ANA	A	D2	JNC	adr	FD	---	
27	DAA		52	MOV	D,D	7D	MOV	A,L	A8	XRA	B	D3	OUT	byte	FE	CPI	byte
28	---		53	MOV	D,E	7E	MOV	A,M	A9	XRA	C	D4	CNC	adr	FF	RST	7
29	DAD	H	54	MOV	D,H	7F	MOV	A,A	AA	XRA	D	D5	PUSH	D			
2A	LHLD	adr	55	MOV	D,L	80	ADD	B	AB	XRA	E	D6	SUI	byte			

HEX-ASCII TABLE

Hex	Char		Hex	Char	Hex	Char	Hex	Char
00	NUL		21	!	42	B	63	c
01	SOH		22	"	43	C	64	d
02	STX		23	#	44	D	65	e
03	ETX		24	$	45	E	66	f
04	EOT		25	%	46	F	67	g
05	ENQ		26	&	47	G	68	h
06	ACK		27	'	48	H	69	i
07	BEL		28	(	49	I	6A	j
08	BS		29	)	4A	J	6B	k
09	HT		2A	*	4B	K	6C	l
0A	LF		2B	+	4C	L	6D	m
0B	VT		2C	,	4D	M	6E	n
0C	FF		2D	–	4E	N	6F	o
0D	CR		2E	.	4F	O	70	p
0E	SO		2F	/	50	P	71	q
0F	SI		30	0	51	Q	72	r
10	DLE		31	1	52	R	73	s
11	DC1	(X-ON)	32	2	53	S	74	t
12	DC2	(TAPE)	33	3	54	T	75	u
13	DC3	(X-OFF)	34	4	55	U	76	v
14	DC4		35	5	56	V	77	w
15	NAK		36	6	57	W	78	x
16	SYN		37	7	58	X	79	y
17	ETB		38	8	59	Y	7A	z
18	CAN		39	9	5A	Z	7B	{
19	EM		3A	:	5B	[	7C	|
1A	SUB		3B	;	5C	\	7D	} (ALT MODE)
1B	ESC		3C	<	5D	]	7E	~
1C	FS		3D	=	5E	∧(↑)	7F	DEL (RUB OUT)
1D	GS		3E	>	5F	–(←)		
1E	RS		3F	?	60	\		
1F	US		40	@	61	a		
20	SP		41	A	62	b		

Courtesy of Intel Corporation

Bibliography

[Andrews 82] Andrews, M. *Programming Microprocessor Interfaces for Control and Instrumentation*. Englewood Cliffs, NJ: Prentice-Hall, 1982.

[Bartee 85] Bartee, T. C. *Digital Computer Fundamentals*, 6th ed. New York: McGraw-Hill, 1985.

[Blakeslee 79] Blakeslee, T. R. *Digital Design with Standard MSI and LSI*, 2d ed. New York: Wiley, 1979.

[Boole 54] Boole, G. *An Investigation of the Laws of Thought*. New York: Dover, 1954.

[Clare 73] Clare, C. R. *Designing Logic Systems Using State Machines*. New York: McGraw-Hill, 1973.

[Clements 82] Clements, A. *Microcomputer Design and Construction*. Englewood Cliffs, NJ: Prentice-Hall, 1982.

[Conffron 83] Conffron, J. W., and W. E. Long. *Practical Interfacing Techniques for Microprocessor Systems*. Englewood Cliffs, NJ: Prentice-Hall, 1983.

[Cushman 86] Cushman, R. H. ''EDN's Thirteenth Annual μP/μC Chip Directory,'' *EDN*, vol. 31, no. 24 (November 27, 1986): 102–206.

[Doty 79] Doty, K. L. *Fundamentals of Microcomputer Architecture*. Portland, OR: Matrix, 1979.

[Fletcher 80] Fletcher, W. I. *An Engineering Approach to Digital Design*. Englewood Cliffs, NJ: Prentice-Hall, 1980.

[Freedman 83] Freedman, D. M., and L. B. Evans. *Designing Systems with Microcomputers—A Systematic Approach*. Englewood Cliffs, NJ: Prentice-Hall, 1983.

[Gorsline 85] Gorsline, G. W. *16-Bit Modern Microcomputers, the Intel 18086 Family*. Englewood Cliffs, NJ: Prentice-Hall, 1985.

[Hill 81] Hill, F. J., and G. R. Peterson. *Introduction to Switching Theory and Logical Design*, 3d ed. New York: Wiley, 1981.

[Intel 79] Intel Corporation. *MCS-80/85 Family User's Manual*. Santa Clara, CA: Intel, 1979.

[Intel-A] Intel Corporation. *Memory Components Handbook*. Santa Clara, CA: Intel, latest edition.

[Intel-B] Intel Corporation. *Microsystem Components Handbook*. Santa Clara, CA: Intel, latest edition.

[Karnaugh 53] Karnaugh, M. "The Map Method for Synthesis of Combinational Logic Circuits," *Trans. AIEE,* 72 (November 1953): 593–598.

[Kintner 71] Kintner, P. M. "Mixed Logic: A Tool for Design Simplification," *Computer Design,* (August 1971): 55–60.

[Kline 83] Kline, R. M. *Structured Digital Design, Including MSI/LSI Components and Microprocessors*. Englewood Cliffs, NJ: Prentice-Hall, 1983.

[Liu 84] Liu, Y. C., and G. A. Gibson. *Microcomputer Systems: The 8086/8088 Family*. Englewood Cliffs, NJ: Prentice-Hall, 1984.

[Mano 79] Mano, M. M. *Digital Logic and Computer Design*. Englewood Cliffs, NJ: Prentice-Hall, 1979.

[Mano 84] Mano, M. M. *Digital Design*. Englewood Cliffs, NJ: Prentice-Hall, 1984.

[McCluskey 75] McCluskey, E. J. *Introduction to the Theory of Switching Circuits*. New York: McGraw-Hill, 1975.

[Mead 80] Mead, C. A., and L. A. Conway. *Introduction to VLSI Systems*. Reading, MA: Addison-Wesley, 1980.

[Mick 80] Mick, J. R., and J. Brick. *Bit-Slice Microprocessor Design*. New York: McGraw-Hill, 1980.

[Monolithic] Monolithic Memories, Inc. *PAL Programmable Array Logic Handbook*. Sunnyvale, CA: Monolithic, latest edition.

[Motorola] Motorola Semiconductor Products, Inc., *Schottky TTL Data Book*. Phoenix, AZ: Motorola, latest edition.

[Muchow 83] Muchow, K., and B. R. Deem. *Microprocessors: Principles, Programming and Interfacing*. Reston, VA: Reston, 1983.

[Peatman 77] Peatman, J. B. *Microcomputer-Based Design*. New York: McGraw-Hill, 1977.

[Peatman 80] Peatman, J. B. *Digital Hardware Design*. New York: McGraw-Hill, 1980.

[Prosser 87] Prosser, F. P., and D. E. Winkel. *The Art of Digital Design—An Introduction to Top-Down Design*. Englewood Cliffs, NJ: Prentice-Hall, 1987.

[Rafiquzzaman 84] Rafiquzzaman, M. *Microprocessors and Microcomputer Development Systems—Designing Microprocessor-Based Systems*. New York: Harper & Row, 1984.

[Roberts 82] Roberts, S. K. *Industrial Design with Microcomputers*. Englewood Cliffs, NJ: Prentice-Hall, 1982.

[Roth 85] Roth, C. H. *Fundamentals of Logic Design,* 3d ed. St. Paul, MN: West, 1985.

[Shannon 38] Shannon, C. E. "A Symbolic Analysis of Relay and Switching Circuits," *Trans. AIEE,* 57 (1938): 713–723.

[Short 81] Short, K. L. *Microprocessors and Programmed Logic*. Englewood Cliffs, NJ: Prentice-Hall, 1981.

[Signetics] Signetics Corporation. *Integrated Fuse Logic Data Manual*. Sunnyvale, CA: Signetics, latest edition.

[Taub 63] Taub, A. H. (Ed.). *Collected Works of John von Neumann,* vol. 5. New York: Macmillan, 1963.

[Taub 82] Taub, H. *Digital Circuits and Microprocessors*. New York: McGraw-Hill, 1982.

[Texas Instruments] Texas Instruments, Inc. *The TTL Data Book*. Dallas: Texas Instruments, latest edition.

[Wiatrowski 80] Wiatrowski, C. A., and C. H. House. *Logic Circuits and Microcomputer Systems*. New York: McGraw-Hill, 1980.

INDEX